RECENT PROGRESS IN FUNCTIONAL ANALYSIS

NORTH-HOLLAND MATHEMATICS STUDIES 189

(Continuation of the Notas de Matemática)

Editor: Saul LUBKIN
University of Rochester
New York, U.S.A.

2001
ELSEVIER
Amsterdam - London - New York - Oxford - Paris - Shannon - Tokyo

RECENT PROGRESS IN FUNCTIONAL ANALYSIS

Proceedings of the International Functional Analysis Meeting on the Occasion of the 70th Birthday of Professor Manuel Valdivia, Valencia, Spain, July 3–7, 2000

Edited by

Klaus D. BIERSTEDT
University of Paderborn
Paderborn, Germany

José BONET
Technical University of Valencia
Valencia, Spain

Manuel MAESTRE
University of Valencia
Valencia, Spain

Jean SCHMETS
University of Liège
Liège, Belgium

N·H

2001
ELSEVIER
Amsterdam - London - New York - Oxford - Paris - Shannon - Tokyo

ELSEVIER SCIENCE B.V.
Sara Burgerhartstraat 25
P.O. Box 211, 1000 AE Amsterdam, The Netherlands

First edition 2001

Library of Congress Cataloging in Publication Data
A catalog record from the Library of Congress has been applied for.

ISBN: 0-444-50219-X
ISSN: 0304-0208

∞ The paper used in this publication meets the requirements of ANSI/NISO Z39.48-1992 (Permanence of Paper).
Transferred to Digital Printing 2006

PREFACE

During its meeting before the Second European Congress in Mathematics in Budapest in 1996, the Council of the European Mathematical Society decided that the Third European Congress of Mathematics would be held in Barcelona in July 2000. Klaus D. Bierstedt, a member of this Council, who had also been one of the organizers of the International Functional Analysis Meeting on the Occasion of the 60th Birthday of Professor M. Valdivia at Peñíscola, Spain, October 22–27, 1990, immediately suggested to organize a satellite meeting on functional analysis in Valencia during the week before the Third European Congress. José Bonet and Manuel Maestre from the two universities of Valencia agreed with this suggestion and decided to hold such a meeting, ten years after the one in Peñíscola, and now on the occasion of the 70th birthday of Professor Valdivia. The preparations for the conference started in 1998, the Scientific Committee was formed, and the first Plenary Speakers were invited.

The Proceedings of the International Functional Analysis Meeting of 1990 had been published as volume 170 (1992) in the series North-Holland Math. Studies. During the International Congress of Mathematicians 1998 in Berlin, Bierstedt and Bonet asked Drs. Arjen Sevenster, Associate Publisher of Elsevier Science, if the Proceedings of the meeting in Valencia in 2000 could again be published in the series North-Holland Math. Studies, one year after the conference. The reaction of Sevenster was very positive; the contract was signed some months later. We thank Drs. Sevenster and the Elsevier/North-Holland company for publishing this book.

The preface of the Proceedings of the Peñíscola meeting contained a short summary of the merits of Professor Valdivia, and the first article in the book, "The mathematical works of Manuel Valdivia" by J. Horváth, gave a report on his books and articles up to 1990. Since then Valdivia had gone on to publish many important results, some of them in joint work with the President of the Belgian Mathematical Society, Professor Jean Schmets. Hence it was natural to ask Schmets to report on the recent work of Valdivia at the meeting in Valencia. Schmets accepted; his article "The mathematical works of Manuel VALDIVIA, II" is the first one in the present Proceedings volume.

Soon it became clear that the International Functional Analysis Meeting in Valencia in 2000 would be much bigger and broader in scope than the one in Peñíscola. In addition to the special talk of Jean Schmets, there were 16 Plenary Lectures of 50 minutes, by well-known specialists from 8 countries, as originally planned. However, the organizers had not anticipated in the beginning that there would be more than 300 participants in the end. And so many abstracts were submitted that, only a few weeks before the meeting, it was finally decided to group the 176 parallel talks of 25 minutes in 10 Special Sessions and to let each Special Session have its own organizers. In addition, 24 posters were presented in three Poster Sessions.

We take the opportunity to thank the people of the Atlas Mathematical Conference Abstracts (AMCA) server at York University, Canada, and here especially Elliott Pearl, for providing an excellent service for the submission of the abstracts of the participants via WWW and for the preparation of the booklet. Announcements of the meeting and several circular letters were distributed by e-mail and by Internet. The homepages of the meeting, with much pertinent information and many useful links, were designed and maintained in the Department of Mathematics and Computer Science at the University of Paderborn by Dr. Silke Holtmanns to whom special thanks are due.

We thank various sponsors, in particular the Universidad Politécnica de Valencia, for contributing money and the facilities for the meeting. Among other things, these contributions made it possible to waive the conference fee and to offer free accomodation and lunches in Valencia to 29 participants from Eastern Europe, from countries of the former Soviet Union, as well as from countries of the Third World. More than twice as many mathematicians had applied for such grants, and the selection process was painful for the Scientific Committee since a much larger number of them would have deserved grants. – Finally, this may be the right point to also thank the members of the Scientific Committee and of various other committees and the organizers of the Special Sessions for their efficient help. (A list of the sponsors, of all the committees, of all Special Sessions and their organizers, the schedule of the meeting, as well as the schedules of the Special Sessions, can be found in the editorial part of this book.)

The organization took much more time and energy than expected. The arrival day was hectic. The week of the meeting turned out to be one of the hottest of the summer of 2000 in Valencia. On Monday, the highest temperature was 39 degrees, and even after midnight there remained 31 degrees. The air condition cooled down the main lecture hall in the Rectorado so much that some participants ended up with a cold. On the other hand, some of the smaller lecture rooms for the Special Sessions did not have any air condition at all, and the speakers (and the audience) suffered from the heat. Fortunately enough, Wednesday afternoon was not as hot as Monday, or else the excursion to Xàtiva would have ended in a virtual disaster. (Temperatures in Xàtiva during the summer are usually five degrees higher than in Valencia.) In fact, both excursions to Xàtiva and to the Hemisferic and Science Museum (on Friday) turned out fine. The participants will remember the Castillo of Xàtiva with its beautiful scenic views, the picturesque center of this town, and the spectacular new buildings of the City of Arts and Sciences in Valencia.

But for the success of a meeting in mathematics, it is mathematics which is by far the most important thing, and in this respect the meeting was very successful indeed. Many interesting lectures and posters were presented during the meeting, and it was reported on plenty of deep and important theorems: the mathematics was first-rate. It remains to thank the speakers for their excellent, well-prepared and inspiring talks, the chairpersons for their help, and the audience for persistent interest and stimulating discussions.

All the Plenary Speakers had been invited, and all the other participants had the opportunity, to submit an article to this Proceedings volume. The original deadline was November 30, 2000, but the last paper published here, by one of the Plenary Speakers, arrived at Paderborn only in early March 2001. Part of the Scientific Committee also served as editors of the book. 12 contributions by Plenary Speakers and 37 articles by other participants were submitted and refereed. We thank the referees very much; some of them took their job very seriously. Many comments and criticisms were made which definitely helped to improve several articles, sometimes not only in the exposition or by removing misprints. 17 articles had to be rejected, revisions of some of which, according to the remarks of referees, would actually have been publishable in a good mathematical journal. The editors regret the limitations of space (the book was supposed to have about 400 pages) and time. Still, the original deadline for the submission of the final manuscript to Elsevier (March 1, 2001) was exceeded by more than two months. We thank Ms. Duddeck for compiling the editorial part of the book in Paderborn.

As a glance at the table of contents shows, the present Proceeedings volume contains 32 articles on various interesting areas of present-day functional analysis and its applications: Banach spaces and their geometry, operator ideals, Banach and operator algebras, operator and spectral theory, Fréchet spaces and algebras, function and sequence spaces. The reports we received from the (sometimes very prominent) referees confirmed our impression that the authors have taken much care with their articles and that many of the papers present important results and methods in active fields of research. Several survey type articles (at the beginning and the end of the book) will be very useful for mathematicians who wants to learn "what's going on" in some particular field of research. We hope, and are quite confident, that this collection of papers, dedicated to Professor Manuel Valdivia on the occasion of his 70th birthday, with best wishes for the future, will prove helpful, valuable and inspiring for anybody interested in functional analysis and related fields.

Paderborn/Valencia/Liège, May 2001

Klaus D. Bierstedt, José Bonet, Manuel Maestre, Jean Schmets

INTERNATIONAL FUNCTIONAL ANALYSIS MEETING on the Occasion of the 70th Birthday of Professor M. Valdivia

Valencia, Spain, 3 - 7 July 2000

(a Satellite Conference to the Third European Congress of Mathematics in Barcelona)

SPONSORS

Universidad Politécnica de Valencia

Dirección General de Enseñanza Superior e Investigación Científica (DGESIC) del Ministerio de Educación y Cultura (MEC)

Conselleria de Cultura i Educació, Generalitat Valenciana

BANCAJA

Facultat de Matemàtiques, Universitat de València

Facultat de Farmàcia, Universitat de València

Real Sociedad Matemática Española

HONOUR COMMITTEE

SCIENTIFIC COMMITTEE

Prof. Richard M. Aron, Kent State University, Ohio, USA
Prof. Klaus D. Bierstedt, Univ. Paderborn, Germany
Prof. José Bonet, Univ. Politécnica de Valencia, Spain
Prof. Joan Cerdà, Univ. Barcelona, Spain
Prof. Hans Jarchow, Univ. Zürich, Switzerland
Prof. Manuel Maestre, Univ. de Valencia, Spain
Prof. Jean Schmets, Univ. Liège, Belgium

LOCAL ORGANIZING COMMITTEE

Prof. Carmen Fernández
Prof. Antonio Galbis
Prof. Pablo Galindo
Prof. Domingo Garcia
Prof. Manuel López Pellicer
Prof. Vicente Montesinos
Prof. Alfred Peris

The Local Committee would like to extend special thanks to the following colleagues of the Universidad Politécnica who have been very helpful with the organization of the meeting:

- Alberto Conejero,
- David Jornet,
- Félix Martínez Giménez,
- María José Rivera,
- Enrique Sánchez.

Bust of Don Manuel Valdivia, a gift of his friends, unveiled during the Opening Ceremony
Sculptor: Danuta Pustkowska, wife of V. Montesinos, graduate in History of Art (Université de Liège, Belgium) and graduate in Fine Arts (Universidad Politécnica de Valencia)

INTERNATIONAL FUNCTIONAL ANALYSIS MEETING
on the Occasion of the 70th Birthday of Professor M. Valdivia
Valencia, Spain, 3 - 7 July 2000

SCHEDULE

Monday, July 3

9.30	E. Effros, On the local structure of non-commutative L_1-spaces Chair: R. Payá
10.30	D. Vogt, The space of real analytic functions has no basis Chair: K. D. Bierstedt
11.30	Coffee break
12.00 - 13.30	**Opening Ceremony, including:** J. Schmets, The recent mathematical work of Manuel Valdivia
14.00	Lunch
15.30	J. D. M. Wright, Wild factors linked to generic dynamics Chair: D. Garling
16.30	Coffee break
17.00 - 20.30	Special sessions [1]
18.00 - 20.00	Poster session [I]

Tuesday, July 4

9.00	J. Lindenstrauss, Affine approximation of Lipschitz mappings between some classes of infinite dimensional Banach spaces Chair: A. Lazar
10.00	N. J. Kalton, Applications of Banach space theory to sectorial operators Chair: L. Drewnowski
11.00	Coffee break
11.30 - 13.30	Special sessions [2]
14.00	Lunch
15.30	A. Defant, Almost everywhere convergence of series in non-commutative L_p-spaces Chair: M. Maestre
16.30	Coffee break
17.00 - 20.30	Special sessions [3]

Wednesday, July 5

9.00	H. G. Dales, Derivations from Banach algebras Chair: J. Schmets
10.00	N. Tomczak-Jaegermann, Geometry, linear structure and random phenomena in finite-dimensional normed spaces Chair: C. Finet
11.00	Coffee break
11.30	K. Seip, Analysis at the Nyquist rate Chair: J. Cerdà
12.30	T. W. Gamelin, Homomorphisms of uniform algebras Chair: K. Floret
14.00	Lunch
16.00	Excursion by bus to Xàtiva
21.00	Dinner at Xàtiva

Thursday, July 6

9.00	R. Meise, Solution operators for linear partial differential operators Chair: J. Bonet
10.00	J. H. Shapiro, The numerical range of a composition operator Chair: R. M. Timoney
11.00	Coffee break
11.30 - 13.30	Special sessions [4]
14.00	Lunch
15.30	J. Eschmeier, Invariant subspaces for commuting contractions Chair: M. Langenbruch
16.30	Coffee break
17.00 - 20.30	Special sessions [5]
18.00 - 20.00	Poster session [II]

Friday, July 7

9.00	A. Pełczyński, Sobolev spaces as Banach spaces
	Chair: H. Jarchow
10.00	Coffee break
10.30 - 13.30	Special sessions [6]
11.00 - 13.00	Poster session [III]
14.00	Lunch
15.30	P. Wojtaszczyk, Greedy and quasi-greedy bases in Banach spaces
	Chair: W. Lusky
16.30	Coffee break
17.00 - 18.00	G. Godefroy, The Szlenk index and its applications Chair: M. Valdivia
18.30 - 19.30	Excursion by bus to Hemisferic and Science Museum

Saturday, July 8

11.00 - 13.30	**Closing Ceremony**

Professors D. García, K. Floret, M. Valdivia after the Closing Ceremony

LIST OF THE SPECIAL SESSIONS
with the Session Organizers

- **Geometry and Structure of Banach spaces**
  H. Jarchow, V. Montesinos
- **Weak topologies in Banach spaces, renormings**
  V. Zizler, B. Cascales
- **Operator algebras, operator spaces**
  E. Sánchez
- **Operator theory, spectral theory, Banach algebras**
  W. Zelazko, P. Paúl
- **Chaotic behaviour of operators and universality**
  K. G. Grosse-Erdmann, A. Peris
- **Non-linear functional analysis**
  M. Poppenberg, M. J. Rivera
- **Fréchet spaces, with applications to complex analysis and (linear) partial differential operators**
  P. Domański, C. Fernández
- **Function spaces and their duals**
  J. Cerdà, A. Galbis
- **Topological vector spaces, duality theory**
  J. Schmets, M. López Pellicer
- **Holomorphy, polynomials (in infinite dimensions)**
  R. Aron, D. García.

General Schedule of the Special Sessions

Monday, 17 - 20.30 h	**(number of talks in each Special Session: 7)**
Weak topologies in Banach spaces, renormings	(talks 1 - 7)
Function spaces and their duals	(talks 1 - 7)
Operator algebras, operator spaces	(talks 1 - 7)
Topological vector spaces, duality theory	(talks 1 - 7)
Holomorphy, polynomials	(talks 1 - 7)
Poster Session I (18 - 20 h)	

Tuesday 11.30 - 13.30 h (number of talks in each Special Session: 4) and 17.00 - 20.20 h (7 talks)	
Geometry and structure of Banach spaces	(talks 1 - 11)
Operator theory, spectral theory, Banach algebras	(talks 1 - 11)
Function spaces and their duals	(talks 8 - 18)
Topological vector spaces, duality theory / Fréchet spaces and applications	(talk 8) (talks 1 - 10, end)
Weak topologies in Banach spaces, renormings / Topological vector spaces, duality theory	(talks 8 - 15, end) (talks 9 - 11)

Thursday, 11.30 - 13.30 h (number of talks in each Special Session: 4) and 17.00 - 20.30 h (7 talks)	
Geometry and structure of Banach spaces	(talks 12 - 22)
Operator theory, spectral theory, Banach algebras	(talks 12 - 22)
Function spaces and their duals / Topological vector spaces, duality theory	(talks 19 - 28, end) (talk 12)
Holomorphy, polynomials	(talks 8 - 18)
Chaotic behaviour of operators and universality	(talks 1 - 9, end)
Poster Session II (18 - 20 h)	

Friday, 10.30 - 13.30 h (number of talks in each Special Session: 6)	
Geometry and structure of Banach spaces	(talks 23 - 27, end)
Topological vector spaces, duality theory	(talks 13 - 18, end)
Operator algebras, operator spaces	(talks 8 - 13, end)
Operator theory, spectral theory, Banach algebras / Holomorphy, polynomials	(talk 23) (talks 19 - 23, end)
Non-linear functional analysis / Operator theory, spectral theory, Banach algebras	(talks 1 - 5, end) (talk 28, end)
Operator theory, spectral theory, Banach algebras	(talks 24 - 27)
Poster Session III (11 - 13 h)	

Participants of the meeting during the excursion to the City of Arts and Sciences on Friday afternoon

Schedule of the Special Session
GEOMETRY AND STRUCTURE OF BANACH SPACES

Tuesday, July 4	**Chair: C. Finet**
11.30 h - 11.55 h	C. Michels, *Summing inclusion maps between symmetric sequence spaces*
12.00 h - 12.25 h	K. Bolibok, *Minimal displacement and retraction problems*
12.30 h - 12.55 h	G. Arango, *Factorization of* (∞, σ)*-integral operators*
13.00 h - 13.25 h	F. Oertel, *Extension of finite rank operators and local structures in operator ideals*
	Chair: D. Werner
17.00 h - 17.25 h	N. Randrianantoanina, *Operators on* C^**-algebras*
17.30 h - 17.55 h	J. Castillo, *Twisted sums of* $C(K)$*-spaces*
18.00 h - 18.25 h	Y. Moreno, *König-Wittstock norms and twisted sums of quasi-Banach spaces*
18.30 h - 18.55 h	F. Albiac, *Some geometric properties of quasi-Banach spaces*
19.00 h - 19.25 h	A. Pelczar, *Remarks on Gowers' dichotomy*
19.30 h - 19.55 h	J. Amigó, *On copies of* c_0 *in the bounded linear operator space*
20.00 h - 20.25 h	J. López-Abad, *A new Ramsey property for Banach spaces*

Thursday, July 6	**Chair: M. Gonzales**
11.30 h - 11.55 h	L. Drewnowski, *Continuity of monotone functions with values in Banach lattices*
12.00 h - 12.25 h	W. Wnuk, *Locally solid Riesz spaces with certain Lebesgue topologies*
12.30 h - 12.55 h	I. Plyrakis, *Minimal lattice-subspaces*
13.00 h - 13.25 h	A. Martínez-Abejón, *On the local dual spaces of a Banach space*
	Chair: H. Jarchow
17.00 h - 17.25 h	B. Randrianantoanina, *A note on the Banach-Mazur problem*
17.30 h - 17.55 h	A. Ulger, *The Phillips properties*
18.00 h - 18.25 h	H.-O. Tylli, *Non-self duality of the weak essential norm*

Thursday, July 6	**Chair: V. Zizler**
18.30 h - 18.55 h	D. Yost, *Projections on big Banach spaces*
19.00 h - 19.25 h	D. Werner, *Narrow operators and the Daugavet property*
19.30 h - 19.55 h	M. Petrakis, *Discrepancy norms*
20.00 h - 20.25 h	A. Naor, *Isomorphic embedding of* l_p^m *into* l_1^n

Friday, July 7	**Chair: H.-O. Tylli**
10.30 h - 10.55 h	C. Finet, *Vector-valued perturbed minimization principles*
11.00 h - 11.25 h	W. Lusky, *On the isomorphic classification of weighted spaces of holomorphic functions*
11.30 h - 11.55 h	Y. Raynaud, *Ultrapowers of non commutative* L_p*-spaces*
	Chair: N. Randrianantoanina
12.00 h - 12.25 h	O. Blasco, (p, q)*-summing sequences*
12.30 h - 12.55 h	V. Zizler, *The structure of uniformly Gâteaux smooth spaces*
13.00 h - 13.25 h	P. Maritz, *Lindenstrauss' short proof of the Lyapunov convexity theorem: Impact and citation analysis*

Schedule of the Special Session

WEAK TOPOLOGIES IN BANACH SPACES; RENORMINGS

Monday, July 3	**Chair: S. Troyanski**
17.00 h - 17.25 h	M. D. Acosta, *Reflexivity and norm attaining functionals*
17.30 h - 17.55 h	B. Cascales, *Measurable selectors for the metric projection*
18.00 h - 18.25 h	V. Montesinos, *Lower semicontinuous smooth norms*
18.30 h - 18.55 h	M. Raja, *Descriptive compact spaces and W*LUR renorming*
19.00 h - 19.25 h	O. Kalenda, *Valdivia compact spaces, renormings and Asplund spaces*
19.30 h - 19.55 h	D. J. Ives, $C(K)$ *spaces that could be Gateaux differentiability spaces and which are not weak Asplund*
20.00 h - 20.25 h	E. Nieto, *On M-structure and the asymptotic-norming property*

Tuesday, July 4	**Chair: V. Montesinos**
11.30 h - 11.55 h	E. Matouskova, *Translating finite sets into convex sets*
12.00 h - 12.25 h	J. Orihuela, *Renormings and coverings in Banach spaces*
12.30 h - 12.55 h	S. Troyanski, *A class of maps for the non separable Banach space renorming theory*
13.00 h - 13.25 h	A. Moltó, *A non linear transfer technique for LUR renorming*
	Chair: M.D. Acosta
17.00 h - 17.25 h	N. Ribarska, *A stability property of LUR renorming*
17.30 h - 17.55 h	S. Falcon, *On non-compact convexity in the James space*
18.00 h - 18.25 h	A. Kryczka, *Measure of weak noncompactness under complex interpolation*
18.30 h - 18.55 h	A. Wisnicki, *On connections between the fixed point property and the* (S_m) *property*

Schedule of the Special Session
OPERATOR ALGEBRAS, OPERATOR SPACES

Monday, July 3	**Chair: E.A. Sánchez-Pérez**
17.00 h - 17.25 h	A. M. Peralta, *Grothendieck's inequalities for real and complex* JBW^**-triples*
17.30 h - 17.55 h	A. Y. Helemskii, *Wedderburn-type theorems for operator algebras and moduls: the traditional and 'quantized' homological approaches*
18.00 h - 18.25 h	V. M. Manuilov, *Asymptotically split extensions of* C^**-algebras and E-theory*
	Chair: A.Y. Helemskii
18.30 h - 18.55 h	A. Morales, *Prime non-commutative algebras*
19.00 h - 19.25 h	A. Rakhimov, *Actions of compact abelian groups on a real semifinite factor*
19.30 h - 19.55 h	R. M. Timoney, *An internal characterization of complete positivity for elementary operators*
20.00 h - 20.25 h	M. I. Berenguer, *Lie derivations and Lie isomorphisms on Banach algebras*

Friday, July 7	Chair: E.A. Sánchez-Pérez
10.30 h - 10.55 h	B. Jefferies, *Spectral theory of noncommutative systems of operators*
11.00 h - 11.25 h	A. Rodríguez Palacios, *A holomorphic characterization of C^*-algebras*
11.30 h - 11.55 h	S. Ayupov, *Non commutative Arens algebras*
12.00 h - 12.25 h	F. F. Nassopoulos, *Involutive and C^*-complexifications: Commutative real and complex involutive complete algebras in effective perspective*
12.30 h - 12.55 h	W. Werner, *Heat kernel expansion and functional calculus*
13.00 h - 13.25 h	B. V. Loginov, *Canonical Jordan sets and Andronov-Hopf bifurcation*

Schedule of the Special Session

OPERATOR THEORY, SPECTRAL THEORY, BANACH ALGEBRAS

Tuesday, July 4	Chair: J.C. Díaz
11.30 h - 11.55 h	M. V. Velasco, *The second transpose of a derivation*
12.00 h - 12.25 h	M. Mathieu, *The norm problem for elementary operators*
12.30 h - 12.55 h	J. Kim, *Spectral interpolation and amenability of symmetric discrete groups*
13.00 h - 13.25 h	A. Villena, *Uniqueness of the norm topology on commutative Banach algebras*
	Chair: W. Ricker
17.00 h - 17.25 h	M. S. Agranovich, *Spectral properties of potential type integral operators on smooth and Lipschitz surfaces*
17.30 h - 17.55 h	I. Domanov, *Invariant and hyperinvariant subspaces of some Volterra operators in Sobolev spaces and related operator algebras*
18.00 h - 18.25 h	I. Karabash, *J-selfadjoint ordinary differential operators similar to selfadjoint operators*
	Chair: W. Lusky
18.30 h - 18.55 h	N. Ivanovski, *Operator valued weighted sequence spaces*
19.00 h - 19.25 h	L. Volevich, *Resolvent of a mixed order system on a manifold with boundary*
19.30 h - 19.55 h	V. Adamyan, *Averaged evolution on a Markov background*
20.00 h - 20.25 h	J. Wosko, *Constrictive Markov operators*

Thursday, July 6	Chair: T. Tonev
11.30 h - 11.55 h	E. Briem, *Operating functions and subspaces of* $C_0(X)$
12.00 h - 12.25 h	V. Kisil, *Spectral theory of operators and group representations*
12.30 h - 12.55 h	Z. Cuckovic, *Products of Toeplitz operators on the Bergman space*
13.00 h - 13.25 h	P. J. Paúl, *Properties of the generalized Toeplitz operators*
	Chair: B. Gramsch
17.00 h - 17.25 h	G. H. Esslamzadeh, *Ideal structure and representation of* L_1*-Munn algebras and semigroup algebras*
17.30 h - 17.55 h	M. Filali, *On some ideal structure of the second conjugate of a group algebra with an Arens product*
18.00 h - 18.25 h	A. J. Calderón-Martin, *Functional analysis in Lie theory*
	Chair: H.G. Dales
18.30 h - 18.55 h	J. Flores, *Domination by strictly-singular and disjointly strictly-singular operators*
19.00 h - 19.25 h	E. Semenov, *Disjoint strict singularity and inclusions of rearrangement invariant spaces*
19.30 h - 19.55 h	N. J. Laustsen, *The Banach algebra of operators on James's quasi-reflexive Banach spaces*
20.00 h - 20.25 h	V. Gorbachuk, *Operator approach to direct and inverse theorems in the approximation theory of functions*

Friday, July 7	Chair: M. Fragoulopoulou
13.00 h - 13.25 h	M. B. Ghaemi *AC-operators and well-bounded operators with dual of scalar-type*

Schedule of the Special Session
CHAOTIC BEHAVIOUR OF OPERATORS AND UNIVERSALITY

Thursday, July 6	**Chair: J.H. Shapiro**
11.30 h - 11.55 h	J. Bes, *Approximation by chaotic operators on a Hilbert space*
12.00 h - 12.25 h	H. Emamirad, *Linear chaos and approximation*
12.30 h - 12.55 h	F. Martínez-Giménez, *On the existence of chaotic operators on Banach spaces*
13.00 h - 13.25 h	A. Peris, *Hypercyclic and chaotic polynomials on Banach spaces*
	Chair: K. Chan
17.30 h - 17.55 h	L. Bernal-González, *Several kinds of strongly omnipresent operators*
18.00 h - 18.25 h	M. C. Calderón-Moreno, *Wild behavior via plane sets: dense-image operators*
	Chair: K.G. Grosse-Erdmann
18.30 h - 18.55 h	E. Gallardo-Gutiérrez, *Supercyclic operators*
19.00 h - 19.25 h	A. Montes-Rodríguez, *Recent development in supercyclic operators*

Schedule of the Special Session
NON-LINEAR FUNCTIONAL ANALYSIS

Friday, July 7	**Chair: M. Poppenberg**
10.30 h - 10.55 h	A. Cañada, *Nonlinear periodic perturbations of linear boundary value problems at resonance*
11.00 h - 11.25 h	G. Grillo, *On the time decay of solutions to classes of quasilinear evolution equations*
	Chair: M.J. Rivera
11.30 h - 11.55 h	M. Poppenberg, *Nash-Moser methods for nonlinear boundary value problems*
12.00 h - 12.25 h	I. M. Tkachenko, *Matrix orthogonal polynomials and the truncated moment problem*
12.30 h - 12.55 h	V. A. Trenogin, *Adjoint operators to nonlinear operators*

Schedule of the Special Session
FRÉCHET SPACES,
WITH APPLICATIONS TO COMPLEX ANALYSIS
AND (LINEAR) PARTIAL DIFFERENTIAL OPERATORS

Tuesday, July 4	**Chair: R. Braun**
11.30 h - 11.55 h	W. Zelazko, *When does a topological algebra have all ideals closed?*
12.00 h - 12.25 h	B. Gramsch, *Fréchet operator algebras with spectral invariance*
12.30 h - 12.55 h	B. Schreiber, *Stochastic continuity algebras*
13.00 h - 13.25 h	A. Pirkovskii, *Relation between Hochschild homology and cohomology of locally convex algebras, and applications to computing injective homological dimensions*
	Chair: B. Gramsch
17.30 h - 17.55 h	M. Langenbruch, *Surjective partial differential operators on spaces of real analytic functions*
	Chair: J.C. Díaz
18.00 h - 18.25 h	R. Braun, *Geometry of the local Phragmén-Lindelöf condition in three variables*
18.30 h - 18.55 h	A. Goncharov, *Bases in the spaces of infinitely differentiable and Whitney functions*
19.00 h - 19.25 h	Ch. Bouzar, *A generalization of the problem of elliptic iterates*
19.30 h - 19.55 h	F. Blasco, *Some properties of sequence spaces*
20.00 h - 20.25 h	P. Sevilla, *Spectra in tensor products of lmc-algebras*

Schedule of the Special Session
FUNCTION SPACES AND THEIR DUALS

Monday, July 3	**Chair: A. Galbis**
17.00 h - 17.25 h	I. Cioranescu, *On the equivalence of the ultradistribution theories*
17.30 h - 17.55 h	C. Fernández-Rosell, *Regularity of solutions of convolution equations*
18.00 h - 18.25 h	J. Bastero, *Commutators for the Hardy-Littlewood maximal function*
18.30 h - 18.55 h	M. Nawrocki, *On the Smirnov class defined by the maximal function*
19.00 h - 19.25 h	M. Lindström, *Factorization of weakly compact homomorphisms on (URM) algebras*
19.30 h - 19.55 h	O. Nygaard, *Slices in the unit ball of a uniform algebra*
20.00 h - 20.25 h	T. Tonev, *Inductive limits of classical uniform algebras*

Tuesday, July 4	Chair: O. Blasco
11.30 h - 11.55 h	P. Domański, *Composition operators on spaces of real analytic functions*
12.00 h - 12.25 h	A. Siskakis, *The Hilbert matrix and composition operators*
12.30 h - 12.55 h	D. Vukotic, *Bergman space operators and the Berezin transform. Old and new*
13.00 h - 13.25 h	N. Zorboska, *Berezin Transform and compactness of operators*
	Chair: J. Bastero
17.00 h - 17.25 h	G. Dolinar, *Stability of disjointness preserving mappings*
17.30 h - 17.55 h	J. A. Jaramillo, *Lattices of uniformly continuous functions and Banach-Stone theorems*
18.00 h - 18.25 h	I. Novikov, *Compactly supported wavelet bases in function spaces*
18.30 h - 18.55 h	F. Bastin, *Riesz bases of spline wavelets in periodic Sobolev spaces*
	Chair: M. Lindström
19.00 h - 19.25 h	J. Taskinen, *On the continuity of Bergman and Szegö projections*
19.30 h - 19.55 h	K. Bogalska, *Multiplication operators on weighted Banach spaces which are an isomorphism into*
20.00 h - 20.25 h	S. Holtmanns, *Biduals of weighted spaces of holomorphic or harmonic functions*

Thursday, July 6	Chair: F. Bastin
11.30 h - 11.55 h	J. M. Ahn, *L_p Fourier-Feyman transform and the first variation on the Fresnel class*
12.00 h - 12.25 h	K. S. Chang, *Feyman's operational calculus for an operator-valued function space integral*
12.30 h - 12.55 h	K. Sik Ryu, *On a measure in Wiener space induced by the sum of measures associated with arbitrary numbers and applications*
13.00 h - 13.25 h	I. Yoo, *Convolution and Fourier-Feynman transforms*

Thursday, July 6	**Chair: J.A. Jaramillo**
17.00 h - 17.25 h	J. M. Almira, *Bernstein theorems in an abstract setting*
17.30 h - 17.55 h	B. Rodríguez Salinas, *On the Radon-Nikodym theorem*
18.00 h - 18.25 h	F. J. Freniche, *Poisson integrals of Pettis integrable functions*
18.30 h - 18.55 h	A. Michalak, *On the Fubini theorem for the Pettis integral for bounded functions*
	Chair: Y. Raynaud
19.00 h - 19.25 h	M. J. Rivera, *The de la Vallée Poussin theorem for vector valued measure spaces*
19.30 h - 19.55 h	E. A. Sánchez-Pérez, *Spaces of integrable functions with respect to a vector measure G: several properties related to the range of G*

Schedule of the Special Session

TOPOLOGICAL VECTOR SPACES, DUALITY THEORY

Monday, July 3	**Chair: C. Stuart**
17.00 h - 17.25 h	S. Saxon, *Fréchet-Urysohn topological vector spaces*
17.30 h - 17.55 h	M.V. Gregori, *On fuzzy metric spaces*
	Chair: L.M. Sánchez-Ruiz
18.00 h - 18.25 h	J. Kakol, *Strongly Hewitt spaces and applications to spaces $C(K)$*
18.30 h - 18.55 h	R. Miralles-Rafart, *The generalized Minkowski functional*
	Chair: I. Tweddle
19.00 h - 19.25 h	W. Ricker, *Locally convex spaces and Boolean algebras of projections*
19.30 h - 19.55 h	L. Oubbi, *P- and Q-properties in weakly topologized algebras*
20.00 h - 20.25 h	M.A. Sofi, *Factoring operators and embedding into product spaces*

Tuesday, July 4	**Chair: B. Gramsch**
17.00 h - 17.25 h	J. Wengenroth, *A splitting theorem for subspaces and quotients of $\mathcal{D}'$*
	Chair: A. Peris
19.00 h - 19.25 h	K. G. Grosse-Erdmann, *On weak criteria for vector-valued holomorphy*
19.30 h - 19.55 h	E. Martín Peinador, *On locally quasi-convex groups*
20.00 h - 20.25 h	S. Hernández, *Pontryagin duality for spaces of continuous functions*

Thursday, July 6 20.00 h - 20.25 h	M. Mauer, *Domains of analytic existence in real locally convex spaces*

Friday, July 7	**Chair: S. Saxon**
10.30 h - 10.55 h	I. Tweddle, *Barrelled countable enlargements for ultrabornological spaces*
11.00 h - 11.25 h	L. M. Sánchez Ruiz, *Cardinals and metrizable barrelled spaces*
	Chair: I. Tweddle
11.30 h - 11.55 h	C. Stuart, *Some gliding hump conditions for barrelledness*
12.00 h - 12.25 h	M. López Pellicer, *Strong barrelledness*
	Chair: J. Kakol
12.30 h - 12.55 h	J. Gómez-Pérez, *The m-topology on algebras between* $C^*(X)$ *and* $C(X)$
13.00 h - 13.25 h	J. Rodríguez-López, *On the relation between uniform convergence and some hypertopologies on semicontinuous function spaces*

Schedule of the Special Session

HOLOMORPHY, POLYNOMIALS (IN INFINITE DIMENSIONS)

Monday, July 3	**Chair: A. Defant**
17.00 h - 17.25 h	K. Floret, *Minimal ideals of n-homogeneous polynomials*
17.30 h - 17.55 h	C. Boyd, *Products of polynomials and the geometry of Banach spaces*
18.00 h - 18.25 h	L. A. de Moraes, *Boundaries for algebras of holomorphic functions*
	Chair: R. Alencar
18.30 h - 18.55 h	G. A. Muñoz, *Bernstein-Markov type inequalities in real Banach spaces*
19.00 h - 19.25 h	Y. S. Choi, *Geometric properties of* δ_k^m *in the space* $P(^mC(K))$
19.30 h - 19.55 h	F. Cabello-Sánchez, *Complemented subspaces of spaces of multilinear forms and tensor products*
20.00 h - 20.25 h	M. L. Lourenço, *A class of polynomials*

Thursday, July 6	**Chair: L.A. de Moraes**
11.30 h - 11.55 h	J. M. Ansemil, *On $(BB)_n$ properties on Fréchet spaces*
12.00 h - 12.25 h	M. Venkova, *On reflexive spaces*
	Chair: F. Bombal
12.30 h - 12.55 h	G. S. Kim, *Exposed points of the unit balls of the spaces $P(^2l_p^2)$ $(p = 1, 2, \ldots, \infty)$*
13.00 h - 13.25 h	B. Grecu, *Geometry of 2-homogeneous polynomials on Hilbert spaces*
	Chair: R. Ryan
17.00 h - 17.25 h	A. Gretsky, *On differentiability of mappings: A modern treatment of classical theorems and criteria of holomorphy*
17.30 h - 17.55 h	A. Zagorodnyuk, *Some results concerning spaces of analytic functions*
18.00 h - 18.25 h	Y. Sarantopoulos, *Some interesting polynomial inequalities on Hilbert spaces*
	Chair: Y.S. Choi
18.30 h - 18.55 h	R. Gonzalo, *Block diagonal polynomials*
19.00 h - 19.25 h	F. Arranz, *Three-space problems for polynomial properties in Banach spaces*
19.30 h - 19.55 h	R. García, *The bidual of a tensor product of Banach spaces*
20.00 h - 20.25 h	C. S. Fernandez, *Counterexample to the Bartle-Graves selection theorem for multilinear maps*

	Chair: K.G. Grosse-Erdmann
19.30 h - 19.55 h	T. Honda, *Duality theory for the space of polynomials*

Friday, July 7	**Chair: Y. Sarantopoulos**
10.30 h - 10.55 h	M. Mackey *The Lindelöf-Cirka theorem on bounded symmetric domains*
11.00 h - 11.25 h	C. Taylor, *Spectra of tensor product elements*
11.30 h - 11.55 h	F. Bombal, *Integral operators on the product of $C(K)$ spaces*
	Chair: J.M. Ansemil
12.00 h - 12.25 h	I. Villanueva, *On the containment of l_1^n's in the projective tensor product of Banach spaces*
12.30 h - 12.55 h	R. Alencar, *Algebras of symmetric holomorphic functions on l_p*

POSTER SESSION I

Monday, July 3	**Chair: M.J. Rivera**
Poster 1	E. Capobianco, *Multiresolution properties of volatility models*
Poster 2	F. J. Naranjo Naranjo, *A characterization of nuclear Fréchet lattices*
Poster 3	S. G. Kim, *Classification of polynomials attaining their norms at a selected unit vector*
Poster 4	S. Falcon, *On weakly locally nearly uniformly convex Banach spaces*
Poster 5	R. Haller, *Quantitative versions of hereditary results on M-ideals of compact operators*
Poster 6	M. Poldvere, *On Phelps' uniqueness property for $K(X,Y)$ in $L(X,Y)$*
Poster 7	M. Romance, *On maximal volume positions and John's decompositions in the non convex case*
Poster 8	D. L. Salinger, J. D. Stegemann, *Fourier analysis on two groups associated with the Cantor set*

POSTER SESSION II

Thursday, July 6	**Chair: P. Galindo**
Poster 1	M. Fragoulopoulou, *Contractible locally C^*-algebras*
Poster 2	J. A. Conejero, *The set of bounded below operators and almost open operators between locally convex spaces*
Poster 3	M. Friz, *Wedge operators between locally convex spaces*
Poster 4	M. Haralampidou, *Tensor products of locally m-convex H^*-algebras. Structure theorems*
Poster 5	R. Rubio-Diaz, *The Hahn-Banach theorem with infinite values*
Poster 6	J. Rubio-Massegu, *Pretopologies and geometric Hahn-Banach theorems*
Poster 7	A. Uglanov, *Surface integrals in infinite-dimesional spaces and their applications*
Poster 8	G. Mora-Martínez, *A characterization of space-filling curves*

POSTER SESSION III

Friday, July 7	**Chair: C. Fernández**
Poster 1	M. L. Gorbachuk, *Bounded values of semigroups of linear operators*
Poster 2	P. J. Miana, *Integrated trigonometric sine functions*
Poster 3	T. Bermudez, *On hypercyclic operators*
Poster 4	M. C. Gómez-Collado, *Almost periodic ultradistributions of Beurling and of Roumieu type*
Poster 5	E. Jordá, *Spaces of vector-valued meromorphic functions*
Poster 6	D. Jornet, *A perturbation result for surjective convolution operators on the space of all non-quasianalytic functions*
Poster 7	M. Urrea, *Matrix orthogonal polynomials and reconstruction of matrix distributions*
Poster 8	X. Domínguez, *The Schur property for topological groups*

INTERNATIONAL FUNCTIONAL ANALYSIS MEETING on the Occasion of the 70th Birthday of Professor M. Valdivia

Valencia, Spain, 3 - 7 July 2000

LIST OF PARTICIPANTS

M.D. Acosta (Granada, Spain)
V. Adamyan (Odessa, Ukraine)
M.S. Agranovich (Moscow State Institute of Electronics and Mathematics, Russia)
J.M. Ahn (Konkuk University, Seoul, Korea)
F. Albiac (Universidad Pública de Navarra, Spain)
R.L. Alencar (Campinas, Brazil)
J.M. Almira (Jaén, Spain)
J.M. Amigo (Elche, Spain)
J.M. Ansemil (Universidad Complutense de Madrid, Spain)
G. Arango (Medellín, Colombia)
P. Arjona (Badajoz, Spain)
F. Arranz (Badajoz, Spain)
J.L. Arregui (Zaragoza, Spain)
S. Ayupov (Institute of Mathematics of the Academy of Sciences, Uzbekistan)
J.J. Bañas (Badajoz, Spain)
J. Bastero (Zaragoza, Spain)
F. Bastin (Liège, Belgium)
M.I. Berenguer (Granada, Spain)
T. Bermúdez (La Laguna, Spain)
L. Bernal (Sevilla, Spain)
J. Bes (Bowling Green State University, USA)
K.D. Bierstedt (Paderborn, Germany)
F. Blasco Contreras (Universidad Politécnica de Madrid, Spain)
O. Blasco de la Cruz (Universidad de Valencia, Spain)
J.L. Blasco Olcina (Universidad de Valencia, Spain)
G. Blower (Lancaster, United Kingdom)
K. Bogalska (Poznań, Poland)
K. Bolibok (Lublin, Poland)
F. Bombal (Universidad Complutense de Madrid, Spain)
J. Bonet (Universidad Politécnica de Valencia, Spain)
A. Bonilla (La Laguna, Spain)
C. Bouzar (Oran, Algeria)
C. Boyd (University College Dublin, Ireland)
R.W. Braun (Düsseldorf, Germany)
M.P. Bravo (Universidad Politécnica de Valencia, Spain)
E. Briem (Reykjavik, Iceland)
F. Cabello Sánchez (Badajoz, Spain)

J.M. Calabuig (Universidad de Valencia, Spain)
A.J. Calderón Martín (Cádiz, Spain)
M.C. Calderón Moreno (Sevilla, Spain)
A. Cañada (Granada, Spain)
B. Cascales (Murcia, Spain)
J.M.F. Castillo (Badajoz, Spain)
J. Cerdà (Universidad de Barcelona, Spain)
K.C. Chan (Bowling Green State University, USA)
K.S. Chang (Yonsei University, Seoul, Korea)
M.J. Chasco (Universidad Pública de Navarra, Spain)
Y.S. Choi (Pohang, Korea)
R. Cilia (Catania, Italy)
I. Cioranescu (University of Puerto Rico, Puerto Rico)
J.A. Conejero (Universidad Politécnica de Valencia, Spain)
Z. Cuckovic (University of Toledo, USA)
H.G. Dales (Leeds, United Kingdom)
A. Defant (Oldenburg, Germany)
J.C. Díaz (Córdoba, Spain)
G. Dolinar (Ljubljana, Slovenia)
I. Domanov (Donetsk, Ukraine)
P. Domański (Poznań, Poland)
X. Domínguez (La Coruña, Spain)
L. Drewnowski (Poznań, Poland)
E. Effros (University of California, Los Angeles, USA)
H. Emamirad (Poitiers, France)
J. Eschmeier (Saarbrücken, Germany)
S. Falcón (Las Palmas de Gran Canaria, Spain)
C. Fernández Rosell (Universidad de Valencia, Spain)
F.J. Fernández y Fernández Arroyo (UNED, Spain)
J. Ferrer Llopis (Universidad de Valencia, Spain)
J.R. Ferrer Villanueva (Universidad Politécnica de Valencia, Spain)
M. Filali (Oulu, Finland)
C. Finet (Mons-Hainaut, Belgium)
J. Flores (Universidad Complutense de Madrid, Spain)
K. Floret (Oldenburg, Germany)
M. Fragoulopoulou (University of Athens, Greece)
W. Freedman (KOC University, Istanbul, Turkey)
F.J. Freniche (Sevilla, Spain)
L. Frerick (Wuppertal, Germany)
M. Friz (Universidad Politécnica de Valencia, Spain)
A. Galbis (Universidad de Valencia, Spain)
P. Galindo (Universidad de Valencia, Spain)
E. Gallardo (Cádiz, Spain)
T.W. Gamelin (University of California, Los Angeles, USA)
J.A. García del Amo (Salamanca, Spain)

D. García Rodríguez (Universidad de Valencia, Spain)
R. García González (Badajoz, Spain)
L.M. García Raffi (Universidad Politécnica de Valencia, Spain)
J.C. García Vázquez (Sevilla, Spain)
D. Garling (London Mathematical Society, United Kingdom)
M.B. Ghaemi (Glasgow, United Kingdom)
F. Ghahramani (University of Manitoba, Canada)
G. Godefroy (Paris VI, France)
M.C. Gómez Collado (Universidad Politécnica de Valencia, Spain)
J. Gómez Pérez (León, Spain)
A. Goncharov (Bilkent University, Ankara, Turkey)
J.L. González Llavona (Universidad Complutense de Madrid, Spain)
M. González Ortiz (Cantabria, Spain)
R. Gonzalo (Universidad Politécnica de Madrid, Spain)
E. Gootman (University of Georgia, USA)
B. Gramsch (Mainz, Germany)
E.E. Granirer (University of British Columbia, Canada)
B.C. Grecu (Galway, Ireland)
V. Gregori (Universidad Politécnica de Valencia, Spain)
P. Gregori (Universidad de Valencia, Spain)
G. Grillo (Torino, Italy)
K.G. Grosse-Erdmann (Hagen, Germany)
T. Gulinskiy (Leeds, United Kingdom)
R. Haller (Tartu, Estonia)
K.H. Han (Pohang, Korea)
M. Haralampidou (University of Athens, Greece)
A.Y. Helemskii (Moscow State University, Russia)
J. Hernández (Universidad Politécnica de Madrid, Spain)
S. Hernández Múñoz (Castellón, Spain)
F. Hernández Rodríguez (Universidad Complutense de Madrid, Spain)
K. Hoe Woon (Kyungpook National University, Taegu, Korea)
S. Holtmanns (Paderborn, Germany)
T. Honda (Ariake National College of Technology, Japan)
D. Ives (University College London, United Kingdom)
J.A. Jaramillo (Universidad Complutense de Madrid, Spain)
H. Jarchow (Universität Zürich, Switzerland)
B. Jefferies (University of New South Wales, Sidney, Australia)
E. Jordà (Universidad Politécnica de Valencia, Spain)
D. Jornet (Universidad Politécnica de Valencia, Spain)
H. Junek (Potsdam, Germany)
U. Kahre (Tartu, Estonia)
J. Kakol (Poznań, Poland)
O. Kalenda (Prague, Czech Republic)
N.J. Kalton (University of Missouri, USA)
I. Karabash (Donetsk, Ukraine)

J.C. Kelly (Hull University, United Kingdom)
J. Kim (Pohang, Korea)
V. Kisil (Leeds, United Kingdom)
A. Kryczka (Lublin, Poland)
S. Lajara (Murcia, Spain)
M. Langenbruch (Oldenburg, Germany)
N.J. Laustsen (Leeds, United Kingdom)
A.J. Lazar (Tel Aviv, Israel)
C. Leranoz (Universidad Pública de Navarra, Spain)
J. Lindenstrauss (The Hebrew University of Jerusalem, Israel)
M. Lindström (Åbo Akademi University, Finland)
B.V. Loginov (Ulyanovsk State Technical University, Russia)
J. López Abad (Universidad Autónoma de Barcelona, Spain)
B. López Brito (Las Palmas de Gran Canaria, Spain)
M. López Martín (Universidad Politécnica de Valencia, Spain)
J.A. López Molina (Universidad Politécnica de Valencia, Spain)
M. López Pellicer (Universidad Politécnica de Valencia, Spain)
M.L. Lourenço (Universidade de São Paulo, Brazil)
W. Lusky (Paderborn, Germany)
M. Mackey (University College Dublin, Ireland)
M. Maestre (Universidad de Valencia, Spain)
G. Manjabacas (Albacete, Spain)
V.M. Manuilov (Moscow State University, Russia)
P. Maritz (Stellenbosch, South Africa)
A. Martínez Abejon (Oviedo, Spain)
F. Martínez Giménez (Universidad Politécnica de Valencia, Spain)
E. Martín-Peinador (Universidad Complutense de Madrid, Spain)
M. Mathieu (Belfast, Ireland)
E. Matouskova (Czech Academy of Sciences)
M. Mauer (Liège, Belgium)
R. Meise (Düsseldorf, Germany)
P. Miana (Zaragoza, Spain)
A. Michalak (Poznań, Poland)
C. Michels (Oldenburg, Germany)
J.A. Mira (Alicante, Spain)
J.M. Mira Ros (Murcia, Spain)
R. Miralles-Rafart (UNED, Spain)
A. Moltó (Universidad de Valencia, Spain)
A. Montes (Universidad de Sevilla, Spain)
V. Montesinos (Universidad Politécnica de Valencia, Spain)
G. Mora (Alicante, Spain)
L.A. de Moraes (Universidade Federal do Rio de Janeiro, Brazil)
A. Morales Campoy (Almería, Spain)
I. Morales González (Universidad Politécnica de Valencia, Spain)
Y. Moreno (Badajoz, Spain)

J. Motos (Universidad Politécnica de Valencia, Spain)
G.A. Muñoz Fernández (Universidad Complutense de Madrid, Spain)
M. Muñoz Guillermo (Murcia, Spain)
A.B. Muravnik (Moscow State Aviation Institute, Russia)
A. Naor (The Hebrew University of Jerusalem, Israel)
F.J. Naranjo (Sevilla, Spain)
L. Narici (St. John's University, New York, USA)
G.F. Nassopoulos (University of Athens, Greece)
M. Nawrocki (Poznán, Poland)
N.J. Nielsen (Odense, Denmark)
E. Nieto (Granada, Spain)
I. Novikov (Voronezh State University, Russia)
O. Nygaard (Agder University College, Norway)
F. Oertel (Bonn, Germany)
J. Orihuela (Murcia, Spain)
L. Oubbi (Rabat, Morocco)
V. Palamodov (Tel Aviv, Israel)
A. Pallarés (Murcia, Spain)
P.L. Papini (Bologna, Italy)
J. Parcet (Universidad Autónoma de Madrid, Spain)
P.J. Paúl (Sevilla, Spain)
R. Payá (Granada, Spain)
A. Pelczar (Kraków, Poland)
A. Pełczyński (Polish Academy of Sciences)
A.M. Peralta (Granada, Spain)
A. Peris (Universidad Politécnica de Valencia, Spain)
M. Petrakis (Crete, Greece)
F. Piquard (Cergy, France)
A. Pirkovskii (Moscow State University, Russia)
M. Pöldvere (Tartu, Estonia)
M. Poppenberg (Dortmund, Germany)
J.A. Prado-Tendero (Sevilla, Spain)
A. Prieto (Universidad Complutense de Madrid, Spain)
M. Raja (Murcia, Spain)
A. Rakhimov (University of World Economy and Diplomacy, Uzbekistan)
B. Randrianantoanina (Miami University, Oxford, Ohio, USA)
N. Randrianantoanina (Miami University, Oxford, Ohio, USA)
Y. Raynaud (Paris VI, France)
N. Ribarska (Sofia University, Bulgaria)
W. Ricker (University of New South Wales, Sidney, Australia)
R. Ríos (Sevilla, Spain)
M. Rivas (Universidad de los Andes, Venezuela)
M.J. Rivera (Universidad Politécnica de Valencia, Spain)
J. Rodríguez López (Universidad Politécnica de Valencia, Spain)
A. Rodríguez Palacios (Granada, Spain)

B. Rodríguez Salinas (Universidad Complutense de Madrid, Spain)
M. Romance (Zaragoza, Spain)
P. Rubio Díaz (Universidad Politécnica de Cataluña, Spain)
J. Rubio Massegú (Universidad Politécnica de Cataluña, Spain)
M.P. Rueda (Universidad de Valencia, Spain)
R. Ryan (Galway, Ireland)
D.L. Salinger (Leeds, United Kingdom)
M.G. Sánchez Lirola (Almería, Spain)
E.A. Sánchez Pérez (Universidad Politécnica de Valencia, Spain)
L.M. Sánchez Ruíz (Universidad Politécnica de Valencia, Spain)
Y. Sarantopoulos (National Technical Univerity of Athens, Greece)
S. Saxon (University of Florida, USA)
J. Schmets (Liège, Belgium)
B.M. Schreiber (Wayne State University, USA
and The Hebrew University of Jerusalem, Israel)
K. Seip (Trondheim, Norway)
E. Semenov (Voronezh State University, Russia)
P. Sevilla (Universidad de Valencia, Spain)
J.H. Shapiro (Michigan State University, USA)
T. Signes (C.E.S. Felipe II, Madrid, Spain)
A. Siskakis (Thessaloniki, Greece)
A. Sofi (Kashmir, India)
H.G. Song (Pohang, Korea)
J. Stegeman (Utrecht, Netherlands)
C. Stuart (Eastern New Mexico University, USA)
A. Suárez Granero (Universidad Complutense de Madrid, Spain)
J. Suárez Ruíz (Murcia, Spain)
K. Sung Guen (Kyungpook National University, Taegu, Korea)
V. Tarieladze (Vigo, Spain and Georgian Academy of Sciences)
J. Taskinen (Joensuu, Finland)
C. Taylor (Tallaght, Ireland)
R.M. Timoney (Trinity College Dublin, Ireland)
I. Tkachenko (Universidad Politécnica de Valencia, Spain)
N. Tomczak-Jaegermann (University of Alberta, Canada)
T. Tonev (University of Montana, Missoula, USA)
V.A. Trenogin (Moscow Steel and Alloys Institute, Russia)
S. Troyanski (Murcia, Spain and Sofia University, Bulgaria)
I. Tweddle (University of Strathclyde, Glasgow, United Kingdom)
H.O.J. Tylli (Helsinki, Finland)
A. Ulger (KOC University, Istanbul, Turkey)
M. Urrea (Universidad Politécnica de Valencia, Spain)
M. Valdivia (Universidad de Valencia, Spain)
M.V. Velasco (Granada, Spain)
M. Venkova (University College Dublin, Ireland)
G. Vera (Murcia, Spain)

R. Vidal (Vigo, Spain)
R. Villa (Universidad de Sevilla, Spain)
I. Villanueva (Universidad Complutense de Madrid, Spain)
A. Villena (Granada, Spain)
D. Vogt (Wuppertal, Germany)
L. Volevich (Keldysh Institute of Applied Mathematics, Russia)
D. Vukotic (Universidad Autónoma de Madrid, Spain)
J. Wengenroth (Trier, Germany)
D. Werner (Galway, Ireland)
W. Werner (Paderborn, Germany)
A. Wiśnicki (Lublin, Poland)
W. Wnuk (Poznán, Poland)
P. Wojtaszczyk (Warszaw, Poland)
J. Wośko (Lublin, Poland)
J.D.M. Wright (Reading, United Kingdom)
I. Yoo (Yonsei University, Wonju, Korea)
D. Yost (King Saud University, Saudi Arabia)
A. Zagorodnyuk (Ukranian Academy of Sciences, Ukraine)
W. Zelazko (Polish Academy of Siences)
V. Zizler (Czech Academy of Sciences)
N. Zorboska (University of Manitoba, Canada)

LIST OF CONTRIBUTORS

M.D. Acosta, J. Becerra Guerrero and M. Ruiz Galán, Departamento de Análisis Matemático, Universidad de Granada, E-18071 Granada, Spain, dacosta@goliat.ugr.es

V.M. Adamyan, Department of Theoretical Physics, Odessa National University, 65026 Odessa, Ukraine

O. Blasco, Departamento de Análisis Matemático, Universidad de Valencia, E-46100 Burjassot (Valencia), Spain, Oscar.Blasco@uv.es

R.W. Braun and R. Meise, Mathematisches Institut, Heinrich-Heine-Universität, Universitätsstrasse 1, D-40225 Düsseldorf, Germany, braun@cs.uni-duesseldorf.de and meise@cs.uni-duesseldorf.de

F. Cabello Sánchez and R. García, Departamento de Matemáticas, Universidad de Estremadura, Avenida de Elvas, E-06071 Badajoz, Spain, fcabello@unex.es and rgarcia@unex.es

A.J. Calderón Martin, Departamento de Matemáticas, Universidad de Cádiz, E-11510 Puerto Real (Cádiz), Spain, ajesus@uca.es

J.M.F. Castillo and Y. Moreno, Departamento de Matemáticas, Universidad de Extremadura, Avenida de Elvas, E-06071 Badajoz, Spain, castillo@unex.es and y.moreno@unex.es

H.G. Dales, Department of Pure Mathematics, University of Leeds, Leeds, LS2 9JT, UK, pmt6hgd@amsta.leeds.ac.uk

A. Defant, Fachbereich Mathematik, Carl von Ossietzky Universität Oldenburg, Postfach 2503, D-26111 Oldenburg, Germany, defant@mathematik.uni-oldenburg.de

P. Domański, Faculty of Mathematics and Computer Science, A. Mickiewicz University, ul. Matejki 48/49, PL-60-769 Poznań, Poland, domanski@amu.edu.pl

L. Drewnowski, Faculty of Mathematics and Computer Science, A. Mickiewicz University, ul. Matejki 48/49, PL-60-769 Poznań , Poland, drewlech@amu.edu.pl

K. El Amin and A. Morales Campoy, Departamento de Algebra y Análisis Matemático, Facultad de Ciencias Experimentales, Universidad de Almeria, E-04120 Almaria, Spain, elamin@ual.es and amorales@ual.es

J. Eschmeier, Fachbereich Mathematik, Universität des Saarlandes, Postfach 15 11 50, D-66041 Saarbrücken, Germany, eschmei@math.uni-sb.de

M. Filali, Department of Mathematical Sciencies, University of Oulu, FIN-90014 Oulu, Finland, Mahmoud.Filali@oulu.fi

K. Floret, Fachbereich Mathematik, Carl von Ossietzky Universität Oldenburg, Postfach 2503, D-26111 Oldenburg, Germany, floret@mathematik.uni-oldenburg.de

F.J. Freniche, J.C. García-Vázquez and L. Rodríguez-Piazza, Departamento de Análisis Matemático, Universidad de Sevilla, P.O. Box 41080, Sevilla, Spain, freniche@cica.es

T.W. Gamelin, Mathematics Department, U.C.L.A., 405 Hilgard Avenue, Los Angeles, CA 90095-1555, USA, twg@math.ucla.edu

N. Kalton, Department of Mathematics, University of Missouri, Columbia, Missouri 65211, USA, nigel@math.missouri.edu

J. Lindenstrauss, Department of Mathematics, The Hebrew University, Jerusalem, Israel, joram@math.huji.ac.il

C. Martín Gonzáles, Departamento de Algebra, Geometría y Topologia, Universidad de Málaga, Apartado 59, E-29080 Málaga, Spain

M. Mastyło, Faculty of Mathematics and Computer Science, A. Mickiewicz University, ul. Matejki 48/49, PL-60-769 Poznań, Poland, mastylo@amu.edu.pl

M. Mathieu, Department of Pure Mathematics, Queen's University Belfast, Belfast BT7 1NN, Northern Ireland, m.m@qub.ac.uk

P.J. Miana, Departamento de Matemáticas, Universidad de Zaragoza, E-50009 Zaragoza, Spain, pjmiana@posta.unizar.es

C. Michels, Department of Pure Mathematics, University of Leeds, Leeds, LS2 9JT, UK, balin@amsta.leeds.ac.uk

A.B. Muravnik, Moscow Aviation Institute, Department of Differential Equations, Volokolamskoe shosse 4, 125871, Moscow, Russia, muravnik@postmaster.co.uk

E. Nieto, Departamento de Análisis Matemático, Facultad de Ciencias, Universidad de Granada, E-18071 Granada, Spain

A.M. Pelczar, Institute of Mathematics, Jagiellonian University, Reymonta 4, PL-30-059 Kraków, Poland, apelczar@im.uj.edu.pl

A. Pełczyński and M. Wojciechowski, Institute of Mathematics, Polish Academy of Sciencies, Śniadeckich 8, PL-00-950 Warszawa, Poland, olek@impan-gor.pl and miwoj@impan.gor.pl

A.M. Peralta and A. Rodríguez Palacios, Departamento de Análisis Matemático, Facultad de Ciencias, Universidad de Granada, E-18071 Granada, Spain, aperalta@goliat.ugr.es and apalacio@goliat.ugr.es

M. Poppenberg, Fachbereich Mathematik, Universität Dortmund, D-44221 Dortmund, Germany, poppenberg@math.uni-dortmund.de

D. Preiss, Department of Mathematics, University College London, London WC1E6BT, UK, dp@math.ucl.ac.uk

W. Ricker, School of Mathematics, University of New South Wales, Sydney, NSW, 2052, Australia, werner@maths.unsw.edu.au

J. Schmets, Institut de Mathématique, Université de Liège, Sart Tilman Bât. B 37, B-4000 Liège 1, Belgium, j.schmets@ulg.ac.be

B.M. Schreiber, Department of Mathematics, Wayne State University, Detroit, MI 48202, USA, berts@math.wayne.edu

J.H. Shapiro, Department of Mathematics, Michigan State University, East Lansing, MI 48824-1027, USA, shapiro@math.msu.edu

J. Taskinen, Department of Mathematics, University of Joensuu, P.O. Box 111, FIN-80101 Joensuu, Finland, jtaskine@joyx.joensuu.fi

B.A. Taylor, Department of Mathematics, University of Michigan, Ann Arbor, MI 48109, USA, taylor@math.lsa.umich.edu

I.M. Tkachenko, Departamento de Matemática Aplicada, Universidad Politécnica de Valencia, E-46022 Valencia, Spain, imtk@mat.upv.es

D. Vogt, Bergische Universität Wuppertal, FB Mathematik, Gaußstr. 20, D-42097 Wuppertal, Germany, dvogt@math.uni-wuppertal.de

J.D.M. Wright, Mathematics Department, University of Reading, Reading RG 6 6AX, UK, j.d.m.wright@reading.ac.uk

TABLE OF CONTENTS

Recent Progress in Functional Analysis
K.D. Bierstedt, J. Bonet, M. Maestre and J. Schmets

The Mathematical Works of Manuel VALDIVIA, II

Jean SCHMETS

Institut de Mathématique, Université de Liège, Sart Tilman Bât. B 37, B-4000 Liège 1, Belgium; e-mail: j.schmets@ulg.ac.be

1. Introduction

At the International Functional Analysis Meeting on the Occasion of the 60th Birthday of Professor M. Valdivia, which took place at Peñíscola on October 22–27, 1990, John Horváth has had the great honour and pleasure to present the mathematical works of Manuel Valdivia.

Since then this presentation has appeared as [Ho] at the beginning of the Proceedings of the Meeting [FA], a volume of the well-known North-Holland Mathematics Studies series. It certainly was quite a tour de force to write down a remarkable presentation of 114 publications in some 44 pages, i.e. the scientific production of Manuel Valdivia from his first paper back in 1963 up to early 1989. If you give a look at these Proceedings, you will remark that on page 55, the editors have added a list of 13 more articles which appeared or were to appear in the short period in between 1989 and 1992. The last sentence of John Horváth's presentation was indeed prophetic. It was saying "I wish" Manuel Valdivia "many more years of happy and fruitful research activity". This certainly has been the case since by now on top of those 114 publications, some 42 new ones are born and ... more are coming.

In these notes I try to assume the responsibility to present these new mathematical works of Manuel Valdivia.

While preparing this address I really measured the quality and the amount of research done by Manuel Valdivia.

2. Foreword

Let me take some caution; about the same as the one John Horváth took.

It would make no sense to try to present each new result of each new paper of M. Valdivia one by one; this would require an enormous amount of space and of time. I have had to make a severe selection. This leaves away numerous deep theorems but there was no way around. What is worse is that even so there is no place either for describing any proof although it is there that you find what John Horváth called the "stupendous ingenuity" of Manuel. I will just advise the following: go to the papers, read them and work on them, then you will realize what "stupendous ingenuity" really means.

Another caution deals with the language: in places, it will get a bit loose. For instance, the word "space" may often be used instead of "Hausdorff topological space". Another possibility is that I may omit from time to time some obvious hypotheses in order to make the presentation friendlier. In case of a doubt, please go to the paper and check.

The list of the "Publications of Manuel Valdivia" at the end of this presentation starts with the item [115]; the 114 first ones refer of course to the corresponding list of [Ho].

Now we are ready to start the survey of the mathematical works of Manuel Valdivia during the period starting in 1989 and going to early 2000, developed by themes.

3. Banach spaces

In the late 80's and early 90's, a large part of the research of M. Valdivia deals with the theory of Banach spaces. It mainly concerns the existence of projective resolutions of the identity, specialized notions of compactness, Markushevich bases, basic sequences, rotundness, ...

Unless specifically otherwise stated, X denotes a Banach space, X^* its topological dual and $B(X^*_\sigma)$ the closed unit ball of X^* endowed with the weak* topology $\sigma(X^*, X)$.

3.1. Projective resolution of the identity

For a set A, let $|A|$ denote the cardinal number of A and, for a Banach space X, let then $\text{dens}(X)$ be the smallest cardinal number λ for which there is a dense subset D of X verifying $|D| = \lambda$. Moreover ω_0 is the first infinite ordinal and ω_1 the first uncountable ordinal.

A *projective resolution of the identity* — for short, a PRI — in X is a well ordered family $\{P_\alpha \colon \omega_0 \leq \alpha \leq \mu\}$ of continuous linear projections in X, where μ is the first ordinal such that $|\mu| = \text{dens}(X)$, satisfying the following conditions:
(1) $\|P_\alpha\| = 1$,
(2) $\text{dens}(P_\alpha(X)) \leq |\alpha|$,
(3) $P_\alpha P_\beta = P_\beta P_\alpha = P_\beta$ if $\omega_0 \leq \beta \leq \alpha \leq \mu$,
(4) $P_\mu = I_X$,
(5) for every limit ordinal α such that $\omega_0 \leq \alpha \leq \mu$, $\cup_{\omega_0 \leq \beta < \alpha} P_\beta(X)$ is dense in $P_\alpha(X)$.

D. Amir and J. Lindenstrauss have shown [AL] that the construction of a PRI is an important tool in the theory of Banach spaces. However getting a PRI mostly is quite a difficult task. To partially overcome this difficulty, S. Gul'ko introduced [Gu] the notion of conjugate pairs of topological spaces. But there are many Banach spaces with a PRI that fail to have such a conjugate pair.

In [115], J. Orihuela and M. Valdivia overcome this problem. In fact they adapt a method developed earlier by M. Valdivia in ([108], [113], [114] and [117]) to construct projections. This leads them to the introduction of a more flexible notion to obtain a PRI in Banach spaces, the notion of *projective generator*. This is a set-valued function φ defined on a norming subset of X^* such that $\varphi(f)$ is a countable subset of X for every element f of the domain of φ, verifying some additional technical conditions. The key lies in the notion of *norming pairs*, a particular kind of the preconjugate pairs of Gul'ko [Gu], that leads naturally to norm one projections hence to projective generators. Then inspired by an idea of M. Fabian and G. Godefroy [FG], they prove that a Banach space is an Asplund space (i.e. the dual of every separable subspace of X is separable or equivalently X^* has the Radon-Nikodym property) if and only if X^* has a projective generator. Next they get the following basic and deep property: a Banach space with a projective generator has a particular PRI and derive therefrom the following known results as direct corollaries: every weakly countably determined Banach space has a PRI [Va] and every dual Banach space with the Radon-Nikodym property has a PRI [FG].

Given a compact subset K of $[0,1]^I$, let $K(I)$ denote the set of the elements x of K such that $\{i \in I: x_i \neq 0\}$ is countable. Then let $\mathcal{A}$ be the family of the now so-called Valdivia compact spaces, i.e. of the topological spaces homeomorphic to a compact subset K of some $[0,1]^I$ such that $K(I)$ is dense in K. This family $\mathcal{A}$ contains all the Corson compact spaces since H. H. Corson proved [Co] that a topological space is Corson compact if and only if it is homeomorphic to a compact subspace of $\mathbb{R}^I$ the elements of which have countably many non zero components. In [117] M. Valdivia proves the existence of a particular PRI in $C(K)$ for every $K \in \mathcal{A}$. His main result is as follows. Let K be an infinite element of $\mathcal{A}$ and let μ be the first ordinal number such that $|\mu| = \text{dens}(K)$. Then there is a family $\{K_\alpha : \omega_0 \le \alpha \le \mu\}$ of compact subsets of K belonging to $\mathcal{A}$ and, for every $\alpha \in [\omega_0, \mu]$, a continuous linear extension map T_α from $C(K_\alpha)$ into $C(K)$ such that $\{T_\alpha(\cdot|_{K_\alpha}) : \omega_0 \le \alpha \le \mu\}$ is a PRI in $C(K)$. As a consequence, for every continuous image K of an element of $\mathcal{A}$, the space $C(K)$ has an equivalent locally uniformly rotund norm (this notion is defined in 3.5).

R. Deville, G. Godefroy and V. Zisler asked in [DGZ], whether this class $\mathcal{A}$ is stable under continuous maps. In [147], M. Valdivia provides a negative answer to this question: indeed he produces a compact space $K \notin \mathcal{A}$ which is a continuous image of $[0, \omega_1]$ and such that $C(K)$ is isometric to a hyperplane of $C([0, \omega_1])$ and isomorphic to $C([0, \omega_1])$.

3.2. Compact spaces

Over the years the notion of compactness has been refined a lot.

For instance a topological space is *Eberlein* (resp. *Radon-Nikodym*; *Gul'ko*; *Talagrand*) *compact* iff it is homeomorphic to a weakly compact subset of a Banach space (resp. to a weak* compact subset of the dual of an Asplund space; to a weak* compact subset of the dual of a weakly compactly generated space; to a weak* compact subset of the dual of a weakly K-analytic Banach space). These notions have been quite deeply investigated. In [DFJP], W. J. Davis, T. Figiel, W. B. Johnson and A. Pełczynski have proved that every Eberlein compact space is Radon-Nikodym compact. Results of S. Gul'ko [Gu] and M. Talagrand [Ta] assert that the following implications:

$$\text{Eberlein} \Rightarrow \text{Talagrand} \Rightarrow \text{Gul'ko} \Rightarrow \text{Corson}$$

hold. In [Na], I. Namioka dealt with the Radon-Nikodym compact spaces and proved that this notion is equivalent to the fragmentation by a lower semi-continuous metric, hence implies the fragmentation by a metric. So two lines of implications were at hand and he asked three questions about their links.

The first answer is due to E. A. Reznichenko [Ke] who provided a Talagrand compact space which is not Radon-Nikodym compact.

J. Orihuela, W. Schachermayer and M. Valdivia give a full answer in [120] to these three questions. They first establish that Talagrand's initial example [Ta] of a Talagrand compact space failing to be Eberlein compact already is not Radon-Nikodym compact.

They also prove what is announced in the title: every Radon-Nikodym and Corson compact space is Eberlein compact. The paper also contains a Banach space version of this result, i.e. a Banach space X is weakly compactly generated if (and only if) its dual unit ball is Corson compact and if there is a continuous linear map $T\colon Y \to X$ with dense range, where Y is an Asplund space. The proof deeply relies on the existence of a PRI.

3.3. Markushevich bases

A biorthogonal system $(x_i, u_i)_{i\in I}$ in a Banach space X is *complete* if the set $\{x_i\colon i \in I\}$ is total in X, *total* if the set $\{u_i\colon i \in I\}$ is total in X^*_σ and a *Markushevich basis* if it is complete and total. A Markushevich basis is *associated to the* PRI $\{P_\alpha\colon \omega_0 \leq \alpha \leq \mu\}$ if there is a partition $\{I_{\omega_0}\}\cup\{I_{\alpha+1}\colon \omega_0 \leq \alpha < \mu\}$ of the index set I such that $(x_i, u_i|_{P_{\omega_0}(X)})_{i\in I}$ is a Markushevich basis in $P_{\omega_0}(X)$ and, for every $\omega_0 \leq \alpha < \mu$, $(x_i, u_i|_{(P_{\alpha+1}-P_\alpha)(X)})_{i\in I}$ a Markushevich basis in $(P_{\alpha+1} - P_\alpha)(X)$.

In [121], M. Valdivia constructs a PRI and a Markushevich basis in certain Banach spaces. More explicitely, let X be a Banach space and M be a total subset of X such that the intersection of

$$S(M) := \{u \in X^*\colon \{x \in X\colon \ < x, u > \neq 0\} \text{ is countable}\}$$

with $B(X^*)$ is dense in $B(X^*_\sigma)$. Then if μ is the first ordinal number such that $|\mu| = \text{dens}(X)$, there is a PRI $\{P_\alpha\colon \omega_0 \leq \alpha \leq \mu\}$ in X and a partition $\{M_{\omega_0}\} \cup \{M_{\alpha+1}\colon \omega_0 \leq \alpha < \mu\}$ of M such that

$$M_{\omega_0} \subset P_{\omega_0}X \text{ and } M_{\alpha+1} \subset (P_{\alpha+1} - P_\alpha)(X), \quad \forall \alpha \in [\omega_0, \mu[.$$

This extends a result of A. N. Plichko [Pl]. There also is a Markushevich basis $(x_i, u_i)_{i\in I}$ associated to a PRI $\{P_\alpha\colon \omega_0 \leq \alpha \leq \mu\}$ such that the linear hulls of M and of $\{x_i\colon i \in I\}$ coincide and such that $S(M) = S(\{x_i\colon i \in I\})$.

In [125], M. Valdivia refines the construction set up in [117]. This leads him to results on the existence of a PRI in $C(K)$ spaces that induce resolutions of the identity simultaneously on countably many subspaces. He also gets the existence of a Markushevich basis associated with a PRI on a subspace of $C(K)$. As corollaries, he obtains the following consequences. Let X be a Banach space such that $B(X^*_\sigma)$ is Corson compact. Then every closed subspace L of X has a quasicomplement in X (i.e. there is a closed subspace M of X such that $L \cap M = 0$ and $L + M$ is dense in X) and X itself has a quasicomplement in $C(B(X^*_\sigma))$.

M. Valdivia derives consequences of the existence of a Markushevich basis in [137]. One of the statements deals with a Banach space X such that $B(X^*_\sigma)$ is of weak*-countable tightness, i.e. every element u of the weak*-closure of a subset of $B(X^*_\sigma)$ belongs to the weak*-closure of a countable subset of $B(X^*_\sigma)$. In such a case, if X^* has a Markushevich basis $(u_i, z_i)_{i\in I}$ such that the closed linear hull of $\{z_i\colon i \in I\}$ contains X, then X is weakly compactly generated. This leads to an example of a scattered compact space K and of an equivalent norm $|\cdot|$ on $C(K)$ such that $(C(K), |\cdot|)^*$ has no 1-norming Markushevich basis. Another result concerns Banach spaces X such that X^* is non separable and weakly countably determined. It states the existence of $\delta \in]0, 1[$ and of a Banach space Y such that Y^* is isomorphic to $X^* \oplus \ell_1$ and has no α-norming Markushevich basis for $\delta \leq \alpha \leq 1$.

In [130], M. Valdivia uses and refines methods developed in [114], with the idea of obtaining total biorthogonal systems. The main result asserts the following. Let Y be an infinite dimensional closed subspace of a Banach space X such that $\text{dens}(Y) \geq \text{dens}(X^*_\sigma)$. If $B(Y^*_\sigma)$ is Corson compact, then there is a total biorthogonal system $(x_i, u_i)_{i\in I}$ in X such that $\{x_i : i \in I\}$ is total in Y.

In [141], M. Valdivia firstly considers Asplund spaces X and gets as a corollary that X admits a total biorthogonal system $(x_i, u_i)_{i\in I}$ such that the closed linear hull of $\{x_i : i \in I\}$ is weakly compactly generated. He next investigates the weakly countably convex determined spaces defined in [113] and obtains the following statement, by use of a method similar to the one developed in [130]. Let Y be a normed subspace of a Banach space X such that $\text{dens}(Y) \geq \text{dens}(X^*_\sigma)$. If Y is weakly countably convex determined, then there is a total biorthogonal system $(x_i, u_i)_{i\in I}$ in X such that the linear hull of $\{x_i : i \in I\}$ is a dense subspace of Y.

3.4. Basic sequences

In [126], M. Valdivia deals with basic sequences in Banach spaces and essentially establishes the following results. If X is a Banach space with a shrinking basis and a separable bidual, then for every closed subspace Z of X^{**} containing X, there is a shrinking basis $(x_n)_{n\in\mathbb{N}}$ in X and a partition $\{N_1, N_2\}$ of $\mathbb{N}$ such that the closed linear hull of $\{x_n : n \in N_1\}$ is reflexive and $X+Y = Z$ where Y is the weak*-closure of the linear hull of $\{x_n : n \in N_2\}$ in X^{**}. If $(x_n)_{n\in\mathbb{N}}$ is a normalized sequence of a Banach space X with separable bidual, then $\{x_n : n \in \mathbb{N}\}$ is not weakly relatively compact if and only if there is a subsequence such that the closed linear hull of each of its subsequences has codimension 1 in X^{**}. If the Banach space X has a basis, then every basic sequence contains a subsequence which extends to a basis of X.

Basic sequences are also revisited in [151]. As a corollary of a deep result, M. Valdivia gets the following property. Let X be a Banach space with separable dual and let Y, Z be two norming closed subspaces of X^* such that $Y \subset Z$. Then there is a basic sequence $(x_n)_{n\in\mathbb{N}}$ such that $Y + L = Z$ and $(Y \cap L)^{\sim} = L$ where $L = \{x_n : n \in \mathbb{N}\}^{\perp}$ and $A^{\sim}$ is the weak*-closure of $A \subset X^*$. This article also contains properties unifying or refining properties of W. B. Johnson and H. P. Rosenthal [JR] such as if X is a Banach space and if $(u_n)_{n\in\mathbb{N}}$ is a weak*-Cauchy sequence in X^* equivalent to the unit vector basis of ℓ_1, then X/L is isomorphic to c_0 where $L = \{u_n : n \in \mathbb{N}\}_{\perp}$. As a corollary based on a result of J. Hagler and W. B. Johnson [HJ], M. Valdivia also gets that if the Banach space X contains no copy of ℓ_1 and if X^* contains a copy of ℓ_1, then there is a quotient of X isomorphic to c_0.

3.5. Uniform rotundness

A normed space $(X, \|\cdot\|)$ is *uniformly rotund* (resp. *weakly uniformly rotund*) — in short UR (resp WUR) — if given sequences $(x_n)_{n\in\mathbb{N}}$ and $(y_n)_{n\in\mathbb{N}}$ of the unit sphere of X such that $\|x_n + y_n\| \to 2$, the sequence $x_n - y_n$ converges (resp. weakly converges) to 0. It is *locally uniformly rotund* (resp. *weakly locally uniformly rotund*) — in short LUR (resp WLUR) — if given a point x and a sequence $(x_n)_{n\in\mathbb{N}}$ of the unit sphere of X such that $\|x + x_n\| \to 2$, the sequence $x - x_n$ converges (resp. weakly converges) to 0. Of course the following implications hold:

$$\begin{array}{ccc} \text{UR} & \Rightarrow & \text{WUR} \\ \Downarrow & & \Downarrow \\ \text{LUR} & \Rightarrow & \text{WLUR} \end{array}$$

The interest of these notions in renorming theory comes from the fact that they may characterize geometric or topological properties of normed spaces. For instance a result of R. C. James, P. Enflo and G. Pisier states that a Banach space is superreflexive if and only if it has an equivalent UR norm.

It was well known that, with the exception of LUR $\Rightarrow$ WLUR, the inverse implications do not hold for equivalent norms. So several authors investigated conditions under which WLUR implies LUR. For instance in [DGZ], R. Deville, G. Godefroy and V. Zisler proved that a WLUR Banach space with a Fréchet differentiable norm is LUR renormable. In the same vein, R. Haydon [Ha] established that if T is a tree and if $C(T)$ is WLUR, then $C(T)$ has an equivalent LUR norm.

In [156], with A. Moltó, J. Orihuela and S. Troyanski, M. Valdivia solves completely the problem: every WLUR normed space has an equivalent LUR norm. The proof uses an elegant technique of countable covers of a topological space by sets of small local diameters. This paper also contains the fact that the unit sphere of a WUR normed space endowed with the weak topology is metrizable under a metric that may differ from the norm-metric.

3.6. Direct sum decomposition of spaces

In [119] and [128], M. Valdivia continues previous investigations on the direct sum decompositions of locally convex spaces.

The main result of [119] deals with two closed subspaces Y and Z of a Banach space X. If $Y \neq \{0\}$, $X = Y + Z$ and Z is weakly countably determined, then there is a continuous linear projection P on X such that $\|P\| = 1$, $PX \supset Y$, $\ker(P) \subset Z$ and $\text{dens}(TX) = \text{dens}(Y)$. It leads to the fact that every Banach space is a topological direct sum $X = X_1 \oplus X_2$ with X_1 reflexive and $\text{dens}(X_2^{**}) = \text{dens}(X^{**}/X)$, a result that he had already established under the assumption that X^{**}/X is separable [54].

In [128], M. Valdivia considers the case of a Fréchet space E such that $(E', \mu(E', E''))$ is barrelled. He then proves that E is the direct sum of two closed subspaces F and G such that G is reflexive and $\text{dens}(F'', \beta(F'', F')) \leq \text{dens}(E', \beta(E', E'')/E)$, generalizing the result of [119] mentioned above, as well as the one of [110].

4. Real analyticity

4.1. Some historical background

In the 1990's, M. Valdivia has been very much interested in generalizations of the Borel, Mityagin, Ritt and Whitney theorems. The historical setting can roughly be presented in the following way.

In [Bo], E. Borel proved that for every sequence $(c_n)_{n \in \mathbb{N}_0}$ of complex numbers, there is a C^∞-function f on $\mathbb{R}$ such that $f^{(n)}(0) = c_n$ for every $n \in \mathbb{N}_0$. The first deep improvement of this result is due to J. F. Ritt [Ri]: the function f may be supposed real analytic outside the origin, indeed holomorphic on an open sector of $\mathbb{C}$ with $\sum_{n=0}^{\infty} c_n z^n/n!$ as asymptotic behaviour at 0. Next comes the striking Whitney generalization [Wi] of these properties, characterizing the jets $\varphi = (\varphi_\alpha)_{\alpha\in\mathbb{N}_0^n}$ on a closed subset F of $\mathbb{R}^n$ coming from a function $f \in C^\infty(\mathbb{R}^n)$, i.e. such that $\varphi_\alpha = f^{(\alpha)}|_F$ for every $\alpha \in \mathbb{N}_0^n$. In fact, for these jets (since then known as Whitney jets), H. Whitney also proved that the function f may be supposed real analytic on $\mathbb{R}^n \setminus F$, indeed holomorphic on some open subset of $\mathbb{C}^n$ containing $\mathbb{R}^n \setminus F$. Moreover H. Whitney introduced the space $\mathcal{E}(\mathcal{F})$ of the Whitney jets on F, endowed it with a Fréchet structure and asked the following question: when does the continuous linear surjective restriction map $R\colon C^\infty(\mathbb{R}^n) \to \mathcal{E}(\mathcal{F})$ have a continuous linear right inverse? In other words, when is there a continuous linear extension map from F to $\mathbb{R}^n$ in the C^∞-setting? The first answers are due to B. Mityagin who proved in [Mi] that there is no continuous linear extension map from 0 to $\mathbb{R}$ but there is one from $[-1, 1]$ to $\mathbb{R}$.

These results have been extended in many different ways. The contribution of M. Valdivia deals mainly with the real-analyticity property of the extensions, a property which was not considered previously.

4.2. The Borel theorem in real Banach spaces

A first generalization of the Borel theorem with real-analytic extension outside the origin in a real normed space appears in [132]. This 'one direction' result can be stated as follows. Let X be a real normed space satisfying the Kurzweil condition (i.e. there is a polynomial P on X such that $P(0) = 0$ and $\inf\{P(x)\colon \|x\| = 1\} > 0$). Then for every direction ϵ and sequence $(a_n)_{n\in\mathbb{N}_0}$ of real numbers, there is a real C^∞-function f on X which is real-analytic on $X \setminus \{0\}$ and such that $D_\epsilon^n f(0) = a_n$ for every $n \in \mathbb{N}_0$.

Next in [134], M. Valdivia deals with real Hilbert spaces X and proves the following result for every real number A_0 and sequence $(A_n)_{n\in\mathbb{N}}$ of continuous symmetric n-linear functionals on X^n. There always is a holomorphic function on a domain of the complexification of X, containing $X \setminus \{0\}$, which has a real C^∞-extension f on X, bounded on the bounded subsets of X and such that $f^{(n)}(0) = A_n$ for every $n \in \mathbb{N}_0$.

This result is finally generalized to the setting of the real Banach spaces X in [146]. Let A_0 be a real number and for every $n \in \mathbb{N}_0$, let A_n be a n-linear symmetric and approximable real functional on X^n. Then there is a real C^∞-function f on X such that

a) $f^{(n)}(0) = A_n$ for every $n \in \mathbb{N}_0$,

b) $f^{(n)}(x)$ is approximable for every $x \in X$ and $n \in \mathbb{N}$,

c) $f^{(n)}$ is bounded on the bounded subsets of X for every $n \subset \mathbb{N}_0$,

d) f is real-analytic on $X \setminus \{0\}$ endowed with the topology of uniform convergence on the compact subsets of X^*.

4.3. Generalizing the Mityagin results

Since the Mityagin results appeared, intensive research has been going on to find examples of closed subsets F of $\mathbb{R}^n$ for which there is (is not) a continuous linear extension map, to characterize them by means of properties of the boundary of F or by means of locally convex properties of the Fréchet space $\mathcal{E}(\mathcal{F})$. This literature is very rich (cf. [150] for an attempt to describe the situation in the C^∞-setting around 1997).

This research has also been extended from the C^∞-setting to the Beurling and Roumieu type spaces of ultradifferentiable jets and functions. These can be defined by use of a weight w — then the appropriate definition of w is due to R. Meise and B. A. Taylor on the basis of one going back to A. Beurling (cf. [BMT]). They also can be defined by means of a normalized, logarithmically convex and non quasi-analytic sequence $\boldsymbol{M}$ of positive numbers. Let us designate by $\mathcal{E}_*(\mathbb{R}^n)$ (resp. $\mathcal{E}_*(\mathcal{F})$) the corresponding space of functions on $\mathbb{R}^n$ (resp. of jets on F). In this vast literature, the real-analyticity part of the Borel-Ritt theorem [Pe] or of the Whitney theorem ([BBMT], [BMT], [MT], ...) had not been considered.

M. Valdivia has been the first to investigate the possibility to get continuous linear extension maps from $\mathcal{E}_*(\mathcal{F})$ to $\mathcal{E}_*(\mathbb{R}^n)$, with real-analyticity on $\mathbb{R}^n \setminus F$. He first published two papers [142], [145] where he solves the problem when F is compact. His results brought more light and new importance to the results obtained by the previous authors. Here is the main property he got.

Let K be a compact subset of $\mathbb{R}^n$.

a) If the jet $\varphi \in \mathcal{E}_*(\mathcal{K})$ comes from a $\mathcal{E}_*(\mathbb{R}^n)$-function, then it also comes from an element of the same space which moreover is real-analytic on $\mathbb{R}^n \setminus K$, indeed holomorphic on an open subset of $\mathbb{C}^n$ containing $\mathbb{R}^n \setminus K$.

b) If there is a continuous linear extension map from $\mathcal{E}_*(\mathcal{K})$ into $\mathcal{E}_*(\mathbb{R}^n)$, then there also is such an extension map E such that $E\varphi$ is real-analytic on $\mathbb{R}^n \setminus K$ for every $\varphi \in \mathcal{E}_*(\mathcal{K})$.

The part a) of this result can be extended to the case of a closed subset of $\mathbb{R}^n$; the method is basically the same but requires quite refined arguments. M. Valdivia has done this successively for the Beurling type with $* = \boldsymbol{M}$ in [154], for the Roumieu type with $* = \boldsymbol{M}$ in [148], and for the Beurling and Roumieu types with $* = w$ in [155]. The very same method [150], but simpler, also leads to parts a) and b) of the theorem for $\mathcal{E}(\mathcal{K})$ instead of $\mathcal{E}_*(\mathcal{K})$ and $\mathrm{BC}^\infty(\mathbb{R}^n)$ instead of $\mathcal{E}_*(\mathbb{R}^n)$, where $\mathrm{BC}^\infty(\mathbb{R}^n)$ stands of course for the Fréchet space of the C^∞-function on $\mathbb{R}^n$ which are bounded on $\mathbb{R}^n$ as well as all their derivatives. Since then, M. Langenbruch [La] has set up a unified approach to get the $*$-cases, up to a condition when $* = \boldsymbol{M}$; it is based on the existence of a special function obtained in [150].

4.4. The Ritt theorem and the interpolation property

The Ritt theorem leads naturally to the following question: on which subsets D of the boundary $\partial\Omega$ of a non void domain Ω of $\mathbb{C}$ is it possible to fix arbitrarily the asymptotic behaviour of some holomorphic function? It is a direct matter to check that such a set D may not have any accumulation point and that no element of D may be an isolated point of $\partial\Omega$. Known results go back to the work of T. Carleman [Ca] who proved that the answer is positive in the following two cases:

(a) D is finite and Ω is convex and bounded,

(b) $D = \{0\}$ and $\Omega = \{z \in \mathbb{C} \colon |z| < R\} \setminus \{(x,0) \colon x \leq 0\}$,

and also to the work of Ph. Franklin [Fr] dealing with a case when D is infinite.

In [118], M. Valdivia proves that the answer is positive if D is finite and such that the answer is positive at each single point of D separately.

In [123], M. Valdivia completes this result in the following way: the answer is positive for $D = \{z_0\}$ if the connected component of z_0 in $\partial\Omega$ has more than one point. The proof lies on a deep study of the space of the holomorphic functions on Ω which have an asymptotic behaviour at the point z_0, endowed with an appropriate locally convex topology.

The next and final step is contained in [138] where the main statement contains a rather technical condition leading to the following corollary which generalizes all previous known results. If $D \subset \partial\Omega$ (finite or not) has no accumulation point and if the connected component in $\partial\Omega$ of any point of D contains more than one point, then the answer is positive.

4.5. Domains of real analytic existence

Let X be a real normed space. A domain Ω of X is of *real-analytic existence* if there is a real-analytic function f on Ω such that, for every domain Ω_1 of X verifying $\Omega_1 \not\subset \Omega \not\subset X \setminus \Omega_1$ and every connected component Ω_0 of $\Omega \cap \Omega_1$, the restriction $f|_{\Omega_0}$ has no real-analytic extension onto Ω_1. It is a *real-analytic domain* if, for every domain Ω_1 of X verifying $\Omega_1 \not\subset \Omega \not\subset X \setminus \Omega_1$ and every connected component Ω_0 of $\Omega \cap \Omega_1$, there is a real-analytic function f on Ω such that $f|_{\Omega_0}$ has no real-analytic extension onto Ω_1. Of course every real-analytic existence domain is a real-analytic domain.

M. Valdivia has considered the characterization of such domains twice: in [132] and [136]. The final result is as follows. For every non void domain Ω of a separable real normed space, there is a C^∞-function on X which is real-analytic on Ω and has Ω as domain of real-analytic existence. The separability hypothesis is compulsory since he also proved that if A is an uncountable set, then the open unit ball of $c_0(A;\mathbb{R})$ is a real-analytic domain but not a domain of real-analytic existence.

5. Spaces of polynomials and multilinear forms

The study of the reflexivity of the spaces of polynomials and multilinear forms has received much interest over the last years. The relation with weak continuity was discovered by R. Ryan [Ry], the connection with weak sequential continuity by R. Alencar, R. Aron and S. Dineen [AAD], the link with the use of upper and lower estimates as well as of spreading models by J. Farmer [Fa] and the reference to the existence of a basis by R. Alencar [Al]. The study of the bidual of the space $\mathcal{P}(^mX)$ was investigated by R. Aron and S. Dineen in [AD].

M. Valdivia has been very active in this area too.

For a Banach space X and a positive integer m, $\mathcal{P}(^mX)$ designates the space of the m-homogeneous polynomials on X endowed with the uniform norm on $B(X)$. The interest lies in the Banach space $\mathcal{P}_{w^*}(^mX^*)$, i.e. the vector subspace of $\mathcal{P}(^mX^*)$ whose elements are continuous on $B(X^*_\sigma)$ endowed with the sup-norm on $B(X^*_\sigma)$.

In [140] M. Valdivia first establishes that if X is an Asplund space, then $\mathcal{P}_{w^*}(^mX^*)$ is an Asplund space too. He then proves that if $\mathcal{P}_{w^*}(^mX^*)$ contains no copy of ℓ_1 (hence in particular if X is an Asplund space), then $\mathcal{P}_{w^*}(^mX^*)$ is a closed subspace of $\mathcal{P}(^mX^*)$ endowed with the compact-open topology. He also gets that if X is an Asplund space such that X^* has the approximation property, then $\mathcal{P}_{w^*}(^mX^*)^{**}$ coincides with $\mathcal{P}(^mX^*)$ and that if X is a weakly compactly generated Asplund space, then so is $\mathcal{P}_{w^*}(^mX^*)$.

Let us note that in [143], M. Valdivia gets similar results to those developed in [140], in the setting of holomorphic functions. Let X be a Banach space and designate by $\mathcal{H}_b(X)$ the Fréchet space of the holomorphic functions on X which are bounded on the bounded subsets of X, endowed with the fundamental system of the norms $\|\cdot\|_{mB(X)}$ for $m \in \mathbb{N}$. Then $\mathcal{H}_{w^*}(X^*)$ is the subspace of $\mathcal{H}_b(X^*)$ whose elements are weak*-continuous on $mB(X^*)$ for every $m \in \mathbb{N}$ and $\mathcal{H}_{(w^*)}(X^*)$ is the closure of $\mathcal{H}_{w^*}(X^*)$ in $\mathcal{H}_b(X^*)$ for the compact-open topology. Now come the results. If $\mathcal{H}_{w^*}(X^*)$ contains no copy of ℓ_1, then its strong dual coincides canonically with $\mathcal{H}_{(w^*)}(X^*)$ and even with $\mathcal{H}_b(X^*)$ if moreover X^* has the approximation property. If X is an Asplund space, then $\mathcal{H}_{w^*}(X^*)$ contains no copy of ℓ_1; if moreover X is weakly compactly generated, then so is $\mathcal{H}_{w^*}(X^*)$.

In [149], M. Valdivia gets the following result as a corollary dealing with finite tensor products. Let $p \in]2, \infty[$ and $m \in \mathbb{N}$ be such that $1 < m < p$, and designate by s the conjugate number of p/m. Let moreover $(e_n)_{n\in\mathbb{N}}$ be the unit vector basis of ℓ_p. Then T defined by $TP = (P(e_n))_{n\in\mathbb{N}}$ is a continuous linear surjection from $\mathcal{P}(^m\ell_p)$ onto ℓ_s and S defined by

$$(SP)(x) = \sum_{n=1}^{\infty} P(e_n) < x, e_n^* >^m$$

is a continuous linear projection whose range is isomorphic to ℓ_s. He also gets an analogous result for $\mathcal{P}(^m c_0)$ and $\mathcal{P}(^m X)$ instead of $\mathcal{P}(^m\ell_p)$, if X is a Banach space such that X^* has type $p \in]1, 2]$.

The Proposition 2 of [149] runs as follows. Let $m \in \mathbb{N}$ and $p_1, \ldots, p_m \in]1, \infty[$ be such that $1/p_1 + \ldots + 1/p_m = 1/p < 1$. Let moreover for every $j \in \{1, \ldots, m\}$, $(e_{j,n})_{n\in\mathbb{N}}$ be a sequence of the Banach space X having an upper p_j-estimate. Then the sequence $(e_{1,n} \otimes \ldots \otimes e_{m,n})_{n\in\mathbb{N}}$ has an upper p_j-estimate in $X_1 \tilde{\otimes}_\pi \cdots \tilde{\otimes}_\pi X_m$, i.e. there is a constant $c_j > 0$ such that

$$\|\sum_{n=1}^{p} a_n e_{j,n}\| \leq c_j(\sum_{n=1}^{p} |a_n|^p)^{1/p}$$

for every scalars $a_1, \ldots, a_p$.

Using this proposition, M. Valdivia establishes in [152] results that imply the following consequences. If X and Z are Banach spaces, then for every compact linear map T from X to Z, there is a reflexive separable Banach space Y containing no copy of ℓ_p for any p and compact linear maps $T_1 : X \to Y$ and $T_2 : Y \to Z$ such that $T = T_2T_1$. There is a separable reflexive Banach space X without the approximation property such that for every closed subspace Y of X and every $m \in \mathbb{N}$, the spaces $\mathcal{L}(^mY)$ and $\mathcal{P}(^mY)$ are reflexive. Here $\mathcal{L}(^mX)$ is the space of the continuous m-linear functionals on X^m endowed with the norm

$$\|f\| = \sup\{|f(x_1, \ldots, x_m)| : \|x_1\|, \ldots, \|x_m\| \leq 1\}.$$

In [153], M. Valdivia investigates the reflexivity of the spaces $\mathcal{P}(^mX)$. Refining methods developed in [140], he proves that the approximation property can be avoided in some known results and gives new proofs and presentations of known results.

6. Fréchet spaces

It was well known that a Fréchet space E is a Montel space (resp. a Schwartz space) if and only if it is separable and such that every $\sigma(E', E)$-null sequence is $\beta(E', E)$-null (resp. converges uniformly to 0 on some zero-neighbourhood in E). Then the Josefson-Nissenzweig theorem led H. Jarchow [Ja] to ask whether the separability condition is superfluous. The Schwartz case received a positive answer by M. Lindström and Th. Schlumprecht [LS] and by J. Bonet [BoJ]. In a very short joint paper [131] with J. Bonet and M. Lindström, M. Valdivia solves the two questions positively.

Totally reflexive Fréchet spaces, i.e. Fréchet spaces of which every Hausdorff quotient is reflexive, have received much attention from M. Valdivia. In [109] already, he had proved the following deep characterization: a Fréchet space is totally reflexive if and only if it is isomorphic to a closed subspace of a countable product of reflexive Banach spaces. He comes back to this property in [124]. This time he considers sequences of the dual of a Fréchet space E and proves that E is totally reflexive if and only if the following two conditions hold:
(a) every $\mu(E', E)$-null sequence is $\beta(E', E)$-null and
(b) every $\sigma(E', E)$-null sequence is a weak-null sequence in E'_p for some continuous semi-norm p.
As a step of the proof, M. Valdivia gets that this property (a) characterizes the Fréchet spaces E which contain no subspace isomorphic to ℓ_1, a result obtained independently by P. Domański and L. Drewnovski [DD].

Another major concern of M. Valdivia is the study of the locally convex spaces which (do not) contain a copy of ℓ_1. This appears in several of his articles. For instance in [133], he studies specifically those Fréchet spaces E which contain no subspace isomorphic to ℓ_1. The basic result relies on the notion of a convex block sequence $(v_k)_{k\in\mathbb{N}}$ of a sequence $(u_j)_{j\in\mathbb{N}}$ of E, i.e. there is a sequence $(A_k)_{k\in\mathbb{N}}$ of subsets of $\mathbb{N}$ such that $r < s$ for every $k \in \mathbb{N}$, $r \in A_k$ and $s \in A_{k+1}$ as well as positive numbers a_j such that $\sum_{j\in A_k} a_j \leq 1$ and $v_k = \sum_{j\in A_k} a_j u_j$ for every $k \in \mathbb{N}$. M. Valdivia proves the following results: if the Fréchet space E contains no subspace isomorphic to ℓ_1, then
(a) every equicontinuous sequence of E' has a convex block sequence which $\sigma(E', E)$-converges;
(b) if E is separable, then E'_μ is ultrabornological;
(c) each separable closed subspace of $E_{\mathcal{M}}$ is complete, where $E_{\mathcal{M}}$ denotes the space E endowed with the topology of uniform convergence on the absolutely convex, compact and metrizable subsets of E'_σ.
Therefore a Fréchet space E is reflexive if and only if it contains no copy of ℓ_1 and verifies the Grothendieck condition (i.e. every $\sigma(E', E)$-null sequence is $\sigma(E', E'')$ null).

7. The Zahorski theorem

The Zahorski theorem shows how flexible the C^∞-functions are. If f belongs to $C^\infty(\mathbb{R}^n)$, let us say that a point x of $\mathbb{R}^n$ is real-analytic (resp. defect; divergent) if the Taylor series of f at x represents f on a neighbourhood of x (resp. has a positive radius but represents f on no neighbourhood of x; has radius of convergence equal to 0). It is a direct matter to check that the set Ω (resp. F; G) of the real-analytic (resp. defect; divergent) points is open (resp. F_σ and of first category; G_δ) and that $\{\Omega, F, G\}$ is a partition of $\mathbb{R}^n$. The Zahorski theorem [Za] first proved with $[0, 1]$ instead of $\mathbb{R}^n$, says that each such partition comes from a C^∞-function. It was extended by J. Siciak [Si] to $\mathbb{R}^n$. In [144], M. Valdivia has generalized this result to the setting of the Gevrey classes on $\mathbb{R}^n$.

8. Infinite dimensional complex analysis

Let K be a compact subset of a complex Fréchet space and designate by $\mathcal{H}(\mathcal{K})$ the space of the holomorphic germs on K, i.e. the inductive limit ind$\mathcal{H}^\infty(U_n)$, where $(U_n)_{n\in\mathbb{N}}$ is any decreasing fundamental sequence of open neighbourhoods of K and where $\mathcal{H}^\infty(U_n)$ is the Banach space of the bounded holomorphic functions on U_n. In [BM], K. D. Bierstedt and R. Meise proved that $\mathcal{H}(K)$ is compact if and only if E is a Fréchet-Schwartz space and asked for a characterization when $\mathcal{H}(K)$ is weakly compact.

In [139], J. Mujica and M. Valdivia prove in particular that if K is a compact subset of the Tsirelson Banach space X, then $\mathcal{H}(K)$ is weakly compact. Moreover if U is an open subset of X and if $\mathcal{H}_b(U)$ is defined as the projective limit of the spaces $\mathcal{H}^\infty(U_n)$ with $U_n = \{x \in U\colon\ \|x\| < n, \mathrm{d}_U(x) > 1/n\}$, then $\mathcal{H}_b(U)$ is weakly compact too.

9. Conclusion

Now has come the time to say a few words about the mathematician Manuel Valdivia.

The consideration of the mathematical works of Manuel Valdivia brings two characteristics into evidence. The first one is the diversity of the subjects he investigates: geometry of Banach spaces, Fréchet spaces and locally convex spaces have no hidden corner; compact spaces and real analyticity receive much interest; polynomials and multilinear forms are developed, ... The second is that he comes back again and again to his previous research, refines methods and finally gets a unifying perspective, a master piece of work covering many known results.

His influence on mathematics is great. In Spain it can be described in a few figures. He has directed 31 Ph. D. thesis. We all know many of his students: 15 have become Catedráticos de Universidad, 13 Profesores Titulares de Universidad and 2 Catedráticos de Escuela Universitaria.

He has been investigador principal of several DGICYT projects. ¿From 1993 to 1997, he has been the investigador principal of the unique proyecto de elite de la DGICYT in mathematics and this project has been renewed for another period of 5 years.

He is Dr. h. c. mult.: in 1993 at the Universidad Politécnica de Valencia and at the Universidad Jaime I de Castellon; in 1995 at the Université de Liège and in 2000 at the Universidad de Alicante.

Let me recall that in 1975 he was elected Académico Numerario de la Real Academia Espagnola de Ciencias Exactas, Físicas y Naturales. In 1996, he is Hijo Adoptivo de Valencia and Académico de Número de la Academia de Ingeniera; in 1997, he becomes Colegiado de Honor del Colegio de Ingenieros Agrónomos de Centro y Canarias; in 1999, he is nominated Académico Correspondiente de la Academia Canaria de las Ciencias as well as premio de la Confederacion Española de Organizacion de Empresas a las Ciencias, a distinctive prize dedicated to Spanish scientific researchers of particular merit.

These facts are important but should not hide the man. If you consider the list of publications of Manuel Valdivia, you will immediately realize that for several years, Manuel has been single author and that by now most of his papers come from joint research.

I am one of these co-authors. I always appreciated Manuel Valdivia's articles with their fine and delicate ideas, with their intricate constructions. So I knew the work of the mathematician when, about ten years ago, I got to meet the mathematician when we started our joint research. Soon afterwards, I discovered the man: a scholar who readily became a friend.

At the end of this presentation, allow me to say the following about the mathematician. Manuel has a tremendous memory and an enormous ability to do research. When you see him drawing curves all over a page, be careful: do not disturb! He is in deep thought and be not surprised if suddenly stopping drawing, he starts writing or explaining an idea or telling he has a proof. In the evening when he decides to stop doing mathematics, the scholar appears with a deep knowledge of the literature, about music and ... an unforgettable moment is coming.

Let me take this opportunity to emphasize the quality of the help brought by his wife, Maria Teresa.

In my own name and in the name of all the participants in this International Functional Analysis Meeting in honour of the 70th birthday of Professor Manuel Valdivia, let me renew the words pronounced 10 years ago by John Horváth and say: Dear Manuel, I wish you many more years of happy and fruitful research activity.

Publications of Manuel Valdivia (continued)

[115] Projective generators and resolutions of identity in Banach spaces (with J. Orihuela). Congress on Functional Analysis (Madrid, 1988). *Rev. Mat. Univ. Complut. Madrid* **2** (1989), suppl., 179–199. MR 91j:46021. ZBL 717.46009.

[116] Some properties of Banach spaces Z^{**}/Z. Dedicated to Professor A. Plans; *Geometric aspects of Banach spaces*, 169–194, *London Math. Soc. Lecture Notes Ser.,* **140.** *Cambridge Univ. Press, Cambridge*, 1989. MR 91d:46019.

[117] Projective resolution of identity in $C(K)$ spaces. *Arch. Math. (Basel)* **54** (1990), 493–498. MR 91f:46036. ZBL 707.46009.

[118] Una propriedad de interpolación en espacios de funciones holomorfas con desarrollos asintóticos. Homenaje Prof. N. Hayek Kalil. *Publ. Univ. de la Laguna (Tenerife),* (1990), 351–360. ZBL 759.30017.

[119] Topological direct sum decompositions of Banach spaces. *Israel J. Math.* **71** (1990), 286–296. MR 92d:46047. ZBL 822.46016.

[120] Every Radon-Nikodým Corson compact space is Eberlein compact (with J. Orihuela and W. Schachermayer). *Studia Math.* **98** (1991), 157–174. MR 92h:46025. ZBL 771.46015.

[121] Resoluciones proyectivas del operador identidad y bases de Markushevich en ciertos espacios de Banach. *Rev. Real Acad. Cienc. Exact. Fís. Natur. Madrid* **84** (1990), 23–34. MR 92m:46032. ZBL 822.46011.

[122] On the extent of the (non) quasi-analytic classes (with J. Schmets). *Arch. Math. (Basel)* **56** (1991), 593–600. MR 92f:46021. ZBL 719.46015.

[123] Interpolation in spaces of holomorphic mappings with asymptotic expansions. *Proc. Roy. Irish Acad. Sect. A* **91** (1991), 7–38. MR 93h:46056. ZBL 769.46033.

[124] On totally reflexive Fréchet spaces. Dedicated to Professor Giovanni Aquaro on the occasion of his 70th birthday. Recent developments in mathematical analysis and

its applications (Bari, 1990). *Confer. Sem. Mat. Univ. Bari* (1991), 39–55. MR 93j:46003. ZBL 804.46006.

[125] Simultaneous resolutions of the identity operator on normed spaces. *Collect. Math.* **42** (1991), 265–284. MR 94e:46047. ZBL 788.47024.

[126] On basic sequences in Banach spaces. Dedicated to the memory of Professor Gottfried Kőthe. *Note Mat.* **12** (1992), 245–258. MR 95b:46016. ZBL 811.46006.

[127] Complemented subspaces of certain Banach spaces. *Seminar on Functional Analysis Univ. Murcia.*

[128] Decomposiciones de espacios de Fréchet en ciertas sumas topologicas directas. *Papers in honor of Pablo Bobillo Guerrero*, 59–72, *Univ. Granada, Granada*, 1992. MR 94g:46003. ZBL 792.46002.

[129] Mathematical analysis. *History of mathematics in the XIXth century, Part I (Madrid, 1991)*, 157–174, *Real Acad. Cienc. Exact. Fís. Natur. Madrid*, 1992. MR 1 469 504.

[130] On certain total biorthogonal systems in Banach spaces. *Generalized functions and their applications (Varanasi, 1991)*, 271–280, *Plenum, New York*, 1993. MR 94i:46015. ZBL 845.46005.

[131] Two theorems of Josefson-Nissenzweig type for Fréchet spaces (with J. Bonet and M. Lindstrőm). *Proc. Amer. Math. Soc.* **117** (1993), 363–364. MR 93d:46005. ZBL 785.46002.

[132] Domains of analyticity in real normed spaces (with J. Schmets). *J. Math. Anal. Appl.* **176** (1993), 423–435. MR 94c:46084. ZBL 811.46030.

[133] Fréchet spaces with no subspaces isomorphic to ℓ^1. *Math. Japon.* **38** (1993), 397-411. MR 94f:46006. ZBL 778.46001.

[134] Sobre el teorema de interpolación de Borel en espacios de Hilbert. *Rev. Colombiana Mat.* **27** (1993), 235–247. MR 95j:26027. ZBL 802.41001.

[135] Boundaries of convex sets. *Rev. Real Acad. Cienc. Exact. Fís. Natur. Madrid* **87** (1993), 177–183. MR 96a:46035.

[136] Domains of existence of $\mathbb{R}$-analytic functions in real normed spaces (with J. Schmets). *Bull. Polish Acad. Sci. Math.* **41** (1993), 131–137. MR 97g:46061. ZBL 799.46048.

[137] On certain classes of Markushevich bases. *Arch. Math. (Basel)* **62** (1994), 445–458. MR 95k:46017. ZBL 808.46010.

[138] On the existence of holomorphic functions having prescribed asymptotic expansions (with J. Schmets). *Z. Anal. Anwendungen* **13** (1994), 307–327. MR 96f:30039. ZBL 816.46019.

[139] Holomorphic germs on Tsirelson's space (with J. Mujica). *Proc. Amer. Math. Soc.* **123** (1995), 1379–1384. MR 95f:46074. ZBL 823.46047.

[140] Banach spaces of polynomials without copies of ℓ^1. *Proc. Amer. Math. Soc.* **123** (1995), 3143–3150. MR 95m:46070. ZBL 848.46030.

[141] Biorthogonal systems in certain Banach spaces. Dedicated to Professor Baltasar Rodriguez Salinas. Meeting on Mathematical Analysis (Avila, 1995). *Rev. Mat. Univ. Complut. Madrid* **9** (1996), Special issue, suppl., 191–220. MR 97m:46020. ZBL 872.46007.

[142] On certain linear operators in spaces of ultradifferentiable functions. *Result. Math.* **30** (1996), 321–345. MR 97m:46032. ZBL 867.46028.

[143] Fréchet spaces of holomorphic functions without copies of ℓ^1. *Math. Nachr.* **181** (1996), 277–287. MR 97m:46060. ZBL 860.46026.

[144] The Zahorski theorem is valid in Gevrey classes (with J. Schmets). *Fund. Math.* **151** (1996), 149–166. MR 98a:26026. ZBL 877.26014.

[145] On certain analytic function ranged linear operators in spaces of ultradifferentiable functions. *Math. Japon.* **44** (1996), 415–434. MR 98b:46052. ZBL 874.46027.

[146] On the Borel theorem in real Banach spaces (with J. Schmets). *Functional analysis (Trier, 1994)*, 399–412, *de Gruyter, Berlin*, 1996. MR 98e:46056. ZBL 891.46015.

[147] On certain compact topological spaces. *Rev. Mat. Univ. Complut. Madrid* **10** (1997), 81–84. MR 98d:54046. ZBL 870.54025.

[148] Analytic extension of non quasi-analytic Whitney jets of Roumieu type (with J. Schmets). *Result. Math.* **31** (1997), 374–385. MR 98g:46027. ZBL 877.26015.

[149] Complemented subspaces and interpolation properties in spaces of polynomials. *J. Math. Anal. Appl.* **208** (1997), 1–30. MR 1 440 340. ZBL 890.46034.

[150] On the existence of continuous linear analytic extension maps for Whitney jets (with J. Schmets). *Bull. Polish Acad. Sci. Math.* **45** (1997), 359–367. MR 1 489 879. ZBL 980.23060.

[151] Some properties of basic sequences in Banach spaces. *Rev. Mat. Univ. Complut. Madrid* **10** (1997), 331–361. MR 1 605 662. ZBL 980.15524.

[152] Certain reflexive Banach spaces with no copy of ℓ_p. *Funct. Anal. Select Topics*, Ed. K. Jain (1998), 89–96. MR 2000a:46022.

[153] Some properties in spaces of multilinear functionals and spaces of polynomials. *Proc. Roy. Acad. Sc.* **98A** (1998), 87–106. ZBL 933.46042.

[154] Analytic extension of non-quasi analytic Whitney jets of Beurling type (with J. Schmets). *Math. Nachr.* **195**(1998), 187–197. MR 99m:46064. ZBL 926.26014.

[155] Analytic extension of ultradifferentiable Whitney jets (with J. Schmets). *Collect. Math.* **50**(1999), 73–94. MR 2000i:58015. ZBL 937.26014.

[156] On weakly locally uniformly rotund Banach spaces (with A. Moltó, J. Orihuela and S. Troyanski). *J. Funct. Anal.* **163**(1999), 252–271. MR 2000b:46031. ZBL 927.46010.

References

[AAD] Alencar R., Aron R., Dineen S., A reflexive space of holomorphic functions in infinitely many variables. *Proc. Amer. Math. Soc.* **90**(1984), 407–411. MR 85b:46050. ZBL 536.46015.

[AD] Aron R., Dineen S., Q-reflexive Banach spaces. *Rocky Mountain J. Math.*, **27**(1997), 1009–1025. MR 99h:46010. ZBL 916.46011.

[Al] Alencar R., On reflexivity and basis for $\mathcal{P}(^{\Updownarrow}\mathcal{E})$. *Proc. Roy. Irish Acad.* **85A**(1985), 135–138. MR 87i:46101. ZBL 594.46043.

[AL] Amir L., Lindenstrauss J., The structure of weakly compact sets in Banach spaces. *Ann. of Math.* **88**(1968), 35–46. MR 37.4562. ZBL 164.14903.

[BBMT] Bonet J., Braun R., Meise R., Taylor B. A., Whitney's extension theorem for non quasi-analytic classes of ultradifferentiable functions. *Studia Math.* **99**(1991), 155–184. MR 93e:46030. ZBL 738.46009.

[BM] Bierstedt K. D., Meise R., Nuclearity and the Schwartz property in the theory of holomorphic functions on metrizable locally convex spaces. *Infinite Dimensional Holomorphy and Applications* (Proc. Internat. Sympos., Univ. Estadual de Campinas, Sao Paulo 1975) (M. Matos, ed.), Notas de Mat. **54** *North-Holland Math. Studies* **12**, *North-Holland, Amsterdam,* 1977, 93–129. MR 58#30236. ZBL 409.46051.

[BMT] Braun R., Meise R., Taylor B. A., Ultradifferentiable functions and Fourier analysis. *Result. Math.* **17**(1990), 206–237. MR 91h:46072. ZBL 735.46022.

[Bo] Borel E., Sur quelques points de la théorie des fonctions. *Ann. Ec. Norm. Sup.* **XII**(1895), 9–55.

[BoJ] Bonet J., A question of Valdivia on quasi-normable Fréchet spaces. *Canad. Math. Bull.* **34**(1991), 301–304. MR 92k:46004. ZBL 747.46003.

[Ca] Carleman T., Les fonctions quasi-analytiques. *Collection de monographies sur la théorie des fonctions (E. Borel)*, Gauthier-Villars et Cie, Paris, 1926.

[Co] Corson H. H., Normality in subsets and product spaces. *Amer. J. Math.* **81**(1959), 785–796. MR 21#5947. ZBL 095.37302.

[DD] Domański P., Drewnovski L., Fréchet spaces of continuous vector-valued functions: complementability in dual Fréchet spaces and injectivity. *Studia Math.* **102**(1992), 257–267. MR 93d:46061. ZBL 811.46001.

[DFJP] Davis W. J., Figiel T., Johnson W. B., Pełczyński, Factoring weakly compact operators. *J. Funct. Anal.* **17**(1974), 311–327. MR 50#8010. ZBL 306.46020.

[DGZ] Deville R., Godefroy G., Zizler V., Smoothness and renorming in Banach spaces. *Longman Scientific and Technical. John Wiley and Sons Inc., New York*, 1993. MR 94d:46012. ZBL 782.46019.

[Fa] Farmer J., Polynomial reflexivity in Banach spaces. *Israel J. Math.* **87**(1994), 257–273. MR 95h:46021. ZBL 819.46006.

[FA] Progress in Functional Analysis. Proceedings of the International Functional Analysis Meeting on the Occasion of the 60th Birthday of Professor M. Valdivia, Peñíscola, 22–27 October, 1990. Edited by Klaus D. Bierstedt, José Bonet, John Horváth and Manuel Maestre. *North Holland Mathematics Studies* **170**. *North Holland Publishing Co., Amsterdam*, 1992. xxviii+431 pp. ISBN: 0-444-89378-4. MR 92i:46002. ZBL 745.00031.

[FG] Fabian M., Godefroy G., The dual of every Asplund space admits a projectional resolution of identity. *Studia Math.* **91**(1988), 141–151. MR 90b:46032. ZBL 692.46012.

[Fr] Franklin Ph., Functions of a complex variable with assigned derivatives at an infinite number of points, and an analogue of Mittag-Leffler theorem. *Acta Math.* **47**(1926), 371–385.

[Gu] Gul'ko S., The structure of spaces of continuous functions and their hereditary paracompactness. *Uspekhi Mat. Nauk* **34**(1979), 33–40; translated in *Russian Math. Surv.* **34**(1979), 36–44. MR 81b:54017. ZBL 446.46014.

[Ha] Haydon R., Trees in renorming theory. *Proc. London Math. Soc.* **78**(1999), 541–584. MR 2000d:46011.

[HJ] Hagler J., Johnson W. B., On Banach spaces whose dual ball are not weak*-sequentially compact. *Israel J. Math.* **28**(1977), 325–330. MR 58#2173. ZBL 365.46019.

[Ho] Horváth, John, The mathematical works of Manuel Valdivia. *Progress in functional analysis (Peñíscola, 1990)*, 1–55, *North-Holland Math. Stud.* **170**, *North-Holland, Amsterdam*, 1992. MR 93c:01043. ZBL 776.46001.

[Ja] Jarchow H., Locally convex spaces. *Teubner, Stuttgart*, 1981. MR 83h:46008. ZBL 466.46001.

[JR] Johnson W. B., Rosenthal H. P., On ω^*-basic sequences and their applications to the study of Banach spaces. *Studia Math.* **43**(1972), 77–92. MR 46#9696. ZBL 213.39301.

[Ke] Kenderov P. S., A Talagrand compact space which is not a Radon-Nikodym space (An example of E. A. Reznichenko). (manuscript).

[La] Langenbruch M., Analytic extension of smooth functions. *Result. Math.* **36**(1999), 281–296. MR. 2000i:46026. ZBL 945.26029.

[LS] Lindström M., Schlumprecht Th., A Josefson-Nissenzweig theorem for Fréchet spaces. *Bull. London Math. Soc.* **25**(1993), 55–58. MR 94a:46005. ZBL 940.38257.

[Mi] Mityagin B., Approximate dimension and bases in nuclear spaces. *Uspehi Mat. Nauk* **16**(1961), 63-132; translated as *Russian Math. Surveys* **16**(1961), 59–127. MR 27#2837. ZBL 104.08601.

[MT] Meise R., Taylor B. A., Whitney's extension theorem for ultradifferentiable functions of Beurling type. *Ark. Math.* **26**(1988), 265–287. MR 91h:46074. ZBL 683.46020.

[Na] Namioka I., Radon-Nikodym compact spaces and fragmentability. *Mathematika* **34**(1987), 258–281. MR 89i:46021. ZBL 654.46017.

[Pe] Petzsche H.-J., On E. Borel's theorem. *Math. Ann.* **282**(1988), 299–313. MR 89m:46076. ZBL 633.46033.

[Pl] Plichko A. N., On projective resolutions of the identity operator and Markushevich bases. *Dokl. Akad. Nauk SSSR* **263**(1982), 543–546; *Soviet Math. Dokl.* **25**(1982), 386–389. MR 83d:46014. ZBL 531.46010.

[Ri] Ritt J. F., On the derivatives of a function at a point. *Annals of Math.* **18**(1916), 18–23.

[Ry] Ryan R., Ph.D. Thesis, Trinity College, Dublin, 1980.

[Si] Siciak J., Punkty regularne i osobliwe funkcji klasy C^∞ [Regular and singular ponits of C^∞-functions]. *Zeszyty Nauk. Polit. Śląsk. Ser. Mat.-Fiz.* **48**(1986), 127–146.

[Ta] Talagrand M., Espaces de Banach faiblement K-analytiques. *Ann. of Math.* **110**(1979), 407–438. MR 81a:46021. ZBL 414.46015.

[Va] Vaš'ak L., On one generalization of weakly compactly generated Banach spaces. *Studia Math.* **70**(1981), 11–19. MR 83h:46028. ZBL 466.46017.

[Wi] Whitney H., Analytic extensions of differentiable functions defined on closed sets. *Trans. Amer. Math. Soc.* **36**(1934), 63–89. ZBL 008.08501.

[Za] Zahorski Z., Sur l'ensemble des points singuliers d'une fonction d'une variable réelle admettant les dérivées de tous les ordres. *Fund. Math.* **34**(1947), 183–245. MR 10,23c. ZBL 033.25504.

Recent Progress in Functional Analysis
K.D. Bierstedt, J. Bonet, M. Maestre and J. Schmets

Fréchet differentiability of Lipschitz functions (a survey)

J. Lindenstrauss[a] and D. Preiss[b]

[a]Department of Mathematics, The Hebrew University, Jerusalem, Israel

[b]Department of Mathematics, University College London, London WC1E6BT, England

To Professor Manuel Valdivia on the occasion of his seventieth birthday.

Abstract

We present a survey of what is known concerning existence of points of Fréchet differentiability of Lipschitz maps between Banach spaces. The emphasis is on more recent results involving such topics as ϵ-Fréchet differentiability, Γ-null sets and the validity of the mean value theorem for Fréchet derivatives.
MCS 2000 Primary 46G05; Secondary 46B22, 28B05

1. Preliminaries

Our aim in this survey is to give an outline of the results and examples related to Fréchet differentiability as they are known today. There is (fortunately) constant progress in this direction so that in a few more years more will probably be known. Nevertheless, the situation at present is already quite involved and complicated and thus a survey of what is known may be helpful.

In this preliminary section we present the basic notions and some general remarks concerning them. This section is followed by:

2. Results on Gâteaux differentiability. We discuss briefly the main known results in this direction and some notions involved in their formulation.

3. Examples and theorems showing the non existence of Fréchet derivatives in certain cases.

4. Existence results for ϵ-Fréchet derivatives.

5. Existence results for Fréchet derivatives.

6. Questions related to the mean value theorem for Fréchet derivatives.

7. Some open questions.

The material in this section as well as that of Sections 2 and 3 is mostly discussed in the book [5] with more details, background and additional references. Most of the material in sections 4, 5 and 6 is not discussed in the book [5] and a large part of it is rather recent.

We start with the basic definitions. We consider here Lipschitz functions f from a Banach space X into a Banach space Y. We shall always assume that the domain space X is separable (and therefore it is of no loss of generality to assume that also Y is separable). Since differentiability is a local property all we shall say applies also to functions which are locally Lipschitz or to functions defined only on an open set in X. For the purpose of convenience we shall not work or state our results in this more general context.

The function f is said to have a derivative at x in the direction v if

$$f'(x,v) = \lim_{t\to 0} \frac{f(x+tv) - f(x)}{t} \tag{1}$$

exists.

The function f is said to be Gâteaux differentiable at x if there is a bounded linear operator $T:\ X \to Y$ so that for all $v \in X$

$$Tv = \lim_{t\to 0} \frac{f(x+tv) - f(x)}{t} \tag{2}$$

The operator T (which is obviously unique if it exists) is called the Gâteaux derivative of f at x and is denoted also by $D_f(x)$. Clearly f is Gâteaux differentiable at x iff $f'(x,v)$ exists for all v and depends linearly on v (the boundedness of T is automatic for Lipschitz functions f, always $\|T\| \le Lip\ f$ the Lipschitz constant of f). Note that with our notations $f'(x,v) = D_f(x)v$ if the right hand side exists.

If the limit in (2) exists uniformly in v with $\|v\| = 1$ then we say that f is Fréchet differentiable at x and $D_f(x)$ is then called the Fréchet derivative of f at x. Alternatively, $D_f(x)$ is the Fréchet derivative of f at x iff

$$f(x+u) = f(x) + D_f(x)u + o(\|u\|).$$

If $dim\ X < \infty$ then the notions of Gâteaux derivatives and Fréchet derivatives coincide for Lipschitz functions f. For X with $dim\ X = \infty$ it is easy to see that this is not the case (see Section 3) and this is the fact that makes the question of existence of Fréchet derivatives hard (but, we think, interesting).

If a function $f:\ X \to Y$ is Gâteaux differentiable in a neighborhood of a point x and if the Gâteaux derivative is norm continuous at x, meaning that $\|D_f(z) - D_f(x)\| \to 0$ as $\|z - x\| \to 0$, then f is actually Fréchet differentiable at x. This remark is a trivial consequence of the mean value theorem (applied to $g = f - D_f(x)$) which states that if g is Gâteaux differentiable on the interval I connecting x and z then

$$\|g(x) - g(z)\| \le \|x - z\|\ sup\ \{\|D_g(u)\| :\ u \in I\}.$$

This mean value theorem is, in turn, a direct consequence of the mean value theorem for functions from I to R (consider $y^*g(tz + (1-t)x)$ for a suitable $y^* \in Y^*$). A function f

is said to be of class C^1 in an open set if it is continuously differentiable there. In view of the remark above it does not matter whether we take in this context Fréchet or Gâteaux derivatives.

As we shall see in Section 2 there are strong general existence theorems for Gâteaux derivatives. This is not so for Fréchet derivatives. On the other hand Fréchet derivatives are more useful than Gâteaux derivatives when they exist. We shall discuss presently two examples which illustrate this.

There is a notion which is weaker than Fréchet derivative which can however replace the Fréchet derivative in many arguments and for which it is often much easier to prove existence theorems. This notion is the following.

A Lipschitz map $f: X \to Y$ is said to be ϵ-Fréchet differentiable at x for some $\epsilon > 0$ if there is a bounded linear operator $T: X \to Y$ and a $\delta > 0$ so that

$$||f(x+u) - f(x) - Tu|| \leq \epsilon ||u|| \quad if \quad ||u|| < \delta. \tag{3}$$

Clearly f is Fréchet differentiable at x if and only if f is ϵ-Fréchet differentiable at x for every $\epsilon > 0$. Any operator T which satisfies (3) is called an ϵ-Fréchet derivative at x. An operator which satisfies (3) is not determined uniquely by this equation. If f is Gâteaux differentiable at x we can take as T the operator $D_f(x)$ after replacing ϵ by 2ϵ.

We shall consider now two examples where Fréchet derivatives (or ϵ-Fréchet derivatives) give an important information which we cannot deduce from Gâteaux derivatives.

Let f be a Lipschitz equivalence between X and Y. This means that f is a one to one mapping from X onto Y so that for some $m < M$ and all $u, v \in X$

$$m||u - v|| \leq ||f(u) - f(v)|| \leq M||u - v||.$$

A natural question is whether under these circumstances there is a linear map from X onto Y which has the same property (i.e. whether X is linearly isomorphic to Y). If f is Gâteaux differentiable at some $x \in X$ then it follows directly from our assumption that $m||u|| \leq ||D_f u|| \leq M||u||$, i.e. $D_f(x)$ is an isomorphism from X into Y. The question whether $D_f(x)$ is surjective is harder. In [13] and [22] examples are presented of Lipschitz equivalences $f: \ell_2 \to \ell_2$ so that on a "large" set of points x, $D_f(x)$ exists but $D_f(x)\ell_2$ is a proper subspace of ℓ_2. If however $D_f(x)$ is also an ϵ-Fréchet derivative of f (for $\epsilon < \frac{m}{2}$) then it is easy to show that $D_f(x)$ is surjective. Indeed, we may assume without loss of generality that $x = f(x) = 0$. Assume that the range of $D_f(x)$ is contained in a proper closed subspace Z of Y. Choose $u \in Y$ with $||u|| = 1$ and $d(u, Z) \geq 1/2$. For $0 < t < 1$ let $v_t \in X$ be such that $f(v_t) = tu$. Then $m||v_t|| \leq t$. If $t/m < \delta$ we get

$$||tu - D_f(0)v_t|| < \frac{m}{2}||v_t|| \leq \frac{t}{2}$$

and thus $||u - t^{-1}D_f(0)v_t|| < 1/2$ contradicting the choice of u.

As a matter of fact, partly because we have no strong enough existence theorem for Fréchet or ϵ-Fréchet derivatives, the question we stated above is still open for separable (and in particular reflexive separable) spaces X. We will have more to say on this topic at the end of Section 7.

A related example concerns Lipschitz quotient maps. A map f from X onto Y is called a Lipschitz quotient map if there exist constants m and M so that for every $x \in X$ and every $r > 0$

$$B_Y(f(x), mr) \subset f(B_X(x, r)) \subset B_Y(f(x), Mr)$$

where $B(u, \alpha)$ denotes the ball with center u and radius α in the appropriate space.

Here again the question is: If there is a Lipschitz quotient map f from X onto Y does there exist a linear quotient map from X onto Y? Like above it is easy to show that if for some x, the Gâteaux derivative $D_f(x)$ exists and if it is also an ϵ-Fréchet derivative for ϵ sufficiently small, then $D_f(x)$ is a linear quotient map from X onto Y. Using this remark, in the proper setting, it is proved in [6] that the only infinite-dimensional Banach space which is a Lipschitz quotient of ℓ_2 is (isomorphic to) ℓ_2. In this context Gâteaux derivatives by themselves are not useful at all. In [6] it is shown that there is a Lipschitz quotient map from ℓ_2 onto ℓ_2 whose Gâteaux derivative at some point is 0 (a stronger result in which "some point" is replaced by "many points" can be deduced from [22]). In [15] it is shown that there is a Lipschitz quotient map f from $C(0, 1)$ onto ℓ_1 such that whenever $D_f(x)$ exists it is an operator of rank ≤ 1 (the general existence theorem on Gâteaux derivatives presented in Section 2 ensures that such derivatives exist "almost everywhere"). In this case we have an example that a Lipschitz quotient map exists but there is no linear quotient map from $C(0, 1)$ onto ℓ_1. There are however many other interesting situations where one can ask about the existence of linear quotient maps once one has a Lipschitz quotient map. The answer to these questions often depends on existence theorems for Fréchet or ϵ-Fréchet derivatives.

2. Gâteaux differentiability

The oldest and perhaps most simple result on Gâteaux differentiability in infinite dimensional spaces is the result of Mazur[26] which states that every continuous convex function on a separable Banach space X is Gâteaux differentiable on a dense G_δ subset of X. A more precise theorem, which reflects the well known fact that a continuous convex function on the line can have only a countable number of points where it is nondifferentiable, is the following.

Theorem 1 ([35]) . *Let A be a subset in a separable Banach space X. There is a convex continuous real-valued function on X which is nowhere Gâteaux differentiable on A if and only if A is contained in a countable union of graphs of δ-convex functions.*

A subset B of X is called a graph of a real-valued function ϕ if there is a closed hyperplane Z in X, a vector $u \in X \setminus Z$ (thus $X = Z \oplus \{span\ u\}$) and

$$B = \{z + \phi(z) \cdot u : \ z \in Z\}.$$

A function ϕ is called δ-convex if it is a difference of two convex Lipschitz functions. In particular such a ϕ is itself a Lipschitz function.

In order to formulate the general existence theorem for the Gâteaux differentiability of Lipschitz functions we need first two concepts.

A Banach space Y is said to have the Radon-Nikodým property (RNP) if every Lipschitz (actually every absolutely continuous) function f from R into Y is differentiable almost everywhere. It is noted in [6] that if Y fails the RNP there is a Lipschitz function $f: R \to Y$ and an $\epsilon > 0$ such that f is nowhere ϵ-differentiable (we omitted here the word "Fréchet" since for every f on R, Fréchet differentiability and Gâteaux differentiability mean the same thing).

It is known and easy to see that a separable conjugate space Y (and therefore any reflexive Y) has the RNP. On the other hand a space containing c_0 or $L_1(0,1)$ fails to have the RNP (for more details see [5] Chapter 5).

A Borel set A in a separable space X is called an Aronszajn null set if for every sequence $\{x_i\}_{i=1}^{\infty}$ in X whose closed linear span is the whole space we can decompose A as $\bigcup_{i=1}^{\infty} A_i$ where each A_i is a Borel set which intersects every line in the direction of x_i by a set which has (linear) Lebesgue measure 0.

In the definition above it is important that we consider all sequences $\{x_i\}_{i=1}^{\infty}$ as above (and not only one such sequence or a small set of sequences).

There is a deep characterization of Aronszajn null sets which is due to Csörnyei [8]. A Borel set A is Aronszajn null if and only if for every nondegenerate Gaussian measure μ on X we have $\mu(A) = 0$. A Gaussian measure is nondegenerate if it is not supported on a closed proper hyperplane of X. In view of this characterization Aronszajn null sets are also called Gauss null sets. In the proof of the theorem below it is convenient to use the original definition of Aronszajn null sets instead of the more elegant definition as Gauss null sets. We mention that there is also a useful notion of Haar null sets in X. We do not give here the definition of this concept since Haar null sets will not be used in this survey. We just mention for the purpose of orientation that every Gauss null set is a Haar null set but the converse is false.

The following is a nice generalization of the classical Rademacher theorem (on differentiation of Lipschitz functions from R^n to R^m) to infinite dimensional spaces.

Theorem 2 ([3,7,23]) . *A Lipschitz function f from a separable Banach space into a space Y having the RNP is Gâteaux differentiable outside a Gauss null set.*

In view of the definition of the RNP and the remark we made just after the definition, the assumption that Y has the RNP is essential here.

In [3] Theorem 2 was proved as stated here. In [7] what is proved is a weaker version of Theorem 2 in which Haar null sets replace Gauss null sets. In [23] Theorem 2 is proved with yet another natural class of null sets, the so called cube null sets.The proof of Csörnyei [8] shows however that the class of cube null sets coincides with Gauss null sets.

Because so many classes of null sets turn out to be identical to (or containing) the class of Gauss null sets it was thought for some time that Theorem 2 cannot be strengthened by replacing Gauss null sets by a smaller natural class of exceptional sets. This, however, turned out to be false:

For an $x \in X$ and $\epsilon > 0$ denote by $\tilde{A}(x,\epsilon)$ the system of all Borel sets B in X such that $\{t \in [0,1] : \phi(t) \in B\}$ has Lebesgue measure 0 whenever $\phi : [0,1] \to X$ is such that $\phi(t) - tx$ has Lipschitz constant at most ϵ. Let $\tilde{A}$ be the class of all Borel sets B so that whenever $\overline{span}\{x_i\}_{i=1}^{\infty} = X$, B can be represented as a union $\bigcup_{i=1}^{\infty}\bigcup_{k=1}^{\infty} B(i,k)$ of Borel sets so that $B(i,k) \in \tilde{A}(x_i, 1/k)$ for all i and k. In [32] it is proved that $\tilde{A}$ is a σ-ideal which is properly contained in the class of Gauss null sets and that Theorem 2 remains valid if we demand that exceptional sets belong to $\tilde{A}$. The paper [32] contains several other examples of such σ-ideals.

In view of the examples in [32] as well as some previous examples it seems now to be very difficult to get a characterization of sets of non Gâteaux differentiability of Lipschitz functions which is as precise as Theorem 1.

It seems reasonable to conjecture that if $dim\ X < \infty$ then the σ-ideal generated by the set of points of nondifferentiability of Lipschitz functions from X to R coincides with the class of sets of Lebesgue measure 0. This conjecture is known to be true if $dim\ X = 1$ (this is an easy classical result) and if $dim\ X = 2$ (this is a recent unpublished result of the second author). However the conjecture is still open if $2 < dim\ X < \infty$.

3. Non existence of points of Fréchet differentiability

We start this section with a few simple examples which show that Theorem 2 fails badly if Gâteaux differentiability is replaced by Fréchet differentiability. Actually in most examples we show that points of ϵ-Fréchet differentiability fail to exist if $\epsilon > 0$ is small enough. We later state some theorems which put the examples in a more general context. We end this section with a discussion of yet another concept of exceptional sets which enters already into one of the examples here and which is of basic importance in some results discussed later on in this survey.

Example 1. The norm in ℓ_1 is nowhere Fréchet differentiable. If

$$x = (\lambda_1, \lambda_2, \ldots, \lambda_n, \ldots) \in \ell_1$$

then $\|x\| = \sum_{n=1}^{\infty} |\lambda_n|$ is easily seen to be Gâteaux differentiable exactly at those points x for which $\lambda_n \neq 0$ for all n. The norm in ℓ_1 is not ϵ-Fréchet differentiable with $\epsilon < 1$ at any point $x \in \ell_1$. This follows from the fact that for every n

$$|\|x + te_n\| - (\|x\| + |t|)| < |\lambda_n| \quad , \quad t \in R$$

where e_n is the n'th unit vector in ℓ_1, and that $\lambda_n \to 0$ as $n \to \infty$.

Example 2. The function $x \to |x|$ from ℓ_2 to itself, where for $x = (\lambda_1, \lambda_2, \ldots, \lambda_n, \ldots)$ we put $|x| = (|\lambda_1|, |\lambda_2|, \ldots, |\lambda_n|, \ldots)$, is not ϵ-Fréchet differentiable with $\epsilon < 1$ at any point in ℓ_2. The argument is identical to that of Example 1. Again, the map $x \to |x|$ is Gâteaux differentiable at x if and only if $\lambda_n \neq 0$ for every n.

Example 3. Consider the map $f : L_2[0,1] \to L_2[0,1]$ defined by $f(x)(t) = \sin x(t)$. This map is Gâteaux differentiable at every point x with $D_f(x) = \cos x$ (meaning that $D_f(x)y = \cos x \cdot y$). This map is not ϵ-Fréchet differentiable at any point in $L_2[0,1]$ with $\epsilon < \epsilon_0$ for some ϵ_0. Consider e.g. $x = 0$ and let $u_t = \chi[0,t]$ (the indicator function of $[0,t]$). Then $\sin u_t = \sin 1 \cdot u_t$ and thus

$$\sin u_t - D_{\sin u}(0)u_t = u_t(\sin 1 - 1)$$

and so one can take $\epsilon_0 = |1 - \sin 1|$ (at least for $x = 0$, the computation for all other points is similar). The formula $D_f(x) = \cos x$ may give the impression that $D_f(x)$ depends continuously on x in contradiction to the observation made in Section 1. However $D_f(x)$ is not continuous in the right topology; $\|D_f(x) - D_f(y)\| = \|x - y\|_\infty$ (the maximum norm) and clearly $\|x - y_n\|_2 \to 0$ does not imply $\|x - y_n\|_\infty \to 0$.

Example 4. Let C be a closed convex set with empty interior in a Banach space X. The convex function $f : X \to R$ defined by $f(x) = d(x, C)$ is not ϵ-Fréchet differentiable at any point x in C and for any $\epsilon < 1$. This follows from the fact that for every $x \in C$ (which by definition coincides with *bd* C) there is for every $\rho < 1$ and every $\delta > 0$ a point $y \in X$ so that $\|y - x\| < \delta$ and $d(y, C) \geq \rho\|y - x\|$.

Let μ be any regular probability measure on an infinite dimensional space X. It follows from the regularity of the measure that for every $\delta > 0$ there is a compact convex set C in X so that $\mu(X\backslash C) < \delta$. By taking e.g. μ a Gaussian measure we deduce that the set of points where the convex function $f(x) = d(x, C)$ is not Fréchet differentiable cannot be a Gauss null set.

Let $\{C_n\}_{n=1}^\infty$ be any collection of closed convex sets with empty interior in X, and let $\{\lambda_n\}_{n=1}^\infty$ be positive numbers such that $\sum_{n=1}^\infty \lambda_n d(x, C_n) = f(x)$ exists for all $x \in X$. The function $f(x)$ above is a convex function which is not Fréchet differentiable at any point in $\bigcup_{n=1}^\infty C_n$, in fact it is not ϵ_n-Fréchet differentiable on any point of C_n, for a suitable sequence of $\epsilon_n \downarrow 0$. This follows from the fact that if a sum of convex functions is say, Fréchet differentiable at some point, so must every summand be.

In order to state a more general version of Example 4 we introduce now the following important notion. A set A in a Banach space X is said to be c-porous with $0 < c < 1$ if for every $x \in C$ and $\delta > 0$ there is a $y \in X$ with $\|y - x\| < \delta$ and $B(y, c\|y - x\|) \cap A = \emptyset$. A set is called porous if it is c-porous for some $0 < c < 1$. A countable union of porous sets is called a σ-porous set. (All the definitions make sense in the setting of general metric spaces; we shall, however, deal with them here only in the setting of Banach spaces.) A convex set with empty interior is a simple example of a c-porous set for every $c < 1$.

Example 4'. Let A be a c-porous set in a Banach space. Then the Lipschitz function $f(x) = d(x, A)$ from A to R is not c-Fréchet differentiable at any point in A. Indeed, if $x \in A$ and $B(y, c\|x - y\|) \cap A = \emptyset$ then $|f(y) - f(x)| \geq c\|x - y\|$. The only possible value for a Gâteaux derivative of f at a point of A is evidently 0.

We shall say more on porous sets below. We want first to state theorems which put the examples above into a more general framework.

In the direction of Example 1 there is the following theorem.

Theorem 3 ([16]) . *Assume X is a separable Banach space with X^* non separable. Then there is an equivalent norm $\|\ \|$ on X and an $\epsilon > 0$ so that $f(x) = \|x\|$ has at no point an ϵ-Fréchet derivative.*

In Theorem 10 in Section 5 it will be shown that the assumption in Theorem 3 that X^* is non separable is essential. It is also known that if X^* is separable then X can be renormed so that its norm is Fréchet differentiable for $x \neq 0$. This is impossible if X is separable but X^* non separable.

In the direction of Example 2 we have the following general theorem.

Theorem 4 ([14]) . *Assume that X or Y have an unconditional basis and that there is a non compact linear operator from X to Y. Then there is a Lipschitz function f from X to Y which is nowhere ϵ-Fréchet differentiable for some $\epsilon > 0$.*

The proof of Theorem 4 is quite simple. If we denote by $|x|$ the vector obtained from x by replacing its coordinates with respect to an unconditional basis by their absolute value and if T is a non compact linear operator from X to Y then a function with the desired property is $f(x) = T|x|$ (if X has an unconditional basis) or $f(x) = |Tx|$ (if Y has an unconditional basis).

The following theorem contains as special cases Example 2 (in the continuous case, i.e. the map $x \to |x|$ in $L_2(0,1)$) as well as Example 3.

Theorem 5 ([2]) . *Let Ω be a metric space with a locally finite non-atomic measure μ such that every non-empty open set has strictly positive measure and let $F(t,s)$ ($t \in \Omega$, $s \in R$) be a continuous function such that $x(t) \in L_2(\mu) \Rightarrow F(t,x(t)) \in L_2(\mu)$. Assume that $f(x) = F(t,x)$ is Fréchet differentiable at some x_0 (as a map from $L_2(\mu)$ into itself). Then there are functions $a(t)$ and $b(t)$ such that $F(t,s) = a(t) + b(t)s$.*

The basic idea behind this statement is the observation that the assumptions that $|F(t,s) + F(t,-s) - 2F(t,0)| > c > 0$ for t in some non-empty open set G of suitably small measure and that f is ϵ-Fréchet differentiable at x_0 imply that

$$c\|1_G\| < \|f(x_0+(-x_0+s)1_G)+f(x_0+(-x_0-s)1_G)-2f(x_0-x_0 1_G)\| \leq 4\epsilon(\|x_0\|+s)\|1_G\|$$

which is a contradiction if ϵ is small enough.

The next theorem, which is the most delicate of the theorems in this section, is somewhat connected to Example 4.

Theorem 6 ([25,24]) . *Let X be a uniformly convex space. Then there is an equivalent norm $\|\ \|$ on X so that $f(x) = \|x\|$ is Fréchet differentiable only on a Gauss null set.*

We return now to the discussion of porous sets. A survey on this topic is in [35] where more details and references can be found. If A is a σ-porous Borel set then for any $c < 1$ we can represent A as $\bigcup_{n=1}^{\infty} \tilde{A}_n$ where each $\tilde{A}_n$ is a Borel c-porous set. In any Banach space X a porous set is meager and hence a σ-porous set is of the first category. If *dim* $X < \infty$ then a porous set and thus a σ-porous set is of measure 0 by Lebesgue's density theorem. Actually σ-porous sets form a proper subset of the sets of measure 0 and I'st category in R^n. For example, there is a graph of a continuous function in the plane which is not σ-porous.

In Banach spaces with *dim* $X = \infty$ the situation is different. In [30] it is proved that every infinite dimensional Banach space X can be decomposed into a union of two Borel sets $A \cup B$ so that A is σ-porous and B meets any line in X in a set of measure 0 (and thus in particular B is Aronszajn or Gauss null). Hence, in infinite dimensional spaces we can no longer consider σ-porous sets to be small in the sense of "measure".

By using the decomposition of X which we just mentioned the following theorem, which is a weaker form of Theorem 6 for general separable Banach apaces, is proved. (It is conceivable that Theorem 6 itself holds for a general separable Banach space.)

Theorem 7 ([30]) . *For every separable Banach space X there is a Lipschitz function $f : X \to R$ which is Fréchet differentiable only on a Gauss null set.*

Clearly by Theorem 3 we need to consider only spaces X with X^* separable. (By Theorem 6 we can consider just as well only spaces X which are not superreflexive, but this is not helpful for the proof of the theorem.) If $A = \bigcup_{n=1}^{\infty} A_n$ with $\{A_n\}_{n=1}^{\infty}$ porous it is easy to construct for every n a Lipschitz function $f_n : X \to R$ so that f_n is not Fréchet differentiable on any point of A_n (see e.g. Example 4'). The proof in [30] consists of choosing the $\{f_n\}_{n=1}^{\infty}$ so that $f = \sum_{n=1}^{\infty} \lambda_n f_n$ (for suitable positive $\{\lambda_n\}_{n=1}^{\infty}$) will still be Lipschitz and not Fréchet differentiable at any point of A. This construction is somewhat delicate since we are not dealing here with convex functions $\{f_n\}_{n=1}^{\infty}$. The situation is easiest to handle if all sets $\{A_n\}_{n=1}^{\infty}$ are also closed. This is the case which actually comes out from the decomposition of X constructed in [30].

There is a variant of the notion of porous or σ-porous sets which also in infinite dimensions produces sets which are small in the sense of measure. A set is said to be c-directionally porous if for every $x \in A$ there is a $u \in X$ with $\|u\| = 1$ and a sequence of scalars $\lambda_n \downarrow 0$ so that $B(x + \lambda_n u, c\lambda_n) \cap A = \emptyset$. The notion of σ-directionally porous sets is defined in an obvious way. The same argument as that of Example 4' shows that if A is c-directionally porous for some $0 < c < 1$ then the Lipschitz function $f(x) = d(x, A)$ is nowhere Gâteaux differentiable on A. It follows from Theorem 2 that a directionally porous (and thus also σ-directionally porous) set is always Gauss null. If $\dim X < \infty$ then a simple compactness argument shows that a set is directionally porous if and (clearly) only if it is porous.

It is evident that a graph of a Lipschitz function is a directionally porous set. Thus all the exceptional sets appearing in Theorem 1 are σ-directionally porous.

By the regularity argument appearing in Example 4 not every closed convex set C with empty interior (even compact convex set C) in a Banach space is σ-directionally porous (it need not be Gauss null).

4. Existence theorems for almost Fréchet derivatives

We start this section by presenting a proof of the existence of ϵ-Fréchet derivatives for Lipschitz functions $f : X \to R$ if X^* is uniformly convex. This is perhaps the simplest result on the existence of Fréchet derivatives of non necessarily convex Lipschitz functions on infinite dimensional spaces, but the idea in the proof is basic for the proof of several far more complicated results.

We assume that f is Gâteaux differentiable at x and that $\|D_f(x)\|$ is "very close" to its maximal possible value (which is, by Theorem 2, $Lip\ f$). We prove that then f is ϵ-Fréchet differentiable at x. The amount of closeness of $\|D_f(x)\|$ to $Lip\ f$ which is needed depends on ϵ and on the modulus of uniform convexity of X^*.

We shall normalize f so that $Lip\ f = 1$ and let $\epsilon > 0$. If X^* is uniformly convex, $e \in X$ with $\|e\| = 1$ and $x^* \in X^*$ with $x^*(e) = \|x^*\| = 1$ then for every $\eta > 0$ there is a $\delta = \delta(\eta) > 0$ so that if $\|v\| \le \delta$ and $x^*(v) = 0$ then $\|e + v\| \le 1 + \eta\|v\|$. Another simple observation we need is that if e and x^* are as above, $\eta > 0$ is sufficiently small and $z^* \in X^*$ satisfies $\|z^*\| = 1$ and $z^*(e) \ge 1 - \eta$ then $\|z^* - x^*\| < \epsilon/6$.

We start the proof by letting $x \in X$ and $e \in X$ with $\|e\| = 1$ be such that $D_f(x)e \geq 1 - \delta\eta$ where $\eta > 0$ will be chosen below to be small enough and $\delta = \delta(\eta)$. Put $z^* = D_f(x)$ and let $x^* \in X$ with $\|x^*\| = x^*(e) = 1$.

Since f is Gâteaux differentiable at x there is a $\sigma > 0$ so that

$$|f(x + \gamma e) - f(x) - z^*(e)\gamma| \leq \delta\eta|\gamma| \quad if \ \ |\gamma| < \sigma. \tag{4}$$

Let $y \in X$ and write it as $y = u + v$ where $u = x^*(y)e$. Then $\|u\| \leq \|y\|$, $\|v\| \leq 2\|y\|$ and $x^*(v) = 0$. Put $r = \|v\|/\delta$.

If $\|y\| \leq \delta\sigma/3$ then by (4)

$$\begin{aligned} |f(x + u + re) - f(x + u - re)| &\geq 2rz^*(e) - \delta\eta(3\|y\|/\delta) \\ &\geq 2r(1 - \delta\eta) - 3\|y\|\eta \geq 2r - 2\|v\|\eta - 3\|y\|\eta \geq 2r - 7\|y\|\eta. \end{aligned} \tag{5}$$

Since $Lip\ f = 1$ and $\|v/r\| = \delta$

$$|f(x + y) - f(x + u + re)| \leq \|re - v\| = r\|e - \frac{v}{r}\| \leq r(1 + \eta\delta) \leq r + 2\|y\|\eta. \tag{6}$$

Similarly,

$$|f(x + y) - f(x + u - re)| \leq r + 2\|y\|\eta. \tag{7}$$

We use now the trivial inequality that $|a| \leq \max(|b + a|, |b - a|) - b$ for all real a and b and obtain from (5), (6) and (7) that

$$\left| f(x + y) - \frac{f(x + u + re) + f(x + u - re)}{2} \right| \leq r + 2\|y\|\eta - (r - \frac{7}{2}\|y\|\eta) = 11\eta\|y\|/2. \tag{8}$$

From (4) and (8) and the fact that

$$|z^*(v)| = |(z^* - x^*)(v)| \leq 2\|z^* - x^*\|\|y\| \leq \epsilon\|y\|/3$$

we get that

$$\begin{aligned} &|f(x + y) - f(x) - z^*(y)| \leq \\ &\quad \left| f(x + y) - \frac{f(x + u + re) + f(x + u - re)}{2} \right| + |z^*(v)| + \\ &\quad |f(x + u + re) - f(x) - z^*(u + re)|/2 + |f(x + u - re) - f(x) - z^*(u - re)|/2 \leq \\ &\quad \frac{11\eta\|y\|}{2} + \frac{\epsilon\|y\|}{3} + \frac{3\delta\eta\|y\|}{\delta} + \frac{3\delta\eta\|y\|}{\delta} \leq 18\eta\|y\| + \epsilon\|y\|/3 \leq \epsilon\|y\| \end{aligned}$$

if η is chosen so that $\eta < \epsilon/27$ and satisfies the requirement posed on it at the beginning of the proof.

The proof above leads naturally to the question if it can be modified so as to yield stronger results.

1. If we want to get points of Fréchet differentiability of f by this method we would need points x and e such that $\|D_f(x)e\| = Lip\ f$. This is however impossible in general. To get points of Fréchet differentiability one has to do a tedious and careful process of successive approximation. This is explained in the next section.

2. If we are given two Lipschitz functions $f, g : X \to R$ (or equivalently, a Lipschitz function from X to R^2) we would like to find a point x where both f and g are ϵ-Fréchet differentiable. For this we need, if we use the proof above, a point x so that $\|D_f(x)\|$ and $\|D_g(x)\|$ are both close to *Lip* f and *Lip* g respectively. This again is impossible to achieve. However quite complicated arguments involving passage from global to local maxima of $\|D_f(x)\|$ or $\|D_g(x)\|$ allow us to overcome this difficulty. This is behind the proof of Theorem 8 below.

3. The natural assumption on X for the result proved above is that X^* be separable. In the proof above strong quantitative use was made of the fact that X^* is uniformly convex. We do not know how to carry out the proof above without quantitative information on X^*. There is however a class of spaces more general than super-reflexive spaces so that the proof above can be modified so as to apply to this more general class.

We proceed now to define this more general class. Recall that the modulus of smoothness $\rho_X(t)$ and convexity $\delta_X(t)$ of a Banach space X are defined by

$$\rho_X(t) = \sup_{\|x\|=1, \|y\| \le t} \frac{\|x+y\| + \|x-y\|}{2} - 1 ,$$

$$\delta_X(t) = \inf_{\|u\|, \|v\| \le 1, \|u-v\| \ge t} 1 - \frac{\|u+v\|}{2} .$$

In order to define the new "asymptotic" moduli we put for $Y \subset X$, $x \in S_X$ (the unit sphere of X)

$$\bar{\rho}_X(t, x, Y) = \sup_{y \in Y, \|y\| \le t} \|x+y\| - 1 .$$

Let

$$\bar{\rho}_X(t, x) = \inf_{dim\ X/Y < \infty} \bar{\rho}_X(t, x, Y)$$

and

$$\bar{\rho}_X(t) = \sup_{\|x\|=1} \bar{\rho}_X(t, x).$$

The function $\bar{\rho}_X(t)$ is called the modulus of asymptotic smoothness of X. The space X is said to be asymptotically uniformly smooth if $\lim_{t \to 0} \bar{\rho}_X(t)/t = 0$.

For $Y \subset X$, $x \in S_X$ put

$$\bar{\delta}_X(t, x, Y) = \inf_{y \in Y, \|y\| \ge t} \|x+y\| - 1.$$

Let

$$\bar{\delta}_X(t, x) = \sup_{dim\ X/Y < \infty} \bar{\delta}_X(t, x, Y)$$

and

$$\bar{\delta}_X(t) = \inf_{\|x\|=1} \bar{\delta}_X(t, x).$$

The function $\bar{\delta}_X(t)$ is called the modulus of asymptotic convexity of X. X is said to be asymptotically uniformly convex if $\bar{\delta}_X(t) > 0$ for $t > 0$.

These moduli have quite a long history and much is known about them. A survey of known results and several references are contained in Section 2 of [14].

The asymptotic moduli are especially easy to compute if X is a subspace of ℓ_p, $1 \leq p < \infty$, or c_0. For a subspace X of ℓ_p, $1 \leq p < \infty$, one gets

$$\bar{\rho}_X(t) = \bar{\delta}_X(t) = (1+t^p)^{1/p} - 1 \ , \quad 0 < t < 1$$

For a subspace X of c_0 one gets

$$\bar{\rho}_X(t) = \bar{\delta}_X(t) = 0 \ , \quad 0 < t < 1$$

In particular c_0 is (the most) asymptotically smooth space while ℓ_1 is (the most) asymptotically convex space.

It is also easy to check that for every X and $0 < t < 1$, $\bar{\delta}_X(t) \leq \bar{\rho}_X(t)$, $\delta_X(t) \leq \bar{\delta}_X(t)$ and $\rho_X(t) \geq \bar{\rho}_X(t)/2$. Hence a uniformly smooth (resp. convex) space is asymptotically uniformly smooth (resp. convex).

Another fact which we need to mention here (which is simple but not trivial) is that if $\bar{\rho}_X(t) < t$ for some $0 < t < 1$ and some separable space X then X^* is also separable. In particular an asymptotically uniformly smooth separable space X has a separable dual.

The proof we presented in the beginning of this section for spaces X which are uniformly smooth (i.e. X^* uniformly convex) carries over with not much difficulty to the setting in which one assumes only that X is asymptotically uniformly smooth.

The passage to maps from X to R^n, $n > 1$, is more tricky. In [18] we introduced the notion of a density sequence $\{X_n\}_{n=1}^{\infty}$ of Borel sets in X. We do not repeat the definition here but just mention that the idea is to produce a family of rich subsets in X so as to enable us to imitate arguments which in the setting of finite-dimensional spaces are proved by the usual density points argument.

Roughly speaking the idea in [18] was to prove that for a Lipschitz function f from a uniformly smooth space X to R the set of points where $\|D_f(x)\|$ is close to the local Lipschitz constant of f (and thus f is ϵ-Fréchet differentiable at x for a suitable $\epsilon > 0$) gives a density sequence. The details were however quite complicated.

In [14] a simpler inductive argument is given (which is still quite involved) which proves the same in the more general setting of asymptotically smooth spaces. What is proved in [14] is

Theorem 8 . *Assume that X is an asymptotically uniform smooth space and that f is a Lipschitz function from X to R^n. Then given any set X_0 in X such that $X \backslash X_0$ is Gauss null and $\epsilon > 0$ there exists an $x \in X_0$ such that f is Gâteaux differentiable at x and also ϵ-Fréchet differentiable at that point.*

Though the notion of density set is no longer needed for the proof of Theorem 8 it is likely that this notion may be useful in future applications.

By using Theorem 8 it is possible to deduce the existence of points of ϵ-Fréchet differentiability for Lipschitz maps between certain infinite dimensional Banach spaces.

Theorem 9 ([14]) . *Let X be a separable Banach space and let Y have the RNP. Let $f : X \to Y$ be a Lipschitz function. Assume that for all $t > 0$, $\bar{\delta}_Y(t) > 0$ and that for*

all $c > 0$, $\lim_{t \to 0} \bar{\rho}_X(t)/\bar{\delta}_Y(ct) = 0$. *Then for every* $\epsilon > 0$ *and every subset* X_0 *of* X *with* $X \backslash X_0$ *Gauss null there is an* $x \in X_0$ *such that* f *is Gâteaux differentiable at* x *and also* ϵ*-Fréchet differentiable there.*

An interesting special case where Theorem 9 applies is X a subspace of ℓ_r and $Y = \ell_p$ with $r > p \geq 1$.

For the most asymptotically uniform smooth space c_0 we have even a stronger result.

Theorem 10a ([14]) . *Let* X *be a subspace of* c_0, *let* Y *have the RNP and* $f : X \to Y$ *a Lipschitz function. Then for every* $\epsilon > 0$ *and* X_0 *a subset of* X *with* $X \backslash X_0$ *Gauss null there is an* $x \in X_0$ *such that* f *is Gâteaux differentiable at* x *as well as* ϵ*-Fréchet differentiable there.*

By using a transfinite iteration procedure one gets from Theorem 10a

Theorem 10b ([14]) . *Let* K *be a countable compact space, let* Y *have the RNP and let* f *be a Lipschitz function from* $C(K)$ *into* Y. *Then for every* $\epsilon > 0$ *there is a point in* $C(K)$ *where* f *is* ϵ*-Fréchet differentiable.*

Unlike the previous theorem there is no assertion here that the point of ϵ-Fréchet differentiability can be chosen to be in a given conull set. This assertion might be true also in this case but the iteration process used in proving Theorem 10b does not allow us to carry along this assertion concerning conull sets.

Unlike Theorem 9 and Theorem 10a we cannot take in Theorem 10b a subspace of $C(K)$. Indeed let S be the Schreier space, i.e. the completion of the space of sequences $\{\lambda_i\}_{i=1}^{\infty}$ which are eventually 0 with respect to the norm

$$\|\{\lambda_i\}\| = \sup\{\sum_{k=1}^{n} |\lambda_{i_k}|,\ n \in N,\ n \leq i_1 < i_2 < \ldots < i_n\} .$$

Then it is easily checked that S is isomorphic to a subspace of $C(\omega^\omega)$, the unit vectors form an unconditional basis in S and the formal identity map from S to ℓ_2 is bounded. Hence by Theorem 4 there is a Lipschitz map $f : S \to \ell_2$ which is nowhere ϵ-Fréchet differentiable for ϵ small enough.

A stronger version of both parts of Theorem 10 will be stated in the next section. The proof of Theorem 10 is however simpler than that of the corresponding result in the next section.

5. Existence theorems for Fréchet derivatives

The first results of this nature were obtained naturally in the context of continuous convex functions where the situation is much easier than that of general Lipschitz functions.

A direct verification shows that the points where a convex continuous function is Fréchet differentiable is a G_δ set. In [17] it was proved that if X is reflexive this set is dense. This result was generalized by Asplund [4] where it is shown to hold whenever X^* is separable. In view of Theorem 3 the assumption that X^* is separable cannot be weakened. In view of this result spaces with X^* separable (or more generally spaces X such that for every separable $Y \subset X$, Y^* is separable) are now called Asplund spaces.

The result that in Asplund spaces every convex and continuous function is Fréchet differentiable outside a dense G_δ set was strengthened in a different direction in [31]. In this paper the following theorem is proved:

Theorem 11 . *Let X^* be separable. Then every convex continuous $f: X \to R$ is Fréchet differentiable outside a σ-porous set.*

Surprisingly the proof of this stronger theorem is simpler than the proofs of the weaker results of [17] and [4].

We recall that by Theorem 6 a convex continuous function on Hilbert space may be Fréchet differentiable only on a Gauss null set. Later on in this section we shall prove that in some other sense convex continuous functions on an Asplund space must be Fréchet differentiable "almost everywhere". All these results make it difficult even to conjecture a precise result on the nature of the sets of points of non Fréchet differentiability of convex continuous functions on Asplund spaces (in the spirit of Theorem 1).

We now turn to results on existence of points of Fréchet differentiability of Lipschitz functions. These results are far more complicated than the results on convex functions. As a matter of fact these results are the hardest results on which we report in this survey.

The first result on Fréchet differentiability of Lipschitz functions from an Asplund space to R was obtained in [28]. It is proved there that a Lipschitz function from an Asplund space X to R which is everywhere Gâteaux differentiable must have a point of Fréchet differentiability. The proof of this result, like the subsequent proofs of the same results where the assumption of the existence everywhere of the Gâteaux derivative is dropped, uses an iteration method. One constructs inductively a sequence of points $\{x_n\}_{n=1}^{\infty}$ which converges to a point $x \in X$ and so that the sequence of Gâteaux derivates $D_f(x_n)$ converges also. What makes the final step of the proof relatively easy is the fact that by assumption the function f is Gâteaux differentiable at x. The fact that X is Asplund is used in the proof via the assumption that the norm in X^* can be assumed to be locally uniformly convex, i.e. that $\|x_n^*\| = \|x^*\| = 1$ and $\|x_n^* + x^*\| \to 2$ implies that x_n^* tends in norm to x^*.

It should be mentioned that unlike Lebesgue's original proof for functions on the line (or the subsequent elegant proofs of this theorem) where one proves directly the existence of the derivative almost everywhere, here one had to construct the point of differentiability by iteration. There are at present no known concepts of "almost everywhere" which can be used in the proof for general Asplund spaces X (and in particular for $X = \ell_2$). Later on in this section we will present such a notion of "almost everywhere" which can be used in certain specific spaces X. Of course the lack of an "almost everywhere" approach has the disadvantage that it does not lead to results on the existence of a common point of Fréchet differentiability for say two Lipschitz functions from X to R.

It is worthwhile to point out here (the non surprising fact) that Baire Category arguments cannot help in this context. In [27] an example of the following type is presented.

Example 5. There is a Lipschitz function $f: \ell_2 \to R$ which is everywhere Gâteaux differentiable but which is Fréchet differentiable only on a set of the first category.

In [29] it is proved that if X is Asplund then any Lipschitz function from X to R has a point of Fréchet differentiability. The proof is again by constructing inductively a sequence of points $\{x_n\}$ which converges to a desired point x of Fréchet differentiability. It is evident from the construction that the points x_n can be chosen so that the Gâteaux derivatives $D_f(x_n)$ exist. However the a priori lack of knowledge that $D_f(x)$ exists made it necessary to use a very delicate and complicated procedure. In particular every step of passing from x_n to x_{n+1} necessitated the construction of a new norm in X for which X^* becomes locally uniformly convex.

A much simpler (but still not simple) proof of the main result of [29] is presented in [19]. In this latter paper a somewhat stronger version is proved in which one does not have to assume that X is Asplund.

Theorem 12 ([29,19]) . *Let f be a Lipschitz function from a separable space X into R. Assume that the w^* closure of the set of all Gâteaux derivatives of f (at the points in which they exist) is a norm separable subset of X^*. Then f has a point of Fréchet differentiability.*

Of course if X^* itself is assumed to be separable then no special assumption on f is needed.

The proof in [19] is done via slicing of sets in X^*. A w^*-slice S of a set $A \subset X^*$ is a set of the form

$$S(x, \alpha) = \{x^* \in A : \ x^*(x) \geq \sigma - \alpha, \ where \ \sigma = \sup\{x^*(x) : \ x^* \in A\}\}$$

for some $x \in X$ and $\alpha > 0$. One says also that S is a w^*-slice determined by the vector x. The separability assumption in the statement of Theorem 12 is used via the following proposition proved in [19]: Let $A \subset X^*$ be bounded and have separable w^*-closure, let $x \in X$ and $\epsilon > 0$ so that $\epsilon < \|x\|$. Then there is a w^*-slice S of A which has diameter less than ϵ and which is determined by a vector y with $\|y - x\| < \epsilon$.

In rough terms the strategy of the proof in [19] is to consider the set A of Gâteaux derivatives of f. This set is norm bounded by $Lip\ f$. By using the proposition above one constructs inductively a sequence $\{x_n\}_{n=1}^{\infty}$ of points in X such that the Gâteaux derivatives $D_f(x_n)$ exist and such that $x = \lim_n x_n$ exists; a sequence of slices $\{S_n\}_{n=1}^{\infty}$ so that $S_n \supset S_{n+1}$, $D_f(x_n) \in S_{n+1}$, $diam\ S_n \to 0$ and S_n is determined by a vector e_n so that $\sum \|e_n - e_{n+1}\| < \infty$. Then clearly $y^* = \lim_n D_f(x_n)$ must exist. The delicate point (which necessitates extra care in the inductive construction) is to show that this y^* is the Fréchet derivative of f at the point x.

We concentrate in this survey on separable spaces X, but let us say a little about nonseparable spaces in connection with Theorem 12. There is an argument which allows to reduce Fréchet differentiability results from the nonseparable case to the separable case. This argument is presented in [27] and reproduced in [19]. Using this argument one gets immediately from Theorem 12 that every Lipschitz function on (a possibly nonseparable) Asplund space X to R has points of Fréchet differentiability. (By the way this proves in particular that such an f has points of Gâteaux differentiability, a fact which does not follow from the usual existence proofs for Gâteaux derivatives like Theorem 2.)

We are going to define now a new class of null sets and in terms of this new class new results on Fréchet differentiability can be formulated (and proved). We let $\Sigma = [0, 1]^N$

be endowed with the product topology and the product Lebesgue measure μ. Let X be a Banach space and let $\Gamma(X)$ be the space of continuous maps $\gamma: \Sigma \to X$ having also continuous partial derivatives $\{D_j\gamma\}_{j=1}^{\infty}$. The elements $\gamma \in \Gamma(X)$ will be called surfaces. We equip $\Gamma(X)$ with the topology of uniform convergence of the surfaces and their partial derivatives. In other words the topology in $\Gamma(X)$ is generated by the semi norms $\|\gamma\|_0 = \sup_{t\in\Sigma} \|\gamma(t)\|$ and $\|\gamma\|_k = \sup_{t\in\Sigma} \|D_k\gamma(t)\|$, $k = 1, 2, \ldots$. In this topology $\Gamma(X)$ is a Fréchet space, in particular it is a Polish space (i.e. metrizable by a complete separable metric).

A Borel set $N \subset X$ is called Γ-null if

$$\mu\{t \in \Sigma : \gamma(t) \in N\} = 0$$

for residually many $\gamma \in \Gamma(X)$. Recall that a set is called residual if its complement is of the first category. Note that the definition of Γ-null sets involves both the concepts of measure and category. A possibly non Borel subset of X is called Γ-null if it is contained in a Borel Γ-null set.

The Γ-null sets clearly form a σ-ideal of subsets of X. It can be verified that if $dim\ X < \infty$ then the Γ-null sets coincide with the sets of Lebesgue measure 0 (like e.g. the classes of Gauss or Haar null sets).

With a proof which is perhaps even simpler than that of Theorem 2 it follows that Theorem 2 remains valid if we require the exceptional set to be Γ-null.

In order to study Fréchet differentiability in the context of Γ-null sets a new concept is needed:

Let $f: X \to Y$ be a map. We say that $x \in X$ is a regular point of f if for every $v \in X$ for which the directional derivative $f'(x, v)$ exists

$$\lim_{t\to 0} \frac{f(x + tu + tv) - f(x + tu)}{t} = f'(x, v)$$

uniformly in u with $\|u\| \le 1$.

It is not hard to prove that if $f: X \to R$ is convex and continuous then every $x \in X$ is a regular point of f. Another easy fact is that if $f: X \to Y$ is Lipschitz then the set of irregular points of f is a σ-porous subset of X.

The main result on Fréchet differentiability in the context of Γ-null sets is the following.

Theorem 13 ([20]) . *Let X and Y be separable Banach spaces with Y having the RNP and let L be a separable subspace of the space of bounded linear operators from X to Y. Then any Lipschitz map $f: X \to Y$ whose Gâteaux derivatives belong to L (whenever they exist) is Fréchet differentiable at Γ-almost every point $x \in X$ at which it is regular and Gâteaux differentiable.*

This is a strong theorem and its proof is hard. It follows in particular that any convex continuous function $f: X \to R$ where X is Asplund is Fréchet differentiable Γ-almost everywhere. Thus in view of Theorem 6 the notions of Γ-null and Gauss null sets are incomparable (at least when X is superreflexive). One can decompose X as a union of disjoint Borel sets $X = A_0 \cup B_0$ with A_0 Γ-null and B_0 Gauss null.

It follows also from Theorem 13 that if X is a separable Asplund space and Y has the RNP then for every sequence of convex continuous functions $\{f_i\}_{i=1}^{\infty}$ from X to R and every sequence $\{g_j\}_{j=1}^\infty$ of Lipschitz functions from X to Y, there is a point x so that all the $\{f_i\}_{i=1}^\infty$ are Fréchet differentiable at x and all the $\{g_j\}_{j=1}^\infty$ are Gâteaux differentiable at x. This result cannot be deduced from the preceding results. Till now we knew only that a Lipschitz $g : X \to Y$ is Gâteaux differentiable outside a Gauss null set while the generic result involving Fréchet differentiability of convex functions involved σ-porous sets.

Another consequence of Theorem 13 is the fact that every Lipschitz $f : X \to R$ where X is a separable Asplund space is Fréchet differentiable Γ-almost everywhere if and only if every porous (and therefore σ-porous) set in X is Γ-null. The "only if" part follows from the simple observation in Example 4'.

In view of this result it is worthwhile to investigate those spaces X such that any σ-porous set in X is Γ-null. For this purpose we introduce the following concepts.

A set $A \subset X$ is said to be c-porous in the direction of a subspace $Y \subset X$ if for every $x \in A$ there is a sequence $\{y_n\}_{n=1}^\infty$ in Y with $\|y_n\| \downarrow 0$ and $B_X(x + y_n, c\|y_n\|) \cap A = \emptyset$ for all n.

A decreasing sequence $\{X_k\}_{k=1}^\infty$ of subspaces of X is said to be asymptotically c_0 if there is a constant $C < \infty$ such that for every integer n

$$(\exists k_1 \in \mathcal{N})(\forall u_1 \in X_{k_1})(\exists k_2 \in \mathcal{N})(\forall u_2 \in X_{k_2}) \cdots (\exists k_n \in \mathcal{N})(\forall u_n \in X_{k_n})$$

$$\|u_1 + u_2 + \cdots + u_n\| \leq C \max(\|u_1\|, \|u_2\|, \ldots, \|u_n\|).$$

The main tool for verifying that in certain spaces every σ-porous set is Γ-null is the following

Theorem 14a ([20]) . *Suppose that the space X has a decreasing sequence of subspaces $\{X_k\}_{k=1}^\infty$ which is asymptotically c_0. Then for every $0 < c < 1$ every set $A \subset X$ which is c-porous in the direction of all the subspaces $\{X_k\}_{k=1}^\infty$ is Γ-null.*

As a consequence of Theorem 14a one deduces

Theorem 14b ([20]) . *If X is a subspace of c_0, or a space $C(K)$ with K countable compact, or the Tsirelson space T, then all the σ-porous subsets of X are Γ-null.*

The Tsirelson space is a reflexive space with an unconditional basis which is asymptotically c_0 but (clearly) does not contain c_0 as a subspace. Such a space was first constructed in [33].

We shall see in the next section that if $X = \ell_p$, $1 < p < \infty$, then X fails to have the property that its σ-porous subsets are Γ-null.

For X a subspace of c_0 or for $X = C(K)$ with K countable compact it is easy to check that whenever Y has the RNP the space of bounded linear operators from X to Y is separable. Hence one gets from Theorem 13 and Theorem 14b

Theorem 14c ([20]) . *If X is a subspace of c_0 or the space $C(K)$ with K countable compact then every Lipschitz function from X to a space Y with the RNP is Fréchet differentiable Γ-almost everywhere.*

6. The mean value theorem

In Section 1 we already used a very simple form of the mean value theorem for functions which are Gâteaux differentiable on a segment. A little more general result whose proof is again very simple is the following.

Let X be a separable Banach space and Y a space having the RNP. Let D be a convex open set in X and let $f : D \to Y$ be a Lipschitz function. Let $D_0 \subset D$ be the subset of points in D where f is Gâteaux differentiable. (We know by Theorem 2 that $D \backslash D_0$ is Gauss null.) Then if we put for $u \in X$

$$R_u = \{\frac{f(x+tu)-f(x)}{t} : \; x, x+tu \in D \; and \; t > 0\} \subset Y$$

$$\tilde{R}_u = \{D_f(x)u : \; x \in D_0\} \subset Y$$

then the closed convex hull of R_u coincides with the closed convex hull of $\tilde{R}_u$.

This observation is an easy consequence of the separation theorem and the property of Gauss null sets which ensures that whenever $x, x+tu \in D$ there is a point x_0 arbitrarily close to x so that $x_0 + su \in D_0$ for almost all $0 \leq s \leq t$.

As in other questions, the situation with Fréchet derivatives is much more delicate. For Lipschitz functions $f : X \to R$ where X is Asplund it follows from the description given above of the proof of Theorem 12 that if S is any slice of the set of Gâteaux derivatives of f then there is a point $x \in X$ such that f is Fréchet differentiable at x and $D_f(x) \in S$. This can be expressed in other words as follows (if we consider just functions defined on an open set). Let D be a convex open subset in X with X Asplund and let f be a Lipschitz function from D to R. Let $u, v \in D$ and let $m < f(v) - f(u)$. Then there is a point $x \in D$ so that f is Fréchet differentiable at x and $D_f(x)(v-u) > m$. (This fact is contained in both proofs of Theorem 12, the one in [29] and the one in [19] which was very briefly outlined in Section 5.)

It follows in particular from the result above that if at all points where f is Fréchet differentiable $D_f(x) = 0$ then f has to be a constant.

The formulation of the result above in terms of slices makes sense also for functions with a range space Y of dimension > 1. For example, the natural formulation of the mean value theorem for Fréchet derivatives for a Lipschitz function $f : X \to R^n$ with X Asplund would be the following: Any slice S of the set of Gâteaux derivatives of f (which is a subset in the set of operators from X to R^n, or equivalently $X^* \oplus \cdots \oplus X^*$ (n summands)) contains an element of the form $D_f(x)$ where f is Fréchet differentiable at x.

At this stage it would be impossible to assert that this mean value theorem is true for every Asplund space X since we do not even know if there are at all points of Fréchet differentiability of f. We know however by Theorem 8 that if X is uniformly smooth (or more generally asymptotically uniformly smooth) then f has points of ϵ-Fréchet differentiability for every $\epsilon > 0$. The following very delicate and surprising example from [30] shows that the mean value theorem for Fréchet derivatives is false even if we talk of ϵ-Fréchet derivatives.

Example 6. Let $1 < p < \infty$ and n be an integer with $n > p$. Then there is a Lipschitz map $f = (f_1, f_2, \ldots, f_n)$ from ℓ_p to R^n such that

$$\sum_{j=1}^{n} D_{f_j}(x)e_j = 0$$

where $\{e_j\}_{j=1}^{\infty}$ is the basis of ℓ_p, whenever f is Fréchet differentiable at x. The function f is Gâteaux differentiable at the origin with $\sum_{j=1}^{n} D_{f_j}(0)e_j = 1$. Whenever f is Gâteaux differentiable at a point x with $\sum_{j=1}^{n} D_{f_j}(x)e_j \neq 0$ the function f fails to be even ϵ-Fréchet differentiable for $\epsilon = \epsilon(x) = c|\sum_{j=1}^{n} D_{f_j}(x)e_j|$ with a suitable $c > 0$.

The construction of this example and the proof that it has the desired properties is very complicated. In order not to make the reading of this proof even more complicated than necessary for the potential reader we point out a bad misprint in [30]. On page 227 in the statement of Lemma 2 and in many places on pages 228 and 229 there is a meaningless symbol $g \simeq$ in the formulas. Whenever this symbol occurs it should be replaced by $g\left(\frac{\|\pi_m z\|}{r_m}\right)$.

In terms of slices Example 6 states that the non empty slice

$$S = \left\{ (D_{f_1}(x), D_{f_2}(x), \ldots, D_{f_n}(x)) : \sum_{j=1}^{n} D_{f_j}(x)e_j > \frac{1}{2} \right\}$$

of the set of Gâteaux derivatives of f contains no $D_f(x)$ at a point x in which f is Fréchet differentiable (or even only ϵ-Fréchet differentiable for a suitable fixed ϵ). It is clear from this that the proof of Theorem 12 in [19], as outlined in Section 5, cannot be generalized in an obvious way to maps from X to R^n with $n \geq 2$.

We return now to the Γ-null sets discussed in the previous section. In this setting we have

Theorem 15 ([20]) . *Let $f : X \to Y$ be a Lipschitz function which is Fréchet differentiable at Γ-almost every point of X. Then for any slice S of the set of Gâteaux derivatives of f the set of points x so that f is Fréchet differentiable at x and $D_f(x) \in S$ is not Γ-null.*

It follows from this result combined with Theorem 13 and Example 6 that in ℓ_p, $1 < p < \infty$, not every porous set is Γ-null.

The next theorem shows that Example 6 is in some sense optimal, at least in the case of ℓ_2.

Theorem 16a ([21]) . *Let f be a Lipschitz map from ℓ_2 to R^2. Then for every slice S of the set of Gâteaux derivatives of f and every $\epsilon > 0$ there is a point $x \in X$ such that f is ϵ-Fréchet differentiable at x and $D_f(x) \in S$.*

The strategy of the proof of Theorem 16a is the following: Given a σ-porous set A in X (which in the application to the proof of Theorem 16a will be the set of irregular points of f) and a 2-dimensional surface $\gamma : [0,1]^2 \to X$ we want to modify γ to a "nearby" surface $\hat{\gamma}$ which does not hit A. This modification is a rather tedious iterative procedure which is done locally at the points where the range of γ hits A. The key ingredient in the proof is that of replacing γ locally, at neighborhoods of appropriate points, by pieces of a catenoid (which is, as well known, a surface with minimal surface area).

The same procedure can be done with curves (one-dimensional surfaces) and this works for a general Asplund space X which has the RNP.

In order to state these modification results formally we first define what exactly we mean by finite-dimensional surfaces and what topology we take on them.

For a Banach space X we define $\Gamma_n(X)$ to be the space of continuous maps $\gamma : [0,1]^n \to X$ having a distributional derivative $\gamma' \in L_2([0,1]^n, Y)$, where Y is the space of operators from R^n to X, with the norm

$$\|\gamma\|_{\Gamma_n(X)} = \sup_{t\in[0,1]^n} |\gamma(t)| + \left(\int_{[0,1]^n} \|\gamma'(t)\|^2 dt\right)^{1/2}.$$

All the $\Gamma_n(X)$ are Banach spaces (i.e. complete).

Theorem 16b ([21]) .

1. *Assume that X is a separable Asplund space with the RNP and let $A \subset X$ be a porous set. Then*
$$\{\gamma \in \Gamma_1(X) : \mu\{t : \gamma(t) \in A\} = 0\}$$
is residual in $\Gamma_1(X)$.

2. *Let $X = \ell_2$, and $A \subset X$ a porous set. Then*
$$\{\gamma \in \Gamma_2(\ell_2) : \mu\{t : \gamma(t) \in A\} = 0\}$$
is residual in $\Gamma_2(X)$.

It is not clear that the assumption that X has the RNP is needed in statement 1 of the theorem.

Example 6 shows that statement 2 of the theorem is no longer valid if $X = \ell_p$ with $1 < p < 2$.

Example 6 shows also that there is no analogue of statement 2 for $\Gamma_3(\ell_2)$.

7. Open problems

There are several (implicit or explicit) open problems which are scattered in the material of the previous sections. Here we shall state four explicit problems which seem to be of central interest in the subject matter of the present survey.

Problem 1: Does there exist an example of a pair of separable Banach spaces X and Y so that every Lipschitz map $f : X \to Y$ has for every $\epsilon > 0$ points of ϵ-Fréchet differentiability but so that there is a Lipschitz map $f : X \to Y$ which has no point of Fréchet differentiability?

We do not know of any criterion for the non existence of points of Fréchet differentiability which does not automatically show that points of ϵ-Fréchet differentiability fail to exist for ϵ small enough. The most evident special case of the problem is the case where $X = \ell_2$ and $Y = R^n$ with $1 < n < \infty$.

Problem 2: Is it true that every Lipschitz map $f: X \to Y$ for a pair of separable spaces has points of ϵ-Fréchet differentiability for every $\epsilon > 0$ if and only if the space of bounded linear operators from X to Y has the RNP?

This is an attractive problem which shows that there is conceivably an elegant characterization of all such pairs X and Y. At present there are only partial positive results to the "if" or the "only if" part of this question.

It is clear that if the space of bounded linear operators from X to Y has the RNP then both X^* and Y must have the RNP. It is known that for separable X, the dual space X^* has the RNP iff X^* is separable. It was mentioned above that these assertions on X^* and Y are necessary if every $f: X \to Y$ has points of ϵ-Fréchet differentiability for every $\epsilon > 0$.

It is known (see [9]) that if X^* is separable and Y is a separable space with the RNP and if every bounded linear operator from X to Y is compact then the space of bounded linear operators from X to Y has the RNP. It follows from Theorem 4 that if X or Y have an unconditional basis then the assumption that every bounded linear operator from X to Y is compact is also a necessary condition for ϵ-Fréchet differentiability of Lipschitz functions. Thus the missing piece of information for answering the "only if" part of Problem 2 is what happens to Theorem 4 if we drop the unconditionality assumption. In particular assume that X is a hereditary indecomposable space in the sense of [11] (where the existence of such spaces is proved). Does every Lipschitz map from X to itself have points of ϵ-Fréchet differentiability?

On the "if" part of Problem 2 the existing information is even more fragmentary. The main known positive results are presented in Section 4 above.

Problem 3: Assume that X is a separable Asplund space (or even a superreflexive space). Can the set of points of Gâteaux differentiability of a Lipschitz map from X to a space Y with the RNP be a σ-porous set? With X as above can the decomposition result of [30] be strengthened so that X can be decomposed into a union of Borel sets $A \cup B$ with A a σ-porous set and B belonging to the class $\tilde{A}$ (see the end of Section 2) ? Of course a positive answer to the first question implies a positive answer to the second question.

Problem 4: Assume that X^* is separable, $\{f_i\}_{i=1}^{\infty}$ a sequence of Lipschitz functions from X to R with $\{Lip\ f_i\}_{i=1}^{\infty}$ bounded and g a Lipschitz map from X to a space Y with the RNP. Does there exist for every $\epsilon > 0$ a point $x \in X$ such that all the $\{f_i\}$ are ϵ-Fréchet differentiable at x and g is Gâteaux differentiable at x?

We know from Theorem 14b that the answer is positive for some such spaces X (with even ϵ-Fréchet differentiability replaced by Fréchet differentiability). For superreflexive X and more generally spaces X having an asymptotically uniformly smooth norm there is a positive answer to this question if we are given only a finite sequence $\{f_i\}_{i=1}^{n}$ of maps from X to R. Unfortunately the methods of proof in both [18] and [14] do not seem to make it possible to pass from a finite sequence of real-valued maps to an infinite sequence. Thus Problem 2 is open for superreflexive X. Again, the most interesting open case is $X = \ell_2$.

Problem 4 is related to the question of Lipschitz equivalence of Banach spaces discussed in Section 1. It was noted in Section 1 that if g is a Lipschitz equivalence between X and Y and if there is a point x where g is Gâteaux differentiable and ϵ-Fréchet differentiable for ϵ small enough then $D_g(x)$ is a linear isomorphism from X onto Y. A minor change in this proof shows the same conclusion holds if g is Gâteaux differentiable at x and the sequence of functions $\{y_n^* \circ g\}_{n=1}^{\infty}$ are ϵ-Fréchet differentiable at x with ϵ small enough and $\{y_n^*\}_{n=1}^{\infty}$ being a norming sequence of functionals in B_{Y^*}. As a consequence of this remark and Theorem 14b we get that if $g : T \to Y$ is a Lipschitz equivalence (where T is the Tsirelson space) then there is a point $x \in T$ such that $D_g(x)$ is a linear isomorphism from T onto Y. We used here the trivial fact that the RNP is invariant with respect to Lipschitz equivalence.

The Tsirelson space T is the first example of this kind. Previously there were many examples of Banach spaces X such that every space Y which is Lipschitz equivalent to X must be linearly isomorphic to X. This was proved in [12] for $X = \ell_p$ or $L_p(0,1)$ if $1 < p < \infty$ and also for $X = \ell_1$ is case Y is assumed to be a conjugate space. The proof in [12] used the existence of Gâteaux derivatives of the Lipschitz equivalence $g : X \to Y$. However in order to get an isomorphism onto they had to use the decomposition method of Pełczynski. Their proof does not show that $D_g(x)$ is a linear surjective isomorphism for some $x \in X$. In [10] it is proved that if Y is Lipschitz equivalent to c_0 then Y is linearly isomorphic to c_0. Their argument does not even use Gâteaux differentiability since c_0 does not have the RNP. To show the delicacy of this result we point out that if g is a Lipschitz equivalence from X into c_0 we cannot deduce that X is linearly isomorphic to a subspace of c_0 (in case of spaces with the RNP such a result follows trivially, as pointed out in Section 1, by taking the Gâteaux derivative of g at a point). In fact, it is proved in[1] that any separable Banach space is Lipschitz equivalent to a subspace of c_0. Most Banach spaces (like ℓ_p, $1 \le p < \infty$, or $C(0,1)$) are easily seen not to be isomorphic to a subspace of c_0.

We conclude by remarking that by Theorem 14b and the observations made above, it follows that if g is a Lipschitz quotient map from T onto a separable Banach space Y, then by taking $D_f(x)$ at a suitable x we get a linear quotient map from T onto Y.

REFERENCES

1. I. Aharoni, Every separable metric space is Lipschitz equivalent to a subset of c_0, Israel J. Math. 19(1974), 284-291.
2. A. Ambrosetti and G. Prodi, A primer of nonlinear analysis, Cambridge studies in advanced mathematics vol. 34, Cambridge University Press, 1995.
3. N. Aronszajn, Differentiability of Lipschitzian mappings between Banach spaces, Studia Math. 57(1976), 147-190.
4. E. Asplund, Fréchet differentiability of convex functions, Acta Math. 121(1968), 31-47.
5. Y. Benyamini and J. Lindenstrauss, Geometric nonlinear functional analysis Vol. I, Colloquium Publications, Amer. Math. Soc. n. 48 (2000).
6. S. M. Bates, W. B. Johnson, J. Lindenstrauss, D. Preiss and G. Schechtman, Affine approximation of Lipschitz functions and nonlinear quotients, Geom. Funct. Anal. 9(1999), 1092-1127.
7. J. P. R. Christensen, On sets of Haar measure zero in Abelian Polish groups, Israel J. Math. 13(1972), 255-260.
8. M. Csörnyei, Aronszajn null and Gaussian null sets coincide, Israel J. Math. 111(1999), 191-202.

9. J. Diestel and T. J. Morrison, The Radon Nikodým property for the space of operators I, Math. Nachr. 92(1979), 7-12.
10. G. Godefroy, N. J. Kalton and G. Lancien, Subspaces of $c_0(N)$ and Lipschitz isomorphisms, Geom. Funct. Anal. 10(2000), 798-820.
11. W. T. Gowers and B. Maurey, The unconditional basic sequence problem, J. Amer. Math. Soc. 6(1993), 851-874.
12. S. Heinrich and P. Mankiewicz, Applications of ultrapowers to the uniform and Lipschitz classification of Banach spaces, Studia Math. 73(1982), 225-251.
13. D. J. Ives and D. Preiss, Not too well differentiable isomorphisms, Israel J. Math. 115(2000), 343-353.
14. W. B. Johnson, J. Lindenstrauss, D. Preiss and G. Schechtman, Almost Fréchet differentiability of Lipschitz mappings between infinite dimensional Banach spaces, Submitted for publication.
15. W. B. Johnson, J. Lindenstrauss, D. Preiss and G. Schechtman, Lipschitz quotients of spaces containing ℓ_1, in preparation.
16. E. B. Leach and J. H. M. Whitefield, Differentiable norms and rough norms on Banach spaces, Proc. Amer. Math. Soc. 33(1972), 120-126.
17. J. Lindenstrauss, On operators which attain their norm, Israel J. Math. 1(1963), 139-148.
18. J. Lindenstrauss and D. Preiss, Almost Fréchet differentiability of finitely many Lipschitz functions, Mathematika 43(1996), 393-412.
19. J. Lindenstrauss and D. Preiss, A new proof of Fréchet differentiability of Lipschitz functions, J. Eur. Math. Soc. 2(2000), 199-216.
20. J. Lindenstrauss and D. Preiss, New null sets, In preparation.
21. J. Lindenstrauss and D. Preiss, Avoiding σ-porous sets, In preparation.
22. J. Lindenstrauss, E. Matoušková and D. Preiss, Lipschitz image of a measure-null set can have a null complement, Israel J. Math. 118(2000), 207-219.
23. P. Mankiewicz, On the differentiability of Lipschitz mappings in Fréchet spaces, Studia Math. 45(1973), 15-29.
24. E. Matoušková, An almost nowhere Fréchet smooth norm on superreflexive spaces, Studia Math. 133(1999), 93-99.
25. J. Matoušek and E. Matoušková, A highly nonsmooth norm on Hilbert spaces, Israel J. Math. 112(1999), 1-28.
26. S. Mazur, Über konvexe Mengen in linearen normierten Räumen, Studia Math. 4(1933), 70-84.
27. D. Preiss, Gâteaux differentiable Lipschitz functions need not be Fréchet differentiable on a residual set, Supplemento Rend. Circ. Mat. Palermo, serie II, n. 2(1982), 217-222.
28. D. Preiss, Gâteaux differentiable functions are somewhere Fréchet differentiable, Rend. Circ. Mat. Palermo 33(1984), 122-133.
29. D. Preiss, Differentiability of Lipschitz functions on Banach spaces, J. Funct. Anal. 91(1990), 312-345.
30. D. Preiss and J. Tišer, Two unexpected examples concerning differentiability of Lipschitz functions on Banach spaces, GAFA Israel Seminar 92-94, Birkhäuser (1995), 219-238.
31. D. Preiss and L. Zajíček, Fréchet differentiation of convex functions in a Banach space

with separable dual, Proc. Amer. Math. Soc. 91(1984), 202-204.
32. D. Preiss and L. Zajíček, Directional derivatives of Lipschitz functions, Israel J. Math. (2001).
33. B. S. Tsirelson, Not every Banach space contains an imbedding of ℓ_p or c_0, Funct. Anal. Appl. 8(1974), 138-141.
34. L. Zajíček, On the differentiability of convex functions in finite and infinite-dimensional spaces, Czechoslovak Math. J. 29(1979), 340-348.
35. L. Zajíček, Porosity and σ-porosity, Real Anal. Exchange 13(1987-88), 314-350.

Recent Progress in Functional Analysis
K.D. Bierstedt, J. Bonet, M. Maestre and J. Schmets

Summing inclusion maps between symmetric sequence spaces, a survey

Andreas Defant, Mieczysław Mastyło[*] and Carsten Michels

Fachbereich Mathematik, Carl von Ossietzky University of Oldenburg, Postfach 2503, D-26111 Oldenburg, Germany, e-mail: defant@mathematik.uni-oldenburg.de

Faculty of Mathematics and Computer Science, Adam Mickiewicz University and Institute of Mathematics, Poznań Branch, Polish Academy of Sciences, Matejki 48/49, 60-769 Poznań, Poland, e-mail: mastylo@amu.edu.pl

School of Mathematics, University of Leeds, Leeds LS2 9JT, United Kingdom, e-mail: balin@amsta.leeds.ac.uk

To Professor Manuel Valdivia on the occasion of his seventieth birthday.

Abstract

For $1 \leq p \leq 2$ let E be a p-concave symmetric Banach sequence space, so in particular contained in ℓ_p. It is proved in [14] and [15] that for each weakly 2-summable sequence (x_n) in E the sequence $(\|x_n\|_p)$ of norms in ℓ_p is a multiplier from ℓ_p into E. This result is a proper improvement of well-known analogues in ℓ_p-spaces due to Littlewood, Orlicz, Bennett and Carl, which had important impact on various parts of analysis. We survey on a series of recent articles around this cycle of ideas, and prove new results on approximation numbers and strictly singular operators in sequence spaces. We also give applications to the theories of eigenvalue distribution and interpolation of operators.
MCS 2000 Primary 47B10; Secondary 46M35, 47B06

1. Introduction

Dirichlet proved that a sequence in a finite-dimensional Banach space is absolutely summable if and only if it is unconditionally summable. For infinite-dimensional Banach spaces this is not true: The Dvoretzky–Rogers Theorem asserts that for every sequence $\lambda = (\lambda_n) \in \ell_2$ and every Banach space X there exists an unconditionally summable sequence (x_n) in X with $\|x_n\|_X = |\lambda_n|$ for all n, hence for any given sequence space E properly contained in ℓ_2 there exists in every Banach space X an unconditionally summable sequence (x_n) such that $(\|x_n\|_X) \notin E$. However, this result is false for $E = \ell_2$: There

[*]Research supported by KBN Grant 2 P03A 042 18

exist Banach spaces X in which every unconditionally summable sequence is absolutely 2-summable, e. g. $X = \ell_p$, $1 \le p \le 2$, a famous result due to Orlicz [37]. Moreover, the collection of all unconditionally summable sequences in a Banach space X may be better than absolutely 2-summable if considered as sequences in a larger Banach space $Y \supset X$; a well-known inequality of Littlewood [31] asserts that every unconditionally summable sequence in $X = \ell_1$ is absolutely 4/3-summable if considered as a sequence in $Y = \ell_{4/3}$.

In terms of absolutely summing operators, a notion introduced in the late 60's by Pełczyński and Pietsch (for the definition see Section 3), Orlicz's and Littlewood's results read as follows: The identity operator $\mathrm{id} : \ell_p \hookrightarrow \ell_p$, $1 \le p \le 2$ is absolutely $(2,1)$-summing, and the identity operator $\mathrm{id} : \ell_1 \hookrightarrow \ell_{4/3}$ is absolutely $(4/3, 1)$-summing. Bennett [2] and (independently) Carl [5] extended these results as follows: For $1 \le p \le q \le 2$ the identity operator $\mathrm{id} : \ell_p \hookrightarrow \ell_q$ is absolutely $(r,1)$-summing, where $1/r = 1/p - 1/q + 1/2$, or, equivalently, absolutely $(s,2)$-summing, where $1/s = 1/p - 1/q$. We remark that Wojciechowski, answering a question of Pełczyński, recently in [46] has proved an analogue for Sobolov embeddings, using the original result of Bennett and Carl.

Motivated by a study of Bennett–Carl type inequalities within the setting of Orlicz sequence spaces in [32], the following proper extension was given in [16] (the case $p = 2$) and [15] (general case): For $1 \le p \le 2$ and a p-concave symmetric Banach sequence space E the identity operator $\mathrm{id} : E \hookrightarrow \ell_p$ is $(M(\ell_p, E), 2)$-summing, where $M(\ell_p, E)$ denotes the space of multipliers from ℓ_p into E (for the notions of p-concavity and $(M(\ell_p, E), 2)$-summability see Section 2 and Section 3, respectively). In particular, for each 2-concave symmetric Banach sequence space E every unconditionally summable sequence (x_n) satisfies $(\|x_n\|_{\ell_2}) \in E$.

As in the classical case of Bennett–Carl, this result has interesting consequences in various parts of analysis. Besides a sketch of the proof of the main result from above, we report on applications to the following topics:

- Strictly singular identity operators
- Approximation numbers of identity operators
- Eigenvalues of compact operators
- Interpolation of operators
- Mixing identities between sequence spaces and unitary ideals

We mainly survey recent results for summing inclusions in sequence spaces from [6], [16], [15], [17], [32], [14] and [35]; the results from the first two topics seem to be new. We do not consider Grothendieck and Kwapień type results on summing operators defined on ℓ_1 or ℓ_∞; we only remark that ideas similar to those used here in [8] lead to a recent extension within the framework of Orlicz sequence spaces.

2. Preliminaries

If f and g are real-valued functions we write $f \prec g$ whenever there is some $c \ge 0$ such that $f(t) \le c\, g(t)$ for t in the domain of f and g, and $f \asymp g$ whenever $f \prec g$ and $g \prec f$.

We use standard notation and notions from Banach space theory, as presented e. g. in [29], [30] and [45]. If E is a Banach space, then B_E denotes its (closed) unit ball and E' its dual space. For all information on Banach operator ideals and s-numbers see [19], [25], [42] and [43]. As usual $\mathcal{L}(E,F)$ denotes the Banach space of all (bounded and linear) operators from E into F endowed with the operator norm. For basic results and notation from interpolation theory we refer to [4] and [3].

Throughout the paper by a Banach sequence space we mean a real Banach lattice E which is modelled on the set $\mathbb{J}$ and contains an element x with $\operatorname{supp} x = \mathbb{J}$, where $\mathbb{J} = \mathbb{Z}$ is the set integers or $\mathbb{J} = \mathbb{N}$ is the set of positive integers. A Banach sequence space E modelled on $\mathbb{N}$ is said to be symmetric provided that $\|(x_n)\|_E = \|(x_n^*)\|_E$, where (x_n^*) denotes the decreasing rearrangement of the sequence (x_n). A Banach sequence space E is said to be maximal if the unit ball B_E is closed in the pointwise convergence topology induced by the space ω of all real sequences, and σ-order continuous if $x_n \downarrow 0$ in E pointwise implies $\lim_n \|x_n\|_E = 0$. Note that this condition is equivalent to $E^\times = E'$, where as usual

$$E^\times := \{x = (x_n) \in \omega\,;\, \textstyle\sum_{n\in\mathbb{J}} |x_n y_n| < \infty \text{ for all } y = (y_n) \in E\}$$

is the Köthe dual of E. Note that $E^\times$ is a maximal (symmetric, provided that E is) Banach sequence space under the norm

$$\|x\| := \sup\{\textstyle\sum_{n\in\mathbb{J}} |x_n y_n|\,;\, \|y\|_E \le 1\}.$$

The fundamental function λ_E of a symmetric Banach sequence space E is defined by

$$\lambda_E(n) := \|\textstyle\sum_{i=1}^n e_i\|_E, \quad n \in \mathbb{N};$$

throughout the paper (e_n) will denote the standard unit vector basis in c_0 and E_n the linear span of the first n unit vectors. By $(\xi_i)_{i=1}^n$ we denote the sequence $\sum_{i=1}^n \xi_i \cdot e_i$.

The notions of p-convexity and q-concavity of a Banach lattice are crucial throughout the article. For $1 \le p, q < \infty$ a Banach lattice X is called p-convex and q-concave, respectively, if there exist constants $C_p > 0$ and $C_q > 0$ such that for all $x_1, \dots, x_n \in X$

$$\left\| \left(\sum_{i=1}^n |x_i|^p \right)^{1/p} \right\|_X \le C_p \cdot \left(\sum_{i=1}^n \|x_i\|_X^p \right)^{1/p} \tag{2.1}$$

and

$$\left(\sum_{i=1}^n \|x_i\|_X^q \right)^{1/q} \le C_q \cdot \left\| \left(\sum_{i=1}^n |x_i|^q \right)^{1/q} \right\|_X, \tag{2.2}$$

respectively. We denote by $\mathbf{M}^{(\mathbf{p})}(X)$ and $\mathbf{M}_{(\mathbf{q})}(X)$ the smallest constants C_p and C_q which satisfy (2.1) and (2.2), respectively. Each Banach function space X is 1-convex with $\mathbf{M}^{(\mathbf{1})}(X) = 1$, and the properties "$p$-convex" and "$q$-concave" are "decreasing in p" and "increasing in q". Recall that for $1 \le p < \infty$ the space $L_p(\mu)$ is p-convex and p-concave with constants equal to 1. It can be easily seen that a maximal Banach sequence space E which is p-convex and q-concave satisfies $\ell_p \hookrightarrow E \hookrightarrow \ell_q$.

For two Banach sequence spaces E and F the space $M(E,F)$ of multipliers from E into F consists of all scalar sequences $x = (x_n)$ such that the associated multiplication operator $(y_n) \mapsto (x_n y_n)$ is defined and bounded from E into F. $M(E,F)$ is a (maximal and symmetric provided that E and F are) Banach sequence space equipped with the norm

$$\|x\|_{M(E,F)} := \sup\{\|xy\|_F\,;\, y \in B_E\}.$$

Note that if E is a Banach sequence space then $M(E,\ell_1) = E^\times$.

3. Summing identity maps

The following definition is a natural extension of the notion of absolutely (r,p)-summing operators. For two Banach spaces E and F we mean by $E \hookrightarrow F$ that E is contained in F, and the natural identity map is continuous; in this case we put $c_E^F := \|\mathrm{id} : E \hookrightarrow F\|$ and $c_p^F := c_{\ell_p}^F$ whenever $\ell_p \hookrightarrow F$. If F is a Banach sequence space with $\|e_n\|_F = 1$ for all $n \in \mathbb{J}$, then obviously $\ell_1 \hookrightarrow F$ and $c_1^F = 1$, and in particular $E^\times \hookrightarrow M(E,F)$ and $c_{E^\times}^{M(E,F)} = 1$ for each Banach sequence space E.

Definition 3.1. Let E and F be Banach sequence spaces on $\mathbb{J}$ such that $F \hookrightarrow E$. Then an operator $T : X \to Y$ between Banach spaces X and Y is called (E,F)-summing (resp., (E,p)-summing whenever $F = \ell_p$, $1 \le p \le \infty$) if there exists a constant $C > 0$ such that for all $n \in \mathbb{N}$ and $x_i \in X$, $i \in A_n$ with $A_n := \{-n,\dots,n\}$ in the case when $\mathbb{J} = \mathbb{Z}$, and $A_n := \{1,\dots,n\}$ when $\mathbb{J} = \mathbb{N}$, the following inequality holds:

$$\Big\| \sum_{i\in A_n} \|Tx_i\|_Y\, e_i \Big\|_E \le C \cdot c_F^E \cdot \sup_{x'\in B_{X'}} \Big\| \sum_{i\in A_n} \langle x', x_i\rangle\, e_i \Big\|_F .$$

We write $\pi_{E,F}(T)$ for the smallest constant C with the above property. The space of all (E,F)-summing (resp., (E,p)-summing) operators between Banach spaces X and Y is denoted by $\Pi_{E,F}(X,Y)$ (resp., $\Pi_{E,p}(X,Y)$). If $\|e_n\|_E = \|e_n\|_F = 1$, we obtain the Banach operator ideal $(\Pi_{E,F},\pi_{E,F})$, in particular for $E = \ell_r$ $(r \ge p)$ the well-known Banach operator ideal $(\Pi_{r,p},\pi_{r,p})$ of all absolutely (r,p)-summing operators. In a different context than the one considered here summing norms with respect to sequence spaces appear also in [8] and [34].

We note an obvious fact, however useful in the sequel, that an operator $T : X \to Y$ is (E,p)-summing if and only if

$$\sup_m \|\widehat{T}_m\| < \infty, \tag{3.1}$$

where

$$\widehat{T}_m : \mathcal{L}(\ell_{p'}^m, X) \to E(Y), \quad \widehat{T}_m(S) := \sum_{j\in A_m} (TSe_j)\, e_j;$$

here as usual, $E(Y)$ stands for the vector space of all sequences (y_n) in Y such that $(\|y_n\|) \in E$, which together with the norm (resp., quasi-norm) $\|(y_n)\|_{E(Y)} := \|(\|(y_n)\|)\|_E$ forms a Banach space (resp., a quasi-Banach space whenever E or Y is a quasi-Banach space).

As already explained in the introduction the following result from [16] and [15] extends results of Littlewood, Orlicz, Bennett and Carl as well as of Maligranda and Mastyło, and it is crucial for all our further considerations.

Theorem 3.2. *For* $1 \leq p \leq 2$ *let* E *be a* p*-concave symmetric Banach sequence space. Then the identity map* $\mathrm{id} : E \hookrightarrow \ell_p$ *is* $(M(\ell_p, E), 2)$*-summing.*

The proof follows by abstract interpolation theory. In order to show that

$$\sup_m \|\widehat{\mathrm{id}}_m : \mathcal{L}(\ell_2^m, E) \to M(\ell_p, E)(\ell_p)\| < \infty,$$

we interpolate between the trivial case

$$(\mathrm{id} : \ell_p \hookrightarrow \ell_p) \in \Pi_{\infty,2}$$

and the fact that

$$(\mathrm{id} : \ell_1 \hookrightarrow \ell_p) \in \Pi_{p',2},$$

which is due to Kwapień [28] (global case) and Grothendieck ($p = 2$). By (3.1) this means that

$$\sup_m \|\widehat{\mathrm{id}}_m : \mathcal{L}(\ell_2^m, \ell_p) \to \ell_\infty(\ell_p)\| < \infty$$

and

$$\sup_m \|\widehat{\mathrm{id}}_m : \mathcal{L}(\ell_2^m, \ell_1) \to \ell_{p'}(\ell_p)\| < \infty.$$

Now the aim is to find an exact interpolation functor $\mathcal{F}$ such that

$$M(\ell_p, E) = \mathcal{F}(\ell_{p'}, \ell_\infty);$$

based on the well known result of Calderón-Mityagin which yields that each maximal and symmetric Banach sequence space is an interpolation space with respect to the couple (ℓ_1, ℓ_∞), it is shown in [16] and [15] that this is possible. By interpolation we obtain

$$\sup_m \|\widehat{\mathrm{id}}_m : \mathcal{F}(\mathcal{L}(\ell_2^m, \ell_1), \mathcal{L}(\ell_2^m, \ell_p)) \to M(\ell_p, E)(\ell_p)\| < \infty.$$

Finally it remains to prove that

$$\sup_m \|\mathrm{id} : \mathcal{L}(\ell_2^m, E) \hookrightarrow \mathcal{F}(\mathcal{L}(\ell_2^m, \ell_1), \mathcal{L}(\ell_2^m, \ell_p))\| < \infty,$$

a Kouba type interpolation formula (see [26], and for more recent development [18] and [9]). This crucial step is established with a variant of the Maurey–Rosenthal factorization theorem: Since E is p-concave, each operator T allows a factorization

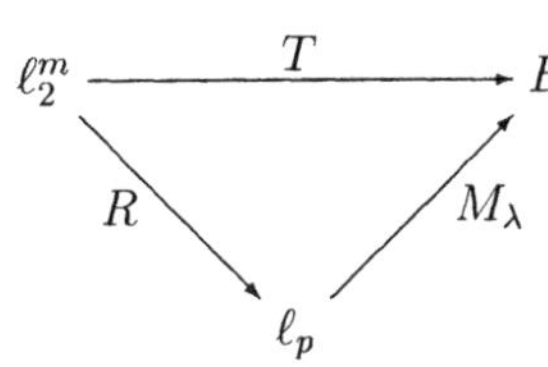

with $\|R\| \cdot \|\lambda\|_{M(\ell_p,E)} \leq \sqrt{2} \cdot \mathbf{M}_{(\mathbf{p})}(E) \cdot \|T\|$ (see [12], [16] and [15]).

Extending well-known formulas by Kwapień and Tomczak-Jaegermann, it was shown in [16, 3.3] that $\Pi_{E,1}(X,Y) = \Pi_{M(\ell_2,E),2}(X,Y)$ whenever E is 2-concave and X is of cotype 2. Using this, we obtain as a corollary of Theorem 3.2 the following consequence which turns out to be of particular interest (see also [16, 4.1]):

Corollary 3.3. *Let E be a 2-concave symmetric Banach sequence space. Then the identity map* $\mathrm{id} : E \hookrightarrow \ell_2$ *is $(E,1)$-summing. In other words, for every unconditionally summable sequence (x_n) in E the scalar sequence $(\|x_n\|_{\ell_2})$ is contained in E.*

This result is optimal in the following sense ([16, 4.6 and 4.7]):

Corollary 3.4. *Let E and F be symmetric Banach sequence spaces such that E is 2-concave.*

(i) *Let F also be 2-concave. Then* $\mathrm{id} : E \hookrightarrow \ell_2$ *is $(F,1)$-summing if and only if $E \hookrightarrow F$.*

(ii) *Let F be maximal with $E \hookrightarrow F$. Then* $\mathrm{id} : E \hookrightarrow F$ *is $(E,1)$-summing if and only if $\ell_2 \hookrightarrow F$.*

For the computation of spaces of multipliers we use powers of sequence spaces: Let E be a (maximal) symmetric Banach sequence space and $0 < r < \infty$ such that $\mathbf{M}^{(\max(\mathbf{1},\mathbf{r}))}(E) = 1$. Then

$$E^r := \{x \in \ell_\infty \,;\, |x|^{1/r} \in E\}$$

endowed with the norm

$$\|x\|_{E^r} := \||x|^{1/r}\|_E^r, \quad x \in E^r$$

is again a (maximal) symmetric Banach sequence space which is $1/\min(1,r)$-convex. With this a straightforward computation shows the following:

Proposition 3.5. *For $1 < p < \infty$ let E and F be symmetric Banach sequence spaces such that E is p-concave with $\mathbf{M}_{(\mathbf{p})}(E) = 1$ and F is p-convex with $\mathbf{M}^{(\mathbf{p})}(F) = 1$. Then*

$$M(\ell_p, E) = (((E^\times)^{p'})^\times)^{1/p'} \quad \textit{and} \quad \lambda_{M(\ell_p,E)}(n) = \frac{\lambda_E(n)}{n^{1/p}}, \tag{3.2}$$

and

$$M(F, \ell_p) = ((F^p)^\times)^{1/p} \quad \textit{and} \quad \lambda_{M(F,\ell_p)}(n) = \frac{n^{1/p}}{\lambda_F(n)}. \tag{3.3}$$

Proof. We start with (3.2). Since all involved spaces are maximal, it is obviously enough to show equality of norms for all $\lambda \in \omega$ with finite support for the first part, and we can also restrict ourselves to the case $\lambda = |\lambda|$:

$$\begin{aligned}\|\lambda\|_{(((E^\times)^{p'})^\times)^{1/p'}} &= \|\lambda^{p'}\|_{((E^\times)^{p'})^\times}^{1/p'} = \sup_{\|\mu\|_{(E^\times)^{p'}} \leq 1} \|\lambda^{p'}\mu\|_{\ell_1}^{1/p'} = \sup_{\|\nu\|_{E^\times} \leq 1} \|\lambda\nu\|_{\ell_{p'}} \\ &= \sup_{\|\nu\|_{E^\times} \leq 1} \sup_{\|\mu\|_{\ell_p} \leq 1} \left| \sum_{i=1}^\infty \lambda_i \nu_i \mu_i \right| = \sup_{\|\mu\|_{\ell_p} \leq 1} \|\lambda\mu\|_E = \|M_\lambda : \ell_p \to E\|.\end{aligned}$$

For the second part of (3.2) note first that

$$\lambda_{F^\times}(n) = n/\lambda_F(n) \tag{3.4}$$

for any symmetric Banach sequence space F (see e.g. [30, 3.a.6]). Hence,

$$\lambda_{M(\ell_p,E)}(n) = \|\sum_1^n e_i\|_{(((E^\times)^{p'})^\times)^{1/p'}} = \|\sum_1^n e_i\|^{1/p'}_{((E^\times)^{p'})^\times}$$
$$= \frac{n^{1/p'}}{\|\sum_1^n e_i\|^{1/p'}_{(E^\times)^{p'}}} = \frac{n^{1/p'}}{\|\sum_1^n e_i\|_{E^\times}} = \frac{\|\sum_1^n e_i\|_E}{n^{1/p}}.$$

Now (3.3) follows by the simple fact that $M(F,\ell_p) = M(\ell_{p'},F^\times)$, the duality relation between convexity and concavity (see e.g. [30, 1.d.4]) and (3.4). □

Example 3.6. Fix $1 < p \le 2$.

(a) Let $1 < p_1 < p$ and $1 \le p_2 \le p$. Then by Creekmore [10] (see also [12]) the Lorentz sequence space ℓ_{p_1,p_2} is p-concave. In this case $M(\ell_p,\ell_{p_1,p_2}) = \ell_{r_1,r_2}$, where $1/r_1 = 1/p_1 - 1/p$ and $1/r_2 = 1/p_2 - 1/p$, hence the identity operator $\mathrm{id} : \ell_{p_1,p_2} \hookrightarrow \ell_p$ is $(\ell_{r_1,r_2},2)$-summing.

(b) For $1 \le q < p$ and a Lorentz sequence ω the Lorentz sequence space $d(\omega,q)$ is p-concave whenever the sequence ω is $q/(q-p)$-regular, i.e., $n\cdot\omega_n^{q/(q-p)} \asymp \sum_{i=1}^n \omega_i^{q/(q-p)}$ (see [44, Theorem 2]). Hence, in this case the identity operator $\mathrm{id} : d(\omega,q) \hookrightarrow \ell_p$ is $(M(\ell_p,d(\omega,q)),2)$-summing.

(c) By [24] an Orlicz sequence space ℓ_φ is 2-concave if and only if the function $t \mapsto \varphi(\sqrt{t})$ is equivalent to a concave function. Hence, in this case the identity operator $\mathrm{id} : \ell_\varphi \hookrightarrow \ell_2$ is $(\ell_\varphi,1)$-summing. This result was first proved in [32], and we note that the proof presented there is based on the following inequality which is interesting in its own right: If $\varphi : \mathbb{R}_+ \to \mathbb{R}_+$ is an Orlicz function, then

$$\sum_{k=1}^n \frac{u_k^2}{v_k^2}\varphi(v_k) \le C\varphi((\sum_{k=1}^n u_k^2)^{1/2}) + C\varphi(\sum_{k=1}^n \frac{u_k^2}{v_k})$$

holds for some $C > 0$ and for all $u_j \ge 0$, $v_j > 0$, $1 \le j \le n$ and $n \in \mathbb{N}$ if and only if $t \mapsto \varphi(\sqrt{t})$ is equivalent to a concave function. In [32] the following more general result on summing inclusions between Orlicz sequence spaces is proved: Let φ_0 and φ_1 be Orlicz functions. Then the following statements hold true:

(i) If $t \mapsto \varphi_0(\sqrt{t})$ is equivalent to a concave function, then for any $0 < \theta < 1$ and any Orlicz function φ_1 such that $\varphi_1^{-1}(t) \asymp \varphi_0^{-1}(t)^{1-\theta}\,(\sqrt{t})^\theta$ the inclusion map $\ell_{\varphi_0} \hookrightarrow \ell_{\varphi_1}$ is $(\ell_\phi,1)$-summing, where $\phi^{-1}(t) = (\sqrt{t})^{1-\theta}\,\varphi_0^{-1}(t)^\theta$ for $t \ge 0$.

(ii) If either $t \mapsto \varphi_0(\sqrt{t})$ is equivalent to a concave function and $\varphi_1 \prec t^2$ or $t \mapsto \varphi_0(\sqrt{t})$ is equivalent to a convex function, φ_0 is supermultiplicative and $\varphi_1 \prec \varphi_0$, then the inclusion map $\ell_{\varphi_0} \hookrightarrow \ell_{\varphi_1}$ is $(\ell_{\varphi_0},1)$-summing.

For other concrete examples of spaces of multipliers see also [33].

4. Strictly singular identity operators

We give an immediate application of Theorem 3.2 to the theory of strictly singular operators; moreover we will also show that it is not possible to avoid the assumption on p-convexity in Theorem 3.2. Recall that an operator between Banach spaces is called strictly singular if it is no isomorphism on any infinite-dimensional closed subspace.

Theorem 4.1. *For $1 < p \leq 2$ let E be a p-concave symmetric Banach sequence space not isomorphic to ℓ_p. Then the identity map $\mathrm{id} : E \hookrightarrow \ell_p$ is strictly singular.*

Proof. Assume that there exists an infinite-dimensional subspace F of E such that the induced norm on F is equivalent to the norm on ℓ_p; by [30, 2.a.2] we can choose F to be complemented in ℓ_p and also isomorphic to ℓ_p. Hence, it follows from Theorem 3.2 that the identity map on ℓ_p is $(M(\ell_p, E), 2)$-summing. But by the Dvoretzky–Rogers Lemma [19, 1.3] there exist $x_1, \ldots, x_n \in \ell_p$ with $\sup_{x' \in B_{\ell_{p'}}} (\sum_{k=1}^n |x'(x_k)|^2)^{1/2} \leq 1$ and $\|x_k\|_p \geq 1/2$, $1 \leq k \leq n$. Since then by definition

$$\|(\|\mathrm{id}_{\ell_p}(x_i)\|_p)_{i=1}^n\|_{M(\ell_p,E)} \leq \pi_{M(\ell_p,E),2}(\mathrm{id}_{\ell_p})$$

and

$$\frac{1}{2}\lambda_{M(\ell_p,E)}(n) = \|\sum_{i=1}^n \frac{1}{2} e_i\|_{M(\ell_p,E)} \leq \|\sum_{i=1}^n \|x_i\|_p e_i\|_{M(\ell_p,E)},$$

we obtain

$$\sup_n \lambda_{M(\ell_p,E)}(n) \leq 2 \cdot \pi_{M(\ell_p,E),2}(\mathrm{id}_{\ell_p}),$$

a contradiction to the assumption $E \neq \ell_p$. □

Note that by [23] and [20] for every $p > 1$ there exists an Orlicz sequence space E which is properly contained in ℓ_p and such that the embedding $\mathrm{id} : E \hookrightarrow \ell_p$ is not strictly singular. The proof of the preceding result then shows that $\mathrm{id} : E \hookrightarrow \ell_p$ is not $(M(\ell_p, E), 2)$-summing, hence our assumption on p-concavity in Theorem 3.2 is essential.

5. Approximation numbers of identity operators

For an operator $T : X \to Y$ between Banach spaces recall the definition of the k-th approximation number

$$a_k(T) := \inf\{\|T - T_k\| \,;\, T_k \in \mathcal{L}(X,Y) \text{ has rank } < k\}$$

and the k-th Gelfand number

$$c_k(T) := \inf\{\|T_{|G}\| \,;\, G \subset X, \operatorname{codim} G < k\}.$$

Of special interest for applications (e. g. in approximation theory) are formulas for the asymptotic behavior of approximation numbers of finite-dimensional identity operators. One of the first well-known results in this direction is due to Pietsch [41]: For $1 \leq k \leq n$ and $1 \leq q \leq p \leq \infty$

$$a_k(\mathrm{id} : \ell_p^n \hookrightarrow \ell_q^n) = (n - k + 1)^{1/q - 1/p},$$

which clearly can be rewritten as follows:

$$a_k(\mathrm{id} : \ell_q^n \hookrightarrow \ell_p^n) = \frac{\lambda_{\ell_q}(n-k+1)}{\lambda_{\ell_p}(n-k+1)}.$$

This leads us to conjecture that for "almost all" pairs (E,F) of symmetric Banach sequence spaces such that $E \hookrightarrow F$

$$a_k(\mathrm{id} : F_n \hookrightarrow E_n) \asymp \frac{\lambda_E(n-k+1)}{\lambda_F(n-k+1)}.$$

Using Theorem 3.2, we establish this formula for all p-concave symmetric Banach sequence spaces E and $F = \ell_p$, where $1 < p \leq 2$:

Theorem 5.1. *For $1 < p \leq 2$ let E be a p-concave symmetric Banach sequence space. Then for all $1 \leq k \leq n$*

$$a_k(\mathrm{id} : \ell_p^n \hookrightarrow E_n) \asymp c_k(\mathrm{id} : \ell_p^n \hookrightarrow E_n) \asymp \frac{\lambda_E(n-k+1)}{\lambda_{\ell_p}(n-k+1)}.$$

To prove the lower estimate we need the following lemma:

Lemma 5.2. *Let F be a maximal symmetric Banach sequence space such that $\ell_2 \hookrightarrow F$. Then for every invertible operator $T : X \to Y$ between two n-dimensional Banach spaces and all $1 \leq k \leq n$*

$$c_k(T) \geq C^{-1} \cdot \frac{\lambda_F(n-k+1)}{\pi_{F,2}(T^{-1})}, \tag{5.1}$$

where $C := 2\sqrt{2e} \cdot c_2^F$.

Proof. We follow the proof of [7, p. 231] for the 2-summing norm. Take a subspace $M \subset X$ with $\operatorname{codim} M < k$. Then

$$n - k + 1 \leq \dim M,$$

hence by [16, (6.2)]

$$\|\textstyle\sum_1^{n-k+1} e_i\|_F \leq \|\sum_1^{\dim M} e_i\|_F \leq C \cdot \pi_{F,2}(\mathrm{id}_M).$$

Clearly (by the injectivity of $\Pi_{F,2}$)

$$\pi_{F,2}(\mathrm{id}_M) = \pi_{F,2}(\mathrm{id} : M \hookrightarrow X),$$

therefore the commutative diagram

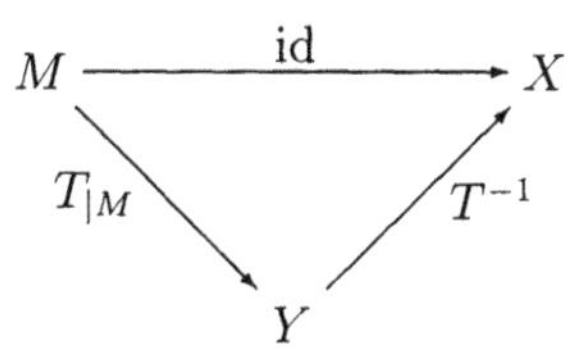

gives, as desired, $\|\sum_1^{n-k+1} e_i\|_F \leq \|T_{|M}\| \cdot C \cdot \pi_{F,2}(T^{-1})$. □

Proof of Theorem 5.1: The upper estimate is easy:

$$a_k(\mathrm{id} : \ell_p^n \hookrightarrow E_n) \leq \|\mathrm{id} : \ell_p^{n-k+1} \hookrightarrow E_{n-k+1}\| = \lambda_{M(\ell_p,E)}(n-k+1).$$

For the lower estimate we obtain from (5.1) together with Theorem 3.2

$$a_k(\mathrm{id} : \ell_p^n \hookrightarrow E_n) \geq c_k(\mathrm{id} : \ell_p^n \hookrightarrow E_n) \succ \lambda_{M(\ell_p,E)}(n-k+1).$$

The conclusion now follows by (3.2). □

Example 5.3. (a) For a Lorentz sequence space $d(\omega, q)$ with finite concavity the sequence ω is 1-regular (see again [44, Theorem 2]), hence $\lambda_{d(\omega,q)}(n) \asymp n^{1/q}\omega_n^{1/q}$. Consequently, under the assumptions of Example 3.6 (b)

$$a_k(\mathrm{id} : \ell_p^n \hookrightarrow d_n(\omega, q)) \asymp (n-k+1)^{1/q-1/p}\omega_{n-k+1}^{1/q}.$$

(b) For an Orlicz sequence space ℓ_φ a straightforward computation shows $\lambda_{\ell_\varphi}(n) = 1/\varphi^{-1}(1/n)$, hence, under the assumptions of Example 3.6 (c),

$$a_k(\mathrm{id} : \ell_p^n \hookrightarrow \ell_\varphi^n) \asymp ((n-k+1)^{1/p}\varphi^{-1}(1/(n-k+1)))^{-1}.$$

Clearly, Theorem 5.1 also has consequences for Lorentz sequence spaces $E = \ell_{p_1,p_2}$, but in this case a direct argument shows that the restriction on p and the assumption on the concavity of ℓ_{p_1,p_2} can be dropped:

Proposition 5.4. *Let $1 < p_1 < p \leq \infty$ and $1 \leq p_2 \leq \infty$. Then for $1 \leq k \leq n$*

$$a_k(\mathrm{id} : \ell_p^n \hookrightarrow \ell_{p_1,p_2}^n) \asymp (n-k+1)^{1/p_1-1/p}.$$

Proof. For the upper estimate define $0 \leq \theta < 1$ by $\theta := p_1/p$. Then, since $\ell_p = (\ell_\infty, \ell_{p_1,p_2})_{\theta,p}$ and the fact that $(\cdot,\cdot)_{\theta,p}$ is an interpolation functor of power type θ, we have

$$\begin{aligned} a_k(\mathrm{id} : \ell_p^n \hookrightarrow \ell_{p_1,p_2}^n) &\leq \|\mathrm{id} : \ell_p^{n-k+1} \hookrightarrow \ell_{p_1,p_2}^{n-k+1}\| \\ &\prec \|\mathrm{id} : \ell_\infty^{n-k+1} \hookrightarrow \ell_{p_1,p_2}^{n-k+1}\|^{1-\theta} \\ &\asymp (n-k+1)^{1/p_1-1/p}. \end{aligned}$$

For the lower estimate choose arbitrary $0 < \theta < 1$, and let $p_1 < r < p$ be defined by $1/r = (1-\theta)/p + \theta/p_1$. Then by the interpolation property of the Gelfand numbers (see e.g. [42, 11.5.8])

$$a_k(\mathrm{id} : \ell_p \hookrightarrow \ell_{p_1,p_2}^n) \geq c_k(\mathrm{id} : \ell_p^n \hookrightarrow \ell_{p_1,p_2}^n) \succ c_k(\mathrm{id} : \ell_p^n \hookrightarrow \ell_r^n)^{1/\theta} = (n-k+1)^{1/p_1-1/p},$$

which gives the claim. □

6. Eigenvalues of compact operators

Recall that for an operator $T : X \to Y$ between Banach spaces the k-th Weyl number $x_k(T)$ is defined as

$$x_k(T) := \sup\{a_k(TS)\,;\, S \in \mathcal{L}(\ell_2, X) \text{ with } \|S\| \leq 1\}.$$

The Weyl–König inequality shows that each Riesz operator T on a Banach space X (in particular, each power compact operator) with Weyl numbers $(x_n(T))$ in ℓ_p $(1 \leq p < \infty)$ has its sequence $(\lambda_n(T))$ of eigenvalues in ℓ_p,

$$\|(\lambda_n(T))\|_{\ell_p} \leq 2\sqrt{e} \cdot \|(x_n(T))\|_{\ell_p}. \tag{6.1}$$

On the other hand König also proved that every $(p, 2)$-summing operator T defined on a Hilbert space H has its sequence of Weyl numbers in ℓ_p; in consequence, for all $T \in \Pi_{p,2}$ and $k \in \mathbb{N}$

$$x_k(T) \leq \lambda_{\ell_p}(n)^{-1}\pi_{p,2}(T). \tag{6.2}$$

In combination with the classical Bennett–Carl/Grothendieck inequalities this lead König to the following two important eigenvalue results for operators in ℓ_p-spaces (see [25, 2.b.11]):

- Each operator $T \in \mathcal{L}(\ell_p)$, $1 \leq p \leq 2$ with values in ℓ_q, $1 \leq q < p$ is a Riesz operator, and for all n

$$|\lambda_n(T)| \leq c \cdot n^{1/p - 1/q}, \tag{6.3}$$

  where c is some uniform constant.

- Each operator $T \in \mathcal{L}(\ell_p)$, $2 \leq p < \infty$ with values in ℓ_q, $1 \leq q < 2$ is a Riesz operator with

$$(\lambda_n(T)) \in \ell_r, \qquad 1/r = 1/q - 1/p. \tag{6.4}$$

Here the case $p = 2$ is of particular interest.

König's techniques show that (6.1) and (6.2) even hold if ℓ_p is replaced by an arbitrary maximal and symmetric Banach sequence space E (for (6.1) see [25, 2.a.8] and for (6.2) analyze [25, 2.a.3]; here one has to assume additionally that $\ell_2 \hookrightarrow E$). Together with Theorem 3.2 this in [15] leads to natural and proper extensions of (6.3) and (6.4):

Theorem 6.1. *For $1 \leq p \leq 2$ let E be a p-concave symmetric Banach sequence space and $T \in \mathcal{L}(\ell_p)$ a Riesz operator with values in E. Then for all n*

$$|\lambda_n(T)| \leq c \cdot \frac{1}{n} \sum_{k=1}^{n} \lambda^{-1}_{M(\ell_p, E)}(k).$$

Theorem 6.2. *Let E and F be maximal symmetric Banach sequence spaces not both isomorphic to ℓ_2 such that E is 2-concave and F is 2-convex and σ-order continuous. Then every operator $T \in \mathcal{L}(F)$ with values in E satisfies*

$$(\lambda_n(T)) \in M(F, E).$$

Again the case $F = \ell_2$ is of particular interest. Note that $M(\ell_p, \ell_q) = \ell_r$ for $1 \leq q \leq p \leq \infty$ and $1/r = 1/q - 1/p$, in particular, $\frac{1}{n}\sum_{k=1}^{n} \lambda^{-1}_{M(\ell_p, \ell_q)}(k) \asymp n^{1/q - 1/p}$.

In [15] we moreover give an extension of a well-known ℓ_p-estimate for the eigenvalues of $n \times n$ matrices due to Johnson, König, Maurey and Retherford [21].

7. Interpolation of operators

In [35] summing inclusion maps between Banach sequence spaces are used to study interpolation of operators between spaces generated by the real method of interpolation. As an application an extension of Ovchinnikov's [39] interpolation theorem from the context of classical Lions-Peetre spaces to a large class of real interpolation spaces is presented.

In [14], we continue the study of interpolation of operators between abstract real method spaces. The results obtained in this fashion applied to L_p allow us to recover a remarkable recent result of Ovchinnikov [40]. In order to present certain results from [35] and [14] we recall some fundamental definitions.

A couple $\overline{\Phi} = (\Phi_0, \Phi_1)$ of quasi-Banach sequence spaces on $\mathbb{Z}$ is called a *parameter of the $\mathcal{J}$-method* if $\Phi_0 \cap \Phi_1 \subset \ell_1$. The $\mathcal{J}$-method space $\mathcal{J}_{\overline{\Phi}}(\overline{X}) = \mathcal{J}_{\Phi_0,\Phi_1}(\overline{X})$ consists of all $x \in X_0 + X_1$ which can be represented in the form

$$x = \sum_{n=-\infty}^{\infty} u_n \quad (\text{convergence in } X_0 + X_1) \tag{7.1}$$

with $u = (u_n) \in \Phi_0(X_0) \cap \Phi_1(X_1)$. Similarly as in the case of Banach sequence spaces Φ_0 and Φ_1, we easily show that $\mathcal{J}_{\overline{\Phi}}(\overline{X})$ is a quasi-Banach space under the quasi-norm

$$\|x\| = \inf \max \{\|u\|_{\Phi_0(X_0)}, \|u\|_{\Phi_1(X_1)}\},$$

where the infimum is taken over all representations (7.1) (cf. [4], [27]). In the case if $\overline{E} = (E_0, E_1)$ is a couple of quasi-Banach sequence spaces on $\mathbb{Z}$ so that $(\Phi_0, \Phi_1) = (E_0(2^{-n\theta}), E_1(2^{n-n\theta}))$ is a parameter of the $\mathcal{J}$-method with $0 \le \theta \le 1$, then the space $\mathcal{J}_{\overline{\Phi}}(\overline{X})$ (resp., $\mathcal{J}_{\Phi_0,\Phi_1}(\overline{X})$) is denoted by $\mathcal{J}_{\theta,\overline{E}}(\overline{X})$ (resp., $\mathcal{J}_{\theta,E_0,E_1}(\overline{X})$). In the particular case $E = E_0 = E_1$ and $\theta = 0$ the space $\mathcal{J}_{0,E,E}(\overline{X})$ is the classical space $\mathcal{J}_E(\overline{X})$ (see [11], [27]).

If E is a (quasi-)Banach lattice on $\mathbb{Z}$ intermediate with respect to $(\ell_\infty, \ell_\infty(2^{-n}))$, then the $\mathcal{K}$-method space $\mathcal{K}_E(\overline{X}) := \overline{X}_E$ is a (quasi-)Banach space which consists of all $x \in X_0 + X_1$ such that $\big(K(2^n, x; \overline{X})\big)_{-\infty}^{\infty} \in E$ with the associated (quasi-)norm

$$\|x\| := \|\big(K(2^n, x; \overline{X})\big)\|_E,$$

where as usual K denotes the K-functional (see [3]).

It is easy to see that similar as in the Banach case $\mathcal{K}_E$ as well as $\mathcal{J}_E$ are exact interpolation functors. Moreover, if in addition a quasi-Banach lattice E is a *parameter of the real method*, i.e. $\ell_\infty \cap \ell_\infty(2^{-n}) \subset E \subset \ell_1 + \ell_1(2^{-n})$ and $T : E \to E$ for any operator $T : (\ell_1, \ell_1(2^{-n})) \to (\ell_\infty, \ell_\infty(2^{-n}))$, then for any Banach couple (X_0, X_1)

$$\mathcal{K}_E(X_0, X_1) = \mathcal{J}_E(X_0, X_1)$$

up to equivalence of norms (see [4], [38]).

Examples of real parameters are spaces $E(2^{-n\theta})$ for any $0 < \theta < 1$, where E is any quasi-Banach space on $\mathbb{Z}$ which is *translation invariant*, i.e., $\|(\xi_{n-k})\|_E = \|(\xi_n)\|_E$ for all $k \in \mathbb{Z}$. In the sequel for such real parameters the space $\overline{X}_{E(2^{-n\theta})}$ is denoted by $\overline{X}_{\theta,E}$ (resp., $\overline{X}_{\theta,p}$ whenever $E = \ell_p$, $0 < p \le \infty$).

The following result (see [35]) shows that under certain additional conditions an interpolation theorem holds with the range space generated by the $\mathcal{J}$-abstract method of interpolation.

Theorem 7.1. *Suppose that E_j, F_j, G_j for $j = 0, 1$ are Banach sequence spaces on $\mathbb{Z}$, and further that Φ_j for $j = 0, 1$ and E are quasi-Banach sequence spaces on $\mathbb{Z}$ satisfying the following conditions:*

(i) $\ell_1 \hookrightarrow E_j$, $\ell_1 \hookrightarrow F_j$, $\Phi_j \hookrightarrow \ell_\infty$ *and* $E \hookrightarrow \ell_\infty$,

(ii) $M(E_j, F_j) \hookrightarrow M(E, \Phi_j)$,

(iii) The inclusion map $F_j \hookrightarrow G_j$ *is an* $(F_j, 1)$*-summing operator.*

If $T : (E_0, E_1(2^{-n})) \to (F_0, F_1(2^{-n}))$, *then T is bounded from* $E(2^{-n\theta})$ *into* $(G_0, G_1)_{\theta,\overline{\Phi}}$ *for any* $0 < \theta < 1$.

Combining this result with the reiteration theorem as well as an orbital equivalence of the real method spaces with the parameter spaces generating these spaces, the following interpolation theorem has been shown in [35].

Theorem 7.2. *Assume that the assumptions of the preceding theorem hold, and in addition that the E_j are translation invariant and for all* $0 < \theta < 1$

(iv) $(F_0, F_1(2^{-n}))_{\theta,\overline{\Phi}} = (G_0, G_1(2^{-n}))_{\theta,\overline{\Phi}}$.

Then each operator $T : (\overline{X}_{\alpha_0,E_0}, \overline{X}_{\alpha_1,E_1}) \to (\overline{Y}_{\beta_0,F_0}, \overline{Y}_{\beta_1,F_1})$ *with* $0 < \alpha_j < 1$, $0 < \beta_j < 1$, $\alpha_0 \neq \alpha_1$ *and* $\beta_0 \neq \beta_1$ *is bounded from* $\overline{X}_{\alpha,E}$ *into* $\overline{Y}_F$, *where* $\alpha = (1-\theta)\alpha_0 + \theta\alpha_1$, $0 < \theta < 1$ *and* $F = (F_0(2^{-n\beta_0}), F_1(2^{-n\beta_1}))_{\theta,\overline{\Phi}}$.

An immediate consequence of the above theorem is the following theorem of Ovchinnikov [39] (for details we refer to [35]).

Theorem 7.3. *Let $\overline{X}$ and $\overline{Y}$ be any Banach couples and let* $T : (\overline{X}_{\alpha_0,p_0}, \overline{X}_{\alpha_1,p_1}) \to (\overline{Y}_{\beta_0,q_0}, \overline{Y}_{\beta_1,q_1})$, $0 < \alpha_j < 1$, $0 < \beta_j < 1$, $\alpha_0 \neq \alpha_1$, $\beta_0 \neq \beta_1$ *and* $1 \leq p_i \leq \infty$, $1 \leq q_j \leq \infty$, $j = 0, 1$. *Then T is bounded from* $\overline{X}_{\alpha,p}$ *into* $\overline{Y}_{\beta,q}$, *where* $\alpha = (1-\theta)\alpha_0 + \theta\alpha_1$, $\beta = (1-\theta)\beta_0 + \theta\beta_1$, $0 < p \leq \infty$, *and* $1/q = 1/p + (1-\theta)(1/q_0 - 1/p_0)_+ + \theta(1/q_1 - 1/p_1)_+$, *where* $x_+ := \max\{0, x\}$ *for* $x \in \mathbb{R}$.

The next theorem is essential in the proof of the main result in [14].

Theorem 7.4. *Suppose that* E_j, F_j *for* $j = 0, 1$ *are Banach sequence spaces on* $\mathbb{Z}$ *satisfying the following conditions:*

(i) $\ell_1 \hookrightarrow F_j \hookrightarrow E_j \hookrightarrow \ell_\infty$,

(ii) $M(E_0, F_0) = M(E_1, F_1) =: F$.

(iii) The inclusion map $F_j \hookrightarrow \ell_\infty$ *is an* $(F_j, 1)$*-summing operator,*

If $T : (E_0, E_1(2^{-n})) \to (F_0, F_1(2^{-n}))$, *then T is bounded from E into* $E \odot F$ *for any quasi-Banach lattice E on $\mathbb{Z}$ which is a parameter of the real method.*

Here, for given quasi-Banach lattices X and Y on a measure space (Ω, μ), we define a quasi-Banach lattice $X \odot Y := \{x \cdot y \,;\, x \in X,\ y \in Y\}$ equipped with the quasi-norm

$$\|z\|_{X \odot Y} := \inf\{\|x\|_X \, \|y\|_Y \,;\, z = x \cdot y,\, x \in X,\, y \in Y\}.$$

Now we can state the main result proved in [14].

Theorem 7.5. *Let E_j, F_j be translation invariant Banach sequence spaces on $\mathbb{Z}$ satisfying the assumptions of Theorem 7.4 and let E be a parameter of the real method. Assume further that $\overline{X}$ and $\overline{B}$ are Banach couples, $0 < \alpha_j < 1$, $0 < \beta_j < 1$, $j = 0, 1$ with $\alpha_0 \neq \alpha_1$, $\beta_0 \neq \beta_1$ and $\Phi := (E_0(2^{-n\alpha_0}), E_1(2^{-n\alpha_1}))_E$. Then the following statements hold true for any operator $T : (\overline{X}_{\alpha_0,E_0}, \overline{X}_{\alpha_1,E_1}) \to (\overline{B}_{\beta_0,F_0}, \overline{B}_{\beta_1,F_1})$.*

(i) An operator T is bounded from $\overline{X}_\Phi$ into $\overline{B}_\Psi$, where $\Psi = (F_0(2^{-n\beta_0}), F_1(2^{-n\beta_1}))_{E \odot F}$.

(ii) If $F_j = G_j \odot F$ for some translation invariant Banach sequence space G_j on $\mathbb{Z}$ and $\overline{B} := (Y_0 \odot Y, Y_1 \odot Y)$ is a couple of Banach lattices with Y a quasi-Banach lattice satisfying an upper F-estimate, then T is bounded from $\overline{X}_\Phi \to \overline{Y}_G \odot Y$, where $G = (G_0(2^{-n\beta_0}), G_1(2^{-n\beta_1}))_E$.

We say (analogously to [30], pp. 82–84) that a quasi-Banach lattice X on (Ω, μ) satisfies an *upper F-estimate* if the following continuous inclusion $F(X) \hookrightarrow X[\ell_\infty]$ holds, where $X[\ell_\infty]$ denotes the mixed quasi-Banach lattice of all sequences (x_n) in $L^0(\mu)$ such that $\sup_{n \in \mathbb{Z}} |x_n| \in X$ with the associated quasi-norm

$$\|(x_n)\|_{X[\ell_\infty]} := \|\sup_{n \in \mathbb{Z}} |x_n|\|_X.$$

Recall that if X is a symmetric space on $(0, a)$, $0 < a \leq \infty$, then the dilation operators D_s on X are defined by $D_s f(t) = f(t/s)$ (we let $x(t) = 0$ for $t > 0$ in the case $a < \infty$). We can then define the *Boyd indices* α_X and β_X of X by (see [27])

$$\alpha_X := \lim_{s \to 0} \frac{\log \|D_s\|_X}{\log s}, \qquad \beta_X := \lim_{s \to \infty} \frac{\log \|D_s\|_X}{\log s}.$$

We finish this section by presenting a corollary of Theorem 7.5 which is a remarkable result of Ovchinnikov [40].

Theorem 7.6. *Assume that T is a linear and bounded operator from a Banach scale of L_{p_θ}-spaces into a Banach scale of L_{q_θ}-spaces, where $0 < \theta < 1$, $1/p_\theta = (1-\theta)/p_0 + \theta/p_1$, $1/q_\theta = (1-\theta)/q_0 + \theta/q_1$ and $1/q_j = 1/r + 1/p_j$ for $j = 0, 1$. If X is any symmetric space with Boyd indices satisfying $1/p_0 < \alpha_X \leq \beta_X < 1/p_1$, then T is bounded from X into $X \odot L_r$.*

Proof. Take $\theta_0, \theta_1 \in (0,1)$ such that $1/p_0 < 1/p_{\theta_0} < \alpha_X \leq \beta_X < 1/p_{\theta_1} < 1/p_1$. It is well known (see e.g. [1]) that there exists a real parameter E such that $X = (L_{p_{\theta_0}}, L_{p_{\theta_1}})_E$.

Now we let $\overline{X} = (L_{p_0}, L_{p_1})$, $E_j = \ell_{p_{\theta_j}}$, $F_j = \ell_{q_{\theta_j}}$ for $j = 0, 1$. Since $F_j = E_j \odot \ell_r$, we get (see [3], Theorem 5.2.1) $\overline{X}_{\theta_j, E_j} = L_{p_{\theta_j}}$. Taking $Y = L_r$ and $Y_j = L_{p_j}$ $(j = 0, 1)$, we have for $\overline{B} = (Y_0 \odot Y, Y_1 \odot Y)$ that $\overline{B}_{\theta_j, F_j} = (L_{q_0}, L_{q_1})_{\theta_j, F_j} = L_{q_{\theta_j}}$.

Since $M(E_0, F_0) = M(E_1, F_1) = \ell_r$ the result is an immediate consequence of Theorem 7.5(ii) (with $\alpha_j = \beta_j = \theta_j$ and $G_j = F_j$ for $j = 0, 1$), the reiteration formula and the well known Bennett-Carl result on $(p,1)$-summability of the inclusion map $\ell_p \hookrightarrow \ell_\infty$ for any $1 \le p \le \infty$. □

8. Mixing identities

An operator $T \in \mathcal{L}(E,F)$ is called (s,p)-mixing $(1 \le p \le s \le \infty)$ whenever its composition with an arbitrary operator $S \in \Pi_s(F,Y)$ is absolutely p-summing; with the norm

$$\mu_{s,p}(T) := \sup\{\pi_p(ST)\,;\, \pi_s(S) \le 1\}$$

the class $\mathcal{M}_{s,p}$ of all (s,p)-mixing operators forms again a Banach operator ideal. Obviously, $(\mathcal{M}_{p,p}, \mu_{p,p}) = (\mathcal{L}, \|\cdot\|)$ and $(\mathcal{M}_{\infty,p}, \mu_{\infty,p}) = (\Pi_p, \pi_p)$.

Recall that due to [36] (see also [13, 32.10–11]) summing and mixing operators are closely related:

$$(\mathcal{M}_{s,p}, \mu_{s,p}) \subset (\Pi_{r,p}, \pi_{r,p}) \qquad \text{for} 1/s + 1/r = 1/p,$$

and "conversely"

$$(\Pi_{r,p}, \pi_{r,p}) \subset (\mathcal{M}_{s_0,p}, \mu_{s_0,p}) \qquad \text{for } 1 \le p \le s_0 < s \le \infty \text{ and } 1/s + 1/r = 1/p.$$

Moreover, it is known that each $(s,2)$-mixing operator on a cotype 2 space is even $(s,1)$-mixing (see again [36] and [13, 32.2]).

An extension of the original Bennett–Carl result for mixing operators was given in [6]:

Theorem 8.1. *Let $1 \le p \le q \le 2$. Then the identity map $\mathrm{id} : \ell_p \hookrightarrow \ell_q$ is $(s,2)$-mixing, where $1/s = 1/2 - 1/p + 1/q$.*

While Carl and Defant used a certain tensor product trick, in [17] a proof by complex interpolation was given. By factorization we obtain the following asymptotic formula:

Corollary 8.2. *For $1 \le p \le q \le 2$ let E and F be 2-concave symmetric Banach sequence spaces such that E is p-convex and F is q-concave. Then for $2 \le r, s \le \infty$ such that $1/r \ge 1/p - 1/q$ and $1/s = 1/2 - 1/r$*

$$\pi_{r,2}(\mathrm{id} : E_n \hookrightarrow F_n) \asymp \mu_{s,2}(\mathrm{id} : E_n \hookrightarrow F_n) \asymp n^{1/r} \cdot \frac{\lambda_F(n)}{\lambda_E(n)}.$$

Proof. For the upper estimate factorize through the identity $\ell_p^n \hookrightarrow \ell_q^n$ and observe that by (3.3) and (3.2) one has $\|\mathrm{id} : E_n \hookrightarrow \ell_p^n\| \asymp n^{1/p}/\lambda_E(n)$ and $\|\mathrm{id} : \ell_q^n \hookrightarrow F_n\| \asymp \lambda_F(n)/n^{1/q}$. The lower estimate can be obtained with the help of Weyl numbers exactly as in [17, (3.12)]. □

Finally, we state analogues of these results for Schatten classes / unitary ideals. For a maximal symmetric Banach sequence space E we denote by $\mathcal{S}_E$ the Banach space of all operators $T : \ell_2 \to \ell_2$ for which the sequence of its singular numbers $(s_n(T))$ is contained in E, endowed with the norm $\|T\|_{\mathcal{S}_E} := \|(s_n(T))\|_E$. If $E = \ell_p$, $1 \le p < \infty$, we write as usual $\mathcal{S}_p$ instead of $\mathcal{S}_{\ell_p}$. By $\mathcal{S}_E^n$ and $\mathcal{S}_p^n$ we denote the space $\mathcal{L}(\ell_2^n, \ell_2^n)$ endowed with the norm induced by $\mathcal{S}_E$ and $\mathcal{S}_p$, respectively.

Using a non-commutative analogue of the Kouba formula due to Junge [22], the following asymptotic formulas were also proved in [17]—note that the lower estimate follows from the fact that ℓ_2^n is naturally contained in each $\mathcal{S}_p$ and $\pi_{r,2}(\mathrm{id}_{\ell_2^n}) = n^{1/r}$.

Theorem 8.3. *Let $1 \le p \le q \le 2$. Then for $2 \le r, s \le \infty$ such that $1/r \ge 1/p - 1/q$ and $1/s = 1/2 - 1/r$*

$$\pi_{r,2}(\mathrm{id} : \mathcal{S}_p^n \hookrightarrow \mathcal{S}_q^n) \asymp \mu_{s,2}(\mathrm{id} : \mathcal{S}_p^n \hookrightarrow \mathcal{S}_q^n) \asymp n^{1/r}.$$

Motivated by the definition of limit orders of Banach operator ideals (see e.g. [42, 14.4]), we define

$$\lambda_{\mathcal{S}}(\Pi_{r,2}, u, v) := \inf\{\lambda > 0 \mid \exists \rho > 0 \,\forall n : \pi_{r,2}(\mathcal{S}_u^n \hookrightarrow \mathcal{S}_v^n) \le \rho \cdot n^\lambda\}.$$

The results for $\lambda_{\mathcal{S}}(\Pi_{r,2}, u, v)$ in [17, Corollary 10] can be summarized in the following picture:

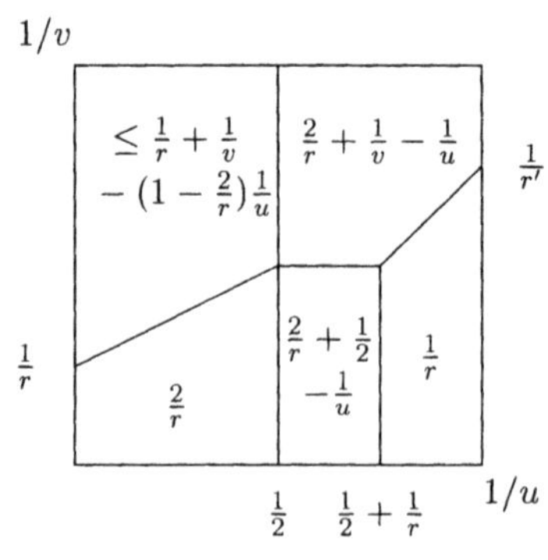

Compared to the original limit order $\lambda(\Pi_{r,2}, u, v)$ of the ideal $\Pi_{r,2}$, this gives for almost all u, v except those in the upper left corner

$$\lambda_{\mathcal{S}}(\Pi_{r,2}, u, v) = \lambda(\Pi_{r,2}, u, v) + 1/r;$$

we conjecture that this equality is true for all u, v.

Combining the proof of the preceding corollary with factorization, we also obtain the following extension of Theorem 8.3:

Corollary 8.4. *For $1 \le p \le q \le 2$ let E and F be 2-concave symmetric Banach sequence spaces such that E is p-convex and F is q-concave. Then for $2 \le r, s \le \infty$ such that $1/r \ge 1/p - 1/q$ and $1/s = 1/2 - 1/r$*

$$\pi_{r,2}(\mathrm{id} : \mathcal{S}_E^n \hookrightarrow \mathcal{S}_F^n) \asymp \mu_{s,2}(\mathrm{id} : \mathcal{S}_E^n \hookrightarrow \mathcal{S}_F^n) \asymp n^{2/r} \cdot \frac{\lambda_F(n)}{\lambda_E(n)}.$$

REFERENCES

1. S.V. Astashkin. On stability of the real interpolation method. *Dokl. Akad. Nauk. Uzb. SSR*, 4:10–11, 1983 (in Russian).
2. G. Bennett. Inclusion mappings between ℓ^p-spaces. *J. Funct. Anal.*, 13:20–27, 1973.
3. J. Bergh and J. Löfström. *Interpolation spaces.* Springer-Verlag, 1978.
4. Yu. A. Brudnyĭ and N. Ya. Krugljak. *Interpolation functors and interpolation spaces.* North-Holland, 1991.
5. B. Carl. Absolut $(p,1)$-summierende identische Operatoren von ℓ_u nach ℓ_v. *Math. Nachr.*, 63:353–360, 1974.
6. B. Carl and A. Defant. Tensor products and Grothendieck type inequalities of operators in L_p-spaces. *Trans. Amer. Math. Soc.*, 331:55–76, 1992.
7. B. Carl and A. Defant. Asymptotic estimates for approximation quantities of tensor product identities. *J. Approx. Theory*, 88:228–256, 1997.
8. J. Cerdà and M. Mastyło. The type and cotype of Calderón–Lozanovskiĭ spaces. *Bull. Pol. Acad. Sci. Math.*, 47:105–117, 1999.
9. F. Cobos and T. Signes. On a result of Peetre about interpolation of operator spaces. *Preprint.*
10. J. Creekmore. Type and cotype in Lorentz L_{pq} spaces. *Indag. Math.*, 43:145–152, 1981.
11. M. Cwikel and J. Peetre. Abstract K and J spaces. *J. Math. Pures Appl.*, 60:1–50, 1981.
12. A. Defant. Variants of the Maurey–Rosenthal theorem for quasi Köthe function spaces. *To appear in Positivity.*
13. A. Defant and K. Floret. *Tensor Norms and Operator Ideals.* North-Holland, 1993.
14. A. Defant, M. Mastyło and C. Michels. Applications of summing inclusion maps to interpolation of operators. *Preprint.*
15. A. Defant, M. Mastyło and C. Michels. Eigenvalues of operators between symmetric sequence spaces. *In preparation.*
16. A. Defant, M. Mastyło and C. Michels. Summing inclusion maps between symmetric sequence spaces. *Report No. 105/2000, 22 pp., Faculty of Math. & Comp. Sci., Poznań.*
17. A. Defant and C. Michels. Bennett–Carl inequalities for symmetric Banach sequence spaces and unitary ideals. *E-print math.FA/9904176.*
18. A. Defant and C. Michels. A complex interpolation formula for tensor products of vector-valued Banach function spaces. *Arch. Math.*, 74:441–451, 2000.
19. J. Diestel, H. Jarchow, and A. Tonge. *Absolutely Summing Operators.* Cambridge Univ. Press, 1995.
20. F. L. Hernández and B. Rodriguez-Salinas. On ℓ^p-complemented copies in Orlicz spaces II. *Israel J. Math.*, 68:27–55, 1989.
21. W. B. Johnson, H. König, B. Maurey and J. R. Retherford. Eigenvalues of p-summing and ℓ_p-type operators in Banach spaces. J. Functional Anal., 32:353–380, 1979.
22. M. Junge. *Factorization theory for spaces of operators.* Univ. Kiel, Habilitationsschrift, 1996.
23. N. J. Kalton. Orlicz sequence spaces without local convexity. *Math. Proc. Camb.*

Phil. Soc., 81:253–277, 1977.
24. I. A. Komarchev. 2-absolutely summable operators in certain Banach spaces. *Math. Notes*, 25:306–312, 1979.
25. H. König. *Eigenvalue distributions of compact operators.* Birkhäuser, 1986.
26. O. Kouba. On the interpolation of injective or projective tensor products of Banach spaces. *J. Functional Anal.*, 96:38–61, 1991.
27. S.G. Krein, Yu.I. Petunin and E.M. Semenov. *Interpolation of Linear Operators.* Nauka, Moscow, 1978 (in Russian; English transl.: Amer. Math. Soc., Providence, 1982.)
28. S. Kwapień. Some remarks on (p,q)-summing operators in ℓ_p-spaces. *Studia Math.*, 29:327–337, 1968.
29. J. Lindenstrauss and L. Tzafriri. *Classical Banach Spaces I: Sequence Spaces.* Springer-Verlag, 1977.
30. J. Lindenstrauss and L. Tzafriri. *Classical Banach Spaces II: Function Spaces.* Springer-Verlag, 1979.
31. J. E. Littlewood. On bounded bilinear forms in an infinite number of variables. *Quart. J. Math.*, 1:164–174, 1930.
32. L. Maligranda and M. Mastyło. Inclusion mappings between Orlicz sequence spaces. *J. Functional Anal.*, 176:264–279, 2000.
33. L. Maligranda and L. E. Persson. Generalized duality of some Banach function spaces. *Indagationes Math.*, 51:323–338, 1989.
34. V. Mascioni and U. Matter. Weakly (q,p)-summing operators and weak cotype properties of Banach spaces. *Proc. R. Ir. Acad.*, 88:169–177, 1988.
35. M. Mastyło and M. Milman. Interpolation of real method spaces via some ideals of operators. *Studia Math.*, 136:17–35, 1999.
36. B. Maurey. *Théorèmes de factorisation pour les opérateurs linéaires à valeurs dans les espaces L^p.* Astérisque 11, 1974.
37. W. Orlicz. Über unbedingte Konvergenz in Funktionenräumen (I). *Studia Math.*, 4:33–37, 1933.
38. V.I. Ovchinnikov. *The methods of orbits in interpolation theory.* Math. Rep., 1:349–516, 1984.
39. V.I. Ovchinnikov. Interpolation theorems for $L_{p,q}$-spaces. *Mat. Sb.*, 136:227–240, 1984 (in Russian).
40. V.I. Ovchinnikov. Interpolation theorem for uniformly deteriorating operators in rearrangement invariant spaces. *Preprint* 7 pp. (in Russian).
41. A. Pietsch. s-numbers of operators in Banach spaces. *Studia Math.*, 51:201–223, 1974.
42. A. Pietsch. *Operator Ideals.* North-Holland, 1980.
43. A. Pietsch. *Eigenvalues and s-numbers.* Cambridge University Press, 1987.
44. S. Reisner. A factorization theorem in Banach lattices and its applications to Lorentz spaces. *Ann. Inst. Fourier*, 31:239–255, 1981.
45. N. Tomczak-Jaegermann. *Banach–Mazur Distances and Finite-Dimensional Operator Ideals.* Longman Scientific & Technical, 1989.
46. M. Wojciechowski. On the summing property of the Sobolev embedding operators. *Positivity*, 1:165-170, 1997.

Recent Progress in Functional Analysis
K.D. Bierstedt, J. Bonet, M. Maestre and J. Schmets

Applications of Banach space theory to sectorial operators

Nigel Kalton *

Department of Mathematics
University of Missouri
Columbia
Missouri 65211

To Professor Manuel Valdivia on the occasion of his seventieth birthday.

Abstract

We will discuss recent joint work of the author with Gilles Lancien and Lutz Weis on the theory of sectorial operators. Our presentation is informal and we hope to show how classical Banach space theory finds applications in this area.
MCS 2000 Primary 47A60; Secondary 47D06

1. Introduction

In this survey we will try to show how Banach space methods can be used in the study of sectorial operators. In particular we will show that the *maximal regularity problem* can be considered as a variant of the *complemented subspace problem* solved thirty years ago by Lindenstrauss and Tzafriri [23]. We then show how the ideas developed in resolving this problem lead to a new approach to the theory of operators with an H^∞-calculus initiated by McIntosh [27].

In general, sectorial operators have a very nice and complete theory on Hilbert spaces. The problem is always to try to extend the theory to more general Banach spaces. Even for applications in partial differential equations it is very natural to, at least, consider the classical spaces L_p when $p \neq 2$. It is exactly in considering such problems that Banach space theory has much to offer, because much of the classical theory has the same basic theme of attempting to find what parts of Hilbert space theory can be carried over to Banach spaces.

We will write this survey from the point-of-view of a Banach space specialist and our aim is partly to advertise the interesting results that can be achieved when Banach space methods are applied externally to other (related) fields.

*The author was supported by NSF grant DMS-9870027

2. Sectorial operators

Let us introduce some notation. Let X be a complex Banach space, and let A be a closed operator on X with domain $\mathcal{D}(A)$ and range $\mathcal{R}(A)$. We say that A is *sectorial* if $\mathcal{D}(A)$ and $\mathcal{R}(A)$ are dense, A is one-one and there exists $0 < \phi < \pi$ and a constant C such that if $\lambda \in \mathbf{C} \setminus \{0\}$ with $|\arg \lambda| > \phi$ then the resolvent $R(\lambda, A) = (\lambda - A)^{-1}$ is bounded and $\|\lambda R(\lambda, A)\| \leq C$. We can then define the angle of sectoriality $\omega(A)$ as the infimum of all angles ϕ for which the above statement holds; clearly $0 \leq \omega(A) < \pi$.

Let us observe at this point an important property of sectorial operators. The resolvent $R(\lambda, A)$ is certainly defined for λ on the negative real axis. Let $S_t = -tR(-t, A) = t(t+A)^{-1}$ for $t > 0$. Then S_t is uniformly bounded. Next note that if $x = Ay$ then

$$\lim_{t\to 0} \|S_t x\| = \lim_{t\to 0} t\|A(t+A)^{-1}y\| = 0$$

while if $x \in \mathcal{D}(A)$

$$\lim_{t\to\infty} \|x - S_t x\| = \lim_{t\to\infty} \|(t+A)^{-1}Ax\| = 0.$$

Since both the domain and range are dense this means that $\lim_{t\to 0} S_t x = 0$ and $\lim_{t\to\infty} S_t x = x$ for all $x \in X$.

Suppose A is sectorial with angle $\omega = \omega(A)$. Then we can define a functional calculus for A as follows. Suppose f is bounded and analytic on a sector $\Sigma_\phi := \{z : |\arg z| < \phi\}$ where $0 < \omega < \phi$. Suppose $\omega < \nu < \phi$ and consider the contour $\Gamma_\nu = \{t \exp(i\nu \mathrm{sgn}\, t) : -\infty < t < \infty\}$. Then we can formally "define"

$$f(A)x = -\frac{1}{2\pi i}\int_{\Gamma_\nu} f(\zeta)R(\zeta, A)x\, d\zeta. \tag{2.1}$$

Of course (2.1) is not well-defined without some restrictions on f. The most natural is that

$$\sup_{\zeta\in\Sigma_\phi} \int_0^\infty |f(t\zeta)|\frac{dt}{t} < \infty \tag{2.2}$$

as this implies that the integral in (2.1) is convergent as a Bochner integral for all $x \in X$, and that $f(A)$ is bounded with

$$\|f(A)\| \leq C \sup_{\zeta\in\Sigma_\phi} \int_0^\infty |f(t\zeta)|\frac{dt}{t} \tag{2.3}$$

where C depends only on A and ϕ.

It is often easier to require that f satisfies an estimate of the form $|f(z)| \leq C|z|^\delta(1 + |z|^2)^{-\delta}$ for $z \in \Sigma_\phi$ where $\delta > 0$. The set of such functions is denoted $H_0^\infty(\Sigma_\phi)$. It is easily established that the map $f \to f(A)$ is an algebra homomorphism on $H_0^\infty(\Sigma_\phi)$. Now if we set

$$\varphi_n(z) = \frac{(n^2-1)z}{(1+nz)(n+z)}$$

then $\varphi_n(A) = S_n - S_{1/n}$. It is then possible to show that if $f \in H^\infty(\Sigma_\phi)$ that $\lim_{n\to\infty}(\varphi_n f)(A)x$ exists for all $x \in X$ if and only if $\sup_n \|(\varphi_n f)(A)\| < \infty$. It is then possible to define $f(A)x$ unambiguously as $\lim_{n\to\infty}(\varphi_n f)(A)x$.

To summarize let us define $\mathcal{H}(A)$ to be the space of all (germs of) functions f which are analytic and bounded on some sector Σ_ϕ where $\phi > \omega(A)$ and such that $\sup_n \|(\varphi_n f)(A)\| < \infty$. Then $\mathcal{H}(A)$ is an algebra and we have constructed an algebra homomorphism $f \to f(A)$ from $\mathcal{H}(A) \to \mathcal{L}(X)$. It may also be shown that if (f_n) is a sequence in $H^\infty(\Sigma_\phi)$, where $\phi > \omega$, such that $\sup_n \|f_n\|_{H^\infty}(\Sigma_\phi) < \infty$ and $f_n(z) \to f(z)$ for all $z \in \Sigma_\phi$ then the conditions $f_n \in \mathcal{H}(A)$ and $\sup_n \|f_n(A)\| < \infty$ imply that $f \in \mathcal{H}(A)$ and $f_n(A)x \to f(A)x$ for all $x \in X$.

Notice that the space $\mathcal{H}(A)$ contains all functions $(\lambda - z)^{-1}$ for $|\arg\lambda| > \omega$ and also $H_0^\infty(\Sigma_\phi)$ for $\phi > \omega$. In particular if $\omega < \pi/2$ then we can see that $e^{-wz} \in \mathcal{H}(A)$ for $|\arg w| + \omega < \pi/2$ since $e^{-wz} - wz(1+wz)^{-1} \in H_0^\infty(\Sigma_\phi)$ whenever $|\arg w| + \phi < \pi/2$. In fact if $\psi + \omega < \pi/2$ the set of operators $\{e^{-wA} : \ w \in \Sigma_\psi\}$ is uniformly bounded and hence we have:

Proposition 2.1 *If A is sectorial with $\omega(A) < \frac{\pi}{2}$ then $-A$ is the generator of a bounded analytic semigroup.*

Conversely suppose $-A$ is the generator of a bounded strongly continuous semigroup. Then the equation

$$R(\lambda, A) = \int_0^\infty e^{\lambda t} e^{-tA} dt$$

shows that A is sectorial with $\omega(A) \leq \frac{\pi}{2}$.

We say that A has an H^∞*−calculus* or is H^∞*−sectorial* if there exists $\phi < \pi$ so that $H^\infty(\Sigma_\phi) \subset \mathcal{H}(A)$. Then we set $\omega_H(A) = \inf\{\phi : \ H^\infty(\Sigma_\phi) \subset \mathcal{H}(A)\}$. This notion was originally introduced by McIntosh [27] for operator son Hilbert spaces, and later an extensive study was undertaken in [9].

Note that A has an H^∞−calculus then in particular it has *bounded imaginary powers (BIP)* i.e. A^{it} is a bounded operator for all real t. Indeed it is further true that if $\phi > \omega_H(A)$ then A satisfies the estimate $\|A^{it}\| \leq Ce^{\phi t}$. If X is a Hilbert space then (BIP) is actually equivalent to an H^∞-calculus, but this is false for the spaces L_p when $p \neq 2$ [9].

Example 1. The primary motivating example to consider here is the case of the differentiation operator $Af = f'$ on the space $L_p(\mathbf{R})$ where $1 \leq p < \infty$. Here $\mathcal{D}(A) = \{f \in L_p : \ f' \in L_p\}$. It is easy to see that $-A$ is the generator of the translation semigroup $(e^{-tA}f)(s) = f(s-t)$. Thus A is sectorial and $\omega(A) = \frac{\pi}{2}$. Now if $f \in H^\infty(\Sigma_\phi)$ where $\phi > \frac{\pi}{2}$ then the boundedness of $f(A)$ is equivalent to the boundedness of the Fourier multiplier $f(i\xi)$. It then follows from the Hörmander-Mikhlin conditions that if $1 < p < \infty$ the operator A has an $H^\infty(\Sigma_\phi)$−calculus where $\phi > \frac{\pi}{2}$. Thus $\omega_H(A) = \frac{\pi}{2}$.

Example 2. Now consider the vector-valued analogue $Af = f'$ on $L_p(\mathbf{R}; X)$ where X is a Banach space. Then the same arguments show that A is sectorial with $\omega(A) = \frac{\pi}{2}$, but for $1 < p < \infty$, one only obtains an H^∞-calculus in those spaces where the vector-valued analogues of the Hörmander-Mikhlin conditions give boundedness of Fourier multipliers. A Banach space X with this property is called a *(UMD)-space* for *unconditional martingale differences.* This class of spaces was introduced by Burkholder [6], and the Fourier multiplier result we need was proved by McConnell [26]. For our purposes it is simplest to use a characterization of Bourgain [4] to define (UMD)-spaces: a Banach space X has (UMD) if and only for some (and hence for every) $1 < p < \infty$ the vector-valued Hilbert transform is bounded on $L_p(\mathbf{R};X)$. Readers who are not so familiar with concept may like to note the important classical spaces L_p for $1<p<\infty$ are (UMD) spaces.

It is worth reproducing a simple argument to show that if A has an $H^\infty(\Sigma_\phi)$-calculus for some $\phi<\pi$ (or even (BIP)) then indeed X is (UMD). In fact it is enough to consider imaginary powers. In fact the boundedness of A^{2it} implies that the multiplier $m_1(\xi)=\exp(-t\pi\text{sgn }\xi)|\xi|^{2it}$ is bounded on $L_p(X)$. However if A has $H^\infty(\Sigma_\phi)$-calculus then so has $-A$ and the same reasoning leads to the boundedness of the multiplier $m_2(\xi)=\exp(t\pi\text{sgn }\xi)|\xi|^{2it}$. Taking $t=1$ we can deduce the boundedness of the multiplier $m_3(\xi)=|\xi|^{2i}\text{sgn }\xi$ while from $t=-1$ we obtain the boundedness of the multiplier $m_4(\xi)=|\xi|^{-2i}$. Combining gives us the boundedness of the multiplier sgn ξ i.e. the Hilbert transform.

Example 3. Now let us consider an example closer in spirit to Banach space theory. Suppose X has a Schauder basis (e_n) and let $(a_n)_{n=1}^\infty$ be an increasing real sequence with $a_1>0$. Let us define

$$A(\sum_{n=1}^\infty c_ne_n)=\sum_{n=1}^\infty a_nc_ne_n.$$

Here the domain of A is the set of $x=\sum_{n=1}^\infty c_ne_n$ so the series $\sum_{n=1}^\infty a_nc_ne_n$ converges. Such examples were first considered by Baillon and Clément [2]. It is easy to show that A is sectorial and $\omega(A)=0$.

If the basis (e_n) is unconditional then A has an H^∞-calculus and $\omega_H(A)=0$. If we take a_n be an interpolating sequence for $H^\infty(\Sigma_\phi)$ where $\phi>0$ then the converse is true; an example is $a_n=2^n$.

It is clear these ideas can be extended to Schauder decompositions. If (E_n) is a Schauder decomposition of X we can define, in an exactly similar way,

$$A(\sum_{n=1}^\infty x_n)=\sum_{n=1}^\infty a_nx_n$$

where $x_n\in E_n$ and $\mathcal{D}(A)$ is again the set where the right-hand series converges,

3. The maximal regularity problem

Let us now suppose that A is sectorial with $\omega(A)<\frac{\pi}{2}$. For $0<T<\infty$ consider the Cauchy problem:

$$\frac{dx}{dt}+Ax=h(t)\quad 0\le t\le T\quad x(0)=0. \tag{3.1}$$

This can be formally solved by

$$x(t)=\int_0^t e^{-A(t-s)}h(s)ds. \tag{3.2}$$

Now suppose $h\in L_p([0,T];X)$ where $1<p<\infty$. Then one may easily check that $x\in L_p([0,T];X)$. We say that A has L_p*–maximal regularity* if it also follows (for any T) that $dx/dt\in L_p(X)$; this clearly equivalent to the L_p-boundedness of the operator

$$h\to\int_0^t Ae^{-A(t-s)}h(s)ds.$$

Although this definition apparently depends on p in fact L_p-maximal regularity for some $1 < p < \infty$ implies L_p-maximal regularity for every p when $1 < p < \infty$; see [13]. It is also not difficult to see that this definition is independent of T. We can therefore refer just to the problem of maximal regularity.

Note that we have restricted our problem to a finite interval. It is possible also to consider the case when $T = \infty$ which leads to a slightly stronger form of maximal regularity; let us refer to this as *strong maximal regularity*.

It is an old result of de Simon [11] that if X is a Hilbert space then *every* sectorial operator A with $\omega(A) < \frac{\pi}{2}$ has maximal regularity. Let us say that a Banach space has the *maximal regularity property* or (MRP) if it satisfies this condition that every sectorial operator with $\omega(A) < \frac{\pi}{2}$ has maximal regularity. It is not too difficult to find counter-examples in certain Banach spaces (cf.[21]), but it was conjectured that *at least* the classical Banach spaces $X = L_p$ where $1 < p < \infty$ have (MRP). This conjecture is usually attributed to Brézis (around 1980) although it may have been around before that time. It is based on the fact that for all concrete examples arising from partial differential equations this seems to be the case. Subsequently this conjecture crystallized into the form that any space with (UMD) has (MRP); we will discuss this further below. It should perhaps be remarked that the space L_∞ has (MRP) for somewhat trivial reasons: any sectorial operator which generates a bounded semigroup is already a bounded operator by a theorem of Lotz [25].

Let us now fix $T = 2\pi$ and suppose that A^{-1} is bounded; this second assumption is not necessary but allows us to make a convenient alternative formulation; in fact we can always reduce to this case by replacing A by $I + A$. It is shown in [17] that, under the this assumption, we can transfer the maximal regularity problem to the circle. Effectively we can replace the boundary condition $x(0) = 0$ by the boundary condition:

$$\int_0^{2\pi} x(s)ds = 0.$$

In this case we can expand the solution in a Fourier series:

$$x(t) \sim -\sum_{n\neq 0} R(-in, A)\hat{h}(n)e^{int}.$$

Then maximal regularity becomes equivalent to the boundedness of the operator

$$h \to \sum_{n\neq 0} AR(-in, A)\hat{h}(n)e^{int} \tag{3.3}$$

on the space $L_p(\mathbf{T}; X)$ where $\mathbf{T}$ denotes the unit circle with normalized Haar measure $dt/2\pi$ and

$$\hat{h}(n) = \frac{1}{2\pi}\int_0^{2\pi} h(t)e^{-int}dt.$$

Taking, as we may, $p = 2$, we summarize this in the statement that A has maximal regularity if and only if there is a constant C so that for all finitely non-zero sequences $(x_n)_{n\in\mathbf{Z}}$ we have:

$$\left(\int_0^{2\pi} \|\sum_{n\in\mathbf{Z}} AR(-in, A)x_n e^{int}\|^2 \frac{dt}{2\pi}\right)^{\frac{1}{2}} \leq C\left(\int_0^{2\pi} \|\sum_{n\in\mathbf{Z}} x_n e^{int}\|^2 \frac{dt}{2\pi}\right)^{\frac{1}{2}}. \tag{3.4}$$

At this point one can see why de Simon's theorem must hold. We can take $p = 2$. By using Parseval's identity when X is a Hilbert space we have:

$$\|\sum_{n\neq 0} AR(in, A)x_n e^{int}\| = (\sum_{n\neq 0} \|AR(in, A)x_n\|^2)^{\frac{1}{2}}.$$

Thus all we need is uniform boundedness of the operators $\{AR(in, A) : n \neq 0\}$ and this is almost exactly the assumption that $\omega(A) < \frac{\pi}{2}$. This also suggests that such a theorem is really rather unlikely to hold in a general Banach space, because such properties are very special.

Let us now explain the role of the (UMD) assumption. (UMD)-spaces became important in Banach space theory during the 1980's after their introduction by Burkholder. Two results from the 1980's strongly suggested this was the right assumption for maximal regularity.

First suppose $L_p(\mathbf{T}; X)$ has (MRP) where $1 < p < \infty$. Then consider the subspace $L_p^0(\mathbf{T}; X)$ of all functions h with mean zero, i.e. $\hat{h}(0) = 0$. Then we have an example of a sectorial operator A by taking

$$Ah \sim \sum_{n\neq 0} |n|\hat{h}(n)e^{int}$$

on a its natural domain. (We need to restrict to functions of mean zero to prevent A having non-trivial kernel). In fact $\omega(A) = 0$. Now by maximal regularity the operator on $L_p(\mathbf{T}^2; X)$ given by

$$\sum_{m,n} \hat{h}(m, n)e^{ims+int} \to \sum_{m,n} \frac{im}{im + |n|}\hat{h}(m, n)e^{ims+int}$$

is bounded. But if we consider the subspace of functions of the form $h(s - t)$ we quickly see that the Hilbert transform is bounded on $L_p(\mathbf{T}; X)$ i.e. X has (UMD). This example is a version of a result of Coulhon-Lamberton [8].

The second result was the spectacular and important result of Dore-Venni (1987)[14]:

Theorem 3.1 *Suppose X is a (UMD)-space and A is a sectorial operator whose imaginary powers A^{it} are bounded and satisfy an estimate*

$$\|A^{it}\| \leq Ke^{\theta|t|} \tag{3.5}$$

where $\theta < \frac{\pi}{2}$. Then X has maximal regularity.

This is deduced from a more general result which we will discuss later. Let us note that (3.5) is stronger than $\omega(A) < \frac{\pi}{2}$. Thus the maximal regularity conjecture essentially reduces to whether the assumption (3.5) is really necessary.

4. Solution of the maximal regularity problem

We will now describe the approach taken by the author and Gilles Lancien in [17] to resolve the maximal regularity problem. The most important observation is that we should look at examples in the spirit of Example 3 of §2, rather than examples arising from partial differential equations, which have a habit of behaving well. In this manner we can essentially transport basis theory in Banach spaces, much of which was developed in the period 1960-1975, and make it immediately applicable in this new setting.

Let us first look more carefully at (3.4). It is immediately tempting to *restrict* the range of summation to some sequence such as $\{2^n\}$ which forms a Sidon set, because then a result of Pisier [29] implies that have an estimate

$$\left(\int_0^{2\pi} \|\sum_{n=1}^{\infty} x_n e^{i2^n t}\|^2 \frac{dt}{2\pi}\right)^{\frac{1}{2}} \approx (\mathbf{E}\|\sum_{n=1}^{\infty} \epsilon_n x_n\|^2)^{\frac{1}{2}}$$

where (ϵ_n) is an independent sequence of Rademacher-type random variables.

We thus see that if A has maximal regularity we have an inequality of the type;

$$(\mathbf{E}\|\sum_{n=1}^{\infty} \epsilon_n R(i2^n, A)x_n\|^2)^{\frac{1}{2}} \leq C(\mathbf{E}\|\sum_{n=1}^{\infty} \epsilon_n x_n\|^2)^{\frac{1}{2}}. \tag{4.1}$$

This inequality turns out to have far reaching ramifications, which we shall return to later. In fact for the results of this section we do not really need to use (4.1), but clearly it motivated our work.

Now let us suppose X is a Banach space with (MRP) and suppose (E_n) is a Schauder decomposition of X. We can use the idea of Example 3. Let us take A corresponding to the sequence

$$(a_n)_{n=1}^{\infty} = \{1, 2, 2, 4, 4, \cdots\}.$$

Suppose $x_n \in E_n$ is finitely non-zero. Then we have

$$\left(\mathbf{E}\|\sum_{n=1}^{\infty} \epsilon_n (\frac{2^{n-1}}{2^{n-1} + i2^n} x_{2n-1} + \frac{2^n}{2^n + i2^n} x_{2n})\|^2\right)^{\frac{1}{2}} \leq C(\mathbf{E}\|\sum_{n=1}^{\infty} \epsilon_n (x_{2n-1} + x_{2n})\|^2)^{\frac{1}{2}}.$$

This inequality clearly implies an estimate

$$(\mathbf{E}\|\sum_{n=1}^{\infty} \epsilon_n x_{2n}\|^2)^{\frac{1}{2}} \leq C'(\mathbf{E}\|\sum_{n=1}^{\infty} \epsilon_n (x_{2n-1} + x_{2n})\|^2)^{\frac{1}{2}}. \tag{4.2}$$

This leads to the following conclusion:

Lemma 4.1 *Let X be a Banach space with (MRP). If X has a Schauder decomposition (E_n) so that the blocked decomposition $(E_{2n_1} + E_{2n})$ is unconditional then the subspace $\sum_{n=1}^{\infty} E_{2n}$ is complemented (and hence (E_n) is unconditional).*

This Lemma follows from (4.2) because unconditionality allows us to estimate

$$\|\sum x_n\| \approx (\mathbf{E}\|\sum_{n=1}^{\infty} \epsilon_n (x_{2n-1} + x_{2n})\|^2)^{\frac{1}{2}}.$$

Once we have Lemma 4.1 the problem reduces to some classical results in Banach space theory. Let us first suppose X has an unconditional basis (e_n) and suppose (y_n) is a normalized block basic sequence i.e.

$$y_n = \sum_{k=r_{n-1}+1}^{r_n} c_k e_k$$

where $0 = r_0 < r_1 < r_2 < \cdots$. Then a Lemma of Zippin [32] allows us to make a new basis (e'_n) so that $[e'_k : r_{n-1} < k \leq r_n] = [e_k : r_{n-1} < k \leq r_n]$ and $e_{r_n} = y_n$. Then let $E_{2n-1} = [e'_k : r_{n-1} < k < r_n]$ and $E_{2n} = [y_n]$. It follows from the Lemma that $[y_n]$ is a *complemented* subspace of X. Thus *every block basic sequence spans a complemented subspace.*

However, more is clearly true: we can apply this argument to any permutation of the original basis, so any block basic sequence of any permutation spans a complemented subspace of x. This is exactly the hypothesis of a theorem of Lindenstrauss and Tzafriri [23],[24] Theorem 2.a.10:

Theorem 4.2 *If X has a normalized unconditional basis (e_n) such that every block basic sequence of every permutation spans a complemented subspace, then (e_n) is equivalent to the canonical basis of one the spaces ℓ_p where $1 \leq p < \infty$ or c_0.*

Thus any Banach space X with (MRP) and an unconditional basis is one of these spaces. But it is clear that we can eliminate the spaces ℓ_p when $p \neq 1, 2$. For in these spaces Pełczyński showed that it is possible to find an unconditional basis which is not equivalent to the standard one [28], and working with that basis leads to a contradiction. We are now left with c_0, ℓ_1 and ℓ_2 ; these are the precise three spaces with a unique unconditional basis by beautiful results of Lindenstrauss-Pełczyński and Lindenstrauss-Zippin, [24] Theorem 2.b.10. It would be nice to conclude by the argument by saying these three spaces have (MRP); but unfortunately that is false. The cases of c_0 and ℓ_1 can be eliminated by working with the summing basis of c_0 (see [17] for details).

Thus we have proved:

Theorem 4.3 *[17] Let X be a Banach space with an unconditional basis and the maximal regularity property. Then X is isomorphic to a Hilbert space.*

Since the spaces L_p when $1 < p < \infty$ have unconditional bases this implies the original conjecture is false. It is possible to refine this argument to give the following results:

Theorem 4.4 *[17] Let X be a Banach space with (MRP). If either (a) X is an order-continuous Banach lattice or (b) $X = L_p(Y)$ for some $1 \leq p < \infty$ then X is a Hilbert space.*

Note here that (b) completes the result of Coulhon-Lamberton cited in §3. It is possible to ask whether every separable Banach space with (MRP) is a Hilbert space or at least every separable Banach space with a basis. This seems harder, but some partial results are obtained in [18]. One final remark is in order: one may wonder why these techniques fail in L_∞; the answer is that $L_\infty \sim \ell_\infty$ has no Schauder decompositions [12].

5. Rademacher boundedness and maximal regularity

Since Theorems 4.3 and 4.4 effectively end the discussion of the maximal regularity property for Banach spaces, one is naturally led to seek conditions on a sectorial operator so that it has maximal regularity. Let us look again at (4.1) which is clearly (at least if

A^{-1} is bounded) a necessary condition. It turns out (although we did not immediately know this) that this condition embodies a concept which has made sporadic appearances in the literature over the last 20 years. Let us say that a family of operators $\mathcal{F} \subset \mathcal{L}(X)$ is *Rademacher-bounded* or *R-bounded* if there is a constant C so that if $T_1, \cdots, T_n \in \mathcal{F}$ then

$$(\mathbf{E}\|\sum_{k=1}^{n} \epsilon_k T_k x_k\|^2)^{\frac{1}{2}} \leq C(\mathbf{E}\|\sum_{k=1}^{n} \epsilon_k x_k\|^2)^{\frac{1}{2}}. \tag{5.1}$$

This concept seems to date implicitly to a paper of Bourgain [4]. It was subsequently used by Berkson and Gillespie [3] and a comprehensive study was undertaken by Clément, de Pagter, Sukochev and Witvliet [7]. Let us observe that in the spaces L_p when $1 \leq p < \infty$ (or indeed in any Banach lattice with nontrivial cotype) R-boundedness is equivalent to a square-function estimate:

$$\|(\sum_{k=1}^{n} |T_k x_k|^2)^{\frac{1}{2}}\| \leq C\|(\sum_{k=1}^{n} |x_k|^2)^{\frac{1}{2}}\|.$$

From (4.1) we obtain the fact that if A has maximal regularity then the sequence $\{AR(\pm i2^n, A)\}_{n=1}^{\infty}$ is R-bounded; a slight variation gives us that for $\frac{1}{2} \leq s \leq 1$ the sequences $\{AR(\pm is2^n, A)\}_{n=1}^{\infty}$ are uniformly Rademacher-bounded (i.e. with the same constant C.). From this it is possible to see that the sets $\{AR(it, A) : \ |t| \geq 1\}$ are R-bounded, based on simple properties of the resolvent. If A^{-1} is bounded then using the analyticity of the resolvent one can actually show that the set $\{AR(\lambda, A) : \ \Re\lambda \leq 0\}$ is R-bounded.

The remarkable fact is that this condition is actually sufficient for maximal regularity in a (UMD)-space. This was discovered by Lutz Weis [31] and myself (unpublished) independently in the early summer of 1999. My own argument was a sledgehammer technique using the unconditionality of the Haar series expansion of any $h \in L_p([0,T]; X)$. Weis instead proved an elegant and more general Fourier multiplier result which implied the same conclusion. Let us state Weis's theorem:

Theorem 5.1 *Let X be a Banach space with (UMD) and suppose A is a sectorial operator with $\omega(A) < \frac{\pi}{2}$. Then A has strong maximal regularity if and only if $\{AR(\lambda, A) : \Re\lambda \leq 0\}$ is Rademacher-bounded.*

Remark. For maximal regularity one only needs that $\{AR(\lambda, A) : \ \Re\lambda \leq -\delta\}$ is Rademacher-bounded for some $\delta > 0$.

This suggest a new concept of *Rademacher-sectoriality.* We say that a sectorial operator is Rademacher-sectorial or R-sectorial for some angle ϕ if the set $\{R(\lambda, A) : \ |\arg\lambda| \geq \phi\}$ is R-bounded. We can denote by $\omega_R(A)$ the infimum of all such angles ϕ. Clearly $\omega_R(A) \geq \omega(A)$. The content of the results of §4 is that a Banach space with unconditional basis always admits a sectorial operator which is not Rademacher-sectorial for any angle.

At this point Weis and I got in contact and decided to pool our resources and work on these ideas together, starting in the fall of 1999 and continuing in the spring and summer of 2000.

6. The joint functional calculus

For specialists in the area, the maximal regularity problem is but one aspect of a more general problem concerning the sum of two closed commuting operators. Let us suppose A, B are two sectorial operators on a Banach space X which commute in the sense that their resolvents commute. We can consider the sum $A+B$ on $\mathcal{D}(A)\cap\mathcal{D}(B)$; however it is not immediately clear that with this domain $A+B$ is closed. One sufficient condition is that $\|Ax\| \le C\|Ax+Bx\|$ for $x \in X$. This can be viewed as the problem of whether $A(A+B)^{-1}$ can be defined as a bounded operator. The problem of maximal regularity is exactly of this type (cf.[10]). We consider D on $L_p([0,T];X)$ to be the operator $Df = f'$ on the set of all $f \in L_p$ of the form $f(t) = \int_0^t g(s)ds$ where $g \in L_p$. We then consider the equation

$$(D+\tilde{A})f = h$$

where $\tilde{A}f(t) = A(f(t))$ and $\tilde{A}$ has domain of all f such that $f(t) \in \mathcal{D}(A)$ almost everywhere. Maximal regularity is exactly the requirement that $A(D+A)^{-1}$ is bounded or that $D+A$ is closed [10].

From this more general viewpoint the Dore-Venni Theorem (Theorem 3.1) reads [14]:

Theorem 6.1 *Suppose X is a Banach space with (UMD) and that A, B are commuting sectorial operators with bounded imaginary powers satisfying*

$$\|A^{it}\| \le Ke^{\theta_A|t|} \qquad \|B^{it}\| \le Ke^{\theta_B|t|}.$$

Then if $\theta_A+\theta_B < \pi$ then $A+B$ is closed on $\mathcal{D}(A)\cap\mathcal{D}(B)$ and $A(A+B)^{-1}$ extends to a bounded operator on X.

To derive Theorem 3.1 one need only observe that when X has (UMD) and $1<p<\infty$ D has (BIP) with any $\theta_D > \frac{\pi}{2}$. In fact, we saw in Example 2 of §2 that X has (UMD) if and only if D has an H^∞-calculus (or even (BIP)), and that in this case it always follows that $\omega_H(D) = \frac{\pi}{2}$.

Thus it is natural for us to consider the general problem of developing a joint functional calculus for two commuting sectorial operators. Suppose $\phi_A > \omega(A)$ and $\phi_B > \omega(B)$. Then if $f \in H^\infty(\Sigma_{\phi_A}\times\Sigma_{\phi_B})$ we make a formal definition of $f(A,B)$ by modifying (2.1):

$$f(A,B)x = \frac{-1}{4\pi^2}\int_{\Sigma_{\nu_A}}\int_{\Sigma_{\nu_B}} f(\zeta_1,\zeta_2)R(\zeta_1,A)R(\zeta_2,A)x\,d\zeta_2 d\zeta_1. \tag{6.1}$$

The general question is then to give conditions so that $f(A,B)$ is a bounded operator. In the particular case when $f(z_1,z_2) = z_1(z_1+z_2)^{-1}$ then we will require of course that $\omega(A)+\omega(B) < \pi$ to avoid the singularities of f. It is natural to assume that one of the operators, say A, has an H^∞-calculus.

In [19] we discovered a very general such theorem:

Theorem 6.2 *Suppose X is any Banach space and that A, B are commuting sectorial operators on X. Suppose that $\sigma > \omega_H(A)$ and $\sigma' > \omega(B)$ and that $f \in H^\infty(\Sigma_\sigma \times \Sigma_{\sigma'})$. Suppose further that the set $\{f(w,B) : w \in \Sigma_\sigma\}$ consists of bounded operators and is R-bounded. Then $f(A,B)$ is bounded.*

We should emphasize that this theorem is really quite easy to prove; this is significant because it can be used to replace quite delicate and involved arguments in some existing theorems in the literature. A second point is that the R-boundedness assumption is too strong; one can replace it by U-boundedness, where we say that $\mathcal{F}$ is U-bounded if for some C and all $T_1, \cdots, T_n \in \mathcal{F}$, $x_1, \cdots, x_n \in X$ and $x_1^*, \cdots, x_n^* \in X^*$ we have

$$\sum_{k=1}^{n} |\langle T_k x_k, x_k^* \rangle| \leq C \max_{\epsilon_k=\pm 1} \| \sum_{k=1}^{n} \epsilon_k x_k \| \max_{\epsilon_k=\pm 1} \| \sum_{k=1}^{n} \epsilon_k x_k^* \|. \tag{6.2}$$

U-boundedness is significantly weaker than R-boundedness, but, unfortunately for many practical examples it yields nothing new.

Let us put Theorem 6.2 to use by considering two problems. First let us look at the results of Lancien, Lancien and Le Merdy [20] on the existence of a full H^∞−joint calculus (these results extended earlier work by Albrecht, Franks and McIntosh [1], [15] who considered only L_p-spaces when $1 < p < \infty$). Suppose B also has an H^∞−calculus and $\phi_A > \omega_H(A)$, $\phi_B > \omega_H(B)$. We say that (A, B) has a joint $H^\infty(\Sigma_{\phi_A} \times \Sigma_{\phi_B})$-functional calculus if for every $f \in H^\infty(\Sigma_{\phi_A} \times \Sigma_{\phi_B})$ the operator $f(A, B)$ is bounded and then one necessarily has an estimate $\|f(A, B)\| \leq C\|f\|_{H^\infty}$. To apply Theorem 6.2 to a particular f one needs that the family $\{f(w, B) : \ w \in \Sigma_{\phi_A}\}$ is R-bounded; to obtain this for every f we really need that the family

$$\{f(B) : \ f \in H^\infty(\Sigma_{\phi_B}), \ \|f\|_{H^\infty} \leq 1\}$$

is R-bounded. This means that B must have a Rademacher-bounded H^∞−calculus. Thus we are led to the question of classifying Banach spaces where an H^∞−calculus already implies a Rademacher-bounded H^∞-calculus. There is a nice example due to Lancien, Lancien and Le Merdy to show that this is not always the case even in UMD-spaces.

Example 4. Consider the Schatten ideal $\mathcal{C}_p$ where $1 \leq p < \infty$. We can consider this as a space of infinite matrices $\mathbf{a} = (a_{jk})_{j,k}$. Now define $A(\mathbf{a}) = (2^j a_{jk})_{j,k}$ and $B(\mathbf{a}) = (2^k a_{jk})_{j,k}$ on their appropriate domains. Both A and B are H^∞−sectorial with $\omega_H(A) = \omega_H(B) = 0$. Now for any suitable bounded analytic f defined on some $\Sigma_{\phi_A} \times \Sigma_{\phi_B}$ it is clear that $f(A, B)$ is simply a Schur multiplier i.e. $f(A, B)\mathbf{a} = (f(2^j, 2^k) a_{jk})_{j,k}$. Now it is pointed out in [20] that $(2^j, 2^k)$ is interpolating for $H^\infty(\Sigma_{\phi_A} \times \Sigma_{\phi_B})$ so that if (A, B) have any joint H^∞−calculus then every Schur multiplier $(a_{jk})_{j,k} \to (m_{jk} a_{jk})_{j,k}$ would have to be bounded. That only happens when $p = 2$. However if $1 < p < \infty$ these spaces even have (UMD) [5].

However, under appropriate conditions, one can get the desired conclusion. We say following Pisier [30] that X has property (α) if for some constant C if (ϵ_j) and (ϵ'_k) are two mutually independent sequences of Rademachers then for any $(x_{jk})_{j,k \leq n}$ in X and scalar $(a_{jk})_{j,k \leq n}$ we have

$$(\mathbf{E}\| \sum_{j=1}^{n} \sum_{k=1}^{n} a_{jk} \epsilon_j \epsilon'_k x_{jk} \|^2)^{\frac{1}{2}} \leq C \max_{j,k} |a_{jk}| (\mathbf{E}\| \sum_{j=1}^{n} \sum_{k=1}^{n} \epsilon_j \epsilon'_k x_{jk} \|^2)^{\frac{1}{2}}. \tag{6.3}$$

Then any subspace of a Banach lattice with non-trivial cotype has (α). Then we have [19]:

Theorem 6.3 *Suppose X has property (α) and that A is an $H^\infty-$sectorial operator on X. Then for any $\phi > \omega_H(B)$ the set $\{f(B) : f \in H^\infty(\Sigma_\phi),\ \|f\|_{H^\infty} \leq 1\}$ is R-bounded.*

From this we deduce immediately by Theorem 6.2, as explained above:

Theorem 6.4 *[20] Suppose X has property (α) and that A, B are $H^\infty-$sectorial operators on X. If $\phi_A > \omega_H(A)$ and $\phi_B > \omega_H(B)$ then (A, B) has an $H^\infty(\Sigma_{\phi_A} \times \Sigma_{\phi_B})-$functional calculus.*

In fact in [20] some weaker conditions on X are given which imply the theorem (e.g. if X is any Banach lattice); these results can be obtained from the more delicate version of Theorem 6.2 using U-boundedness (6.2).

Next we turn to the case when $\omega_H(A) + \omega(B) < \pi$ and consider the function $f(z_1, z_2) = z_1(z_1 + z_2)^{-1}$.

If we check Theorem 6.2 we see that we need that $\{w(w + B)^{-1} : w \in \Sigma_\sigma\}$ is Rademacher-bounded for some $\sigma > \omega_H(A)$. This exactly means that $\omega_R(B) + \omega_H(A) < \pi$.

We can then ask, as in the previous example if it is sufficient that $\omega_H(A) + \omega_H(B) < \pi$. Example 4 above provides again some guidance for if we take A, B as before, we require the boundedness of the Schur multiplier $(2^j(2^j + 2^k)^{-1})_{j,k}$ on $\mathcal{C}_p$. If one spaces out the rows and columns one quickly deduces that if this multiplier is bounded then the lower triangular projection, corresponding to the Schur multiplier which is one below the diagonal and zero elsewhere is also bounded. In the case when $p = 1$ (the trace-class) this is false. Thus there are examples when A, B are both $H^\infty-$sectorial and with $\omega_H(A) = \omega_H(B) = 0$ and yet $A(A + B)^{-1}$ is unbounded.

It turns out that we can prove a result somewhat analogous to Theorem 6.3. We first say that X has property (Δ) if it obeys an inequality somewhat weaker than (α):

$$(\mathbf{E}\|\sum_{j=1}^{n}\sum_{k=j}^{n} \epsilon_j\epsilon'_k x_{jk}\|^2)^{\frac{1}{2}} \leq C(\mathbf{E}\|\sum_{j=1}^{n}\sum_{k=1}^{n} \epsilon_j\epsilon'_k x_{jk}\|^2)^{\frac{1}{2}}. \tag{6.4}$$

Note that this is somewhat like a lower triangular projection as in the preceding discussion. Property (Δ) is not nearly as restrictive as (α). It is enjoyed by any space with (α), and also by (UMD)-spaces and even spaces with analytic (UMD). It fails in $\mathcal{C}_1$ ([16]).

Now in place of Theorem 6.3 we have:

Theorem 6.5 *Let X be a Banach space with property (Δ). Then if A is $H^\infty-$sectorial, then A is also R-sectorial and $\omega_R(A) = \omega_H(A)$.*

As before, we deduce:

Theorem 6.6 *[19] Let X be a Banach space with property (Δ) and suppose A, B are $H^\infty-$sectorial operators on X with $\omega_H(A) + \omega_H(B) < \pi$. Then $A(A + B)^{-1}$ is bounded (and $A + B$ is a closed operator on $\mathcal{D}(A) \cap \mathcal{D}(B)$).*

More recently, Le Merdy has improved Theorem 6.6:

Theorem 6.7 *[22] Under the hypotheses of Theorem 6.6, $A + B$ is $H^\infty-$sectorial and $\omega_H(A + B) \leq \max(\omega_H(A), \omega_H(B))$.*

7. Concluding Remarks

We have attempted to give the flavor of recent work in this area, particularly in [17] and [19]. There are many problems left to resolve. Let us mention just two intriguing questions.

We know examples of sectorial which are not R-sectorial. However we do not know any example of an R-sectorial operator for which $\omega_R(A) > \omega(A)$.

The second question is more vague. The maximal regularity conjecture was made because all natural examples on say the spaces L_p have maximal regularity. The problem is to explain why this phenomenon occurs. This would require isolating the properties of a sectorial operator induced by some differential operator which force it to be R-sectorial.

REFERENCES

1. D. Albrecht, E. Franks and A. McIntosh, Holomorphic functional calculi and sums of commuting operators. Bull. Austral. Math. Soc. 58 (1998), 291–305.
2. J.-B. Baillon and Ph. Clément, Examples of unbounded imaginary powers of operators. J. Funct. Anal. 100 (1991), 419–434.
3. E. Berkson and T.A. Gillespie, Spectral decompositions and harmonic analysis on UMD spaces, Studia Math. 112 (1994) 13-49.
4. J. Bourgain, Some remarks on Banach spaces in which martingale differences are unconditional, Arkiv Math. 21 (1983) 163-168.
5. J. Bourgain, Vector-valued singular integrals and $H^1 - BM)$ duality, 1-19 in *Probability theory and harmonic analysis,* (edited by J.-A. Chao and W. Woyczynski), Marcel Dekker, New York, 1986.
6. D.L. Burkholder, A geometrical characterization of Banach spaces in which martingale difference sequences are unconditional. Ann. Probab. 9 (1981), 997–1011.
7. P. Clément, B. de Pagter, F. Sukochev and H. Witvliet, Schauder decomposition and multiplier theorems, Studia Math. 138 (2000), 135–163.
8. T. Coulhon and D. Lamberton, Régularité L^p pour les équations d'évolution, Séminaire d'Analyse Fonctionnelle Paris VI-VII (1984-85),155-165.
9. M. Cowling, I. Doust, A. McIntosh and A. Yagi, Banach space operators with a bounded H^∞-calculus, J. Austral. Math. Soc. 60 (1996) 51-89.
10. G. Da Prato and P. Grisvard, Sommes d'opérateurs linéaires et équations différentielles opérationnelles, J. Math. Pures Appl. 54 (1975), 305-387.
11. L. De Simon, Un' applicazione della theoria degli integrali singolari allo studio delle equazioni differenziali lineare astratte del primo ordine, Rend. Sem. Mat., Univ. Padova (1964), 205-223.
12. D.W. Dean, Schauder decompositions in (m). Proc. Amer. Math. Soc. 18 (1967) 619–623.
13. G. Dore, L^p-regularity for abstract differential equations, in *Functional analysis and related topics,* H. Komatsu, editor, Springer Lecture Notes 1540, 1993.
14. G. Dore, A. Venni, On the closedness of the sum of two closed operators, Math. Z. 196 (1987), 189-201.
15. E. Franks and A. McIntosh, Discrete quadratic estimates and holomorphic functional calculi in Banach spaces. Bull. Austral. Math. Soc. 58 (1998), 271–290.

16. U. Haagerup and G. Pisier, Factoriztion of analytic functions with values in non-commutative L_1-spaces and applications, Canad. J. Math. 41 (1989) 882-906.
17. N.J. Kalton and G. Lancien, A solution to the problem of L^p-maximal regularity, Math. Zeit. 235 (2000) 559-568.
18. N.J. Kalton and G. Lancien, L^p-maximal regularity on Banach spaces with a Schauder basis, Arch. Math. to appear.
19. N.J. Kalton and L. Weis, The $H^\infty-$calculus and sums of closed operators, Math. Ann. to appear.
20. F. Lancien, G. Lancien and C. Le Merdy, A joint functional calculus for sectorial operators with commuting resolvent, Proc. London Math. Soc. 77 (1998) 387-414.
21. C. Le Merdy, Counterexamples on L^p-maximal regularity, Math. Zeit. 230 (1999), 47-62.
22. C. Le Merdy, Two results about H^∞ functional calculus on analytic UMD Banach spaces, to appear.
23. J. Lindenstrauss and L. Tzafriri, On the complemented subspaces problem, Israel J. Math. 9 (1971), 263-269.
24. J. Lindenstauss and L. Tzafriri, *Classical Banach spaces I*, Springer-Berlin (1977).
25. H.P. Lotz, Uniform convergence of operators on L^∞ and similar spaces, Math. Z. 190 (1985), 207-220.
26. T. McConnell, On Fourier multiplier transformations of Banach-valued functions. Trans. Amer. Math. Soc. 285 (1984), 739–757.
27. A. McIntosh, Operators which have an H_∞ functional calculus. Miniconference on operator theory and partial differential equations (North Ryde, 1986), 210–231, Proc. Centre Math. Anal. Austral. Nat. Univ., 14, Austral. Nat. Univ., Canberra, 1986.
28. A. Pełczyński, Projections in certain Banach spaces, Studia Math. 19 (1960), 209-228.
29. G. Pisier, Les inégalités de Kahane-Khintchin d'après C. Borell, Séminaire sur la géometrie des éspaces de Banach, Ecole Polytechnique, Palaiseau, Exposé VII, 1977-78.
30. G. Pisier, Some results on Banach spaces without local unconditional structure, Comp. Math. 37 (1978) 3-19.
31. L. Weis, Operator-valued Fourier multiplier theorems and maximal regularity, Math. Ann. to appear.
32. M. Zippin, A remark on bases and reflexivity in Banach spaces, Israel J. Math. 6 (1968) 74-79.

Recent Progress in Functional Analysis
K.D. Bierstedt, J. Bonet, M. Maestre and J. Schmets

Derivations from Banach algebras

H. G. Dales

Department of Pure Mathematics,
University of Leeds,
Leeds, LS2 9JT, UK

To Professor Manuel Valdivia on the occasion of his seventieth birthday.

Abstract

This paper is a survey of results on the continuity and structure of derivations from a Banach algebra. We shall recall the basic definitions and some classical results, and highlight some strong, recent advances. A number of questions are raised.
MCS 2000 Primary 46G20: Secondary 46B15, 46B28, 46E15

1. Algebraic background

First we recall some basic algebra. Let A be an algebra, always taken to be complex and associative; if A has an identity, it is denoted by e_A. A linear space E is an *A-bimodule* if there are bilinear maps $(a, x) \mapsto a \cdot x$ and $(a, x) \mapsto x \cdot a$ from $A \times E$ to E such that $a \cdot (b \cdot x) = ab \cdot x$, $(x \cdot a) \cdot b = x \cdot ab$, and $a \cdot (x \cdot b) = (a \cdot x) \cdot b$ for all $a, b \in A$ and $x \in E$. For example, we can take $E = A$, and let $a \cdot x$ and $x \cdot a$ be the product in A. Again, let $E = \mathbb{C}$ and take $\varphi \in \Phi_A$, the character space of A (so that $\varphi : A \to \mathbb{C}$ is a non-zero homomorphism), and set $a \cdot z = z \cdot a = \varphi(a)z$ for $a \in A$ and $z \in \mathbb{C}$. Clearly $\mathbb{C}$ is an A-bimodule for these operations; it is denoted by $\mathbb{C}_\varphi$.

Let A be a commutative algebra. Then a symmetric A-bimodule is said to be an *A-module*. Here E is *symmetric* if $a \cdot x = x \cdot a$ $(a \in A,\ x \in E)$.

The following notion is one way of setting up an abstract version of 'differentiating a function'. Let E be an A-bimodule. Then a linear map $D : A \to E$ is a *derivation* if

$$D(ab) = a \cdot Db + Da \cdot b \quad (a, b \in A).$$

The set of these linear maps is itself a linear space: it is denoted by $Z^1(A, E)$. For example, choose $x \in E$, and define

$$\delta_x(a) = a \cdot x - x \cdot a \quad (a \in A).$$

Then it is quickly checked that δ_x is a derivation; such derivations on A are said to be *inner*. The set of inner derivations from A to E is a linear subspace of $Z^1(A,E)$, called $N^1(A,E)$.

For example, a linear functional $d : A \to \mathbb{C}_\varphi$ is a derivation if

$$d(ab) = \varphi(a)d(b) + \varphi(b)d(a) \quad (a, b \in A).$$

Functionals of this form are called *point derivations on A at the character* φ.

We now give a key definition. Let A be an algebra, and let E be an A-bimodule. Then the *first cohomology group* of the algebra A with coefficients in E is defined to be

$$H^1(A,E) = Z^1(A,E)/N^1(A,E).$$

(This is a specific example of the groups $H^n(A,E)$, which are defined for each $n \in \mathbb{N}$: these are the basic objects of study in the so-called *Hochschild cohomology theory of algebras*. See [He 1] and [Da, §1.9], for example). To say that $H^1(A,E) = \{0\}$ is just to say that every derivation from A into the A-bimodule E is inner.

One of the themes that we shall explore is when an algebra A is 'cohomologically trivial' in some sense. In pure algebra, this question asks for a characterization of the algebras A such that $H^1(A,E) = \{0\}$ for each A-bimodule E. The answer is classical; it shows that such an algebra is 'almost trivial'.

Let A be an algebra. Then the linear space $A \otimes A$ is an A-bimodule for maps $\cdot$ that satisfy the following equations:

$$a \cdot (b \otimes c) = ab \otimes c, \quad (b \otimes c) \cdot a = b \otimes ca \quad (a, b, c \in A). \tag{1.1}$$

There is a linear map $\pi_A : A \otimes A \to A$ defined by the condition that

$$\pi_A(a \otimes b) = ab \quad (a, b \in A);$$

this map is called the *induced product map*. In the case where A has an identity, a *diagonal* in $A \otimes A$ is an element $u \in A \otimes A$ with $\pi_A(u) = e_A$ and $a \cdot u = u \cdot a$ $(a \in A)$. Thus u has the form $\sum_{j=1}^n a_j \otimes b_j$, where $\sum_{j=1}^n a_j b_j = e_A$ and $\sum_{j=1}^n a a_j \otimes b_j = \sum_{j=1}^n a_j \otimes b_j a$ for each $a \in A$.

A full matrix algebra is an algebra $\mathbb{M}_n(\mathbb{C})$ of $n \times n$ matrices over $\mathbb{C}$.

Theorem 1.1 *Let A be an algebra. Then the following conditions on A are equivalent:*

(a) $H^1(A,E) = \{0\}$ *for each A-bimodule E;*

(b) *A has an identity and there is a diagonal in $A \otimes A$;*

(c) *A is finite-dimensional and semisimple;*

(d) *A is a finite direct sum of full matrix algebras.* □

The algebraic theory of derivations and the cohomology of algebras is given in [Da, §§1.8, 1.9] and many standard algebra texts, such as [Pi].

2. Continuous derivations

We now add some topology to our considerations. Thus, let $(A, \|\cdot\|)$ be a Banach algebra. Suppose that E is a Banach space for a norm $\|\cdot\|$ and that E is also an A-bimodule. Then E is said to be a *Banach A-bimodule* if

$$\|a \cdot x\| \leq \|a\| \|x\|, \quad \|x \cdot a\| \leq \|a\| \|x\| \quad (a \in A, x \in E).$$

For example, we can take $E = A$. Again, since every character on A is continuous, the one-dimensional A-bimodule $\mathbb{C}_\varphi$ is also a Banach A-bimodule. Next, let $\|\cdot\|_\pi$ be the projective norm on the space $A \otimes A$, so that, for $z \in A \otimes A$, we have

$$\|z\|_\pi = \inf \left\{ \sum_{j=1}^{n} \|a_j\| \|b_j\| : z = \sum_{j=1}^{n} a_j b_j,\, a_1 \ldots a_n, b_1, \ldots b_n \in A,\, n \in \mathbb{N} \right\}.$$

The completion of $A \otimes A$ with respect to this norm is the *projective tensor product* $A \widehat{\otimes} A$. Now $A \widehat{\otimes} A$ is a Banach A-bimodule for the maps defined in equation (1.1), and there is a continuous linear operator $\widehat{\pi}_A : A \widehat{\otimes} A \to A$ such that $\widehat{\pi}_A(a \otimes b) = ab \;\; (a, b \in A)$; this map is called the *projective induced product map.*

Finally we describe the important class of dual Banach A-bimodules. Let E be a Banach A-bimodule, with dual space E'. Then E' is also a Banach A-bimodule for the maps given by the equations:

$$\langle x, a \cdot \lambda \rangle = \langle x \cdot a, \lambda \rangle, \quad \langle x, \lambda \cdot a \rangle = \langle a \cdot x, \lambda \rangle \quad (a \in A, x \in E, \lambda \in E').$$

The bimodule E' is said to be the *dual* of E. In particular, the dual space A' is a Banach A-bimodule, called the *dual module* of A.

Let E and F be Banach spaces. The Banach space of continuous linear operators from E to F is denoted by $\mathcal{B}(E, F)$.

Let A be a Banach algebra, and let E be a Banach A-bimodule. The space of continuous derivations from A to E is denoted by $\mathcal{Z}^1(A, E)$: it is a closed subspace of $\mathcal{B}(A, E)$, and it is a subspace of $Z^1(A, E,)$. The space of continuous inner derivations is $\mathcal{N}^1(A, E)$; in fact $\mathcal{N}^1(A, E) = N^1(A, E)$ because it is clear that every inner derivation is continuous. In general $\mathcal{N}^1(A, E)$ is not a closed subspace of $\mathcal{B}(A, E)$.

We detour to give an example that will be used later. Let A be a Banach algebra, let $\varphi \in \Phi_A$, and let $d \in \mathcal{Z}^1(A, \mathbb{C}_\varphi)$, so that d is a continuous point derivation on A at the character φ. Define $D = \varphi \otimes d : A \to A'$, so that

$$Da = d(a)\varphi \quad (a \in A).$$

It is immediately checked that $D \in \mathcal{Z}^1(A, A')$. Suppose that $D \in \mathcal{N}^1(A, A')$. Then there exists $\lambda \in A'$ such that $Da = a \cdot \lambda - \lambda \cdot a \;\; (a \in A)$. We evaluate both sides of this equation at the element a of A to see that

$$d(a)\varphi(a) = \langle a, a \cdot \lambda \rangle - \langle a, \lambda \cdot a \rangle = \langle a^2, \lambda \rangle - \langle a^2, \lambda \rangle = 0.$$

It follows that $d(a) = 0$ whenever $\varphi(a) \neq 0$, and so $d = 0$. Thus D is only an inner derivation in the case where $d = 0$.

Let A be a Banach algebra, and let E be a Banach A-bimodule. Then the *first continuous cohomology group of A with coefficients in E* is defined to be

$$\mathcal{H}^1(A, E) = \mathcal{Z}^1(A, E)/\mathcal{N}^1(A, E).$$

Thus $\mathcal{H}^1(A, E)$ is always a seminormed space. (In general, the seminormed spaces $\mathcal{H}^n(A, E)$ are defined for each $n \in \mathbb{N}$: these spaces are the main object of study in the subject which is called *topological homology.*)

Here are the standard questions about derivations that arise in the above setting.

(I) When are all derivations from A into E automatically continuous? That is, when does $\mathcal{Z}^1(A, E) = Z^1(A, E)$? If this is not the case, can an arbitrary derivation $D \in Z^1(A, E)$ be written in the form $D = D_1 + D_2$, where D_1 is a continuous derivation and D_2 is a discontinuous derivation of a special form?

(II) When does $\mathcal{H}^1(A, E) = \{0\}$ for some specific Banach A-bimodule E, or for all such E in a certain class of Banach A-bimodules? If this is the case for a reasonable class of bimodules, A is said to be 'cohomologically trivial'.

(III) Can one calculate the space $\mathcal{H}^1(A, E)$?

There is an enormous literature on questions (I) and (II); a small subset of this material is contained in [Da].There are a few, but not many, examples of calculations of $\mathcal{H}^1(A, E)$ in the literature; see [DaDu] for some results.

3. Contractible algebras

The first way in which a Banach algebra A could be 'cohomologically trivial' is to require that $\mathcal{H}^1(A, E) = \{0\}$ for *all* Banach A-bimodules E. Banach algebras satisfying this condition are said to be *contractible*. This seems to be the natural analogue of the above algebraic condition. However it is likely that this condition already forces A to be finite-dimensional; certainly there are no infinite-dimensional examples in any of the major classes of Banach algebras.

In the case where A has an identity, a *projective diagonal* in $A \widehat{\otimes} A$ is an element $u \in A \widehat{\otimes} A$ such that $a \cdot u = u \cdot a \ (a \in A)$ and $\widehat{\pi}_A(u) = e_A$.

The analogue of Theorem 1.1 is the following.

Theorem 3.1 *Let A be a Banach algebra. Then the following conditions are equivalent:*

(a) $\mathcal{H}^1(A, E) = \{0\}$ *for every Banach A-bimodule E;*

(b) *A has an identity and there is a projective diagonal in $A \widehat{\otimes} A$.* □

For a proof and further equivalent conditions, see [Da, 2.8.48].

A contractible Banach algebra A is known to be finite-dimensional if A satisfies any of the following extra conditions: (i) A is commutative; (ii) A has the compact approximation property as a Banach space; (iii) A is a C^*-algebra. For a discussion and more results, see [Ru 1]. It is a nice theoretical problem to determine if a contractible Banach algebra is always finite-dimensional and semisimple, but this is not a central question.

It is clear that the class of contractible Banach algebras is too small: a much more important class of 'cohomologically trivial' Banach algebras will be described in the next section.

4. Amenable and weakly amenable algebras

In [Jo 1], Johnson introduced the class of amenable Banach algebras. In the subsequent years, it has become clear that this is a most significant and important class: the determination of the amenable algebras in diverse classes of Banach algebras has subsequently been a major theme in Banach algebra theory, throwing much light on the structure of various Banach algebras and generating many beautiful theorems.

Definition 4.1 *A Banach algebra A is* amenable *if $\mathcal{H}^1(A, E') = \{0\}$ for each Banach A-bimodule E.*

Thus we require that every continuous derivation into a dual Banach A-bimodule be inner. There are many intrinsic characterizations of amenable Banach algebras (see [Da, 2.9.65] and [He 1, VII.2.3]). We give one which is analogous to the two theorems that we have already stated. It is due to Johnson [Jo 2].

Let A be a Banach algebra. An *approximate diagonal* for A is a bounded net (u_α) in $(A \otimes A, \|\cdot\|_\pi)$ such that

$$\lim_\alpha (u_\alpha \cdot a - a \cdot u_\alpha) = 0 \quad \text{and} \quad \lim_\alpha \widehat{\pi}_A(u_\alpha)a = a$$

for each $a \in A$. A *virtual diagonal* for A is an element M of $(A \widehat{\otimes} A)''$ such that

$$\mathrm{M} \cdot a = a \cdot \mathrm{M} \quad \text{and} \quad \pi_A''(\mathrm{M}) \cdot a = a$$

for each $a \in A$. Here $\pi_A'' : (A \widehat{\otimes} A)'' \to A''$ is the second adjoint of $\widehat{\pi}_A$, and A'' and $(A \widehat{\otimes} A)''$ are the Banach A-bimodules which are the duals of A' and $(A \widehat{\otimes} A)'$, respectively.

Theorem 4.2 *Let A be a Banach algebra. Then the following conditions are equivalent:*

(a) *$\mathcal{H}^1(A, E') = \{0\}$ for every Banach A-bimodule E;*

(b) *A has an approximate diagonal in $A \otimes A$;*

(c) *A has a virtual diagonal in $(A \widehat{\otimes} A)''$.* □

There is a surprising connection between amenability and bounded approximate identities. A net (e_α) in a Banach algebra A is a *bounded left approximate identity* (BLAI) if $\sup_\alpha \|e_\alpha\| < \infty$ and $\lim_\alpha e_\alpha a = a$ for each $a \in A$. Similarly, we define a bounded right approximate identity (BRAI). The net (e_α) is a *bounded approximate identity* (BAI) it is both a BLAI and a BRAI. The famous Cohen factorization theorem (see, for example, [Da, 2.9.24] and [HR 1, (32.22)]) says that $A^{[2]} = A$ for every Banach algebra A with a BLAI or BRAI. Here $A^{[2]} = \{ab : a, b \in A\}$; later we shall use the space A^2, which is the linear span of $A^{[2]}$. Then every amenable Banach algebra has a BAI, and so $A = A^{[2]}$: we say that A *factors*. In fact, let I be a closed ideal in an amenable Banach algebra A such that I is complemented as a Banach space. Then I has a BAI and $I = I^{[2]}$, and I itself is amenable. Indeed, rather stronger results are true ([Da, §2.9]).

Let A be an algebra. Then $A^{\rm op}$ is the same algebra with the opposite multiplication. The *algebraic enveloping algebra* of A is $A \otimes A^{\rm op}$: this is the space $A \otimes A$ with a product that satisfies the rule

$$(a_1 \otimes b_1)(a_2 \otimes b_2) = a_1 a_2 \otimes b_2 b_1 \quad (a_1, a_2, b_1, b_2 \in A).$$

Let π_A be the induced product map. Then $\ker \pi_{\rm A}$ is a left ideal in $A \otimes A^{\rm op}$. Now suppose that A is a Banach algebra. The *enveloping algebra* A^e of A is defined to be the completion $A \widehat{\otimes} A^{\rm op}$ of $(A \otimes A^{\rm op}, \|\cdot\|_\pi)$, and $\ker \widehat{\pi}_{\rm A}$ is a closed left ideal in A^e.

Theorem 4.3 *Let A be a Banach algebra. Then A is amenable if and only if A has a* BAI *and* $\ker \widehat{\pi}_{\rm A}$ *has a* BRAI. □

For a proof of this result, see [He 1] and [CL].

There is a variant of the notion of amenability. A Banach algebra A is said to be *weakly amenable* if $\mathcal{H}(A, A') = \{0\}$: every continuous derivation from A into its dual module A' is inner. In the case where A is commutative, certainly $\mathcal{N}^1(A, A') = \{0\}$, and so A is weakly amenable if and only if there are no non-zero, continuous derivations from A into A'. This implies that $\mathcal{Z}^1(A, E) = \{0\}$ for every Banach A-module E.

Let A be a Banach algebra. Then the following implications are trivial.

- A amenable $\Rightarrow$ A weakly amenable;
- A weakly amenable $\Rightarrow$ A has no non-zero, continuous point derivations at any character.

The second implication follows from a remark in §2. In general, there is no converse to either of these implications, but the converses may hold in certain important classes of Banach algebras. We are interested in determining which algebras in various major classes satisfy one or more of the three conditions.

The seminal paper on the amenability of Banach algebras is [Jo 1]. Notions of amenability were also developed by Helemskii around the same time;

for an account, see [He 1] and [He 2]. The characterization of amenable Banach algebras in terms of virtual (and approximate) diagonals is in [Jo 2]. The concept of a weakly amenable Banach algebra was first defined in [BCDa]. For a characterization of weakly amenable, commutative Banach algebras, see [Gr] and [Da, 2.8.73].

5. C^*-algebras and their closed subalgebras

Let A be a commutative C^*-algebra with an identity. Then A has the form $C(\Omega)$, the algebra of all continuous, complex-valued functions on a compact space Ω (which is equal to Φ_A). The algebra $C(\Omega)$ has the uniform norm $|\cdot|_\Omega$, where

$$|f|_\Omega = \sup\{|f(x)| : x \in \Omega\} \quad (f \in C(\Omega))$$

Theorem 5.1 *Each algebra $C(\Omega)$ is amenable.* □

For a proof, see [BD, 43.12] and [Da, 5.6.2]. In fact, several different proofs are known; they are discussed in [Da, §5.6].

A *uniform algebra* is a closed subalgebra A of an algebra $C(\Omega)$ such that A contains the constants and separates the points of Ω. For example, let $A = A(\overline{\mathbb{D}})$ be the disc algebra of all continuous functions on the closed unit disc $\overline{\mathbb{D}}$ which are analytic on the open disc $\mathbb{D}$. Then A is a uniform algebra, and the map $f \mapsto f'(0)$ is a non-zero, continuous point derivation at the character $\varepsilon_0 : f \mapsto f(0)$ on A. Thus A is not weakly amenable.

We have the following theorem of Sheinberg [Sh] (see [Da, 5.6.23]).

Theorem 5.2 *Let $A \subset C(\Omega)$ be a uniform algebra. Then A is amenable if and only if $A = C(\Omega)$.* □

However, I suspect that this is not the best-possible theorem. Let A be a uniform algebra on Ω, and suppose that A is just known to be weakly amenable. Does it follow that necessarily $A = C(\Omega)$? For some partial results, see [Fe]. However an example of Wermer [Wer] shows that there are uniform algebras $A \subset C(\Omega)$ (with $\Omega = \Phi_A$) such that $A \neq C(\Omega)$, but such that there are no non-zero, continuous point derivations on A.

We now consider when an arbitrary (non-commutative) C^*-algebra has any of our three properties. This involves us in some very deep mathematics. It is easy to see that there are no non-zero, continuous point derivations on a C^*-algebra. A much deeper result is the following theorem of Haagerup [Ha]. For a proof, see [Da, 5.6.77]; this latter proof is taken from [HaL].

Theorem 5.3 *Every C^*-algebra is weakly amenable.* □

The characterization of amenable C^*-algebras is very deep, and we cannot even explain the terms that are involved. It is the work of several hands, principally Connes, Haagerup, and Effros. For an expository account, with some important simplifications of the original arguments, see [Ru 2].

Theorem 5.4 *Let A be a C^*-algebra. Then the following conditions on A are equivalent:*

(a) *A is amenable;*

(b) *A is nuclear;*

(c) *A'' is amenable as a von Neumann algebra;*

(d) *A'' is semidiscrete.* □

The intuition is that 'fairly small' C^*-algebras are amenable, but larger ones are not. Thus $\mathcal{K}(H)$, the C^*-algebra of all compact operators on a Hilbert space H, is always amenable, but $\mathcal{B}(H)$ is only amenable in the case where H is finite-dimensional. We still do not have an elementary proof of this latter result; one that is more elementary than the original, but still not very elementary, is given in [Ru 2].

A second question above asked when all derivations from various Banach algebras A into an arbitrary Banach A-bimodule are automatically continuous. It is a theorem of Ringrose (see [KR, 4.6.65]) that all derivations from an arbitrary C^*-algebra are automatically continuous: see [Da, 5.3.4] for a more general result. On the other hand there are discontinuous derivations from the disc algebra $A(\overline{\mathbb{D}})$ (despite the fact that all point derivations on this algebra are continuous) and from many other proper uniform algebras. For a construction of such discontinuous derivations, see [Da, §5.6]. However, there is one open question that we find puzzling. Let $A = A(\overline{\mathbb{D}})$ be the disc algebra, let E be a Banach A-bimodule, and let $D : A \to E$ be a derivation. The value of D on the polynomials p of the form $\alpha_0 + \alpha_1 X + \cdots + \alpha_n X^n$ is determined by the value of $D(X)$ in E. Indeed $Dp = p' \cdot D(X)$, where p' is the formal derivative of p. Is the restriction of D to the subalgebra of polynomials necessarily continuous? The known discontinuous derivations D are constructed so that $D(p) = 0$ for each polynomial p, but such that $D(\exp X) \neq 0$. Thus the answer to this is not obvious. For some partial results, see [St].

6. Commutative Banach algebras

Let Ω be a compact space. A *Banach function algebra* on Ω is a subalgebra A of $C(\Omega)$ such that A contains the constants, separates the points of Ω, and is a Banach algebra for some norm $\|\cdot\|$; necessarily $\|f\| \geq |f|_\Omega$ $(f \in A)$. The Banach funtion algebra A is a *natural* if every character on A has the form $\varepsilon_x : f \mapsto f(x)$ for some $x \in \Omega$. See [Da, §4.1].

We give some examples in the case where $\Omega = \mathbb{I}$, the closed internal [0,1]. Let $C^{(1)}(\mathbb{I})$ denote the set of functions f on $\mathbb{I}$ such that the derivative f' exists and is continuous on $\mathbb{I}$, and define

$$\|f\|_1 = |f|_{\mathbb{I}} + |f'|_{\mathbb{I}} \quad (f \in C^{(1)}(\mathbb{I})),$$

Then $(C^{(1)}(\mathbb{I}), \|\cdot\|_1)$ is a natural Banach function algebra on $\mathbb{I}$. There are obvious, non-zero, continuous point derivations on this algebra: the map

$f \mapsto f'(0)$ is such a point derivation at the character ε_0. Thus $C^{(1)}(\mathbb{I})$ is neither weakly amenable nor amenable. It is also easy to see that there are many discontinuous point derivations at each character on $C^{(1)}(\mathbb{I})$. Again, the Banach space $C(\mathbb{I})$ is a Banach $C^{(1)}(\mathbb{I})$-module for the pointwise product, and the map

$$D : f \mapsto f', \quad C^{(1)}(\mathbb{I}) \to C(\mathbb{I}),$$

is a continuous derivation.

Now take α with $0 < \alpha < 1$, and let

$$p_\alpha(f) = \sup \left\{ \frac{|f(s) - f(t)|}{|s-t|^\alpha} : s, t \in \mathbb{I},\ s \neq t \right\}$$

for a function f on $\mathbb{I}$. Define

$$\mathrm{Lip}_\alpha \mathbb{I} = \{ f : p_\alpha(f) < \infty \},$$

so that $\mathrm{Lip}_\alpha \mathbb{I}$ is a *Lipschitz space*. It is easily checked that $\mathrm{Lip}_\alpha \mathbb{I}$ is a Banach function algebra on $\mathbb{I}$ with respect to the norm $\|\cdot\|_\alpha$, where

$$\|f\|_\alpha = |f|_{\mathbb{I}} + p_\alpha(f) \quad (f \in \mathrm{Lip}_\alpha \mathbb{I}).$$

There is a subalgebra $\mathrm{lip}_\alpha \mathbb{I}$ of $\mathrm{Lip}_\alpha \mathbb{I}$: this consists of the functions f in $\mathrm{Lip}_\alpha \mathbb{I}$ such that

$$|f(s) - f(t)| \,/\, |s-t|^\alpha \to 0 \quad \text{as } |s-t| \to 0.$$

This subalgebra is also a Banach function algebra on $\mathbb{I}$; it is the closure in $\mathrm{Lip}_\alpha \mathbb{I}$ of the space of (restrictions to $\mathbb{I}$ of) polynomials. The character space of both $\mathrm{Lip}_\alpha \mathbb{I}$ and $\mathrm{lip}_\alpha \mathbb{I}$ is just $\mathbb{I}$, so that both algebras are natural. See [Da, §4.4] and [Wea]. The algebras $\mathrm{lip}_\alpha \mathbb{I}$ form a chain between $C(\mathbb{I})$ and $C^{(1)}(\mathbb{I})$: if $0 < \alpha < \beta < 1$, then

$$C^{(1)}(\mathbb{I}) \subset \mathrm{lip}_\beta \mathbb{I} \subset \mathrm{lip}_\alpha \mathbb{I} \subset C(\mathbb{I}).$$

There are no non-zero, continuous point derivations on the algebras $\mathrm{lip}_\alpha \mathbb{I}$, but there are many discontinuous point derivatations at each character. Let $M_\alpha = \{ f \in \mathrm{lip}_\alpha \mathbb{I} : f(0) = 0 \}$, a maximal ideal of $\mathrm{lip}_\alpha \mathbb{I}$. Then M_α^2 has infinite codimension in M_α, and this shows that $\mathrm{lip}_\alpha \mathbb{I}$ is not amenable. It remains to determine when $\mathrm{lip}_\alpha \mathbb{I}$ is weakly amenable. The answer is a result of [BCDa]; see [Da, 5.6.14].

Theorem 6.1 *Let* $\alpha \in (0,1)$. *Then* $\mathrm{lip}_\alpha \mathbb{I}$ *is weakly amenable if and only if* $\alpha \leq 1/2$. □

Thus, if $\mathrm{lip}_\alpha \mathbb{I}$ is 'sufficiently close' to $C^{(1)}(\mathbb{I})$ (i.e., if $\alpha > 1/2$), there is a non-zero, continuous derivation on $\mathrm{lip}_\alpha \mathbb{I}$, and so the functions in this algebra have some residual 'differentiability properties'.

Now let us consider derivations on a general commutative Banach algebra A. The (Jacobson) radical of an algebra A is denoted by $\mathrm{rad}\, A$; the algebra is *semisimple* if $\operatorname{rad} A = \{0\}$ and *radical* if $\operatorname{rad} A = A$. For each Banach algebra A, $\operatorname{rad} A$ is a closed ideal in A; for a commutative Banach algebra A, $\operatorname{rad} A$ consists of the quasi-nilpotent elements of A, that is, the elements $a \in A$ such that $\|a^n\|^{1/n} \to 0$ as $n \to \infty$.

The intuition is that the range of a derivation on a commutative Banach algebra A is necessarily 'small'. This is confirmed by the following theorem.

Theorem 6.2 *Let $D : A \to A$ be a derivation on a commutative Banach algebra A. Then $D(A) \subset \operatorname{rad} A$.* □

This theorem, in the case where D is assumed to be continuous, is due to Singer and Wermer [SW]; see [BD, 18.16] and [Da, 2.7.20]. It is a very much deeper result that the theorem also holds in the case where D may be discontinuous: this is the achievement of Thomas [Th]. See [Da, 5.2.48] for a full proof.

A famous open question is to seek a non-commutative version of Theorem 6.2. The conjecture is that $D(P) \subset P$ for each primitive ideal P in a Banach algebra A and each derivation D on A. Partial results towards this conjecture and various equivalent formulations are given in [Da, §5.2].

We return to commutative Banach algebras. It seems from examples that an amenable commutative Banach algebra A should be 'close to $C(\Phi_A)$' in some sense. A specific form of this feeling was the conjecture, open for many years, that there could be no non-zero, commutative, radical, amenable Banach algebra. However this conjecture has recently been refuted by Read [Re] with a dramatic new example that may open a door to new vistas in the theory of commutative, radical Banach algebras.

Theorem 6.3 *There is a non-zero, commutative, radical, amenable Banach algebra.* □

For an exposition of this construction, see [Da, §5.6].

There remains much that is unknown about commutative Banach algebras which are amenable. For example, it is open whether or not there is such an algebra (other than $\mathbb{C}$) which is an integral domain. This takes us close to perhaps the hardest question in Banach algebra theory: is there a commutative Banach algebra A (other than $\mathbb{C}$) such that the only closed ideals of A are $\{0\}$ and A? Such an algebra is said to be *topologically simple*.

It is known that, if there is a primitive ideal P in a Banach algebra A and a derivation D on A such that $D(P) \not\subset P$, then there is a topologically simple Banach algebra, but there has been no progress following this line of development for many years.

7. Group algebras

We now consider the answers to our various questions for several classes of group algebras.

Let G be a locally compact group, with left Haar measure $m = m_G$. The *group algebra* of G is $L^1(G)$; this is the set of (equivalence classes of) complex-valued, measurable functions f on G such that

$$\|f\|_1 = \int_G |f(t)| \,\mathrm{d}m(t) < \infty,$$

taken with the convolution product $\star$, where

$$(f \star g)(t) = \int_G f(s)g(s^{-1}t)\,\mathrm{d}m(s) \quad (f, g \in L^1(G)).$$

We obtain the much-studied Banach algebra $(L^1(G), \star, \|\cdot\|)$; this algebra is commutative if and only if the group G is abelian.

A larger algebra than $L^1(G)$ is the *measure algebra* $M(G)$, which consists of all complex-valued, regular Borel measures on G. Let $\mu \in M(G)$. Then the total variation of μ is denoted by $|\mu|$, and $\|\mu\| = |\mu|(G)$. Thus $(M(G)), \|\cdot\|)$ is a Banach space. For $\mu, \nu \in M(G)$, we define $\mu \star \nu$ by the formula

$$(\mu \star \nu)(E) = \int_G \nu(x^{-1}E)\,\mathrm{d}\mu(x)$$

for each Borel set E in G. With respect to this product, $M(G)$ is a Banach algebra. As a Banach space, $M(G) = C_0(G)'$, and the product $\mu \star \nu$ of $\mu, \nu \in M(G)$ is the element of $C_0(G)'$ such that

$$\langle f, \mu \star \nu \rangle = \int_G \int_G f(st)\,\mathrm{d}\mu(s)\,\mathrm{d}\nu(t) \quad (f \in C_0(G)).$$

Let δ_s denote the point mass at s for $s \in G$. Then $M(G)$ contains the *discrete measures* of the form $\mu = \sum_{s \in G} \alpha_s \delta_s$, where $\|\mu\| = \sum_{s \in G} |\alpha_s| < \infty$. The set of discrete measures is denoted by $M_d(G)$, and it is identified with $\ell^1(G)$. Clearly $M_d(G)$ is a closed subalgebra of $M(G)$, and δ_{e_G} is the identity of $M(G)$. In the case where the group G is discrete (as a topological space), we have $M(G) = \ell^1(G)$. The group algebra $L^1(G)$ has an identity if and only if the group G is discrete; however $L^1(G)$ always has a BAI (of bound equal to 1), and so each algebra $L^1(G)$ factors.

A measure μ is *continuous* if $\mu(s) = 0$ for each $s \in G$. The set of continuous measures in $M(G)$ is denoted by $M_c(G)$: this is a closed ideal in $M(G)$, and we have

$$M(G) = M_d(G) \oplus M_c(G) = \ell^1(G) \oplus M_c(G).$$

In the case where G is not discrete, $M_c(G) \neq \{0\}$.

The algebra $L^1(G)$ is identified (via the Radon-Nikodym theorem) with the closed ideal $M_{ac}(G)$ of *absolutely continuous* measures in $M(G)$. Indeed, for $f \in L^1(G)$, define $\mu_f \in M(G)$ by

$$\langle g, \mu_f \rangle = \int_G f(s)g(s)\,\mathrm{d}m(s) \quad (g \in C_0(G));$$

the map $f \mapsto \mu_f$ is the required idenification.

For a full description of these algebras, see [HR 1] and [Da, §3.3].

There is always one character on $M(G)$. This is the *augmentation character*

$$\varphi : \mu \mapsto \int_G \mathrm{d}\mu(s) = \mu(G), \quad M(G) \to \mathbb{C}.$$

For example, let $\mu = \sum_{s \in G} \alpha_s \delta_s \in \ell^1(G)$. Then $\varphi(\mu) = \sum_{s \in G} \alpha_s$. The intersection of $\ker \varphi$ with $L^1(G)$ gives a closed, maximal modular ideal $L_0^1(G)$ of codimension one in $L^1(G)$. This is the *augmentation ideal* of $L^1(G)$, so that

$$L_0^1(G) = \left\{ f \in L^1(G) : \int_G f(s) \,\mathrm{d}m(s) = 0 \right\}.$$

In the case where G is discrete, there may be no other characters on $M(G)$. (If G is abelian, there are many such characters, and they form the dual group $\widehat{G}$ of G.) However, if G is not discrete, there is another character, which we now describe: it will be required later.

The construction of this new character makes sense in a more general context. Let A be an algebra such that A has a subalgebra B and an ideal I and is such that A is the linear space direct sum of B and I; in this case A is the *semidirect product* of B and I, written $A = B \ltimes I$. Let $\pi : A \to A/I = B$ be the quotient map, and let $\varphi \in \Phi_B$. Then $\widetilde{\varphi} = \varphi \circ \pi$ is a character on A. Let λ be a linear functional on I, and extend λ to a linear functional on A by requiring that $\lambda \mid B = 0$. We ask when λ is a point derivation on A at $\widetilde{\varphi}$. A little calculation shows that this is the case if

$$\lambda \mid I^2 = 0, \quad \lambda(bx) = \lambda(xb) = \varphi(b)\lambda(x) \quad (b \in B,\, x \in I). \tag{7.2}$$

We shall apply this in the case where $A = M(G), B = \ell^1(G)$, and $I = M_c(G)$, so that $A = B \ltimes I$. Now φ is taken to be the augmentation character on B, and the corresponding character $\widetilde{\varphi}$ is the *discrete augmentation character.*

The most dramatic theorem about the amenability of $L^1(G)$ is the seminal result of Johnson from [Jo 1]. Recall that a locally compact group G is *amenable* if there is a continuous linear functional Λ on $L^\infty(G)$ such that $\langle 1, \Lambda \rangle = \|\Lambda\| = 1$ and Λ is translation-invariant. Abelian groups and compact groups are amenable; the standard example of a discrete, non-amenable group is $\mathbb{F}_2$, the free group on two generators.

Theorem 7.1 *Let G be a locally compact group. Then the following conditions are equivalent.*

(a) *the Banach algebra $L^1(G)$ is amenable;*

(b) *the locally compact group G is amenable;*

(c) *the augmentation ideal $L_0^1(G)$ has a* BLAI. □

It was this theorem that suggested the name 'amenable' for the class of Banach algebras that we are discussing. For an exposition of this result, see [Da, 5.7.42].

At a later stage, Johnson also proved that $L^1(G)$ is always weakly amenable. A shorter proof of this result is due to Despic and Ghahramani [DG]; see [Da, 5.6.48].

Theorem 7.2 *Let G be a locally compact group. Then $L^1(G)$ is weakly amenable.* □

In particular, by taking G to be a non-amenable locally compact group, we obtain a group algebra $L^1(G)$ such that $L^1(G)$ is weakly amenable, but not amenable.

Now consider the augmentation ideal $L_0^1(G)$. It is clear from remarks that we have made that $L_0^1(G)$ is also amenable if and only if G is an amenable group. What about the weak amenablility of $L_0^1(G)$? It was suspected by considering Theorem 7.2 that this Banach algebra would always be weakly amenable, and this was proved for many groups in [GrL]. However a recent example of Johnson and White [JoW] shows that, for the group $G = SL(2, \mathbb{R})$, the augmentation ideal $L_0^1(G)$ is *not* weakly amenable. It is surprising that whether or not an algebra is weakly amenable can be changed by moving to a closed ideal that has codimension just 1 in the larger algebra.

Let A be a Banach algebra. As well as considering continuous derivations from A into the dual module A', one can consider derivations from A into its nth dual space $A^{\{n\}}$: the algebra A is *n-weakly amenable* if $\mathcal{H}^1(A, A^{\{n\}}) = \{0\}$. For a study of this notion, see [DaGGr]. It is trivial to see that an algebra which is $(n+2)$-weakly amenable is always n-weakly amenble, but the converse is not true: an example of a 1-weakly amenable Banach algebra which is not 3-weakly amenable has been given recently by Zhang [Z]. It is proved in [DaGGr] that $L^1(G)$ is always $(2n-1)$-weakly amenable for $n \in \mathbb{N}$, but we do not know about the 'even' dual spaces. In particular, we would like to prove that $L^1(G)$ is always 2-weakly amenable. See [Jo 3] for a recent strong partial result: the result is trivial for amenable groups G, and Johnson proves it for all free groups, so there are not many groups for which the question remains open.

A question apparently closely related to that of the amenability of $L(G)$ is whether or not $\mathcal{H}^1(L^1(G), M(G)) = \{0\}$ for every group G. (Recall that $M(G) = C_0(G)'$). This is clearly equivalent to the following question.

• Let G be a locally compact group, and let D be a continuous derivation on the group algebra $L^1(G)$. Does there necessarily exist a measure μ in $M(G)$ such that $Df = \mu \star f - f \star \mu$ for each f in $L^1(G)$?

This question lies at the very beginning of the cohomology theory of Banach algebras; it suggested the theory of amenable Banach algebras to Barry Johnson. The question is still open. It is known [Jo 1] to have a positive answer in many special cases: for example, this is so whenever G is an amenable group, whenever G is a discrete group, and whenever G is a so-called SIN group. See [Da, §5.6] for proofs of these and other results. The following advance has recently been announced by Johnson [Jo 4].

Theorem 7.3 *Let G be a connected locally compact group. Then*

$$\mathcal{H}^1(L^1(G)), M(G) = \{0\}\,.$$

□

We do know explicit groups G for which this question remains open.

Here is the second, substantial question about 'automatic continuity' and group algebras.

• Characterize those locally compact groups G such that every derivation from $L^1(G)$ into an arbitrary Banach $L^1(G)$-bimodule is automatically continuous.

It is a basic *conjecture* that this is true for all locally compact groups. Study of this conjecture has forced us to consider more deeply than heretofore the structure of the algebras $L^1(G)$. For example, let $J = L_0^1(G)$, the augmentation ideal. We should like to know when $J = J^{[2]}$ or $J = J^2$. Certainly $J = J^{[2]}$ in the case where G is amenable, for then J is itself amenable. It is always true that $J = J^2$ [Wi 1], but we need a stronger result to deal with the above continuity question. Such a result has recently been established by Willis [Wi 2] in a major study of the structure of $L^1(G)$, as follows. This work is a striking example of the process whereby a specific question on automatic continuity leads to new insight on the structure of $L^1(G)$: the proof involves the probability theory and 'random walks on groups'.

Theorem 7.4 *Let G be a σ-compact, locally compact group, and let K be a closed ideal of finite codimension in $L^1(G)$. Then there are a closed left ideal L with a* BRAI *and a closed right ideal R with a* BLAI *such that $K = L + R$.*
□

A consequence of this result is the following. Let K be as in the theorem, and let (f_n) be a sequence in K such that $f_n \to 0$. Then there exist sequences (g_n) and (h_n) in K such that $g_n \to 0$ and $h_n \to 0$ and elements h_0 and g_0 in K such that

$$f_n = g_n \star h_0 + g_0 \star h_n \quad (n \in \mathbb{N}).$$

This is a key piece of information for an attack on the above conjecture. Nevertheless, it is not in itself sufficient, and the conjecture has not been resolved in general. Partial results, mostly due to Johnson and Willis, are given below; see [Da, §5.7] for further details.

Theorem 7.5 *Let G be a locally compact group. Then all derivations from the group algebra $L^1(G)$ are continuous in the following cases:*

(i) *G is abelian;*

(ii) *G is compact:*

(iii) *G is discrete and locally finite;*

(iv) *G is soluble;*

(v) *G is connected.* □

The question is open in the case where G is $\mathbb{F}_2$, the free group on two generators.

8. Measure algebras

For rather a long time, the characterization of the locally compact groups such that $M(G)$ is amenable was left open. The point that remained unproved was to establish that $M(G)$ is not amenable in the case where G is not discrete. Recently we have resolved this question in a strong form in joint work with F. Ghahramani and A. Ya. Helemskii.

Let G be a non-discrete locally compact group. We shall show that $M(G)$ always has a non-zero, continuous point derivation. This was shown in the case where G is abelian by Brown and Moran [BM]; see also [GM, 8.5.3]. Remarks in §4 show that it is sufficient to show that there is a non-zero, continuous point derivation at a character of $M(G)$; the character that we utilize is the discrete augmentation character of §7.

Throughout this section, G is a non-discrete locally compact group; we set $B = \ell^1(G)$ and $I = M_c(G)$, so that $M(G) = B \ltimes I$.

The first step is to construct a 'Cantor-type' set in G in the case where H is a non-discrete, metrizable, locally compact group.

Lemma 8.1 *There is a decreasing nest $(K_n : n \in \mathbb{N})$ of compact subsets of H such that the following hold for each $n \in \mathbb{N}$:*

(1)K_n is the disjoint union of 4^n sets $K_n(i_1 \dots, i_n)$, where each i_j belongs to $\{1,2,3,4\}$;

(2) $K_{n+1}(i_1, \dots, i_{n+1}) \subset K_n(i_1 \dots, i_n)$ in each case;

(3) the diameter of each set $K_n(i_1 \dots, i_n)$ is at most 2^{-n} and each such set has non-empty interior;

(4) $x_1x_2^{-1}x_3x_4^{-1} \neq e_H$ whenever x_1, x_2, x_3, x_4 belong to four distinct sets of the form $K_n(i_1 \dots, i_n)$.

Proof Start with a non-empty, open subset U of H such that $\overline{U}$ is compact, and set $V = U \times U \times U \times U$. For each permutation σ of the set $\{1,2,3,4\}$, define

$$T_\sigma = \left\{(x_1, x_2, x_3, x_4) \in V : x_{\sigma(1)} = x_{\sigma(2)}\right\},$$

$$S_\sigma = \left\{(x_1, x_2, x_3, x_4) \in V : x_{\sigma(1)}x_{\sigma(2)}^{-1}x_{\sigma(3)}x_{\sigma(4)}^{-1} = e_H\right\}.$$

Each of these sets is closed and nowhere dense in G, and so their union S is also closed and nowhere dense. Set $W = V \setminus S$, so that W is open and dense in H. Take $(a_1, a_2, a_3, a_4) \in W$. Then $a_i \neq a_j$ whenever $i \neq j$. Next, take $r > 0$ such that the four open balls $B(a_i; r)$ in G are pairwise disjoint and such that $\prod_{i=1}^4 B(a_i; r) \subset W$. Set $K_1(i) = \overline{B(a_i; r/2)}$ for $i = 1,2,3,4$, and set $K_1 = \bigcup_{i=1}^4 K_1(i)$.

This is the first step of the construction. We continue in an obvious way. It is easy to check that we obtain all the required properties. □

We define $K = \bigcap\{K_n : n \in \mathbb{N}\}$, so that K is a non-empty, totally disconnected, perfect, compact set in G: in fact, K has cardinality at least $2^{\aleph_0}$, and K is a Cantor set.

It is clear that $x_1 x_2{}^{-1} x_3 x_4^{-1} \neq e_H$ whenever x_1, x_2, x_3, x_4 are four distinct points of K.

There is a positive measure μ_0 in $M_c(H)$ such that $\|\mu_0\| = \mu_0(K) = 1$.

The above proof required H to be metrizable. Now suppose that G is an arbitrary non-discrete locally compact group. Then results about topological groups to be found in [HR 1] show that G has a closed subgroup G_1 such that G_1 is separable and $H := G_1/N$ is a non-discrete, metrizable group. We construct a compact set K in H as above, and try to lift K to a subset of G. In fact, by using a 'Borel cross-section theorem', we find a Borel subset V of G such that $x_1 x_2^{-1} x_3 x_4^{-1} \neq e_G$ whenever x_1, x_2, x_3, x_4 are four distinct points of V. (We cannot obtain a compact set with this property, but this does not matter.) The measure μ_0 is transferred to a positive measure, also called μ_0, in $M_c(G)$ with $\|\mu_0\| = \mu_0(V) = 1$.

The following key step is an elementary combinatorical argument.

Lemma 8.2 *The set $xV \cap yV$ contains at most three points whenever x and y are distinct points of G.*

Proof We can suppose that $y = e_G$ and that $xV \cap V \neq \emptyset$. Choose an element $x_1 \in xV \cap V$, say $x_1 = x x_2$ with $x_2 \in K$, so that $x_1 \neq x_2$. Assume towards a contradiction that there exists $x_3 \in xK \cap V$ such that x_3 is not equal to any of the three points x_1, x_2, or $x x_1$, say $x_3 = x x_4$ for some $x_4 \in V$ with $x_4 \neq x_3$. We check that $x_4 \neq x_1$ and that $x_4 \neq x_2$. Thus x_1, x_2, x_3, x_4 are four distinct points of K. But $x_1 x_2^{-1} x_4 x_3^{-1} = x x^{-1} = e_G$, a contradiction. The result follows. □

Lemma 8.3 *Take $\mu \in I = M_c(G)$, and set $E(\mu) = \{x \in G : |\mu|(xV) > 0\}$. Then $E(\mu)$ is a countable set.*

Proof It is sufficient to show that, for each $k \in \mathbb{N}$, the set

$$F := \{x \in G : |\mu|(xV) > 1/k\}$$

is finite. Assume towards a contradiction that this is false, and take a set $\{y_n : n \in \mathbb{N}\}$ of distinct points of F. Since $|y_m V \cap y_n V| \leq 3$, the set $W := \bigcup \{y_m V \cap y_n V : m \neq n\}$ is countable, and so $|\mu|(W) = 0$ because the measure μ is continuous. The sets $y_n W \setminus W$ for $n \in \mathbb{N}$ are pairwise disjoint and $|\mu|(y_n W \setminus W) > 1/k$ for each $n \in \mathbb{N}$. This is not possible because $|\mu|(G) < \infty$. Thus the result holds. □

Define $\Lambda \in I'$ by setting $\langle \mu, \Lambda \rangle = \mu(V) \quad (\mu \in I)$. It is clear that $\Lambda \neq 0$ because $\langle \mu_0, \Lambda \rangle = \mu_0(V) = 1$. Indeed $\|\Lambda\| = 1$.

Lemma 8.4 *Let Λ be as above. Then $\Lambda \,|\, I^2 = 0$.*

Proof Take $\mu, \nu \in I$. Then

$$\begin{aligned}\langle \mu \star \nu, \Lambda \rangle &= \int_G \int_G \chi_V(st)\, \mathrm{d}\mu(s)\, \mathrm{d}\mu(t) \\ &= \int_G \nu(x^{-1}V)\, \mathrm{d}\mu(x) = \int_{E(\nu)^{-1}} \nu(x^{-1}V)\, \mathrm{d}\mu(x),\end{aligned}$$

and so $|\langle \mu \star \nu,\, \Lambda \rangle| \leq \|\nu\|\, |\mu|\, (E(\nu)^{-1})$. But $|\mu|\, (E(\nu)^{-1} = 0$ because $E(\nu)^{-1}$ is countable and μ is continuous. The result follows □

We note that we have so far shown that $\overline{I^2} \neq I$. In fact, $\overline{I^2}$ has infinite codimension in I.

We extend Λ to an element of $M(G)$ by requiring that $\Lambda \mid B = 0$.

We need to modify Λ to make it 'translation-invariant'. For $\lambda \in M(G)'$ and $s, t \in G$, we define

$$\langle \mu,\, s \cdot \Lambda \cdot t \rangle = \langle \delta_t \star \mu \star \delta_s,\, \lambda \rangle \quad (\mu \in M(G)).$$

The functional λ is *translation-invariant* if $s \cdot \lambda \cdot t = \lambda$ for each $s, t \in G$.

In fact, $M(G)'$ has the form $C_0(G)'' = C(X)$ for a certain extremely disconnected, compact space X, and $M_{\mathbb{R}}(G)$ is isotonic to $C(X, \mathbb{R})$ when these spaces have their obvious partial orders. The space $C(X, \mathbb{R})$ is a boundedy complete lattice in the sense that any subset which is bounded above has a supremum. By using this, we modifty Λ in the way that we require.

Let Λ be as above. As a subset of $C(X, \mathbb{R})$, the set $\{s \cdot \Lambda \cdot t : s, t \in G\}$ is bounded above by the constant function 1. Then we can define

$$d = \sup \{s \cdot \Lambda \cdot t : s, t \in G\}\ ,$$

taking the supremum in $C(X, \mathbb{R})$ and transferring it to $M_{\mathbb{R}}(G)'$. Clearly $d \in M(G)'$, and it is easy to check that $d\,|B = 0$ and that $d\,|I^2 = 0$. Let $\nu = \sum_{s \in G} \alpha_s \delta_s$ belong to B, and let $\mu \in I$. Then

$$d(\mu \star \nu) = \sum_{s \in G} \alpha_s \langle \delta_s \star \mu, d \rangle = \sum_{s \in G} \alpha_s \langle \mu, d \cdot \delta_s \rangle = \sum_{s \in G} \alpha_s \langle \mu, d \rangle = \varphi(\nu) d(\mu),$$

and similarly, $d(\nu \star \mu) = \varphi(v) d(\mu)$. It follows from our earlier remarks (see (7.2)) that d is a point derivation at the discrete augmentation character. Finally $d \neq 0$ because $\Lambda \neq 0$.

Theorem 8.5 *Let G be a locally compact group. Then the following conditions are equivalent:*

(a) *G is discrete;*

(b) *$M(G)$ is weakly amenable;*

(c) *there is a non-zero, continuous point derivation on $M(G)$.* □

Theorem 8.6 *Let G be a locally compact group. Then the following conditions are equivalent:*

(a) *G is discrete and amenable;*

(b) *$M(G)$ is amenable.* □

The above results, some related theorems, and some open questions are contained in [DaGH 1] and [DaGH 2].

REFERENCES

[BDa] W. G. Bade and H. G. Dales, Discontinuous derivations from Banach algebras of power series, Proc. London Math. Soc. (3), 59 (1989), 133–52.

[BCDa] W. G. Bade, P. C. Curtis, Jr., and H. G. Dales, Amenability and weak amenability for Beurling and Lipschitz algebras, Proc. London Math. Soc. (3), 55 (1987), 359–77.

[BD] F. F. Bonsall and J. Duncan, Complete normed algebras, Springer-Verlag, New York, 1973.

[CL] P. C. Curtis, Jr. and R. J. Loy, The structure of amenable Banach algebras, J. London Math. Soc. (2), 40 (1989), 89-104.

[BM] G. Brown and W. Moran, Point derivations on $M(G)$, Bull. London Math. Soc., 8 (1976), 57-64.

[Da] H. G. Dales, Banach algebras and automatic continuity. London Mathematical Society Monographs, Vol. 24, Clarendon Press, Oxford, 2001.

[DaDu] H. G. Dales and J. Duncan, Second-order cohomology of some semigroup algebras. In Banach algebras '97 (ed. E. Albrecht and M. Mathieu), Walter de Gruyter, Berlin, 1998, 101–117.

[DaGH 1] H. G. Dales, F. Ghahramani, and A. Ya. Helemskii, Amenability of the measure algebra, preprint, 2000.

[DaGH 2] H. G. Dales, F. Ghahramani, and A. Ya. Helemskii, Point derivations on measure algebras, preprint 2000.

[DaGGr] H. G. Dales, F. Ghahramani, and N. Grønbæk, Derivations into iterated duals of Banach algebras, Studia Math., 128 (1998), 19–54.

[DG] M. Despić and F. Ghahramani, Weak amenability of group algebras of locally compact groups, Canadian Math. Bulletin, 37 (1994), 165–7.

[Fe] J. F. Feinstein, Weak (F)-amenability of $R(K)$, Proc. Centre for Mathematical Analysis, Australian National University, 21 (1989), 97-125.

[GM] C. C. Graham and O. C.. McGehee, Essays in commutative harmonic analysis, Springer-Verlag, Berlin, 1979.

[Gr] N. Grønbæk, A characterization of weakly amenable Banach algebras, Studia Math., 94 (1989), 149–62.

[GrL] N. Grønbæk, and A. T.-M. Lau, On Hochschild cohomology of the augmentation ideal of a locally compact group, Math. Proc. Cambridge Philos. Soc., 126 (1999), 139–48.

[Ha] U. Haagerup, All nuclear C^*-algebras are amenable, Inventiones Math., 74 (1983), 305–19.

[HaL] U. Haagerup and N. J. Laustsen, Weak amenability of C^*-algebras and a theorem of Goldstein. In Banach algebras '97, (ed. E. Albrecht and M. Mathieu), Walter de Gruyter, Berlin, 1998, 223-243.

[He 1] A. Ya. Helemskii, The homology of Banach and topological algebras, Kluwer Academic Publishers, Dordrecht, 1989.

[He 2] A. Ya. Helemskii, Banach and locally convex algebras, Clarendon Press, Oxford, 1993.

[HR 1] E. Hewitt and K. A. Ross, Abstract harmonic analysis, Volume I: Structure of topological groups. Integration theory, group representations, Springer-Verlag, Berlin, 1979.

[Jo 1] B. E. Johnson, Cohomology in Banach algebras, Memoirs American Math. Soc., 127, 1972.

[Jo 2] B. E. Johnson, Approximate diagonals and cohomology of certain annihilator Banach algebras, American J. Math., 94 (1972), 685–98.

[Jo 3] B. E. Johnson, Permanent weak amenability of group algebras of free groups, Bull. London Math. Soc., 31, (1999), 569–73.

[Jo 4] B. E. Johnson, The derivation problem for group algebras of connected locally compact groups, preprint, 2000.

[JoW] B. E. Johnson and M. C. White, in preparation.

[KR] R. V. Kadison and J. R. Ringrose, Fundamentals of the theory of operator algebras, Volume 1, Elementary theory, Academic Press, New York and London, 1983.

[Pi] R. S. Pierce, Associative algebras, Springer-Verlag, New York, 1982.

[Re] C. J. Read, Commutative, radical amenable Banach algebras, Studia Math., 140 (2000), 199–212.

[Ru 1] V. Runde, The structure of contractible and amenable Banach algebras. In Banach algebras '97, (ed. E. Albrecht and M. Mathieu), Walter de Gruyter, Berlin, 1998, 415–430.

[Ru 2] V. Runde, Lectures on amenability, in preparation.

[Sh] M. V. Sheinberg, A characterization of the algebra $C(\Omega)$ in terms of cohomology groups, Uspekhi. Mat. Nauk, 32, (1959), 203–204.

[SW] I. M. Singer and J. Wermer, Derivations on commutative normed algebras, Math. Annalen, 129 (1955), 260–264.

[St] H. Steiniger, Derivations which are unbounded on the polynomials. In Banach algebras '97 (ed. E. Albrecht and M. Mathieu), Walter de Gruyter, Berlin, 1998, 461–474.

[Th] M. Thomas, The image of a derivation is contained in the radical, Annals of Maths. (2), 128 (1988), 435–460.

[Wi 1] G. A. Willis, The continuity of derivations from group algebras :

factorizable and connected groups, J. Australian Math. Soc. (Series A), 52 (1992), 185–204.

[Wea] N. Weaver, Lipschitz algebras. World Scientific, Singapore, 1999.

[Wer] J. Wermer, Bounded point derivations on certain Banach algebras, J. Functional Analysis, 1 (1967), 28–36.

[Wi 2] G. A. Willis, Factorization in finite-codimensional ideals of group algebras, Proc. London Math. Soc, (3), 82 (2001).

[Z] Y. Zhang, Weak amenability of module extensions of Banach algebras, preprint 2000.

Recent Progress in Functional Analysis
K.D. Bierstedt, J. Bonet, M. Maestre and J. Schmets

Homomorphisms of Uniform Algebras

T.W. Gamelin

Mathematics Department, U.C.L.A., 405 Hilgard Avenue, Los Angeles, CA 90095-1555, USA

To Professor Manuel Valdivia on the occasion of his seventieth birthday.

Abstract

The aim of this paper is to describe some fairly recent results on homomorphisms of uniform algebras. It includes an exposition of results of Udo Klein obtained in his thesis, and some work of the author on algebras of analytic functions on domains in the plane.
MCS 2000 Primary 46J10, 46J15

1. Unital Homomorphisms and Composition Operators

A *uniform algebra* is a closed subalgebra A of the complex algebra $C(K)$ that contains the constants and separates points. Here K is a compact Hausdorff space, and A is endowed with the supremum norm inherited from $C(K)$. We denote the spectrum (maximal ideal space) of A by M_A. We may regard $f \in A$ as a continuous complex-valued function on M_A, by identifying f with its Gelfand transform.

We consider a homomorphism $T : A \to B$ from a uniform algebra A to another uniform algebra B. Thus T is a continuous linear transformation from A to B that is multiplicative, $T(fg) = T(f)T(g)$ for $f,\ g \in A$. We will focus on unital homomorphisms, that is, homomorphisms that satisfy

$$T(1) = 1. \tag{1}$$

This guarantees that T is nontrivial, and in fact that $||T|| = 1$.

We denote by ϕ the restriction of the adjoint $T^* : B^* \to A^*$ to the spectrum M_B of B. Since T is multiplicative, ϕ maps M_B to M_A, and T coincides with the composition operator

$$T_\phi(f) = f \circ \phi, \qquad f \in A. \tag{2}$$

Thus any unital homomorphism T can be regarded as a composition operator T_ϕ, and conversely, any composition operator from A to B is a unital homomorphism.

There are a number of natural questions about T_ϕ that arise. When is T_ϕ compact? When is T_ϕ weakly compact? When is T_ϕ completely continuous?

Since an operator is compact if and only if its adjoint is compact, it is not difficult to prove the following.

Theorem 1. *A unital homomorphism T_ϕ from A to B is compact if and only if $\phi(M_B)$ is a norm-compact subset of A^*.*

There is an analogous characterization of weakly compact homomorphisms.

Theorem 2. *A unital homomorphism T_ϕ from A to B is weakly compact if and only if $\phi(M_B)$ is a weakly compact subset of A^*, that is, $\phi(M_B)$ is A^{**}-compact.*

In the case that $T_\phi : A \to A$ is a unital homomorphism from A to itself, we may ask: What is the spectrum of T_ϕ? What are the eigenvalues and eigenfunctions of T_ϕ? In this case, observe that 1 is always an eigenvalue of T_ϕ, with eigenfunction the constant function $f \equiv 1$. If $T_\phi(f_j) = \lambda_j f_j$ for $1 \leq j \leq n$, then $T_\phi(f_1 \cdots f_n) = \lambda_1 \cdots \lambda_n f_1 \cdots f_n$. Thus $\lambda_1 \cdots \lambda_n$ is an eigenvalue providing $f_1 \cdots f_n \neq 0$, and we have the following.

Lemma. *Suppose that the uniform algebra A is an integral domain. Then the eigenvalues of a unital homomorphism T_ϕ of A to itself form a unital multiplicative semigroup of complex numbers.*

2. Homomorphisms of the Disk Algebra

We denote by $\mathbf{D}$ the open unit disk in the complex plane, $\mathbf{D} = \{|z| < 1\}$. The prototypical uniform algebra is the disk algebra $A(\mathbf{D})$, which consists of the analytic functions on $\mathbf{D}$ that extend continuously to the boundary circle $\partial\mathbf{D}$. The unital homomorphisms T_ϕ of $A(\mathbf{D})$ are composition operators corresponding to continuous functions ϕ from $\overline{\mathbf{D}}$ to $\overline{\mathbf{D}}$ that are analytic on D. Each such function ϕ, excepting the identity $\phi(z) \equiv z$, has a unique fixed point $z_0 \in \overline{\mathbf{D}}$, that is, $\phi(z_0) = z_0$. If $\phi(z) \equiv z_0$ is constant, then T_ϕ is one-dimensional, its spectrum is $\sigma(T_\phi) = \{0, 1\}$, and there is not much to say.

Suppose that $\phi(z)$ is neither the identity nor constant, and suppose that the fixed point z_0 of $\phi(z)$ lies in $\mathbf{D}$. The eigenvalue equation $T_\phi f = \lambda f$ becomes Schröder's equation,

$$f(\phi(z)) = \lambda f(z), \qquad z \in \mathbf{D}, \tag{3}$$

which arises in the study of complex dynamical systems. A solution f of Schröder's equation satisfying $f(z_0) = 0$ and $f'(z_0) \neq 0$ conjugates the map ϕ in some neighborhood of the fixed point z_0 to the complex dilation $w \to \lambda w$ near $w = 0$. If there is such an f, the corresponding eigenvalue of T_ϕ is necessarily $\lambda = \phi'(z_0)$.

Compact homomorphisms of $A(\mathbf{D})$ and their spectra are characterized as follows.

Theorem 3. *Let T_ϕ be a unital homomorphism of $A(\mathbf{D})$, where $\phi : \overline{\mathbf{D}} \to \overline{\mathbf{D}}$ is not the identity. If the fixed point z_0 of ϕ belongs to $\partial\mathbf{D}$, then T_ϕ is compact if and only if $\phi(z) \equiv z_0$ is constant. If the fixed point z_0 of ϕ belongs to $\mathbf{D}$, then T_ϕ is compact if and only if $\phi(\mathbf{D})$ is a relatively compact subset of $\mathbf{D}$.*

In the case that T_ϕ is compact, the iterates of ϕ converge uniformly to the fixed point z_0. As we shall see shortly, this has been generalized considerably by H. Kamowitz and further by U. Klein. The following result is more specific to the unit disk, though Klein has obtained a partial generalization.

Theorem 4. *Let T_ϕ be a compact unital homomorphism of $A(\mathbf{D})$, and suppose that the fixed point z_0 of ϕ belongs to $\mathbf{D}$. If $\phi'(z_0) \neq 0$, then the spectrum of T_ϕ consists of 0 and* *a sequence of simple eigenvalues which are the powers of* $\phi'(z_0)$,

$$\sigma(T_\phi) = \{0, 1, \phi'(z_0), \phi'(z_0)^2, \phi'(z_0)^3, \cdots\}. \tag{4}$$

If $\phi'(z_0) = 0$, *then the spectrum of* T_ϕ *reduces to two points,* $\sigma(T_\phi) = \{0, 1\}$.

In the case that $\phi'(z_0) \neq 0$, the eigenfunction $f(z)$ corresponding to the eigenvalue $\phi'(0)$ is called the *principal eigenfunction* of T_ϕ. It is a solution of Schröder's equation, and it satisfies $f'(z_0) \neq 0$. If we normalize $f(z)$ by $f'(z_0) = 1$, then it is unique. The power $f(z)^n$ is an eigenfunction with eigenvalue $\phi'(z_0)^n$.

Proof of Theorem 4 (following [12]): We can assume that $z_0 = 0$. Denote $\mu = \phi'(0)$ and $\Delta_\delta = \{|z| \leq \delta\}$. Choose $\delta > 0$ so small that $\phi(\Delta_\delta) \subset \Delta_\delta$. Fix $m \geq 1$, let A_δ be the algebra of continuous functions on Δ_δ that are analytic on the interior of Δ_δ, and let B_m be the subspace of A_δ of functions that satisfy $g(0) = g'(0) = \cdots = g^{(m-1)}(0) = 0$. Thus $B_0 = A_\delta$, and B_m has codimension m in A_δ. Define $S_m(g) = g \circ \phi$ for $g \in B_m$. Let C be the supremum of $|\phi(z)|^m/\delta^m$ on Δ_δ. The Schwarz inequality shows that $||S_m|| \leq C$. Since B_m has codimension m in A_δ, S_0 has at most m linearly independent eigenfunctions corresponding to eigenvalues λ with $|\lambda| > C$. When δ is very small, C is approximately $|\mu|^m$. We conclude that there are at most m linearly independent eigenfunctions of S_0 whose eigenvalues satisfy $|\lambda| > |\mu|^m$. Since eigenfunctions of the operator T_ϕ on $A(\mathbf{D})$ restrict to eigenfunctions of the restricted composition operator S_0 on A_δ, there are at most m linearly independent eigenfunctions of T_ϕ whose eigenvalues satisfy $|\lambda| > |\mu|^m$. To complete the proof, it suffices now to establish the existence of the principal eigenfunction. For this, we take $m = 2$, and we assume that δ chosen so that $C < |\mu|$. Then $S_2 - \mu$ is invertible on B_2, and there is $h \in B_2$ such that $\phi - \mu z = (S_2 - \mu)h$. Then $f(z) = z - h(z)$ satisfies $f(\phi(z)) = \mu f(z)$ for $z \in \Delta_\delta$. Since the iterates of $\overline{\mathbf{D}}$ under ϕ eventually are contained in Δ_δ, we can use this functional equation to extend $f(z)$ to $\overline{\mathbf{D}}$, thus obtaining the principal eigenfunction.

3. The Pseudohyperbolic Metric on the Spectrum

The pseudohyperbolic metric in the unit disk $\mathbf{D}$ is given by

$$\rho(z, w) = \frac{|z - w|}{|1 - \bar{w}z|}, \qquad z, w \in \mathbf{D}. \tag{5}$$

Thus $\rho(z, w) = |F(z)|$, where F is the conformal self-map of $\mathbf{D}$ that sends w to 0. This interpretation makes it clear that the expression $\rho(z, w)$ is invariant under conformal self-maps of $\mathbf{D}$. It is easy to check (using properties of conformal self-maps) that $\rho(z, w)$ satisfies the triangle inequality, so that it is indeed a metric.

The *pseudohyperbolic metric* $\rho_A(x, y)$ on the spectrum M_A of the uniform algebra A is defined by

$$\rho_A(x, y) = \sup\{\rho(f(x), f(y)) : f \in A, ||f|| < 1\}, \qquad x, y \in M_A. \tag{6}$$

The triangle inequality for ρ_A follows immediately from the triangle inequality for ρ.

If $f \in A$ satisfies $||f|| < 1$, and if F is a conformal self-map of $\mathbf{D}$, then $F \circ f \in A$ and $||F \circ f|| < 1$. In other words, the unit ball of A is invariant under composition with conformal self-maps of $\mathbf{D}$. By composing with the self-map that sends $f(y)$ to 0, one sees that $\rho_A(x,y)$ is the supremum of $|f(x)|$ over all $f \in A$ satisfying $||f|| < 1$ and $f(y) = 0$. Since this supremum is the norm of the restriction of the evaluation functional at x to the null space of the evaluation functional at y, we have

$$\rho_A(x,y) = ||x|_{y^{-1}(0)}||, \qquad x, y \in M_A. \tag{7}$$

It is easy to check that

$$\rho_A(x,y) \leq ||x-y|| \leq 2\rho_A(x,y), \qquad x,y \in M_A. \tag{8}$$

Thus convergence in the pseudohyperbolic metric is the same on M_A as convergence in the norm of A^*.

In the case that A is the disk algebra $A(\mathbf{D})$, the pseudohyperbolic metric on $\overline{\mathbf{D}}$ is given by (5) if $z, w \in \mathbf{D}$, and by $\rho_A(z,w) = 1$ if either $|z| = 1$ or $|w| = 1$, $z \neq w$.

Two points $x, y \in M_A$ are said to belong to the same *Gleason part* of M_A if $\rho_A(x,y) < 1$. It is easy to check that this is an equivalence relation, so that the Gleason parts are well-defined. The Schwarz inequality for analytic functions shows that any connected analytic set in M_A is contained in a single Gleason part.

Theorem 5. *The Gleason parts of a uniform algebra are open and closed in the weak topology (A^{**}-topology) of M_A.*

For detailed proofs, see [23] and [8]. The theorem follows directly from the work of B. Cole (see [10]) on idempotents in A^{**}. The double dual A^{**} of A is a uniform algebra, and we may regard M_A as a subset of $M_{A^{**}}$. According to Cole, each Gleason part Q in M_A corresponds to a minimal idempotent F_Q in A^{**}, which satisfies $F_Q = 1$ on Q and $F_Q = 0$ on $M_A \backslash Q$. Since F_Q is A^{**}-continuous on M_A, the Gleason part Q is A^{**}-clopen in M_A.

Corollary. *A weakly compact subset of M_A meets only finitely many Gleason parts.*

One application of this corollary is to show that any weakly compact homomorphism from the disk algebra $A(\mathbf{D})$ to a uniform algebra B is compact. For this, we suppose the homomorphism is the operator of composition with $\phi : M_B \to \overline{D}$. Since $\phi(M_B)$ is weakly compact, it meets only finitely many Gleason parts, each in a weakly compact set. Hence $\phi(M_B)$ consists of a finite number of points of $\partial\mathbf{D}$ and a compact subset of $\mathbf{D}$, so that in fact $\phi(M_B)$ is norm compact in $A(\mathbf{D})^*$, and T is compact.

This argument can be extended somewhat. We say that A is a *unique representing measure algebra*, or *URM-algebra*, if every $x \in M_A$ has a unique representing measure on the Shilov boundary of A. The disk algebra $A(\mathbf{D})$ and the algebra $H^\infty(\mathbf{D})$ are URM-algebras. Each Gleason part of a URM-algebra is either a point or an analytic disk (see [16,9]), and the proof technique used to prove this shows also that the weak and norm topologies coincide on the spectrum of a URM-algebra. The above argument then yields the following result of A. Ülger ([23]; see also [8]).

Theorem 6. *A weakly compact homomorphism from a URM-algebra A to a uniform algebra B is compact.*

4. Homomorphisms and Pseudohyperbolic Contractions

In this section we describe several results obtained by Udo Klein in his thesis [19]. We organize the results as four theorems.

Theorem 7. *Let T_ϕ be a unital homomorphism from A to B, and let C be the ρ_A-diameter of $\phi(M_B)$. Then*

$$\rho_A(\phi(x), \phi(y)) \leq C\rho_B(x,y), \qquad x, y \in M_B. \tag{9}$$

The constant C is sharp.

Proof. Suppose $f \in A$ satisfies $||f|| < 1$ and $f(\phi(y)) = 0$. If $w \in M_B$, then $|f(\phi(w))| = \rho(f(\phi(w)), f(\phi(y))) \leq \rho_A(\phi(w), \phi(y)) \leq C$. Thus $g = (f \circ \phi)/C$ satisfies $||g|| \leq 1$ and $g(y) = 0$. Hence $|g(x)| \leq \rho_B(x,y)$, which yields $|f(\phi(x))| \leq C\rho_B(x,y)$. Taking the supremum over such f, we obtain (9).

The existence of the unique fixed point in the next theorem is due to H. Kamowitz [17], who obtained a result valid for homomorphisms of Banach algebras.

Theorem 8. *Suppose that M_A is connected. Let T_ϕ be a unital homomorphism of A, where $\phi : M_A \to M_A$. Suppose T_ϕ is compact. Then ϕ is a strict contraction with respect to the pseudohyperbolic metric ρ_A, and the iterates of ϕ converge in norm to a (unique) fixed point x_0 of ϕ.*

Proof. In this case, $\phi(M_A)$ is norm compact and connected. Hence it is contained in a single Gleason part and has ρ_A-diameter C strictly less than 1. Thus ϕ is a contraction with respect to the ρ_A-metric, and the contraction mapping principle applies.

For the remaining theorems we introduce some notation. We suppose that T_ϕ is a unital homomorphism of A and that $\phi : M_A \to M_A$ has a fixed point x_0. Let A_0 denote the null-space of x_0, that is, A_0 consists of the functions $f \in A$ such that $f(x_0) = 0$. Let T_0 be the restriction of T_ϕ to A_0. From (7) we have

$$\rho_A(x, x_0) = \sup\{|f(x)| : f \in A_0, ||f|| < 1\}. \tag{10}$$

Since $\sup\{|f(\phi(x))| : f \in A_0, ||f|| < 1\} \leq ||T_0|| \sup\{|g(x)| : g \in A_0, ||g|| < 1\}$, we obtain

$$\rho_A(\phi(x), x_0) \leq ||T_0||\, \rho_A(x, x_0), \qquad x \in M_A. \tag{11}$$

In particular, $\rho_A(\phi(x), x_0) \leq ||T_0||$. If $f \in A_0$ satisfies $||f|| < 1$, then $|(T_0 f)(x)| = |f(\phi(x))| \leq \rho_A(\phi(x), x_0)$. Taking the suprema first over $x \in M_A$ and then over such f, we conclude that

$$||T_0|| = \sup_{x \in M_A} \rho_A(\phi(x), x_0). \tag{12}$$

We will denote by ϕ^k the kth iterate of ϕ, so that $T_\phi^k = T_{\phi^k}$. The iterate ϕ^k also has fixed point x_0, and the associated restriction operator is T_0^k. Finally, we denote the spectral radius $\lim_{n\to\infty} ||S^n||^{1/n}$ of an operator S by $||S||_{sp}$.

Theorem 9. *Let T_ϕ be a unital homomorphism of A. Suppose ϕ has a fixed point x_0, and denote $T_0 = T_\phi|_{A_0}$ as above. If $||T_0||_{sp} < 1$, then*

$$||T_0||_{sp} \leq \limsup_{x \to x_0} \frac{\rho_A(\phi(x), x_0)}{\rho_A(x, x_0)}. \tag{13}$$

Proof. Let α be strictly greater than the lim sup. If $||T_0^N|| < 1$, then ϕ^N is a contraction. By choosing N very large, we can assume that $\phi^N(M_A)$ is contained in a neighborhood of x_0 on which $\rho_A(\phi(x), x_0) \le \alpha\, \rho_A(x, x_0)$. Then

$$\frac{\rho_A(\phi^{N+k}(x), x_0)}{\rho_A(\phi^N(x), x_0)} = \prod_{j=1}^{k} \frac{\rho_A(\phi^{N+j}(x), x_0)}{\rho_A(\phi^{N+j-1}(x), x_0)} \le \alpha^k \tag{14}$$

for all $x \in M_A$. From (12) we then have

$$||T_0^{N+k}|| = \sup_{x \in M_A} \rho_A(\phi^{N+k}(x), x_0) \le \alpha^k \sup_{x \in M_A} \rho_A(\phi^N(x), x_0) = \alpha^k ||T_0^N||. \tag{15}$$

If now we take the $(N+k)$th root and let $k \to \infty$, we obtain $||T_0||_{sp} \le \alpha$, which establishes the estimate.

Recall that a point derivation at x_0 is a linear functional on A that satisfies the Leibniz rule

$$L(fg) = f(x_0)L(g) + g(x_0)L(f), \qquad f, g \in A. \tag{16}$$

Let D_0 denote the subspace of A^* consisting of the continuous point derivations at x_0. Since x_0 is a fixed point of ϕ, D_0 is an invariant subspace of T^*.

Theorem 10. *Let T_ϕ be a unital homomorphism of A. Suppose that M_A is connected and that T_ϕ is compact. Let x_0 be the fixed point of ϕ, let D_0 be the space of continuous point derivations at x_0, and let $T_0 = T_\phi|_{A_0}$ be as above. Then the spectral radius of T_0 coincides with the spectral radius of the restriction of the adjoint operator T_ϕ^* to D_0,*

$$||T_0||_{sp} = ||T_\phi^*|_{D_0}||_{sp}. \tag{17}$$

Proof. Let $R = ||T_0||_{sp}$ denote the spectral radius of T_0. Fix $k \ge 2$. From the preceding theorem, applied to ϕ^k, we have

$$\limsup_{x \to x_0} \frac{\rho_A(\phi^k(x), x_0)}{\rho_A(x, x_0)} \ge ||T_0^k||_{sp} = R^k. \tag{18}$$

Choose $x_n \in M_A$ such that $\rho_A(x_n, x_0) \to 0$ and $R^k \le (1+\varepsilon)\rho_A(\phi^k(x_n), x_0)/\rho_A(x_n, x_0)$. Since $y_n = (x_n - x_0)/\rho_A(x_n, x_0)$ satisfies $||y_n|| \le 2$, by (8), we may pass to a subsequence and assume that $T_\phi^*(y_n) = (\phi(x_n) - x_0)/\rho_A(x_n, x_0)$ converges in norm, say to $L \in A^*$. Then $||L|| \le 2$, and it is straightforward to check that L satisfies the Leibnitz identity, so that $L \in D_0$. Further, $||(T_\phi^*)^{k-1}L|| = \lim ||\phi^k(x_n) - x_0||/\rho_A(x_n, x_0) \ge \lim \rho_A(\phi^k(x_n), x_0)/\rho_A(x_n, x_0) \ge R^k/(1+\varepsilon)$, where we have used (8) again. Thus the norm of $(T_\phi^*)^{k-1}$ on D_0 is at least $R^k/2$. Taking kth roots and letting $k \to \infty$, we obtain $||T_\phi^*|D_0||_{sp} \ge R$. The reverse inequality is easier and does not require the compactness hypothesis. Because derivations kill the constants, one obtains $||T_\phi^* L|| \le 2||T_0^* L||$ for any $L \in D_0$, and consequently $||T_\phi^* L|| \le 2||T_0^*|_{D_0}|| \le 2||T_0||$. Apply this with T_ϕ replaced by T_ϕ^k and send $k \to \infty$, to obtain $||T_\phi^*|D_0||_{sp} \le R$.

The theorem implies in particular that if the spectral radius of T_0 is strictly positive, then there exist nontrivial continuous point derivations at x_0. In some sense these point derivations represent the vestiges of an analytic structure at x_0.

In the case of the disk algebra, with fixed point $z_0 \in \mathbf{D}$ satisfying $\phi'(z_0) \neq 0$, the space D_0 is one-dimensional, spanned by the functional $g \mapsto g'(z_0)$. The operator T_ϕ^* on D_0 is multiplication by $\phi'(z_0)$. Thus the norm of T_ϕ^* on D_0 coincides with the absolute value $|\phi'(z_0)|$ of the principal eigenvalue of T_ϕ.

Question. How much of this analysis can be carried out if it is assumed that T_ϕ is only weakly compact? In particular, if M_A is connected and T_ϕ is weakly compact, does ϕ have a fixed point?

5. Homomorphisms of the Ball Algebra

We turn now to some specific families of uniform algebras. In this section we describe a theorem of Aron, Galindo and Lindström [2]. It extends a theorem on compact composition operators that is known for the disk algebra (Section 2) and the ball algebra in finite dimensions (see [5, Theorem 7.20]) to a setting in which the unit disk is replaced by the open unit ball B of a Banach space X. We consider for simplicity the algebra $H^\infty(B)$ of bounded analytic functions on B.

Theorem 11. *Let B be the open unit ball of a Banach space X. Let T_ϕ be a compact unital homomorphism of $H^\infty(B)$, and suppose that ϕ maps B analytically into the ball rB centered at 0 of radius $r < 1$. Then the iterates of ϕ converge to a fixed point $x_0 \in B$ of ϕ. The spectrum of T_ϕ is the unital semigroup generated by the spectrum of the Fréchet derivative $(d\phi)(x_0)$, together with 0.*

Proof. The existence of the fixed point x_0 follows from Theorem 8. It is straightforward to check that $(d\phi)(x_0)$ is a compact operator.

The affine approximation to the analytic function $f(x)$ at x_0 is given by $f(x) \approx f(x_0) + f'(x_0)(x - x_0)$, where $f'(x_0) \in X^*$. Suppose $g = (\lambda I - T_\phi)f$ is in the range of $\lambda I - T_\phi$. Comparing affine approximations and using the chain rule, we obtain $g(x_0) + g'(x_0)(x - x_0) = \lambda f(x_0) + \lambda f'(x_0)(x - x_0) - f(x_0) - f'(x_0)(d\phi)(x_0)(x - x_0)$. Equating the linear terms, we obtain

$$g'(x_0)(y) = f'(x_0)((\lambda I - (d\phi)(x_0))y), \qquad y \in X. \tag{19}$$

From this identity we see that the null space of $\lambda I - (d\phi)(x_0)$ is annihilated by $g'(x_0)$ for all g in the range of $\lambda I - T_\phi$. If now $\lambda \neq 0$ is in the spectrum of the compact operator $(d\phi)(x_0)$, then the null space of $\lambda I - (d\phi)(x_0)$ is nonzero, the operator $\lambda I - T_\phi$ is not onto, and λ is in the spectrum of T_ϕ. Thus the spectrum of T_ϕ includes the spectrum of $(d\phi)(x_0)$, hence it includes the unital semigroup generated by the spectrum of $(d\phi)(x_0)$.

For the reverse inclusion, we must use the higher order terms of the Taylor series. Each complex-valued analytic function f has a Taylor expansion $f(x) = \sum f_m(x - x_0)$ near x_0, where $f_m(y)$ is an m-homogeneous analytic function on X that is bounded on bounded sets. Such m-homogeneous analytic functions, with the supremum norm over B, form a Banach space P_m. The operator T_ϕ induces an operator on P_m, by composing and neglecting higher order terms. The idea is to show that this operator is compact and has eigenvalues consisting of all m-fold products $\lambda_1 \cdots \lambda_m$, where the λ_j's are eigenvalues of $(d\phi)(x_0)$. One way to go about this is to use the fact that P_m is the dual space of a certain tensor product space, spanned by symmetric tensors of the form $x \otimes \cdots \otimes x$, $x \in X$ (the m-fold symmetric projective tensor product). Under this duality, the operator on P_m is the dual of the operator $x \otimes \cdots \otimes x \to (d\phi)(x_0)x \otimes \cdots \otimes (d\phi)(x_0)x$, whose spectrum is known (see [22]) to consist of m-fold products as above. If now f is a nonzero analytic function that satisfies $(\lambda I - T_\phi)f = 0$, and m is the smallest integer for which the term f_m in the Taylor expansion of f is nonzero, then f_m is an eigenfunction for the corresponding operator on P_m, thus the eigenvalue λ is an m-fold product of eigenvalues of $(d\phi)(x_0)$.

We mention an interesting example, pointed out in [2], of a homomorphism T_ϕ that is weakly compact but not compact. Such an example is obtained by taking X to be the Tsirelson space and ϕ to be the restriction to B of any noncompact linear operator S on X satisfying $||S|| < 1$.

6. The Algebra $A(D)$

Let D be a bounded domain in the complex plane, and let $A(D)$ be the algebra of analytic functions on D that extend continuously to the boundary ∂D of D. The spectrum of $A(D)$ is the closure $\overline{D} = D \cup \partial D$ of D. The Gleason parts of $A(D)$ are the one-point parts consisting of peak points on ∂D, and a single Gleason part consisting of D and the nonpeak points on ∂D. The unital homomorphisms T_ϕ of $A(D)$ to itself are composition operators corresponding to continuous functions ϕ from $\overline{D}$ to $\overline{D}$ that are analytic on D.

The following theorem is a sharpened version of a theorem in [14] (see also [7]).

Theorem 12. *Let S be a subset of $\overline{D}$ that is not precompact with respect to the pseudohyperbolic metric $\rho_{A(D)}$. Then for any $\varepsilon_k > 0$, $1 \le k < \infty$, there are points $z_k \in S$, disjoint subsets E_k of D, and functions $g_k \in A(D)$ such that $|g_k| \le C$ for some universal constant C, $|g_k| \le \varepsilon_k$ off E_k, $g_k(z_k) = 1$, and $g_k(z_n) = 0$ for $n \ne k$. If $\sum \varepsilon_k = \varepsilon$ is small, and if the z_k's converge (in the topology of $\overline{D}$), then the linear operator $a \to \sum a_k g_k$ maps the Banach space c_0 of null sequences bicontinuously onto a closed complemented subspace of $A(D)$.*

The proof is similar to those given in [14] and [7], though there are more technical details. It is easy to construct smooth functions g_k with the properties of the theorem. A key idea in the proof is to modify the smooth functions to obtain analytic functions. This is done by using the Cauchy transform to solve a $\overline{\partial}$-problem, with explicit estimates for the solutions. The projection of $A(D)$ onto the subspace c_0 is given by $f \to \sum f(z_k) g_k$.

The point evaluations at the z_k's span a closed subspace of $A(D)^*$ that is isomorphic to ℓ^1. Consequently the evaluation functionals at the z_k's are not weakly compact in $A(D)^*$, and the set S in the theorem above is not weakly compact. This proves the following theorem and also raises a question.

Theorem 13. *The weak and norm topologies of $\overline{D}$, regarded as a subset of $A(D)^*$, have the same compact sets.*

Question: Do the weak and norm topologies of $A(D)^*$ always coincide on $\overline{D}$?

From the theorem we obtain the following dichotomy, which shows in particular that any weakly compact homomorphism of $A(D)$ is compact.

Theorem 14. *Let T be a homomorphism from $A(D)$ to a uniform algebra B. Either T is compact, or there is an embedding $c_0 \hookrightarrow A(D)$ of c_0 onto a complemented subspace of $A(D)$ on which T is an isomorphism (linear and bicontinuous).*

Proof. Assume that $T = T_\phi$ is unital, and apply the preceding theorem to $S = \phi(M_B)$.

A similar theorem holds for the uniform algebra $R(K)$, where K is a compact subset of the complex plane. The key procedure of solving a $\bar{\partial}$-problem is the same as that used for $A(D)$.

Now the algebras $A(D)$ and $R(K)$ are tight uniform algebras. Tight uniform algebras form a class of algebras for which the $\bar{\partial}$-problem is solvable in some abstract sense (see [CG]). S. Saccone [21] has shown that every tight uniform algebra A has the Pelczynski property: if T is a continuous linear operator from A to a Banach space, then either T is weakly compact, or there is an embedding $c_0 \hookrightarrow A$ of c_0 onto a subspace of A on which T is an isomorphism. The preceding theorem can be viewed as a sharpened form of this dichotomy in the special case at hand. For another version of the dichotomy, which applies to Hankel operators, see [7].

7. The Algebra $H^\infty(D)$

Now we turn to the algebra $H^\infty(D)$ of bounded analytic functions on D. There are analogues for $H^\infty(D)$ of each of the theorems in Section 6. The analogue of Theorem 13 is as follows.

Theorem 15. *Let D be a bounded domain (or open set) in the complex plane. The following are equivalent, for a subset E of D: (i) E is a norm-precompact subset of $H^\infty(D)^*$; (ii) E is a weakly precompact subset of $H^\infty(D)^*$; (iii) E does not contain an interpolating sequence for $H^\infty(D)^*$; (iv) if $\{z_n\}$ is a sequence in E that tends to $\zeta \in \partial D$ (in the topology of the complex plane), then $\{z_n\}$ converges in the norm of $H^\infty(D)^*$ to a "distinguished homomorphism."*

For background on distinguished homomorphisms, see [14]. The equivalence of (iii) and (iv) is proved in [14, Theorems 4.2 and 4.4], and the other equivalences follow easily. The analogue of Theorem 14 is as follows.

Theorem 16. *Let T_ϕ be a unital homomorphism from $H^\infty(D)$ to a uniform algebra B. Suppose that $\phi(M_B)$ is contained in the norm closure of D in the spectrum of $H^\infty(D)$. Then either T is compact, or there is an embedding $\ell^\infty \hookrightarrow H^\infty(D)$ of ℓ^∞ onto a complemented subspace of $H^\infty(D)$ on which T is an isomorphism (linear and bicontinuous).*

Again this raises several questions.

Question: Do the weak and norm topologies of $H^\infty(D)^*$ coincide on D?

Question: Does weak compactness imply compactness for arbitrary homomorphisms from $H^\infty(D)$ to a uniform algebra?

We have seen that the answers to both questions are affirmative in the case of the open unit disk $\mathbf{D}$. We turn to a class of infinitely connected domains for which the answers are also affirmative.

8. Behrens Domains

A *Behrens domain* is an infinitely connected domain in the plane obtained from the open unit disk $\mathbf{D}$ by excising the origin 0 and a sequence of disjoint closed subdisks with centers c_j and radii r_j tending to 0, such that there are annular collars $\{r_j < |z - c_j| < R_j\}$ in D that are pairwise disjoint and that satisfy $\sum r_j/R_j < \infty$. These conditions guarantee that $\sum r_j/|c_j| < \infty$, thus the measure $dz/(2\pi iz)$ on ∂D is finite. It represents the distinguished homomorphism φ_0 of $H^\infty(D)$ at 0. The Gleason part of $H^\infty(D)$ containing D consists of D together with a family of analytic disks with their centers identified to the distinguished homomorphism φ_0. For details, see [3].

Theorem 17. *If D is a Behrens domain, then the weak and norm topologies of the spectrum of $H^\infty(D)$, regarded as a subset of $H^\infty(D)^*$, coincide.*

The proof proceeds along the following lines. The weak and norm topologies on the Gleason part of D can be described concretely using the function theory described in [Be], and they coincide. The remaining Gleason parts of $H^\infty(D)$ are points or analytic disks, and $H^\infty(D)$ is effectively a URM-algebra with respect to the spectrum minus the Gleason part containing D. The proof that the weak and norm topologies on the spectrum for a URM-algebra coincide can be adapted to complete the proof of the theorem.

The algebra $A(D)$ is strongly pointwise boundedly dense in $H^\infty(D)$, that is, each $f \in H^\infty(D)$ can be approximated pointwise on D by a sequence of functions $f_n \in A(D)$ satisfying $||f_n|| \le ||f||$. Consequently the pseudohyperbolic metrics on D determined by the algebras $A(D)$ and $H^\infty(D)$ coincide. It follows that 0 belongs to the norm closure with respect to $A(D)^*$ of a subset E of D if and only if the distinguished homomorphism φ_0 belongs to the norm closure with respect to $H^\infty(D)^*$ of E. Using the function theory, one shows that this occurs if and only if 0 and φ_0 belong respectively to the weak closures of E in $A(D)^*$ and $H^\infty(D)^*$. In particular, we obtain the following.

Theorem 18. *If D is a Behrens domain, then the weak and norm topologies of $\overline{D}$, regarded as a subset of $A(D)^*$, coincide.*

REFERENCES

1. R.M.Aron, B.J.Cole and T.W.Gamelin, *Weak-star continuous analytic functions*, Canad. J. Math., **47** (1995), 673-683.
2. R.Aron, P.Galindo and M.Lindström, *Compact homomorphisms between algebras of analytic functions*, Studia Math., **123** (1997), 235-247.
3. M.Behrens, *The maximal ideal space of algebras of bounded analytic functions on infinitely connected domains*, T.A.M.S., **161** (1971), 359-379.
4. B.J.Cole and T.W.Gamelin, *Tight uniform algebras and algebras of analytic functions*, J. Funct. Anal., **46** (1982), 158–220.

5. C.Cowan and B.MacCluer, *Composition Operators on Spaces of Analytic Functions*, CRC Press, 1995.
6. S.Dineen, *Complex Analysis on Infinite Dimensional Spaces*, Springer-Verlag, 1999.
7. J.Dudziak, T.W.Gamelin and P.Gorkin, *Hankel operators on bounded analytic functions*, Trans. A.M.S., **352** (1999), 363-377.
8. P.Galindo and M.Lindström, *Gleason parts and weakly compact homomorphisms between uniform Banach algebras*, Monatshefte Math. **128** (1999), 89-97.
9. T.W.Gamelin, *Uniform Algebras*, 2nd edition, Chelsea Press, 1984.
10. T.W.Gamelin, *Uniform algebras on plane sets*, in *Approximation Theory*, G.G.Lorentz (ed), Academic Press, 1973, 101-149.
11. T.W.Gamelin, *Analytic functions on Banach spaces*, in *Complex Function Theory*, Gauthier and Sabidussi (ed), Kluwer Academic Publishers, 1994, 187-233.
12. T.W.Gamelin, *Conjugation theorems via Neumann series*, MRS Report 99-1 (1999), UCLA Mathematics Report Series, www.math.ucla.edu/pure/reports/
13. T.W.Gamelin, *Multiplicative operators on spaces of analytic functions*, in preparation.
14. T.W.Gamelin and J.B.Garnett, *Distinguished homomorphisms and fiber algebras*, Amer. J. Math. **92** (1970), 455-474.
15. T.W.Gamelin and S.V.Kislyakov, *Uniform algebras as Banach spaces*, in *Handbook of Banach Spaces*, W.B.Johnson and J.Lindenstrauss (ed), Elsevier Science, to appear.
16. K.Hoffman, *Analytic functions and Gleason parts*, Ann. Math. **86** (1967), 74-111.
17. H.Kamowitz, *Compact endomorphisms of Banach algebras*, Pac. J. Math. **89** (1980), 313-325.
18. H.Kamowitz, S.Scheinberg and D.Wortman *Compact endomorphisms of Banach algebras II*, Proc. A.M.S. **107** (1989), 417-422.
19. U.Klein, *Kompakte multiplikative Operatoren auf uniformen Algebren*, Dissertation, Karlsruhe, 1996.
20. J.Mujica, *Complex Analysis in Banach Spaces*, North-Holland, 1986.
21. S.Saccone, *The Pelczynski property for tight subspaces*, J. Funct. Anal. **148** (1997), 86-116.
22. M.Schechter, *On the spectra of operators on tensor products*, J. Funct. Anal. **4** (1969), 95-99.
23. A.Ülger, *Some results about the spectrum of commutative Banach algebras under the weak topology and applications*, Monatshefte Math., **121** (1996), 353-379.

Recent Progress in Functional Analysis
K.D. Bierstedt, J. Bonet, M. Maestre and J. Schmets

Generic Dynamics and Monotone Complete C*-Algebras

JD Maitland Wright

Mathematics Department
University of Reading
Reading RG 6 6AX
England

To Professor Manuel Valdivia on the occasion of his seventieth birthday.

I wish, first, to offer my felicitations to Professor Valdivia on the occasion of his seventieth birthday and, secondly, to express my gratitude to the organising committee for inviting me and for the efficiency with which they have organised this stimulating conference. It is an especial pleasure to thank Professor Bonet and Professor Bierstedt for their great helpfulness and hospitality.

My talk today is focused on some recent joint work by Professor Kazuyuki Saitô and me [5]. Instead of going into great detail and generality my aim here is to start from an elementary standpoint and give a sketch of some of the basic ideas.

1. Preliminaries

Let us recall that a C^*-algebra is a Banach algebra with an involution * such that $||xx^*|| = ||x||^2$.

Example 1.1. Let H be a Hilbert space. Let $L(H)$ be the C^*-algebra of all bounded operators on H. Then $L(H)$ is a C^*-algebra. Moreover, every closed *-subalgebra of $L(H)$ is, itself, a C^*-algebra. Conversely, given a C^*-algebra A, A can always be embedded in $L(H)$ for some Hilbert space H. In general, A can be embedded in more than one way.

An elegant and concise introduction to the theory of operator algebras is given in [2] and, for a more encyclopaedic account [1] is excellent.

Ordering 1.2. Let B be a C^*-algebra. Let $B_{sa} = \{x \in B \mid x = x^*\}$. Then B_{sa} is a ***real*** Banach space.

Let $B_{sa}^+ = \{zz^* : z \in B\} = \{x^2 : x \in B_{sa}\}$. Then B_{sa}^+ is a cone in B_{sa}.
We partially order B_{sa} by $b \geq c$ when $b - c \in B_{sa}^+$.

All C^*-algebras considered here will be unital, that is, possess a unit element.

Definition 1.3. Let A be a (unital) C^*-algebra. It is **monotone complete** if, whenever Y is an upward directed, upper bounded subset of A_{sa}, then Y has a least upper bound in A_{sa}.

Every von Neumann algebra is monotone complete. The converse is FALSE.

Example 1.4. Let $B^\infty[0,1]$ be the C^*-algebra of bounded, Borel measurable (complex-valued) functions on $[0,1]$. Let $M[0,1]$ be the set of all $f \in B^\infty[0,1]$ such that $\{x \in [0,1] : f(x) \neq 0\}$ is meagre. [Let us recall that a subset of a topological space is meagre, or equivalently, of first Baire category, if it is of the form $\bigcup_{n=1}^\infty E_n$, where $\overline{E_n}$ has empty interior.] Then $M[0,1]$ is an ideal of $B^\infty[0,1]$. Let $D = B^\infty[0,1]/M[0,1]$. Then D is a monotone complete C^*-algebra.

D does not possess ***any*** normal states. So D is not a von Neumann algebra. The algebra D is usually known as the **Dixmier algebra**. See [6].

Remark 1.5. Let A be a monotone complete C^*-algebra which has a maximal abelian *-subalgebra M, where M is *-isomorphic to D. Then A is not a von Neumann algebra.

Proof. Each maximal abelian *-subalgebra of a von Neumann algebra is, itself, a von Neumann algebra. □

A monotone complete C^*-algebra is said to be a ***factor*** if its centre is one dimensional.

Let A and M be (unital) C^*-algebras. Let $E : A \to M$ be a positive linear map.

If A and M are monotone complete (σ-complete) the map E is ***normal*** (σ-normal) if, whenever Y is (a countable) upper bounded, upward directed subset of A_{sa} with least upper bound b then Eb is the least upper bound of $\{Ey : y \in Y\}$.

Let A be a monotone complete C^*-algebra. Let M be a maximal abelian *-subalgebra of A. Then a unitary, u, in A is said to be M-normalising if $uMu^* = M$.

We say that A is countably M-generated if A is σ-generated by M and a countable family of M-normalising unitaries.

Definition 1.6. Let A be a monotone complete factor. Let M be a maximal abelian *-subalgebra of A such that M is *-isomorphic to D, the Dixmier algebra. Let A be countably M-generated. Let E be a faithful σ-normal positive projection from A onto M. Then A is said to be a Generic Dynamics Factor and the pair (E, M) is called a ***Cayley system*** for A [5].

It turns out that E is uniquely determined by M.

Theorem 1.7. *(Uniqueness Theorem) For $j = 1, 2$, let A_j be a Generic Dynamics Factor with a Cayley system (E_j, M_j). Then there exists a *-isomorphism π from A_1 on to A_2 such that $\pi[M_1] = M_2$ and $\pi E_1 \pi^{-1} = E_2$.*

It follows from this theorem that it is reasonable to talk about the Generic Dynamics Factor. A proof may be found in [7]. This result depends ultimately on the dynamical results of [6].

Definition 1.8. (Automorphisms). Let us recall the following elementary notions:

For any unital $C*$-algebra B, an automorphism is ***inner*** if it is of the form $z \to uzu^*$ for some unitary, u, in B.

Let $Aut(B)$ be the group of all automorphisms of B. Let $Inn(B)$ be the group of all inner automorphisms of B. Then $Inn(B)$ is a normal subgroup of $Aut(B)$. We define the outer automorphism group to be $Out(B) = Aut(B)/Inn(B)$.

It may happen that $Out(B)$ is small. For example, $Out(L(H))$ is trivial.

K. Saitô and I are investigating $Out(A)$, where A is the Generic Dynamics Factor [5].

It is not immediately obvious that outer automorphisms exist. But when A is the Generic Dynamics Factor it turns out that $Out(A)$ is rather large. We obtain much more general results, but here I shall confine myself to indicating why $Out(A)$ contains every countable group.

2. Polish Spaces and Group Actions

Although $(-\pi, \pi)$ is not complete in its natural metric, it is homeomorphic to $\mathbb{R}$, which is complete. Thus completeness is not a topological property.

Definition 2.1. A topological space T is ***Polish*** if T is homeomorphic to a complete separable metric space.

So the Baire Category Theorem applies to each Polish space.

Definition 2.2. A topological space is ***perfect*** if it has no isolated points.

Examples of perfect Polish spaces:

(i) $\mathbb{R}, \mathbb{R}^n, \mathbb{C}^n$

(ii) $\mathbb{R} \setminus \mathbb{Q}$

(iii) $\{0,1\}^{\mathbb{N}}, \{0,1,2\}^{\mathbb{N}}$

(iv) $\mathbb{N}^{\mathbb{N}}$

From now onward T is a perfect Polish space.

Let us recall that a subset $S \subset T$ is a **G_δ-set** if S is the intersection of countably many open sets.
Every G_δ-subset of a Polish space is Polish.
A subset Y of T is **generic** if Y is a G_δ-set and $X \setminus Y$ is meagre.

Example 2.3. Let T be the unit circle in $\mathbb{R}^2$ and let θ be an irrational number. Let α be the rotation of T through an angle $2\pi\theta$. Then α is a homeomorphism of T onto T. Define α^0 to be the identity map. Then $n \to \alpha^n$ is an action of $\mathbb{Z}$ onto T. Define α^0 to be the identity map. Then $n \to \alpha^n$ is an action of $\mathbb{Z}$ on T. This action has the following two properties:

(i) For each $x_0 \in T$, the orbit $\{\alpha^n(x_0) : n \in \mathbb{Z}\}$ is dense in T.

(ii) For $n \neq 0$, α^n has no fixed points.

Now let T be any perfect Polish space. Let Γ be a countable infinite group. Let $g \to \alpha^g$ be an action of Γ on T i.e. $g \to \alpha^g$ is a group homomorphism from Γ into the group of all homeomorphisms of T.

(i) This action is **(generically) free** if, for g different from the identity, the closed set $\{x \in T : \alpha^g(x) = x\}$ has empty interior.

(ii) The action is **(generically) ergodic** if, for some $x_0 \in T$, the orbit $\{\alpha^g(x_0) : g \in \Gamma\}$ is dense in T.

This implies the existence of a Γ-invariant generic set $Y \subset T$ such that $\{\alpha^g(y) : g \in \Gamma\}$ is dense for each $y \in Y$.

In the following, Γ is a countable infinite group, T is a perfect Polish space, and $g \to \alpha^g$ is a free, ergodic action of Γ in T. Let $C(T)$ be the C^*-algebra of all bounded continuous functions on T. Let $B^\infty(T)$ be the C^*-algebra of bounded Borel functions on T. Let $M(T)$ be the ideal of $B^\infty(T)$ consisting of all f such that $\{x \in T : f(x) \neq 0\}$ is **meagre**. Then $B^\infty(T)/M(T) \approx D$, the Dixmier algebra.

The Baire Category Theorem implies that the natural map from $C(T)$ into $B^\infty(T)/M(T)$ is injective and, hence, an isometric embedding into the Dixmier algebra.

Let $\gamma : T \to T$ be a homeomorphism. Then $f \to f \circ \gamma$ is an automorphism of $B^\infty(T)$ which maps $C(T)$ onto $C(T)$ and $M(T)$ onto $M(T)$. So the action $g \to \alpha^g$ induces an action $g \to \alpha_g$ of Γ on D, which restricts to an action of Γ on $C(T)$.

By a cross-product construction, we can find a C^*-algebra $B = C(T) \times_{r_\alpha} \Gamma$, in which $C(T) \approx M_0$, where M_0 is a maximal abelian *-subalgebra of B. Furthermore, there exists a group representation of Γ in the unitary group of B, $g \to U_g$ such that

$$U_g m U_g^* = \alpha_g(m)$$

for $m \in M_0 \approx C(T)$.

Also, there exists a faithful conditional expectation E_0 from B onto M_0 with $E_0(U_g) = 0$ for all $g \neq e$, where e is the identity element of Γ.

Let B^∞ be the (Pedersen) Borel* envelope of B. Then E_0 has an extension to a σ-normal map E^∞ from B^∞ onto $B^\infty(T)$. Let $q : B^\infty(T) \to D$ be the natural quotient map onto D. Let $\vartheta^\infty = \{z \in B^\infty : qE^\infty(zz^*) = 0\}$.

Then ϑ is a σ-ideal of B^∞ and $B^\infty/\vartheta^\infty$ can be identified with the Generic Dynamics Factor [5].

Example 2.4. Let $\Gamma = \oplus\mathbb{Z}_2$ and let $T = \Pi\mathbb{Z}_2$. Then Γ has a natural action on T. The (reduced) cross-product construction then gives the Fermion algebra $\mathbb{F}$. See [2]. So $\mathbb{F}^\infty$, the Pedersen Borel* envelope of $\mathbb{F}$, has the Generic Dynamics Factor as a quotient i.e. the Generic Dynamics Factor is "hyperfinite".

Let A be the Generic Dynamics Factor. Since A is hyperfinite it is natural to conjecture that A is **injective**.

This is an open problem, which is intimately related to paradoxical decompositions of solids in $\mathbb{R}^n$ ($n \geq 3$). See [8].

3. Constructing Outer Automorphisms

The arguments outlined below are applicable more widely but are some of the tools used in [5] to obtain information on the outer automorphism group of the Generic Dynamics Factor.

Sketch

Let G be a countable infinite group with a free, ergodic action α on a perfect Polish space T. Let H be an (infinite) **normal** subgroup of G such that α_H, the restriction of α to H, is also ergodic.

Let A be the Generic Dynamics Factor. Then there exists a group representation $(g \to U_g)$ for G into the unitary group of A such that $\{Ug : g \in G\}$ is a countable set of M-normalising unitaries such that $M \cup \{U_g : x \in G\}$ σ-generates A. Now let A_1 be the subalgebra of A σ-generated by $\{U_h : h \in H\} \cup M$. Since $h \to \alpha_h^H$ is a free, ergodic action, A_1 is also isomorphic to the Generic Dynamics Factor.

Let $g \in G/H$. Since H is a normal subgroup of G, for each $h \in H$ there exist $k \in H$ such that

$$U_g U_h U_g^* = U_k.$$

It follows that $U_g A_1 U_g^* = A_1$. So $Ad\, U_g$ induces an automorphism of A_1.

Suppose this is an **inner** automorphism of A_1.

Then there exists $\nu \in A_1$ where ν is unitary and $U_g z U_g^* = \nu z \nu^*$ for every $z \in A_1$.

Then $\nu^* U_g m = m \nu^* U_g$ for every $m \in M$.

Since M is a maximal abelian *-subalgebra of A, and $\nu^* U_g$ commutes with every element of M, it follows that $\nu^* U_g \in M$. So $U_g \in A_1$.

By a certain amount of technical trickery, see [5], $g \notin H$ and $U_g \in A_1$, can be used to show $U_g = 0$, which is impossible. So AdU_g gives an **outer** automorphism of A_1.

From this it can be shown that G/H can be embedded as a subgroup of $Out(A_1)$. Since A_1 and A are both isomorphic copies of the Generic Dynamics Factor, $Out(A_1)$ can be identified with $Out(A)$.

But we want to embed **every** countable group in $Out(A)$.

Let G be any countable group (possibly finite). If $G = \{e\}$ then, trivially, G embeds. So we suppose G has at least 2 elements.

Then ΠG, the product of countably many copies of G, is a perfect Polish space. (It is compact if G is finite and homeomorphic to $\mathbb{N}^{\mathbb{N}}$ if G is infinite.) Let $\oplus G$ be the subgroup of ΠG consisting of all $f : N \to G$ such that $f(n) = e$, the neutral element of G, for all but finitely many n. Then $\oplus G$ is a countable group, which is a normal subgroup of ΠG. For each $g \in G$, let $\underline{g}(n) = g$ for every $n \in \mathbb{N}$.

Let Γ be the subgroup of ΠG generated by $\oplus G$ and $\{\underline{g} : g \in G\}$. Then Γ is a countable group and $\oplus G$ is a normal subgroup of Γ.

Now $\Gamma / \oplus G$ contains a copy of G. So G can be embedded as a subgroup of $Out(A)$.

REFERENCES

1. R.V. Kadison and J.R. Ringrose, *Fundamentals of the theory of operator algebras*, Vol. **I** and **II**, Academic Press, New York and London (1983).

2. G.K. Pedersen, *C^*-algebras and their automorphism groups*, LMS Monographs, Vol. **XIV**, Academic Press, London, (1979).
3. K. Saitô and J.D.M. Wright, "Dynamic systems and duality for monotone complete C^*-algebras", *Quart. J. Math. Oxford* (2), **49** 199-226 (1998).
4. K. Saitô and J.D.M. Wright, "Admissible dynamic systems for monotone complete C^*-algebras", *Quart. J. Math. Oxford* (2), **50** 231-247 (1999).
5. K. Saitô and J.D.M. Wright, "Outer automorphisms of the Generic Dynamics Factor", *J. Math. Anal. Appl.* **248** 41-68 (2000).
6. D. Sullivan, B. Weiss and J.D.M. Wright, "Generic dynamics and monotone complete C^*-algebras", *Trans. Amer. Math. Soc.* **295** 795-809 (1986).
7. J.D.M. Wright, "Hyperfiniteness in wild factors", *J. London Math. Soc.* (2), **38** 492-502 (1988).
8. J.D.M. Wright, "Paradoxical decompositions of the cube and injectivity", *Bull. London Math. Soc.* , **22** 18-24 (1990).

Recent Progress in Functional Analysis
K.D. Bierstedt, J. Bonet, M. Maestre and J. Schmets

Linear topological properties of the space of analytic functions on the real line

P. Domański[a]* and D. Vogt[b]

[a]Faculty of Mathematics and Comp. Sci., Adam Mickiewicz University Poznań; and Institute of Mathematics, Polish Academy of Sciences (Poznań branch), Matejki 48/49, 60-769 Poznań, POLAND
e-mail: domanski@amu.edu.pl

[b]Bergische Universität Wuppertal, FB Mathematik
Gaußstr. 20, D-42097 Wuppertal, GERMANY
e-mail: dvogt@math.uni-wuppertal.de

To Professor Manuel Valdivia on the occasion of his seventieth birthday.

Abstract

The paper gives a survey of the linear topological properties of the space of analytic functions on the real line. Besides the analysis of the classical properties, the emphasis is put on certain, very useful "splitting lemmas". The proofs which are presented here are based on much more elementary tools than the known proofs for the same space on a general domain ω in $\mathbb{R}^d$. We also explain the relation of the reviewed structural results with some results on the convolution equations (differential equations of infinite order).
MSC 2000 Primary 46E10, 34A35; Secondary 46A63, 46E40, 30E05, 46M18, 45E10.

In the recent time much progress was made in the study of the space of real analytic functions $A(\omega)$ on an arbitrary domain $\omega \subset \mathbb{R}^d$. In the present paper we will concentrate on a particular case, namely, the space $A(\mathbb{R})$ of real analytic functions on the real line. In this case, we can use much more elementary tools for the complex analytic, resp. function theoretic parts of the proofs while functional analytic tools remain the same and are just reported, without proof. Of course, the proofs apply then only for $\omega = \mathbb{R}$. We will sketch these simpler proofs and we compare them with the proofs in the case of general $A(\omega)$, published elsewhere, believing that this would put a new light on the whole theory. This is by no means a complete survey, we rather focus our attention on some aspects of the space. In the last section we show how the results on the structure of $A(\mathbb{R})$ imply some results of Langenbruch [14] characterizing the existence of a right inverse for a convolution

*The paper was prepared during a stay in Wuppertal supported by the A. von Humboldt Foundation.

operator $T_\mu: A(\mathbb{R}) \longrightarrow A(\mathbb{R})$ (or more precisely for an ordinary differential operator of infinite order) in terms of the Fourier-Laplace transform of μ. The crucial point is that if such a right inverse exists, then the kernel of T_μ must be a DF-space which follows from our "lack of basis" result (Theorem 3.4 below).

1. General properties of $A(\omega)$.

Clearly, $A(\omega)$ consists of all functions $f: \omega \longrightarrow \mathbb{C}$ which develop at every point $x \in \omega$ into a Taylor series convergent in a neighborhood of x to f. There are at least two natural ways to topologize the space (see [16]). First, every $f \in A(\omega)$ extends to a holomorphic function on some open neighborhood $U \subseteq \mathbb{C}^d$ of ω. Therefore

$$A(\omega) = \mathrm{ind}_U H(U),$$

where the inductive limit runs through all such neighborhoods U and $H(U)$ is the Fréchet space of all holomorphic functions on U with the compact-open topology. This algebraic equality provides $A(\omega)$ with the so-called *inductive topology*, we denote the corresponding locally convex space by $A_i(\omega)$.

On the other hand, for every $f \in A(\omega)$ and $K \subset\subset \omega$ we have $f|_K \in H(K)$ the space of germs of holomorphic functions over K. Therefore, we may equip $A(\omega)$ with the *projective topology* given by

$$A_p(\omega) = \operatorname*{proj}_K H(K),$$

where K runs through all compact subsets of ω. $H(K)$ is a nuclear LB-space when equipped with the standard topology

$$H(K) = \mathrm{ind}_n H(U_n),$$

where (U_n) is a fundamental sequence of open neighborhoods of K in $\mathbb{C}^d$. Of course, we can make the projective limit countable by taking any compact exhaustion $K_1 \subset\subset K_2 \subset\subset \ldots \subset \omega$, $\bigcup_{n\in\mathbb{N}} K_n = \omega$. Then $A_p(\omega) = \mathrm{proj}_{n\in\mathbb{N}} H(K_n)$ topologically.

Since the restriction map $H(U) \longrightarrow H(K)$ is continuous for every $\omega \subset U \subset \mathbb{C}^d$, $K \subset\subset \omega$, the identity map

$$\mathrm{id}: A_i(\omega) \longrightarrow A_p(\omega)$$

is continuous and the inductive topology is not weaker than the projective topology. We observe that the bounded sets in $A_p(\omega)$ are contained in bounded sets of the form

$$B = \{f(z) : \sup_{z\in U_n} |f(z)| < C_n \text{ for all } n\}$$

for some sequence of neighborhoods U_n of K_n and a sequence (C_n) of positive constants, $n \in \mathbb{N}$. Clearly, by the Montel theorem, B is also bounded in $H(\bigcup_{n\in\mathbb{N}} U_n)$ therefore in $A_i(\omega)$. We have proved that $A_i(\omega)$ is the bornological space associated to $A_p(\omega)$.

It was Martineau [16] who observed (under much more general circumstances) that both topologies coincide. To give a proof for the case of $\omega = \mathbb{R}$ we use the following elementary lemma, where $\Re z$ and $\Im z$ denote the real and imaginary part of z, respectively.

1.1 Lemma *If $U \supset [-k,k]$ is open in $\mathbb{C}$, then there exist $\delta > 0$, $R > k$ and continuous linear maps $A : H(U) \to H^\infty(\{z : |\Im z| < \delta\})$ and $B : H(U) \to H^\infty(\{z : |\Re z| < R\})$ so that $Af + Bf = f$ in $\{z : |\Im z| < \delta, |\Re z| < R\} \subset U$ for every $f \in H(U)$.*

Proof. We choose $\delta > 0$ and $R > k$ such that f is holomorphic on a neighborhood of a rectangle P with int$P \supset \{z : |\Re z| \leq R, |\Im z| \leq \delta\}$. By the Cauchy formula

$$f(z) = \frac{1}{2\pi i}\int_{\gamma_1} \frac{f(\zeta)}{\zeta - z} d\zeta + \frac{1}{2\pi i}\int_{\gamma_2} \frac{f(\zeta)}{\zeta - z} d\zeta =: Af(z) + Bf(z),$$

where γ_1 consists of horizontal edges of P and γ_2 of vertical ones. These constitute maps A and B which obviously have the requested properties. □

From now on Banach disc means always a bounded one.

1.2 Theorem (Martineau [16]) *$A_p(\omega)$ is ultrabornological. In particular the inductive topology and the projective topology coincide on $A(\omega)$.*

Proof for $\omega = \mathbb{R}$. Let E be a Banach space and $\varphi : A(\mathbb{R}) \to E$ a linear map which sends Banach discs into bounded subsets of E. Then the restriction of φ to the Fréchet space $H(\mathbb{C})$ is continuous ([19, 24.13]), and therefore there is $k \in \mathbb{N}$ so that $\varphi|_{H(\mathbb{C})}$ factorizes through a map $\varphi_0 \in L(H^\infty(\{z : |z| < k\}), E)$. Let $U \supset [-k,k]$ be an open neighborhood in $\mathbb{C}$. We apply Lemma 1.1, obtain maps A and B and set $\Phi = \varphi \circ A + \varphi_0 \circ B$ on $H(U)$. Clearly $\Phi \in L(H(U), E)$ and $\Phi(f) = \varphi(f)$ for $f \in A(\mathbb{R}) \cap H(U)$. Since this holds for every open neighborhood $U \supset [-k,k]$ we see that φ is continuous in the topology induced by $H([-k,k])$, which means that $\varphi \in L(A_p(\mathbb{R}), E)$. □

Before we continue with a proof for general ω we need some machinery very useful in the theory which is, in fact, an abstract version of the Mittag-Leffler procedure.

Let us consider a sequence $\mathcal{X} = (X_n, i_n^{n+1})$ of linear spaces and operators (a so-called projective spectrum):

$$X_0 \xleftarrow{i_0^1} X_1 \xleftarrow{i_1^2} X_2 \longleftarrow \ldots \longleftarrow X_n \xleftarrow{i_n^{n+1}} X_{n+1} \longleftarrow \ldots$$

We can define, as usual, $X = \operatorname{proj} \mathcal{X} = \{(x_n)_{n\in\mathbb{N}} \in \prod X_n : i_n^{n+1} x_{n+1} = x_n \text{ for } n \in \mathbb{N}\}$ then we have the *fundamental resolution*

$$0 \longrightarrow X \xrightarrow{j} \prod X_n \xrightarrow{\sigma} \prod X_n \longrightarrow 0,$$

$j((x_n)_{n\in\mathbb{N}}) = (x_n)_{n\in\mathbb{N}}$, $\sigma((x_n)_{n\in\mathbb{N}}) = (i_n^{n+1}x_{n+1} - x_n)_{n\in\mathbb{N}}$ and $\operatorname{Proj}^1 \mathcal{X} = \prod X_n / \operatorname{Im}\sigma$. In fact, Proj^1 is the so-called first derived functor of the functor proj, but we will not need this homological definition later on.

If we take X_n to be nuclear LB-spaces (= LN-spaces), then the limit X of the spectrum will be called a PLN-*space.* We call a spectrum $\mathcal{X}$ *reduced* if for every k there is $n > k$ such that the canonical image $i_k X$ of X is dense in the image $i_k^n X_n$, $i_k^n := i_k^{k+1} \circ \ldots \circ i_{n-1}^n$. It can be easily observed (comp. [29]) that if $\mathcal{X}, \mathcal{Y}$ are two reduced spectra of LN-spaces and $\operatorname{proj} \mathcal{X} = \operatorname{proj} \mathcal{Y} = X$, then the spectra are *equivalent* (i.e. $\forall\, k\ \exists\, n > k : (i_x)_k^n, (i_y)_k^n$

factorize through some $(i_y)_m^{m+1}, (i_x)_p^{p+1}$ respectively). In particular, $\mathrm{Proj}^1\mathcal{X} = \mathrm{Proj}^1\mathcal{Y}$ so, in fact, we can write $\mathrm{Proj}^1 X$, as every PLN-space has a representing reduced spectrum of LN-spaces because closed subspaces of LN-spaces are LN-spaces and all such spectra are equivalent [29, Cor. 5.3]. For more information on the functor Proj^1 in the category of locally convex spaces see [22], [28] and [29].

1.3 Theorem (Palamodov [22], Retakh [23]) *$\mathrm{Proj}^1 X = 0$, for a PLN-space $X = \mathrm{proj}\, X_n$, if and only if there is a sequence of bounded Banach discs $(B_n)_{n\in\mathbb{N}}, B_n \subseteq X_n$, such that $i_n^{n+1} B_{n+1} \subset B_n$ for every $n \in \mathbb{N}$ and for every $n \in \mathbb{N}$ there is $k \in \mathbb{N}$ such that*

$$i_n^k X_k \subseteq i_n X + B_n. \tag{1}$$

The functor Proj^1 plays a role in proving surjectivity of various operators but it has also some impact on the linear topological properties of the space itself.

1.4 Theorem (Vogt [28], Wengenroth [30]) *A PLN-space X is ultrabornological if and only if* $\mathrm{Proj}^1 X = 0$.

Now, we are ready to come back to the space $A(\omega)$. For general ω we will show $\mathrm{Proj}^1 A(\omega) = 0$ and use Theorem 1.4 to complete the proof of Theorem 1.2. Again, for $\omega = \mathbb{R}$ the proof is particularly simple (of course, in that case it gives nothing more than the given earlier proof of Theorem 1.2).

1.5 Proposition $\mathrm{Proj}^1 A(\omega) = 0$.

Proof for $\omega = \mathbb{R}$. Let f be a holomorphic function on a neighborhood U of $[-k, k]$. By Lemma 1.1 we decompose $f = Af + Bf$ as described there. If q is the Taylor polynomial of Bf at 0 of sufficiently high order, then $\sup_{|z|\le k} |Bf(z) - q(z)| \le 1$, and we have proved that

$$f = (Af + q) + (Bf - q) \in A(\mathbb{R}) + B_k,$$

where B_k is the unit ball of $H^\infty(\{z : |z| < k\})$. □

Of course, a similar proof can be applied to $A(\mathbb{R}^n)$. Unfortunately for arbitrary domains we have to refer to more sophisticated methods. First we need a representation of $A(\omega)$ of great use later on.

We consider $\mathbb{R}^d \subseteq \mathbb{R}^{d+1} = \{(x,t) : t \in \mathbb{R}, x \in \mathbb{R}^d\}$ and the sheaf $\mathcal{E}$ of C^∞-functions on $\mathbb{R}^{d+1}$. Let us denote by $\mathcal{E}_\bullet$ its restriction to $\mathbb{R}^d$ in the sense of sheaves, i.e., the sheaf of germs of C^∞-functions. That means, for every open set $\omega \subseteq \mathbb{R}^d$, a section $s \in \Gamma(\omega, \mathcal{E}_\bullet)$ corresponds to a function $f \in C^\infty(\Omega), \Omega \subseteq \mathbb{R}^{d+1}$ open, $\Omega \cap \mathbb{R}^d = \omega$. Two such sections are equal if their corresponding $\mathbb{R}^{d+1}$-functions coincide on some neighborhood of $\mathbb{R}^d$.

Clearly, the $d+1$-dimensional Laplace operator Δ_{d+1} operates on the sheaf $\mathcal{E}_\bullet$ and, since Δ_{d+1} is elliptic, for every $\omega \subseteq \mathbb{R}^d$

$$\Delta_{d+1}: \Gamma(\omega, \mathcal{E}_\bullet) \longrightarrow \Gamma(\omega, \mathcal{E}_\bullet)$$

is surjective. Let us denote by $\mathcal{H}_\bullet$ the kernel sheaf of Δ_{d+1}, i.e., the sheaf of germs of $d+1$-dimensional harmonic functions. By the Cauchy-Kowalewska Theorem, every section

$s \in \Gamma(\omega, \mathcal{H}_\bullet)$ is uniquely determined by its Cauchy data, i.e., two real analytic functions on ω. We can identify $\mathcal{H}_\bullet$ with $\mathcal{A}^2$, two copies of the sheaf $\mathcal{A}$ of real analytic functions.

Now we are ready to complete the proof of Proposition 1.5 and, due to Theorem 1.4, also of Theorem 1.2.

Proof of $\mathrm{Proj}^1 A(\omega) = 0$. We have as above the short exact sequence

$$0 \longrightarrow \Gamma(\omega, \mathcal{H}_\bullet) \longrightarrow \Gamma(\omega, \mathcal{E}_\bullet) \xrightarrow{\Delta_{d+1}} \Gamma(\omega, \mathcal{E}_\bullet) \longrightarrow 0.$$

By a partition of unity argument, $H^1(\omega, \mathcal{E}_\bullet) = 0$ for any open set. Hence we get, by the long cohomology exact sequence [11, Theorem 6.2.19],

$$0 \longrightarrow \Gamma(\omega, \mathcal{H}_\bullet) \longrightarrow \Gamma(\omega, \mathcal{E}_\bullet) \xrightarrow{\Delta_{d+1}} \Gamma(\omega, \mathcal{E}_\bullet) \longrightarrow H^1(\omega, \mathcal{H}_\bullet) \longrightarrow 0.$$

This implies $H^1(\omega, \mathcal{H}_\bullet) = 0$ and therefore $H^1(\omega, \mathcal{A}) = 0$ for any open set $\omega \subseteq \mathbb{R}^d$. If $\mathcal{U}$ is an open covering (ω_n) of ω, where $\omega_1 \subset\subset \omega_2 \subset\subset \ldots$, $\bigcup \omega_n = \omega$, then $\mathrm{Proj}^1 A(\omega) = H^1(\mathcal{U}, \mathcal{A})$. Since the natural map $H^1(\mathcal{U}, \mathcal{A}) \longrightarrow H^1(\omega, \mathcal{A})$ is injective (see [11, 6.2.12]), we get the conclusion. □

From now on, we equip the space $A(\omega)$ with its unique natural topology. In the following theorem we summarize its certainly known properties (comp., for instance, [3], [4]).

1.6 Theorem *The space $A(\omega)$ satisfies the following properties:*

(a) it is nuclear and complete;

(b) it is ultrabornological and reflexive;

(c) the polynomials are dense in $A(\omega)$, thus it is separable;

(d) it is webbed, so the open mapping and the closed graph theorems hold for maps $T: A(\omega) \longrightarrow A(\omega)$.

Proof. Clearly (a) follows (see [19, 28.8]) from $A(\omega) = A_p(\omega)$ and the fact that for any compact $K \subset \omega$ the space $H(K)$ is nuclear and complete. Then (b) follows from Theorem 1.2 and (a) which implies that $A(\omega)$ is a subspace of the reflexive space $\prod_{n\in\mathbb{N}} H(K_n)$. (c) follows from the fact that every compact subset of $\mathbb{R}^d$ is polynomial convex in $\mathbb{C}^d$ hence admits a basis of neighborhoods which are polynomial polyhedra and therefore Runge sets, (d) follows from (a), i.e., $A(\omega)$ is a PLN-space (see [19, 24.28, 24.30]). □

We want to describe the dual space of $A(\mathbb{R})$. We start with the following consequence of Theorem 1.2 (see [16, Prop 1.7,1.2], comp. [3, Th. 2], [4, Th. 1]) which we formulate for general ω:

1.7 Proposition *The strong dual $A(\omega)'_\beta$ coincides topologically with $\mathrm{ind}_{n\in\mathbb{N}} H(K_n)'_\beta$. Therefore it has the properties (a), (b) and (d) of Prop. 1.6, and*

(c') the evaluation functionals δ_x, $\delta_x(f) := f(x)$, $x \in \omega$, are linearly sequentially dense in $A(\omega)'_\beta$.

Proof. Clearly, $A(\omega)' = \text{ind}_{n\in\mathbb{N}} H(K_n)'$ algebraically, and the topology of $A(\omega)'_\beta$ is weaker than the inductive one. Since $A(\omega)$ is a complete Schwartz space, $A(\omega)'_\beta$ is ultrabornological (see [19, 24.23]) and must coincide topologically with the inductive limit (see [19, 24.33]).

Since $A(\omega)$ is ultrabornological, $A(\omega)'_\beta$ is complete by [19, 24.11]. The other properties follow from properties of $H(K_n)'_\beta$. Clearly, for int $K_n \neq \emptyset$, the $(\delta_x)_{x\in K_n}$ are linearly dense in the Fréchet space $H([-n,n])'_\beta$, thus (c') follows. □

From now on we restrict ourselves to the case of $\omega = \mathbb{R}$.

There are two standard representations of $A(\mathbb{R})'_\beta$ as a space of functions. The first possibility goes through the Grothendieck-Köthe-Silva duality (see [10, §27 (9)]),

$$H([-n,n])'_\beta = H_0(\mathbb{C}\setminus[-n,n]),$$

where H_0 is the Fréchet space of holomorphic functions vanishing at infinity with the compact open topology. Here we assign to $\varphi \in H([-n,n])'_\beta$ the function

$$f_\varphi(z) = \varphi_\zeta\left(\frac{1}{2\pi i}\frac{1}{\zeta - z}\right), \qquad z \in \mathbb{C}\setminus[-n,n].$$

If, on the other hand, $f \in H_0(\mathbb{C}\setminus[-n,n])$, $g \in H([-n,n])$, then we take a simple closed path γ around $[-n,n]$ lying in the common area of holomorphy of f and g and set

$$< f, g > := \int_\gamma f(z)g(z)dz,$$

by Cauchy's integral theorem we have $< f_\varphi, g > \; = \; \varphi(g)$.

The other possibility goes through the Fourier-Laplace transform and the Paley-Wiener Theorem [9, 4.5.2, 4.5.3]:

$$H([-n,n])'_\beta = A_n(\mathbb{C}),$$

where

$$A_n(\mathbb{C}) = \{f \in H(\mathbb{C}) : ||f||_{n,m} := \sup_{z\in\mathbb{C}} |f(z)| \exp\left(-n|\Im z| - \frac{1}{m}|\zeta|\right) < \infty \text{ for all } m\},$$

and to every $\varphi \in H([-n,n])'_\beta$ we associate the function $f \in A_n(\mathbb{C})$,

$$f(z) = \varphi(\exp i < \cdot, z >).$$

1.8 Corollary *There are natural identifications:*

$$\begin{aligned} A(\mathbb{R})'_\beta &\cong \text{ind}_n H_0(\mathbb{C}\setminus[-n,n]) =: H_0(\mathbb{C}\setminus\mathbb{R}); \\ A(\mathbb{R})'_\beta &\cong \{f \in H(\mathbb{C}) : \exists n \forall m \, ||f||_{n,m} < \infty\}. \end{aligned}$$

It is impossible to transfer these representations to arbitrary ω. The analogue of the first one works on ω being products of one dimensional sets, the second one only for convex ω. For general sets we have to use the "harmonic germs" representation on $A(\omega)$ as in the proof of $\text{Proj}^1 A(\omega) = 0$ and use the Grothendieck-Bengel duality [8, Th. 4], [1, Satz 3] (comp. [13, Satz 2.4]).

2. Fundamental lemmas

We present here a geometrical lemma and its analytic consequences: decomposition results for real analytic functions. Here $\mathbb{D}$ will denote the unit disc.

2.1 Geometrical Lemma *Let $L, K \subseteq \mathbb{C}$ be compact, simply connected sets, $0 \in L \subseteq \operatorname{int} K$. Then there is ρ_0, $0 < \rho_0 < 1$, such that for every ρ, $\rho_0 < \rho < 1$, there is a biholomorphic map $\Psi: \mathbb{D} \longrightarrow \mathbb{C}$ such that*

$$L \subseteq \Psi(\rho_0 \mathbb{D}) \subseteq \Psi(\rho\mathbb{D}) \subset\subset \operatorname{int} K \subseteq K \subseteq \Psi(\mathbb{D}) \subset\subset \mathbb{C}.$$

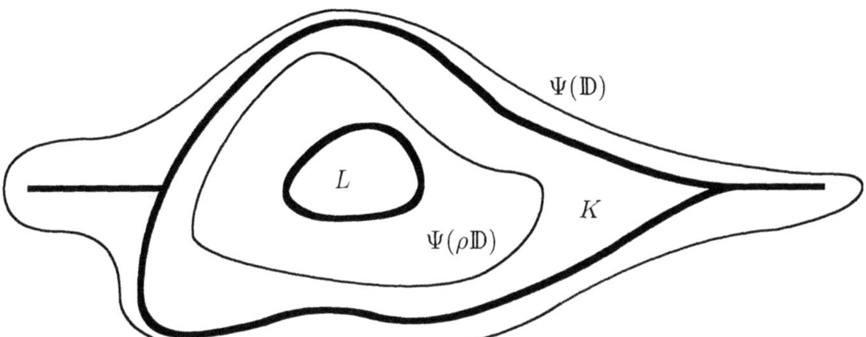

Proof. We choose a decreasing fundamental sequence $(V_k)_{k\in\mathbb{N}}$ of bounded, simply connected open neighborhoods of K and biholomorphic mappings $\psi_k: \mathbb{D} \longrightarrow V_k$, $\psi_k(0) = 0$. Going to a subsequence if necessary, we may assume that (ψ_k) and $(\psi_k^{-1}|_{\operatorname{int} K})$ converge uniformly on compact subsets of $\mathbb{D}$ and $\operatorname{int} K$, respectively. By [2, IV. §2, Satz 11], the compact-open limit ψ of (ψ_k) is a biholomorphic map $\psi: \mathbb{D} \longrightarrow \operatorname{int} K$ and $\varphi_k := \psi_k^{-1}|_{\operatorname{int} K}$ converges to $\varphi := \psi^{-1}$.

Let us take $\rho_0 > \sup_{z\in L} |\varphi(z)|$. Then there is k_0 such that for $k > k_0$ we have $\sup_{z\in L} |\varphi_k(z)| < \rho_0$, i.e., $L \subseteq \psi_k(\rho_0\mathbb{D})$. On the other hand, $M := \psi(\rho\overline{\mathbb{D}}) \subset\subset \operatorname{int} K$. We set $\varepsilon := \operatorname{dist}(M, \partial(\operatorname{int} K))$. There is $k_1 > k_0$ such that for $k > k_1$

$$\sup_{|z|\le\rho} |\psi_k(z) - \psi(z)| \le \frac{\varepsilon}{2}$$

and thus $\psi_k(\rho\overline{\mathbb{D}}) \subset \operatorname{int} K$. It suffices to take $\Psi = \psi_k$ for some $k > k_1$. □

This geometrical lemma will be applied in order to split some real analytic functions into a sum of holomorphic functions on carefully chosen domains.

Let us denote by $\|\cdot\|_\rho$ and B_ρ the norm and its unit ball in the disc algebra $A(\rho\overline{\mathbb{D}})$. We start with a standard and well-known fact:

2.2 Proposition *For $\rho_0 < \rho < 1$, there is a constant C such that*

$$B_\rho \subseteq C(rB_{\rho_0} + \frac{1}{r^a}B_1),$$

for all $r > 0$ *where* $a := \frac{\ln \rho}{\ln \rho_0 - \ln \rho}$.

Proof. It suffices to consider $r < 1$. We take $\nu \in \mathbb{N}$ such that

$$\left(\frac{\rho_0}{\rho}\right)^{\nu} \leq r < \left(\frac{\rho_0}{\rho}\right)^{\nu-1}$$

and split $f(z) = \sum_{n=0}^{\infty} a_n z^n$, $||f||_\rho \leq 1$ as follows:

$$\begin{aligned} g(z) &:= \sum_{n=\nu}^{\infty} a_n z^n = \frac{1}{2\pi i}\int_{|\zeta|=\rho} \frac{z^\nu}{\zeta^\nu}\frac{f(\zeta)}{\zeta - z}d\zeta \quad \text{for } |z| < \rho; \\ h(z) &:= \sum_{n=0}^{\nu-1} a_n z^n. \end{aligned}$$

We get

$$||g||_{\rho_0} \leq \frac{\rho}{\rho - \rho_0}\left(\frac{\rho_0}{\rho}\right)^{\nu} ||f||_\rho \leq C\, r,$$

and, by the Cauchy inequalities

$$||h||_1 \leq \frac{||f||_\rho}{1-\rho} \cdot \frac{1}{\rho^{\nu-1}} \leq C\,\frac{1}{r^a},$$

where $C = \max\left(\frac{\rho}{\rho - \rho_0}, \frac{1}{1-\rho}\right)$. □

Now, we can apply the Geometrical Lemma 2.1 to $L \subset\subset \mathbb{C}$, $K = M \cup \overline{U_0}$ where $U_0 \subset\subset U$ are open neighborhoods of L, $M \subset\subset \mathbb{R}$, then denoting $W := \Psi(\mathbb{D})$, $V = \Psi(\rho_0 \mathbb{D})$ and, by $||\cdot||_V$ the sup-norm on V, we get

$$||\cdot||_U^* \leq C(r||\cdot||_V^* + \frac{1}{r^a}||\cdot||_W^*),$$

where * means the dual norm. Taking the infimum of the right hand side with respect to r we obtain

$$C'||\cdot||_V^{*\,\alpha}||\cdot||_W^{*\,1-\alpha}$$

for $\alpha = \frac{\ln \rho}{\ln \rho_0}$. Therefore we obtain an (Ω)–type condition (see [19, p. 347]):

2.3 (Ω)-Lemma *For arbitrary compact sets* $L \subset\subset \mathbb{C}$, $M \subset\subset \mathbb{R}$, *open neighborhood* U *of* L, $0 < \alpha < 1$, *there are neighborhoods* V *of* L *and* W *of* $\overline{U} \cup M$, *such that*

$$||\cdot||_U^* \leq C||\cdot||_V^{*\,\alpha}||\cdot||_W^{*\,1-\alpha}.$$

The next lemma concerns vector valued functions. We call a function $f: \omega \longrightarrow E$, E a sequentially complete locally convex space, $\omega \subseteq \mathbb{R}^d$, *(topologically) real analytic*, $f \in A(\omega, E)$, if for every point $x \in \omega$ there is a neighborhood of x such that f is a sum of a power series convergent in E around x. For other possible definitions, their relations to the one above and their relevance for various problems of analysis see [3], [4] and [12, Chapter II].

We say that a sequentially complete DF-space E with a fundamental sequence of Banach discs (B_n) has the property $(\underline{A})$ if and only if there is a Banach disc $B \subseteq E$ such that for every n there is k, $\varepsilon_n > 0$ and $C > 0$ so that for every $r > 0$

$$B_n \subseteq C(rB + \frac{1}{r^{\varepsilon_n}} B_k).$$

Without loss of generality we may assume $k = n+1$, $B = B_0$.

The property $(\underline{A})$ is related via duality to the better known property $(\underline{DN})$ (see [19, §29]). The spaces $H(K)$ of germs of holomorphic functions for nice K, as well as duals of finite type power series spaces have the property $(\underline{A})$ (see [25], [19, 29.12]).

2.4 Vector Valued Splitting Lemma *Let $V \subset\subset U$ be open sets in $\mathbb{C}$, E a sequentially complete DF-space with the property $(\underline{A})$, where (B_n) are selected as above. Then for every $l \in \mathbb{N}$ and $f \in H(U, E_{B_l})$, $\varepsilon > 0$ there are $u \in H^\infty(V, E_{B_0})$, $||u||_{V,B_0}^\infty \leq \varepsilon$ and $v \in A(\mathbb{R}, E)$ so that $f = u + v$ on V.*

Here $||\cdot||_{V,B_0}^\infty$ denotes the norm of the space of E_{B_0}-valued bounded holomorphic functions $H^\infty(V, E_{B_0})$.

Proof. Without loss of generality we may assume $l = 1$, $[-1, 1] \subseteq V =: V_1$. We choose U_1 open in $\mathbb{C}$, $V \subset\subset U_1 \subset\subset U$. Then, by the Geometrical Lemma, we define inductively Riemann maps $\Psi_n \colon \mathbb{D} \longrightarrow \mathbb{C}$ and open sets V_n, U_n in $\mathbb{C}$ such that

$$V_{n-1} \subseteq \Psi_n(s_{n-1}\mathbb{D}) \subset\subset \Psi_n(r_{n-1}\mathbb{D}) \subseteq U_{n-1} \cup [-n, n] \subseteq V_n \subset\subset \Psi_n(\mathbb{D}) =: U_n.$$

Here $r_n, s_n < 1$ are chosen in such a way that

$$\frac{s_n^{\varepsilon_n}}{r_n^{1+\varepsilon_n}} \leq 1.$$

We construct inductively sequences of functions $(u_n), (v_n)$ with

$$\begin{aligned} u_n &\in H^\infty(V_{n-1}, E_{B_0}), \quad ||u_n||_{V_{n-1},B}^\infty < 2^{-n}\varepsilon, \\ v_n &\in H(U_n, E_{B_n}) \end{aligned}$$

such that $v_n = u_{n+1} + v_{n+1}$ on V_n.

We start with $u_1 = 0, v_1 = f$. If $v_n \in H(U_n, E_{B_n})$ is defined, then

$$v_n(\Psi_{n+1}(z)) = \sum_{j=0}^{\infty} a_j z^j$$

on some neighborhood of $r_n\mathbb{D}$. By the Cauchy estimates,

$$||a_j||_{B_n} \leq C r_n^{-j},$$

hence (changing C if necessary)

$$a_j \in C r_n^{-j} \left(rB_0 + \frac{1}{r^{\varepsilon_n}} B_{n+1} \right)$$

for all $r > 0$. We apply this to $r = \delta_n C^{-1} r_n^j s_n^{-j}$, obtaining

$$\begin{aligned} b_j &\in \delta_n s_n^{-j} B_0, \\ c_j &\in C^{1+\varepsilon_n} \delta_n^{-\varepsilon_n} \left(\frac{s_n^{\varepsilon_n}}{r_n^{1+\varepsilon_n}} \right)^j B_{n+1} \subseteq C^{1+\varepsilon_n} \delta_n^{-\varepsilon_n} B_{n+1} \end{aligned}$$

with $a_j = b_j + c_j$. Choosing δ_n appropriately and setting

$$u_{n+1}(\Psi_{n+1}(z)) = \sum_{j=0}^{\infty} b_j z^j, \qquad v_{n+1}(\Psi_{n+1}(z)) = \sum_{j=0}^{\infty} c_j z^j$$

we finish our induction. Finally, we define

$$u(z) = \sum_{n=1}^{\infty} u_n(z), \qquad v(z) = \lim_{n \to \infty} v_n(z).$$

Since for $m > n$

$$||v_m - v_n||_{V_n, B_0}^{\infty} = || \sum_{j=n+1}^{m} u_j ||_{V_n, B_0}^{\infty} \leq 2^{-n} \varepsilon$$

the function v is in $A(\mathbb{R}, E)$. □

As we have seen, the proofs of the above results are very much one-dimensional, they can be extended to product sets in $\mathbb{R}^d$ but for general ω the method fails. Nevertheless, in [7, Lemma 3.1] we proved a weaker version of the Geometrical Lemma for general ω, where the role of $\Psi(\mathbb{D})$ is played by an analytic polyhedron Ω, $\Psi(\rho\mathbb{D}), \Psi(\rho_0\mathbb{D})$ are exchanged by sub-level sets Ω_ρ, Ω_{ρ_0} of Ω, but we must assume $L \subseteq \mathbb{R}^d, K = U \cup J, J \subset\subset \mathbb{R}^d$, $U \subset \mathbb{C}^d$ open and $L \subseteq \Omega_{\rho_0} \subseteq \Omega_\rho \subset\subset U, J \subseteq \Omega$. That lemma is not sufficient to prove the analogue of the Vector Valued Splitting Lemma since U need not be contained in Ω. In the forthcoming paper [5] this difficulty is overcome by constructing special strictly pseudoconvex sets Ω.

Having polyhedra Ω instead of biholomorphic images of the unit disc we cannot split functions using Taylor series but one can use a deep result of Zaharjuta [31] as a splitting device to obtain analogues of the (Ω)-Lemma 2.3 (see [7, Lemma 3.3]) and the Vector Valued Splitting Lemma 2.4 ([5]).

3. Applications of the Lemmas

We start with applications to the structure of $A(\mathbb{R})$. Let us recall that a Fréchet space F with the fundamental sequence of seminorms $(|| \cdot ||_n)$ has the property $\left(\overline{\overline{\Omega}}\right)$ if

$$\forall\, k\ \exists\, m\ \forall\, n, \alpha \in]0,1[\ \exists\, C\ \forall\, u \in F' \qquad ||u||_m^* \leq C\,(||u||_k^*)^{\alpha}\,(||u||_n^*)^{1-\alpha},$$

here * denotes as usual the dual norm.

3.1 Corollary ([7], Theorem 3.4) *Every Fréchet quotient F of $A(\mathbb{R})$ has the property* $\left(\overline{\overline{\Omega}}\right)$.

Proof. Let $q: A(\mathbb{R}) \longrightarrow F$ be a quotient map, $F = \operatorname{proj} F_m$, F_m the local Banach space of F with the norm $\|\cdot\|_m$. Clearly, there are $n_p \in \mathbb{N}$ such that q extends to a continuous linear map $q_p: H[-n_p, n_p] \longrightarrow F_p$ for every $p \in \mathbb{N}$. We apply Grothendieck's factorization theorem (see [19, 24.33]) to $i_k F \subset \bigcup_\nu q_k(H^\infty(U_\nu))$, where $(U_\nu)_{\nu \in \mathbb{N}}$ is a basis of open neighborhoods of $[-n_k, n_k]$ in $\mathbb{C}$ and obtain ν so that i_k maps F continuously into $q_k(H^\infty(U_\nu))$. Evaluating the continuity estimates and putting $U = U_\nu$ we get $m \in \mathbb{N}$ and a neighborhood U of $[-n_k, n_k]$ such that $i_k^m: F_m \longrightarrow q_k(H^\infty(U))$ is continuous. Therefore applying the (Ω)-Lemma 2.3 to $L = [-n_k, n_k]$, $M = [-n_n, n_n]$ we get the conclusion. Note that α is arbitrary! □

3.2 Corollary ([7], Theorem 3.7) *Every Fréchet complemented subspace of $A(\mathbb{R})$ is finite dimensional.*

Proof. By [7, Theorem 3.6], every such subspace has $(\underline{DN})$, but $(\overline{\overline{\Omega}})$ plus $(\underline{DN})$ means Banach space (see [27, Satz 4.2] and [26, Satz 3.2], comp. [19, 29.21]). The result follows by nuclearity. □

Let us mention that the Fréchet subspaces of $A(\omega)$ are exactly identified in [6].

The essential functional analytic tool for the proof of the Theorem 3.4 on the nonexistence of a basis in $A(\omega)$ is contained in the following theorem.

3.3 Theorem ([7], Theorem 2.1, 2.2) *Every PLN-space X with basis and* $\operatorname{Proj}^1 X = 0$ *is either a LB-space or it contains an infinite dimensional complemented Fréchet subspace.*

The proof of the following theorem for arbitrary ω is the main result of [7]. As the essential complex analytic tool, namely Corollary 3.1, is proved here only for $\omega = \mathbb{R}$ we formulate the result only for this case.

3.4 Theorem ([7], Theorem 4.1) $A(\mathbb{R})$ *has no basis. Every complemented subspace of* $A(\mathbb{R})$ *with a basis is a DF-space.*

The lemmas from the preceding section allow also to extend some version of the Martineau Theorem to vector-valued analytic functions. Let us observe that

$$A(\mathbb{R}, E) = \operatorname*{proj}_n H([-n, n], E).$$

3.5 Corollary ([5]) $\operatorname{Proj}^1 A(\mathbb{R}, E) = 0$ *for every sequentially complete DF-space E with the property* $(\underline{A})$.

Proof. We prove the result directly from the definition. So let $(S_n) \in \prod H([-n, n], E)$, it is known that each $S_n \in H([-n, n], E_{B_n})$ for some Banach disc B_n. Without loss of generality we may assume that (B_n) is a fundamental sequence from the definition of $(\underline{A})$. Applying verbatim (for Banach-valued integrals) the proof of $\operatorname{Proj}^1 A(\mathbb{R}) = 0$ to the vector case, we get

$$S_n = A_n + H_n,$$

where $H_n \in H(\{z : |\Re z| < n - \frac{1}{2}\}, E_{B_n})$, $A_n \in A(\mathbb{R}, E)$. We apply the Vector-Valued Splitting Lemma 2.4 to H_n obtaining $u_n \in H^\infty((n-1)\mathbb{D}, E_{B_0})$, $||u_n||^\infty_{(n-1)\mathbb{D},B_0} \leq \varepsilon$ and $v_n \in A(\mathbb{R}, E)$ so that

$$H_n = u_n + v_n \quad \text{on } (n-1)\mathbb{D}.$$

Defining

$$T_n = \sum_{j=1}^{n}(v_j + A_j) - \sum_{j=n+1}^{\infty} u_j - S_n$$

we get

$$T_{n+1}|_{[-n,n]} - T_n = S_n.$$

□

The following theorem is proved in Bonet, Domański, Vogt [5] in much greater generality. However, our elementary approach allows already a fairly good interpolation result.

3.6 Theorem (Bonet, Domański, Vogt [5]) *Let E be a sequentially complete DF-space with the property $(\underline{A})$, $\omega \subset \mathbb{R}^d$ open, then for every sequence (w_n) in E and every discrete set $S = \{z_n : n \in \mathbb{N}\}$, $z_n \neq z_m$ for $n \neq m$, there is a real analytic E-valued function $f \in A(\omega, E)$ such that*

$$f(z_n) = w_n \quad \text{for } n \in \mathbb{N}.$$

Proof. Let φ be a real analytic function on ω with $\varphi(z_n) = n$ for all n. If we can find a function $f_0 \in A(\mathbb{R}, E)$ with $f_0(n) = w_n$ for all n then $f = f_0 \circ \varphi$ solves the problem. Therefore it suffices to give the proof of our theorem for $\omega = \mathbb{R}$.

Let $g\colon \mathbb{R} \longrightarrow \mathbb{R}$ be a real analytic function with zeros of order 1 at (z_n) and no other zeros. Clearly, there is a vector-valued polynomial p_n such that for $z_k \in [-n, n]$, $p_n(z_k) = w_k$. Consider $S_n(z) = \frac{p_{n+1}(z) - p_n(z)}{g(z)}$ for z in a neighborhood of $[-n, n]$, $S_n \in H([-n, n], E)$. By the previous Corollary 3.5, there is (T_n), $T_n \in H([-n, n], E)$, such that

$$T_{n+1}|_{[-n,n]} - T_n = S_n.$$

We define $f(z) = p_n(z) - T_n(z)g(z)$ on a neighbourhood of $[-n, n]$, the definitions coincide on the common domain of holomorphy. □

Let us mention that all the presented results are true for $A(\omega)$, ω arbitrary. Analogues of Corollaries 3.1, 3.2 and Theorem 3.4 are proved in [7, Theorem 3.4, 3.7, 4.1] using more sophisticated tools for the preparatory lemmas. The proof of 3.5 cannot be transferred directly because for arbitrary ω we cannot use the Cauchy formula, nevertheless it is true even for coherent sheaves of real analytic functions over arbitrary $\omega \subset \mathbb{R}^d$ instead of $A(\mathbb{R}, E)$ as showed in [5] but we have to refer in the proof to the Cartan-Oka theory. Then one can also prove Theorem 3.6 even for interpolation with multiplicities, i.e., not only values of the function but finitely many derivatives at every point may be prescribed (see [5]). Moreover, it is shown in [5] that $(\underline{A})$ is even necessary for interpolation.

Certainly, Theorem 3.4 is the most striking result giving probably the first example of a natural separable function space without basis (see the discussion in the Introduction of [7]). It is worth pointing out that the step spaces $H([-n, n])$ have very nice bases which unfortunately do not coincide for different n.

3.7 Proposition *The Čebyšev polynomials*

$$T_n(x) = \cos(n \arccos x) = \sum_{j=0}^{[\frac{n}{2}]} (-1)^j \binom{n}{2j} (1-x^2)^j x^{n-2j}$$

are a basis for the following spaces on $[-1,1]$*:*

(i) $L_2((1-x^2)^{-\frac{1}{2}}dx)$*;*

(ii) $C^\infty([-1,1])$*;*

(iii) $H([-1,1])$*;*

(iv) the space of restrictions of $H(\mathbb{C})$*;*

with the corresponding coefficient sequence spaces: $\ell_2, s, \mathbf{\Lambda}_0(n)', \mathbf{\Lambda}_\infty(n)$.

Here we denote by $\mathbf{\Lambda}_r(\alpha_n)$ the power series space of radius r. For the terminology see [19, §29]. Let us note that the real analytic bijection $\varphi\colon(-1,1) \longrightarrow \mathbb{R}$, $\varphi(t) := \tan\frac{\pi}{2}t$, induces a linear topological isomorphism of $A(\mathbb{R})$ and $A(-1,1)$. Thus the system (T_n) cannot be a basis of $A(-1,1)$.

Proof. The first two cases are classical (see [19, 29.5(4)]).
(iii): Using the Joukovski function

$$J(z) = \frac{1}{2}(z+z^{-1}), \quad z \in \mathbb{C} \setminus \{0\}$$

the composition map $C_J : f \mapsto f \circ J$, maps $H([-1,1])$ onto the following subspace of the space $H(\mathbf{T})$ of germs of holomorphic functions over the unit circle $\mathbf{T}$:

$$S = \{g \in H(\mathbf{T}) : g(z) = g(\bar{z}) \text{ for } z \in \mathbf{T}\}.$$

Therefore for $f \in H([-1,1])$ we have

$$C_J f(z) = \sum_{k=-\infty}^{+\infty} b_k z^k \quad \text{with } b_k = b_{-k} \text{ for } k \in \mathbb{N}$$

and $\sum |b_k|\, r^k < \infty$ for some $r > 1$. Thus

$$C_J f(z) = \sum_{k=0}^{\infty} a_k \frac{1}{2}(z^k + z^{-k}) \quad \text{for } a_0 = b_0,\ a_k = 2b_k.$$

Now it is easy to see that $T_k(x) = g_k(J^{-1}(x))$, where

$$g_k(z) = \frac{1}{2}(z^k + z^{-k}).$$

(iv): Let $f(x)=\sum_{j=0}^{\infty}a_jT_j(x)$. If $(a_j)_{j\in\mathbb{N}}\in\Lambda_\infty(n)$ then the series converges on $\mathbb{C}$ because

$$\begin{aligned}\sup_{|z|\le R}|T_n(z)| &\le \sum_{j=0}^{[\frac{n}{2}]}\binom{n}{2j}(1+R^2)^jR^{n-2j}\\ &= R^n\sum_{j=0}^{[\frac{n}{2}]}\binom{n}{2j}(1+\frac{1}{R^2})^j\\ &\le R^n(2+\frac{1}{R^2})^{n/2}.\end{aligned}$$

On the other hand, since $\deg T_n=n$ we have

$$x^k=\sum_{j=0}^{k}b_{k,j}T_j(x).$$

Evaluating

$$\Re\left(\frac{1}{2}\left(e^{it}+e^{-it}\right)\right)^k=\cos^k t=\sum_{j=0}^{k}b_{k,j}\cos jt$$

we obtain

$$b_{k,j}=\frac{1}{2^k}\left(\binom{k}{\frac{1}{2}(k+j)}+\binom{k}{\frac{1}{2}(k-j)}\right)$$

for $0\le j\le k$ and $k+j$ even. Otherwise $b_{k,j}=0$. Since $\sum_{l=0}^{k}\binom{k}{l}=2^k$, in particular, we have $|b_{k,j}|\le 2$ for all j,k.

Now, let b_k be the Taylor coefficients at 0 of an entire function f and a_j the coefficients in its Čebyšev expansion, as before, then clearly

$$a_j=\sum_{k=j}^{\infty}b_kb_{k,j}.$$

Therefore we have for $R>1$ and $|b_k|\le R^{-k}$

$$|a_j|\le 2\sum_{k=j}^{\infty}|b_k|\le 2CR^{-j}\frac{R}{R-1}.$$

So the coefficients (a_j) are in $\Lambda_\infty(n)$ which completes our proof. □

4. One dimensional convolution operators.

In Theorem 3.4 it was stated that a complemented subspace with basis of $A(\mathbb{R})$ must be a DF-space. As $A(I)\cong A(\mathbb{R})$ for any open interval $I\subset\mathbb{R}$ (comp. the remark after Proposition 3.7) this result holds of course also for $A(I)$. Now, kernels of convolution operators T_μ (see below) acting on spaces $A(I)$ do have bases, hence they can be complemented only if they are DF-spaces. It turns out that this yields a condition on the zeros of the Fourier-Laplace transform $\hat{\mu}$ which has been shown by Langenbruch [14] to characterize the convolution operators which admit continuous linear right inverses on spaces $A(I)$. We will exhibit this connection in the present section restricting ourselves for the sake of simplicity to the case of supp $\mu = \{0\}$. One should compare also the analogous results for the nonquasianalytic case obtained in [18]. For parts of the proofs we use a slightly different approach than [14]. Throughout this section we will follow the notation of Langenbruch [14].

We fix $\mu \in A(\mathbb{R})'$ (by Cor. 1.8, μ can be identified with the entire function $\hat{\mu}$) and we assume supp $\mu = \{0\}$. Then μ defines a convolution operator by

$$T_\mu(f)(x) := < \mu, f(x - \cdot) >$$

which acts on $A(I)$ for every open interval $I \subset \mathbb{R}$ and likewise on $A(J)$ for every compact interval $J \subset \mathbb{R}$. From now on I will always denote an open interval, J a compact interval. We set

$$N_\mu(\cdot) := \{f \in A(\cdot) : T_\mu(f) = 0\}.$$

4.1 Lemma *$N_\mu(\cdot)$ has a basis.*

Proof. For compact J this is an immediate consequence of [14, Lemma 1.4] and [17, Proposition 1.4] (use the representation of the dual given in Cor. 1.8). For open I the result is proved in [20, Th. 2.11] since the slowly decreasing assumption is satisfied in our setting by [14, condition (1.8), p. 71]. In fact the claim follows also from the proof of [17, Proposition 1.4] which shows that the choice of basis vectors in $K^J(E)$ can be made independently of J. Just the Köthe matrix depends then on J. □

As an immediate consequence of Lemma 4.1 and Theorem 3.4 we obtain:

4.2 Proposition *If $N_\mu(I)$ is complemented then it is a DF-space.*

To derive consequences from Proposition 4.2 we show the following Lemma. Let $E \subset A(I)$ be a closed subspace. We put

$$V(E) = \{\zeta \in \mathbb{C} : x \mapsto e^{i\zeta x} \in E\}.$$

4.3 Lemma *If $E \subset A(I)$ is a DF-subspace, then the following condition is satisfied*

$$\forall\, \varepsilon\ \exists\, c\ \forall\, \zeta = \xi + i\eta \in V(E) :\ |\eta| \le c + \varepsilon|\xi|.$$

Proof. If I is a bounded interval or $I = \mathbb{R}$, we may assume $I =]-a, a[$, $a > 0$ or $a = +\infty$. From the theory of PDF-spaces we obtain, assuming always $0 < r < a$, $0 < R < a$ and $m, M \in \mathbb{N}$

$$\exists\, r\ \forall\, R, m\ \exists\, M, C \ge 0\ \forall\, f \in E \cap H(\mathbb{C}) :\ \|f\|_{R,M} \le C\, \|f\|_{r,m}. \tag{2}$$

Here

$$\|f\|_{r,m} = \sup\left\{ f(x+iy) : |x| \le r,\ |y| \le \frac{1}{m} \right\}.$$

If we apply this to $f(z) = e^{i\zeta z}$, $\zeta \in V(E)$ then the inequality in (2) takes the form

$$e^{R\,|\eta| + \frac{1}{M}\,|\xi|} \le C\, e^{r\,|\eta| + \frac{1}{m}\,|\xi|}.$$

Hence we obtain from (2)

$$\exists\, r\ \forall\, R, m\ \exists\, M, c\ \forall\, \zeta = \xi + i\eta \in V(E):\ R\,|\eta| + \frac{1}{M}|\xi| \le c + r\,|\eta| + \frac{1}{m}|\xi|.$$

We choose $R > r$ and obtain

$$\forall\, \varepsilon\ \exists\, c\ \forall\, \zeta = \xi + i\eta \in V(E):\ |\eta| \le c + \varepsilon|\xi|.$$

The case where I is a half-line works in a similar fashion. □

If $E = N_\mu(I)$ then, of course

$$V(E) = V(\hat{\mu}) := \{\zeta \in \mathbb{C} : \hat{\mu}(\zeta) = 0\}$$

and we obtain from Lemma 4.3:

4.4 Proposition *If $N_\mu(I)$ is a DF-space then*

$$\forall\, \varepsilon\ \exists\, c\ \forall\, \zeta = \xi + i\eta \in V(\hat{\mu}):\ |\eta| \le c + \varepsilon|\xi|. \tag{3}$$

Therefore, by the sequence of the results above, (3) is a necessary condition for the existence of a right inverse (we always mean a *linear continuous* one) for T_μ. Langenbruch [14] proves the sufficiency of (3) by giving a proper C^∞-extension formula and proving the existence of a solution operator for a certain $\bar{\partial}$-problem (cf. [24]). We may give an alternative proof, at least in our special case. First we conclude from Proposition 4.4:

4.5 Corollary *If condition (3) is satisfied then we have*

$$N_\mu(I) = N_\mu(J) = N_\mu(\{0\})$$

for every open interval I and compact interval J.

Proof. It suffices to show this for $I \supset J \ni 0$. From the concrete representation in [14, Section 1] and condition (3) we conclude that

$$N_\mu(I)' = N_\mu(J)' = N_\mu(\{0\})'.$$

As all these spaces are semi-reflexive and functions in $A(I)$ and $A(J)$ are determined by their germs in 0 we obtain the result. □

From this we conclude easily that for a proof of sufficiency of the condition (3) it is enough to show that $T_\mu : A(J) \to A(J)$ has a right inverse for some compact interval $J \subset I$. As T_μ is translation-invariant and the zeros of $\hat{\mu_s}$, $\mu_s(f) = \mu(f_s)$, $f_s(x) = f(x/s)$ satisfy condition (3) iff those of μ do, it is enough to show it for $J = [-1, 1]$.

For this we need some preparation (cf. [14]). First we note that the Joukovski function from the proof of 3.7 (iii) maps the circle of radius $e^{\frac{1}{n}}$ to the ellipse with the semi-axes $(\cosh(1/n), \sinh(1/n))$, we call its interior U_n. Therefore, due to the proof of Proposition 3.7 (iii), the space $A([-1,1])$ graded by $A([-1,1]) = \mathrm{ind}_n H^\infty(U_n)$, is tamely isomorphic to the dual power series space $\Lambda_0(n)'$. We use the notation "tame" here also for "linear tame" as nothing else will be needed.

By Fourier transformation we may therefore identify $A([-1,1])'$ with

$$\mathcal{H} := \{f \in H(\mathbb{C}) : \ \forall\, n \in \mathbb{N}\ \|f\|_n = \sup_{z\in\mathbb{C}} |f(z)| \exp(-\omega_n(z)) < +\infty\}$$

where $\omega_n(z) = \sqrt{\sinh^2(1/n)\, x^2 + \cosh^2(1/n)\, y^2}$ is the support functional of the ellipse U_n. Therefore $\mathcal{H} \cong \Lambda_0(n)$ tamely.

We set now, following [14],

$$K = \{(f_k) \in \prod E_k : \ \forall\, n \in \mathbb{N}\ \|(f_k)\|_n = \sup_k |f_k|_k \exp(-\omega_n(z_k)) < +\infty\}.$$

Here $(E_k, |\ |_k)$ are certain finite dimensional Banach spaces and $z_k \in \mathbb{C}$ are chosen so that $|\tilde{z}_k - \zeta_k| = o(|\zeta_k|)$, where $(\tilde{z}_k)$ is the sequence of the z_k counted dim E_k times for every k and (ζ_k) is the sequence of zeros of $\hat{\mu}$ counted with multiplicities.

By [14, Lemma 1.4, (1.12")] applied to the ellipses U_n there is a tamely exact sequence

$$0 \longrightarrow \mathcal{H} \xrightarrow{M_F} \mathcal{H} \xrightarrow{\rho} K \longrightarrow 0 \tag{4}$$

where $F(z) := \hat{\mu}(-z)$ and M_F the operator of multiplication with F. Via the identification from Cor. 1.8, this is just the dual sequence to

$$0 \longrightarrow \ker T_\mu \longrightarrow A([-1,1]) \xrightarrow{T_\mu} A([-1,1]) \longrightarrow 0. \tag{5}$$

Let us observe that the surjectivity of T_μ above (or, equivalently, exactness of (5)) is exactly what is called "local surjectivity" in [20, Def. 2.3]. By [20, Th. 2.4] and [14, condition (1.8)], this is always true in our present setting, since we assume supp $\mu = \{0\}$.

We arrive at the following result which is Theorem 4.1 of Langenbruch [14].

4.6 Proposition *The exact sequence (4) splits if and only if (3) is satisfied.*

Proof. Let us assume (3). Then by a tame change of norms we may replace $\omega_n(z_k)$ by $|z_k|/n$.

Notice that in all spaces appearing in (4) the norms can, due to the (even exponential) nuclearity, be changed by a tame change into Hilbertian ones. In the case of K we notice that, assuming the change above has been made, the space $\Lambda_0(|z_k|)$ is a complemented subspace of K hence (exponentially) nuclear hence we may replace the suprema in the definition by sums and then apply Meise's result [17, Proposition 1.4] which makes K tamely isomorphic to a space $\Lambda_0(\alpha)$.

By [21, Corollary 6.3] the sequence (4) now splits.

To prove the reverse direction we refer to the proof of [14, Theorem 4.1]. □

We collect now the results into one theorem. Most of it is, of course, a survey on the results of Langenbruch [14].

4.7 Theorem *For μ with supp $\mu = \{0\}$ the following are equivalent:*

(a) *For some/every open interval $I \in \mathbb{R}$ the operator T_μ has a right inverse in $A(I)$.*

(b) *For some/every compact interval J with non-empty interior the operator T_μ has a right inverse in $A(J)$.*

(c) *There are an open interval I and a compact interval $J \subset I$, so that $N_\mu(I) = N_\mu(J)$, i.e. every zero solution in a neighborhood of J extends to I.*

(d) *$N_\mu(\mathbb{R}) = N_\mu(\{0\})$, i.e. every zero solution in a neighborhood of 0 extends to $\mathbb{R}$.*

(e) *For some/every open interval I the space $N_\mu(I)$ is a DF-space.*

(f) *The zeros of $\hat{\mu}$ satisfy condition (3).*

Let us compare the equivalence of (a) and (e) above to the fact that $T_\mu: A(I) \longrightarrow A(I)$ is surjective if and only if $\ker T_\mu = X_1 \oplus X_2$, where X_1 is a Fréchet space and X_2 is a DF-space. This is proved implicitly in [20, Th. 3.3]. More precisely, it follows from [20, Prop. 3.8, Lemma 3.10] and [29, Cor. 4.4]. For the present state of the art in the problem of surjectivity of convolution operators on $A(\mathbb{R})$ see [15].

Proof. We have just to collect the previous information. If we have (a) for some I, then by 4.2 we have (e) for the same I. By 4.4 we then get (f). Given (f) we get (d) by 4.5. (c) and (e) for every I follow from (d). The condition (c), of course, implies (e) for the same I, hence again (f). The condition (b) is equivalent to (f) by 4.6. So, to complete the proof, it is enough to show that (b) and (d) imply (a) for every I.

For that we choose compact $J \subset I$ with the non-empty interior, and a right inverse R_J in $A(J)$. Let $J \subset M \subset I$ be a compact interval and $f \in A(I)$, then by (b) we find $h \in A(M)$ so that $T_\mu(h) = f$. Clearly $g = R_J(f) - h \in N_\mu(J)$ hence, by (d), it extends to $G \in N_\mu(\mathbb{R})$. Therefore $R_J(f) = g + h$ extends to a neighborhood of M. As this is true for any such M $R_J(f)$ extends to I. So R_J creates a linear map $R : A(I) \longrightarrow A(I)$. It has closed graph, so, by Theorem 1.6 (d), it is continuous. □

REFERENCES

1. G. Bengel, *Das Weylsche Lemma in der Theorie der Hyperfunktionen*, Math. Z. **96** (1967), 373-392.
2. H. Behnke, F. Sommer, *Theorie der analytischen Funktionen einer komplexen Veränderlichen*, 3. Auflage, Springer, Berlin 1965.
3. J. Bonet, P. Domański, *Real analytic curves in Fréchet spaces and their duals*, Mh. Math. **126** (1998), 13-36.
4. J. Bonet, P. Domański, *Parameter dependence of solutions of partial differential equations in spaces of real analytic functions*, Proc. Amer. Math. Soc. **129** (2001), 495-503.
5. J. Bonet, P. Domański, D. Vogt, *Interpolation of vector-valued real analytic functions*, to appear.
6. P. Domański, M. Langenbruch, *Composition operators on spaces of real analytic functions*, to appear.

7. P. Domański, D. Vogt, *The space of real analytic functions has no basis*, Studia Math. **142** (2000), 187-200.
8. A. Grothendieck, *Sur les espaces de solutions d'une classe générale d'équations aux dérivées partielles*, J. Analyse Math **2** (1952/53), 253-280.
9. L. Hörmander, *An Introduction to Complex Analysis in Several Variables*, 3rd ed., North-Holland, Amsterdam 1990.
10. G. Köthe, *Topologische lineare Räume*, Springer, Berlin Göttingen Heidelberg 1960.
11. S. Krantz, *Function Theory of Several Complex Variables*, Wiley, New York 1982.
12. A. Kriegl, P. W. Michor, *The Convenient Setting of Global Analysis*, American Mathematical Society, Providence 1997.
13. M. Langenbruch, *P-Funktionale und Randwerte zu Hypoelliptischen Differentialoperatoren*, Math. Ann. **239** (1979), 55-74.
14. M. Langenbruch, *Continuous linear right inverses for convolution operators in spaces of real analytic functions*, Studia Math. **110** (1994), 65-82.
15. M. Langenbruch, *Hyperfunction fundamental solutions of surjective convolution operators on real analytic functions*, J. Funct. Anal. **131** (1995), 78-93.
16. A. Martineau, *Sur la topologie des espaces de fonctions holomorphes*, Math. Ann. **163** (1966), 62-88.
17. R. Meise, *Sequence space representations for (DFN)-algebras of entire functions modulo closed ideals*, J. reine angew. Math. **363** (1985), 59-95.
18. R. Meise, D. Vogt, *Characterization of convolution operators on spaces of C^∞-functions admitting a continuous linear right inverse*, Math Ann. **279** (1987), 141-155.
19. R. Meise, D. Vogt, *Introduction to Functional Analysis*, Clarendon Press, Oxford 1997.
20. T. Meyer, *Surjectivity of convolution operators on spaces of ultradifferentiable functions of Roumieu type*, Studia Math. **125** (1997), 101-129.
21. M. Poppenberg, D. Vogt, *A tame splitting theorem for exact sequences of Fréchet spaces*, Math. Z. **219** (1995), 141-161.
22. V. P. Palamodov, *Functor of projective limit in the category of topological vector spaces*, Mat. Sb. **75** (1968), 567-603 (in Russian); English transl., Math. USSR Sbornik **4** (1968), 529-558.
23. V. S. Retakh, *Subspaces of a countable inductive limit*, Doklady AN SSSR **194** (1970) (in Russian); English transl., Soviet Math. Dokl. **11** (1970), 1384-1386.
24. B. A. Taylor, *Linear extension operators for entire functions*, Michigan Math. J. **29** (1982), 185-197
25. M. Tidten, *A geometric characterization for the property $(\underline{DN})$ of $\mathcal{E}(K)$ for arbitrary compact subsets K of* $\mathbb{R}$, to appear.
26. D. Vogt, *Charakterisierung der Unterräume eines nuklearen stabilen Potenzreihenraumes von endlichen Typ*, Studia Math. **71** (1982), 251-270.
27. D. Vogt, *Fréchetträume, zwischen denen jede stetige lineare Abbildung beschränkt ist*, J. reine angew. Math. **345** (1983), 182-200.
28. D. Vogt, *Lectures on projective spectra of DF-spaces*, Seminar Lectures, Arbeitsgemeinschaft Funktionalanalysis (1987), Düsseldorf-Wuppertal.
29. D. Vogt, *Topics on projective spectra of LB-spaces*, in: *Advances in the Theory of Fréchet Spaces* (ed. T. Terzioğlu), Proc. of the NATO Adv. Res. Workshop, Istanbul, Turkey 1988, Kluwer Acad. Publ., NATO-ASI Series, Ser. C. Math. Ph. Sci. vol. **87**

Dordrecht 1989, pp. 11-27.

30. J. Wengenroth, *Acyclic inductive spectra of Fréchet spaces*, Studia Math. **120** (1996), 247-258.
31. V. P. Zaharjuta, *Extremal plurisubharmonic functions, Hilbert scales, and the isomorphism of spaces of analytic functions of several variables, I, II*, Teor. Funktsiĭ Funktsional. Anal. Prilozhen. **19** (1974), 133-157; **21** (1974), 65-83 (in Russian).

Recent Progress in Functional Analysis
K.D. Bierstedt, J. Bonet, M. Maestre and J. Schmets

Contribution to the isomorphic classification of Sobolev spaces $L^p_{(k)}(\Omega)$ $(1 \leq p < \infty)$*

Aleksander Pełczyński[a] and Michał Wojciechowski[b]

[a]Institute of Mathematics, Polish Academy of Sciences, Śniadeckich 8, 00-950 Warszawa, Poland, E-mail: olek@impan.gov.pl

[b]Institute of Mathematics, Polish Academy of Sciences, Śniadeckich 8, 00-950 Warszawa, Poland, E-mail: miwoj@impan.gov.pl

To Professor Manuel Valdivia on the occasion of his seventieth birthday.

Abstract

It is proved that if Ω is an open, non-empty subset of $\mathbb{R}^n$ such that for some p with $1 \leq p < \infty$ and some positive integer k there exists a linear extension operator from a Sobolev space $L^p_{(k)}(\Omega)$ into $L^p_{(k)}(\mathbb{R}^n)$ then $L^p_{(k)}(\Omega)$ isomorphic as a Banach space to $L^p_{(k)}(\mathbb{R}^n)$.

MCS 2000 Primary 46B03, 46E35.
Key words and phrases. Sobolev spaces, linear extension operators, isomorphic Banach spaces.

1. Introduction

In this note we prove

Theorem 1 *Let $n = 1, 2, \ldots$. If $\Omega \subset \mathbb{R}^n$ is an $L^p_{(k)}$-linear extension domain for some $1 \leq p < \infty$ and some $k = 1, 2, \ldots$ then $L^p_{(k)}(\Omega)$ is isomorphic to $L^p_{(k)}(\mathbb{R}^n)$.*

For the notation and terminology see Section 2.

Our proof is modelled on Mityagin's [7] proof of his result that for fixed positive integers n and k for every non-empty open set in $\mathbb{R}^n$ the space $C^{(k)}_u(\Omega)$ is isomorphic to $C^{(k)}_u(\mathbb{R}^n)$. His argument heavily depends on the Whitney Extension Theorem (cf. [13]; [3], Theorem 2.3.6) which in our terminology says that every open non-empty subset of $\mathbb{R}^n$ is a $C^{(k)}_u$-linear extension domain ($k = 1, 2, \ldots$). This is not true for $L^p_{(k)}$-extension domains. There are even simply connected domains in $\mathbb{R}^2$ which are not $L^2_{(k)}$-linear extension domains (cf. [6] § 1.5). To overcome this obstruction we use Lemma 3 below to prove infinite divisibility of $L^p_{(k)}(I^n)$ where $I^n = (-1/2, 1/2)^n$ denotes the unit cube in $\mathbb{R}^n$. For $1 < p < \infty$ this fact can be obtained in a simpler way using a direct proof that for the n-dimensional torus $\mathbb{T}^n$

*Supported in part by the Polish KBN grant 2P03A 036 14.

the space $L^p_{(k)}(\mathbb{T}^n)$ is isomorphic to $L^p(0,1)$ for all positive integers k and n (cf. [9]; [10]). For $p = 1$ another possible approach offers the Jones Extension Theorem [4]. However the assumption of the latter theorem is difficult to verify for particular sets.

2. Terminology and notation

Let us recall the basic notation. By $\partial^\alpha f$ and $D^\alpha f$ we denote the α-th partial derivative and the α-th distributional partial derivative of a scalar-valued function f in n variables corresponding to the multiindex $\alpha \in \mathbb{Z}^n_+$, where $\mathbb{Z}_+ := \{0, 1, 2, \ldots\}$.

Recall that given a scalar-valued function f defined on an open set $\Omega \subset \mathbb{R}^n$ a function g on Ω is called the α-th *distributional derivative* of f, in symbols $g := D^\alpha f$ provided

$$\int_\Omega f \partial^\alpha \phi \, dx = (-1)^{|\alpha|} \int_\Omega g \phi \, dx \quad \phi \in \mathcal{D}(\Omega).$$

Here and in the sequel $\mathcal{D}(\Omega)$ stands for the space of scalar-valued infinitely many times differentiable functions ϕ with compact support,

$$\operatorname{supp} \phi := \overline{\{x \in \mathbb{R}^n : \phi(x) \neq 0\}} \subset \Omega;$$

$\overline{A}$ stands for the closure of a subset A of $\mathbb{R}^n$; $\operatorname{bd} A = \overline{A} \setminus A$.

For the multiindex $\alpha = (\alpha_j)_{j=1}^n$ the quantity $|\alpha| := \sum_{j=1}^n \alpha_j$ is called the order of the derivative D^α. For $\alpha = 0 := (0, 0, \ldots, 0)$ we admit for convenience $D^0 f = f$ and $D^\alpha \mu = \mu$. The symbol $\int \ldots dx$ denotes the integral against λ_n—the n-dimensional Lebesgue measure on $\mathbb{R}^n$. By $L^p = L^p(\Omega)$ we denote the Lebesgue space L^p on $\Omega \subset \mathbb{R}^n$ with respect to λ_n. The field of scalars is either real numbers—$\mathbb{R}$ or complex numbers—$\mathbb{C}$.

Let $1 \le p < \infty$. Let $k = 1, 2, \ldots$. The Sobolev space $L^p_{(k)}(\Omega)$ is the Banach space of scalar-valued functions f on Ω such that $D^\alpha f$ exists and belongs to $L^p(\Omega)$ for $|\alpha| \le k$ equipped with the norm

$$||f||_{L^p_{(k)}(\Omega)} = \Big(\sum_{|\alpha| \le k} ||D^\alpha f||^p_{L^p(\Omega)} \Big)^{1/p}.$$

By $C^{(k)}_u(\Omega)$ we denote the space of scalar-valued functions ϕ on Ω, k times differentiable and such that $\partial^\alpha \phi$ is uniformly continuous on Ω and vanishes at infinity for $0 \le |\alpha| \le k$. We admit

$$||\phi||_{C^{(k)}_u(\Omega)} = \max_{|\alpha| \le k} \sup_{x \in \Omega} |\partial^\alpha \phi(x)|.$$

For $\emptyset \neq \Omega_1 \subset \Omega$ and a function f on Ω we denote by $f|\Omega_1$ the restriction of f to Ω_1; recall that f is an extension of f_1 provided $f_1 = f|\Omega_1$.

Fix p with $1 \le p \le \infty$ and $k \in \mathbb{N}$. An open set $\Omega \subset \mathbb{R}^n$ is said to be an $L^p_{(k)}$*-linear extension domain* provided there is a bounded linear operator (called a *linear extension*) $\mathcal{E} = \mathcal{E}^\Omega_{(k),p} : L^p_{(k)}(\Omega) \to L^p_{(k)}(\mathbb{R}^n)$ such that $\mathcal{E}(f)$ is an extension of f for every $f \in L^p_{(k)}(\Omega)$. The definition of $C^{(k)}_u$*-linear extension domain* is analogous.

In the proof of Theorem 1 we make use of the existence of linear extension operators for functions defined on simple domains in $\mathbb{R}^n$ like parallelepipeds, frames and strips. They are constructed in a standard way (cf. [2]; [1], vol. I, Appendix; [12], Chapt. VI) using as building blocks linear extension operators for functions defined in a half-space to functions defined on the whole space $\mathbb{R}^n$. One such operator (cf. e.g. [8]) which we call a *Hestenes type extension operator* is defined by

$$\mathcal{H}f(x) = \begin{cases} f(x), & \text{for } x = (x_j) \in \mathbb{R}^n \text{ with } x_n < 0 \\ \sum_k a_k f(x_1, x_2, \dots x_{n-1}, -2^k x_n)\phi(x_n), & \text{for } x_n > 0; \end{cases}$$

where the a_k's are coefficients of an entire function $F(z) = \sum_k a_k z^k$ such that $F(2^m) = (-1)^m$ for $m = 1, 2, \dots$ and ϕ is a C^∞ function equal 1 at $x_n = 0$ and equal 0 for $|x_n| > 1/2$. The operator $\mathcal{H}$ has the additional property that if $f(x) = 0$ for $(x_j)_{j=1}^n \in \mathbb{R}^n$ with $x_n < 0$ and $x_s > c$ for some $s \in \{1, 2, \dots, n-1\}$ and some $c \in \mathbb{R}$ then $\mathcal{H}f(x) = 0$ for all $(x_j)_1^n \in \mathbb{R}^n$ with $x_s > 1$.

For $f \in L^p_{(k)}(\Omega)$ put

$$\nabla^p_k f := \Big(\sum_{|\alpha|=k} |D^\alpha f|^p\Big)^{1/p} : \Omega \to \mathbb{R}.$$

For $a > 0$ denote also by a the similarity $x \to ax$ for $x \in \mathbb{R}^n$, and by a^o the induced operator on functions (defined on subsets of $\mathbb{R}^n$) given by $a^o(f) = f \circ a$.

Proof of Theorem 1

A set $P \in \mathbb{R}^n$ of the form $P = \bigtimes_{j=1}^n (a_j, b_j)$ with $-\infty < a_j < b_j < +\infty$ for $j = 1, 2, \dots$ is called a *regular parallelepiped.* We put

$$(\complement P)^+ = \{x = (x_j) \in \mathbb{R}^n : x_j > b_j \quad \text{for some} \quad j = 1, 2, \dots, n\}.$$

For a non-empty open $\Omega \subset P$ with $\overline{\Omega} \supset \bigcup_{j'=1}^n \{x = (x_j) \in \operatorname{bd} P : x_{j'} = b_{j'}\}$ we denote by $\overset{+}{L}{}^p_{(k)}(\Omega : P)$ the subspace of $L^p_{(k)}(\Omega)$ consisting of those functions which extend by 0 to functions in $L^p_{(k)}(\Omega \cup \overline{(\complement P)^+})$. We put $\overset{+}{L}{}^p_{(k)}(P) = \overset{+}{L}{}^p_{(k)}(P : P)$. Similarly $L^p_{0,(k)}(P)$ denotes the subspace of those functions in $L^p_k(P)$ which extend by 0 to functions in $L^p_{(k)}(\mathbb{R}^n)$

To see that $\overset{+}{L}{}^p_{(k)}(\Omega : P)$ is closed in $L^p_{(k)}(\Omega)$ pick a norm Cauchy sequence $(f_n) \subset \overset{+}{L}{}^p_{(k)}(\Omega)$. For $n = 1, 2, \dots$ choose $\widetilde{f}_n \in L^p_{(k)}(\Omega \cup \overline{(\complement P)^+})$ so that $\widetilde{f}_n|\Omega = f$ and $\widetilde{f}|(\complement P)^+ = 0$. Then $(\widetilde{f}_n)$ is a norm Cauchy sequence in $L^p_{(k)}(\Omega \cup \overline{(\complement P)^+})$. The desired conclusion easily follows from the completeness of the spaces $L^p_{(k)}(\Omega)$ and $L^p_{(k)}(\Omega \cup \overline{(\complement P)^+})$. We shall need the following essentially known facts.

Lemma 2 (i) *If $P \subset \mathbb{R}^n$ is a regular parallelepiped then there is $C_P = C(P, k, p, n) > 0$ such that*

$$||\nabla^p_k f||_{L^p(P)} \le ||f||_{L^p_{(k)}(P)} \le C_P ||\nabla^p_k f||_{L^p(P)} \quad \textit{for} \quad f \in \overset{+}{L}{}^p_k(P). \quad (\textit{resp. } f \in L^p_{0,(k)}(P)).$$

Moreover the function $P \to C_P$ can be chosen so that if $P_n \to P$ in the Hausdorff metric then $C_{P_n} \to C_P$.

(ii) *Let P and P_1 be regular parallelepipeds in $\mathbb{R}^n$ with $\overline{P} \subset P_1$. Then there is a linear extension operator $\Lambda : L^p_{(k)}(P_1 \setminus \overline{P}) \to L^p_{(k)}(P_1)$ and a constant $A = A(P_1, P, p, n, k)$ such that*

$$||\nabla^p_k(\Lambda f)||_{L^p(P_1)} \leq A||\nabla^p_k f||_{L^p(P_1\setminus\overline{P})} \quad \textit{for} \quad f \in L^p_{(k)}(P_1 \setminus \overline{P}). \tag{1}$$

Moreover, for every $a > 0$,

$$(1/a)^o\Lambda a^o : L^p_{(k)}(a(P_1 \setminus \overline{P})) \to L^p_{(k)}(aP_1)$$

is a linear extension operator such that with the same constant A one has

$$||\nabla^p_k((1/a)^o\Lambda a^o f)||_{L^p(aP_1)} \leq A||\Lambda||\,||\nabla^p_k f||_{L^p(a(P_1\setminus\overline{P}))} \quad \textit{for } f \in L^p_{(k)}(a(P_1 \setminus \overline{P})).$$

Proof. (i) Denote by $\mathcal{P}_{n,k}(P)$ the (finite dimensional!) subspace of $L^p_{(k)}(P)$ consisting of all the polynomials in n variables of degree less than k. Note that $\mathcal{P}_{n,k}(P) \cap \overset{+}{L}{}^p_{(k)}(P) = \{0\}$ because the 0 polynomial is the only polynomial of degree less than k which together with all its partial derivatives of degree less than k vanishes on bd P. (If $f \in L^p_{(k)}(P)$ then $D^\alpha(f)$ has the trace at λ_{n-1} almost every point of bd P for $|\alpha| < k$.) Thus there is a bounded projection $\pi : L^p_{(k)}(P) \to L^p_{(k)}(P)$ such that $\pi(L^p_{(k)}(P)) = \mathcal{P}_{n,k}(P)$ and $\overset{+}{L}{}^p_{(k)}(P) \subset \ker P$. Let us put

$$|||f||| = ||\nabla^p_k f||_{L^p(P)} + ||\pi(f)||_{L^p(P)} \quad (f \in L^p_{(k)}(P)).$$

A regular parallelepiped has the so called cone property (cf. [6], 1.1.9, Definition II). Thus it follows from [6], 1.1.11, Corollary and 1.1.15, Theorem, that the norm $|||\cdot|||$ is equivalent to the original norm $||\cdot||_{L^p_{(k)}}$. Combining this fact with the observation that $|||f||| = ||\nabla^p_k f||_{L^p(P)}$ for $f \in \ker\pi$ we get the right hand side inequality of (i). The left hand side one is trivial. The argument for $L^p_{0,(k)}(P)$ is the same. The left hand side inequality is trivial.

For the moreover part of (i) note that if Q is another regular parallelepiped in $\mathbb{R}^n$ that there is an isomorphism of $\mathbb{R}^n$, say T_Q, which is a composition of a shift with rescaling of each coordinate axis such that $T_Q(P) = Q$. Analyzing how the norms $||\nabla^p_k(\cdot)||_{L^p(P)}$ and $||\cdot||_{L^p_{(k)}(P)}$ depend on T_Q, we infer that they continuously depend on the ratios of the lengths of parallel edges of P and Q. Clearly if P_n tends to P in the Hausdorff metric then T_{P_n} tends to the identity operator on $\mathbb{R}^n$.

(ii) For the frame $P_1 \setminus \overline{P}$ (more generally for every bounded connected domain in $\mathbb{R}^n$) one has

$$\nabla^p_k f \equiv 0 \quad \text{iff} \quad f \in \mathcal{P}_{n,k}(P_1 \setminus \overline{P}) \quad (f \in L^p_{(k)}(P_1 \setminus \overline{P})).$$

Thus $||\nabla^p_k(\cdot)||_{L^p(P_1\setminus\overline{P})}$ induces a norm on the quotient space $L^p_{(k)}(P_1\setminus\overline{P})/\mathcal{P}_{n,k}(P_1\setminus\overline{P})$ which by [6], 1.1.11 Corollary and 1.1.13 Theorem 1 is equivalent to the quotient norm. Hence there is a closed subspace Y of $L^p_{(k)}(P_1 \setminus \overline{P})$ such that codim $Y = \dim(\mathcal{P}_{n,k}(P_1 \setminus \overline{P}))$ and $Y \cup \mathcal{P}_{n,k}(P_1 \setminus \overline{P}) = \{0\}$. Thus there is a projection π from $L^p_{(k)}(P_1 \setminus \overline{P})$ onto $\mathcal{P}_{n,k}(P_1 \setminus \overline{P})$ such that $\ker\pi = Y$. The desired extension operator Λ is defined by

$$\Lambda f = \mathcal{E}(f - \pi(f)) + I(\pi(f)) \quad \text{for} \quad f \in L^p_{(k)}(P_1 \setminus \overline{P}),$$

where $\mathcal{E} : L^p_{(k)}(P_1 \setminus \overline{P}) \to P_1$ is an arbitrary linear extension operator (cf [12], Chapt. VI for the existence of an $\mathcal{E}$) and $I : \mathcal{P}_{n,k}(P_1 \setminus \overline{P}) \to \mathcal{P}_{n,k}(P_1)$ denotes the unique extension operator.

The moreover part of (ii) is a straightforward consequence of the definition of the induced operator a^o, the jacobian formula and the way how partial derivatives of order k transform via a^o. □

Let $J^n = (0,1)^n$. Let us consider in $\mathbb{R}^n$ the regular parallelepipeds $P = (1/2, 3/2)^n$ and $P_1 = (1/4, 7/4)^n$. Let $a_m = 8^{-m}$ for $m = 1, 2, \ldots$. Let us put $\Sigma = J^n \setminus \bigcup_{m=1}^{\infty} a_m P$. The infinite divisibility of $L^p_{(k)}(I^n)$ bases upon the next

Lemma 3 *There exists a linear extension operator* $\mathcal{E} : \overset{+}{L}{}^p_{(k)}(\Sigma) \to \overset{+}{L}{}^p_{(k)}(J^n)$.

Proof. For $m = 1, 2, \ldots$ put

$$R_m = \left(\frac{15}{8} 8^{-m}, 1\right)^n; \quad F_m = a_m(P_1 \setminus \overline{P}); \quad \Sigma_m = R_m \setminus \bigcup_{k=1}^{m-1} \overline{F_k}.$$

Clearly

$$\bigcup_{k=1}^{m-1} \overline{F_k} \subset R_m; \quad \overline{\bigcup_{k=m}^{\infty} F_k} \cap R_m = \emptyset.$$

For $f \in \overset{+}{L}{}^p_{(k)}(\Sigma_m : R_m)$ and for $m = 1, 2, \ldots$ we put

$$\Lambda_m f(x) = \begin{cases} f(x), & \text{for } \in \Sigma_m \\ (1/a_k)^o \Lambda (a_k)^o (f|F_k)(x), & \text{for } x \in F_k \quad (k = 1, 2, \ldots, m-1), \end{cases}$$

where $\Lambda : L^p_{(k)}(P_1 \setminus \overline{P}) \to L^p_{(k)}(P_1)$ is a fixed linear extension operator satisfying (1). It is easy to see that $\Lambda_m : \overset{+}{L}{}^p_{(k)}(\Sigma_m) \to \overset{+}{L}{}^p_{(k)}(R_m)$ is a linear extension operator for $m = 1, 2, \ldots$. Now fix $f \in \overset{+}{L}{}^p_{(k)}(\Sigma)$ and a multiindex $\alpha \in \mathbb{Z}^n_+$ with $|\alpha| \leq k$. Obviously for $m = 1, 2, \ldots$ the derivative $D^\alpha \Lambda_m(f|\Sigma_m) : R_m \to \mathbb{C}$ exists and $D^\alpha \Lambda_{m+1}(f|\Sigma_{m+1})$ is an extension of $D^\alpha \Lambda_m(f|\Sigma_m)$. Thus for $x \in J^n$ a.e. the sequence $(D^\alpha \Lambda_m(f|\Sigma_m)(x))$ is well defined for large m (such that $x \in \Sigma_m$) and it is eventually constant. Let us put $g_\alpha(x) = \lim_m D^\alpha \Lambda_m(f|\Sigma_m)(x)$ for $x \in J^n$ a.e. We define

$$(\mathcal{E} f)(x) = g_{(0,0,\ldots,0)}(x) = \lim_m \Lambda_m(f|\Sigma_m)(x) \quad \text{for } x \in J^n \text{ a.e.} \quad .$$

We first show that $g_\alpha = D^\alpha \mathcal{E} f$. Indeed, pick $\phi \in \mathcal{D}(J^n)$. Then $\operatorname{supp} \phi \subset R_{m_0}$ for some $m_0 \in \mathbb{N}$. Thus

$$\begin{aligned} \int_{J^n} \phi g_\alpha \, dx &= \int_{R_{m_0}} \phi \cdot D^\alpha \Lambda_{m_0}(f|\Sigma_{m_0}) \, dx \\ &= (-1)^{|\alpha|} \int_{R_{m_0}} \partial^\alpha \phi \cdot \Lambda_{m_0}(f|\Sigma_{m_0}) \, dx \\ &= (-1)^{|\alpha|} \int_{J^n} \partial^\alpha \phi \cdot \mathcal{E} f \, dx. \end{aligned}$$

Now we shall show that $D^\alpha \mathcal{E} f \in L^p(J^n)$ for $|\alpha| \le k$. The moreover part of Lemma 2 (ii) yields

$$\begin{aligned}||\nabla_k^p \Lambda_m (f|\Sigma_m)||^p_{L^p(R_m)} &= ||\nabla_k^p (f|\Sigma_m)||^p_{L^p(\Sigma_m)} + \sum_{k=1}^{m-1} ||\nabla_k^p (a_k)^o \Lambda (1/a_k)^o (f|F_k)||^p_{L^p(a_k P)} \\ &\le ||\nabla_k^p (f|\Sigma_m)||^p_{L^p(\Sigma_m)} + (A||\Lambda||)^p \sum_{k=1}^{m-1} ||\nabla_k^p (f|F_k)||^p_{L^p(F_k)} \\ &\le (1 + (A||\Lambda||)^p ||\nabla_k^p (f|\Sigma_m)||^p_{L^p(\Sigma_m)}.\end{aligned}$$

Thus

$$||\nabla_k^p \mathcal{E} f||_{L^p(J^n)} = \lim_m ||\nabla_k^p \Lambda f|\Sigma_m)||_{L^p(R_m)} \le 2A||\Lambda||||\nabla_k^p (f|\Sigma)||_{L^p(\Sigma)} < \infty.$$

Therefore $D^\alpha f \in L^p(J^n)$ for $|\alpha| = k$. Finally if $|\beta| < k$ then it follows from Lemma 2 (i) that

$$||D^\beta \Lambda_m (f|\Sigma_m)||_{L^p(R_m)} \le C_{R_m} ||\nabla_k^p \Lambda_m (f|\Sigma_m)||_{L^p(R_m)}$$

because if $f \in \overset{+}{L}{}^p_{(k)}(\Sigma : J^n)$ then $f|\Sigma_m \in \overset{+}{L}{}^p_{(k)}(\Sigma_m : R_m)$, hence $\Lambda_m (f|\Sigma_m) \in \overset{+}{L}{}^p_{(k)}(R_m)$. Thus, taking into account the moreover part of (i), we get

$$\begin{aligned}||D^\beta \Lambda f||_{L^p(J^n)} &= \lim_m \left(\int_{R_m} |D^\beta f|^p \, dx\right)^{1/p} \\ &= \lim_m ||D^\beta \Lambda_m (f|\Sigma_m)||_{L^p(R_m)} \\ &\le C_{J^n} ||\nabla_k^p \mathcal{E} f||_{L^p(J^n)} < \infty.\end{aligned}$$

Therefore $D^\beta \mathcal{E} f \in L^p(J^n)$ for $0 \le |\beta| < k$. Thus $\mathcal{E}$ is the desired linear extension operator. □

We employ the following notation. Let X and Y be Banach spaces. Then:
"$X \sim Y$" stands for "X is isomorphic to Y",
"$X|Y$" stands for "X is isomorphic to a complemented subspace of Y".

Note that if $\emptyset \ne \Omega \subset \Omega_1 \subset \mathbb{R}^n$, E, E_1 are subspaces of $L^p_{(k)}(\Omega)$ and $L^p_{(k)}(\Omega_1)$ respectively and $\mathcal{E}$ is a linear extension operator from E into E_1 such that the restriction $f|\Omega$ belongs to E for every $f \in E_1$ then $E|E_1$. Indeed $f \to \mathcal{E}(f|\Omega)$ is a projection from E_1 onto E and $f \to \mathcal{E} f$ is an isomorphism from E onto $\mathcal{E}(E)$.

Now we are ready for

Proof of Theorem 1. First we show that $L^p_{(k)}(I^n) \sim L^p_{(k)}(J^n)$ is infinitely divisible, precisely that

$$L^p_{(k)}(J^n) \sim F, \tag{2}$$

where F denotes the infinite l^p sum,

$$F := (L^p_{0,(k)}(I^n) \times L^p_{0,(k)}(I^n) \times \ldots)_p.$$

To establish (2) note that obviously $F \sim (F \times F \times \ldots)_p$ and $L^p_{0,(k)}|F$. Thus, by the decomposition method (cf., e.g., [5]), it is enough to verify that $F|L^p_{0,(k)}(2I^n)$. To this end it suffices to show

(a) $\overset{+}{L}{}^p_{(k)}(J^n)|L^p_{0,(k)}(2I^n)$,

(b) $F|\overset{+}{L}{}^p_{(k)}(J^n)$.

To prove (a) one uses the projection $f \to \mathcal{E}f|J^n$ for $f \in \overset{+}{L}{}^p_{(k)}(J^n)$ where the linear extension operator $\mathcal{E}$ is defined in a standard way by means of Hestenes type extension operators across the hyperplanes $\{x_j = 0\}$ for $j = 1, 2, \ldots, n$ multiplied by a C^∞-function which equals 1 on J^n and equals 0 in the neighborhood of the set $\bigcup_{j'=1}^n \{x = (x_j) \in \mathrm{bd}(2I^n) : x_{j'} = -1\}$.

To establish (b) observe first that if $\mathcal{E} : \overset{+}{L}{}^p_{(k)}(\Sigma) \to \overset{+}{L}{}^p_{(k)}(J^n)$ is the operator constructed in Lemma 3 then the operator $f \to f - \mathcal{E}f|\Sigma$ for $f \in \overset{+}{L}{}^p_{(k)}(J^n)$ is a projection whose range naturally identifies with $L^p_{0,(k)}(J^n \setminus \overline{\Sigma})$. Thus $L^p_{0,(k)}(J^n \setminus \overline{\Sigma})|\overset{+}{L}{}^p_{(k)}(J^n)$. Since $J^n \setminus \overline{\Sigma} = \bigcup_{m=1}^\infty a_m P$ and the sets $\overline{a_m P}$ are mutually disjoint ($m = 1, 2, \ldots$), we have the (isometric) isomorphism

$$L^p_{0,(k)}(J^n \setminus \overline{\Sigma}) \sim (L^p_{0,(k)}(a_1 P) \times L^p_{0,(k)}(a_2 P) \times \ldots)_p. \tag{3}$$

For $m = 1, 2, \ldots$ let us consider on $L^p_{0,(k)}(a_m P)$ (regarded as a subspace of $\overset{+}{L}{}^p_{(k)}(J^n)$) the norm $||\nabla^p_k(\cdot)||_{L^p(a_m P)} = ||\nabla^p_k(\cdot)||_{L^p(J^n)}$. Thus, by Lemma 2 (i), this norm is equivalent on $\overset{+}{L}{}^p_{(k)}(J^n)$ to the original norm $||\cdot||_{L^p_{(k)}(J^n)}$. Thus there is a constant $C > 0$ independent of m such that

$$||\nabla^p_k(f)||_{L^p(a_m P)} \leq ||f||_{L^p_{(k)}(a_m P)} \leq C||\nabla^p_k(f)||_{L^p(a_m P)} \quad \text{for } f \in L^p_{0,(k)}(a_m P).$$

Next observe that the map $f \to (a_m)^{\frac{n}{p}-k}[(a_m)^o f]$ is the isometric isomorphism from the space $(L^p_{0,(k)}(a_m P), ||\nabla^p_k(\cdot)||_{L^p(a_m P)})$ onto the space $(L^p_{0,(k)}(P), ||\nabla^p_k(\cdot)||_{L^p(P)})$. Thus regarding all $L^p_{0,(k)}(a_m P)$ and $L^p_{0,(k)}(P)$ in the original norms we infer that

$$(L^p_{0,(k)}(a_1 P) \times L^p_{0,(k)}(a_2 P) \times \ldots)_p \sim (L^p_{0,(k)}(P) \times L^p_{0,(k)}(P) \times \ldots)_p;$$

the desired isomorphism is given by $(f_m) \to ((a_m)^{\frac{n}{p}-k}[(a_m)^o f_m])$. Since the parallelepipeds P and I^n are isometric, invoking (3) we get the isomorphism $L^p_{0,(k)}(J^n \setminus \overline{\Sigma}) \sim F$. This completes the proof of (b).

Next we prove that $L^p_{(k)}(\mathbb{R}^n) \sim F$. To this end we consider the following "tiling" of $\mathbb{R}^n$ by cubes and strips (here $e_j = (\delta^j_k)^n_{j=1}$ stands for the j-th unit vector and $\operatorname{int} A$ denotes the interior of a set A)

$$Q_0 = I^n; \quad Q_j = \operatorname{int} \overline{\bigcup_{m \in \mathbb{Z}} \{Q_{j-1} + me_j\}};$$

$$Q^e_j = \bigcup_{m \in \mathbb{Z}} \{Q_{j-1} + 2me_j\}; \quad Q^o_j = \bigcup_{m \in Z} \{Q_{j-1} + (2m+1)e_j\}$$

$$\text{for } j = 1, 2, \ldots, n.$$

We also need the following spaces

$$\overset{+}{L}{}^p_{(k)}(Q_j) = \{f \in L^p_{(k)}(Q_j) : \exists \widetilde{f} \in L^p_{(k)}(\mathbb{R}^n) \quad \text{such that} \quad \widetilde{f}|Q_j = f \quad \text{and}$$
$$\widetilde{f}(x) = 0 \quad \text{for} \quad x = (x_s) \quad \text{with} \quad x_s > 1/2 \quad \text{for some } s > j\} \quad (j = 0, 1, \ldots, n);$$
$$\overset{+}{L}{}^p_{(k)}(Q^e_j) = \{f \in L^p_{(k)}(Q^e_j) : \exists \widetilde{f} \in \overset{+}{L}{}^p_{(k)}(Q_j) \quad \text{with } \widetilde{f}|Q^o_j = f\} \quad (j = 1, 2, \ldots, n).$$

Note that the new definition of $\overset{+}{L}{}^p_{(k)}(Q_0)$ is compatible with the definition of $\overset{+}{L}{}^p_{(k)}(P)$ for a regular parallelepiped P.

Using Hestenes type extension operator we construct for $j = 1, 2, \ldots, n$ a linear extension operator $\mathcal{E}_j : L^p_{(k)}(Q^e_j) \to L^p_{(k)}(Q_j)$ such that $\mathcal{E}_j(\overset{+}{L}{}^p_{(k)}(Q^e_j)) \subset \overset{+}{L}{}^p_{(k)}(Q_j)$. The existence of such an $\mathcal{E}_j$ in particular implies

$$\overset{+}{L}{}^p_{(k)}(Q_j) \sim \overset{+}{L}{}^p_{(k)}(Q^e_j) \times L^p_{0,(k)}(Q^o_j).$$

The desired isomorphism is given by $f \to (\mathcal{E}_j f|Q^e_j, f - \mathcal{E}_j f|Q^e_j)$. Clearly $L^p_{0,(k)}(Q^o_j) \sim (L^p_{0,(k)}(Q_{j-1}) \times L^p_{0,(k)}(Q_{j-1}) \times \ldots)_p$ and $\overset{+}{L}{}^p_{(k)}(Q^e_j) \sim (\overset{+}{L}{}^p_{(k)}(Q_{j-1}) \times \overset{+}{L}{}^p_{(k)}(Q_{j-1}) \times \ldots)_p$. It follows from (a) and (b) by the decomposition method that $\overset{+}{L}{}^p_{(k)}(J^n) \sim F$ and obviously $L^p_{0,(k)}(Q^o_1) \sim F$. Thus $\overset{+}{L}{}^p_{(k)}(Q_1) \sim F \times F \sim F$. By induction after n steps we get $\overset{+}{L}{}^p_{(k)}(Q_n) \sim F$. Clearly $Q_n = \mathbb{R}^n$, hence $\overset{+}{L}{}^p_{(k)}(Q_n) \sim L^p_{(k)}(\mathbb{R}^n) \sim F$.

Finally let $\emptyset \neq \Omega \subset \mathbb{R}^n$ be an $L^p_{(k)}$-extension domain. Then $L^p_{(k)}(\Omega)|L^p_{(k)}(\mathbb{R}^n)$. Clearly Ω contains a cube, say $aI^n + x$ for some $a > 0$ and some $x \in \mathbb{R}^n$. Thus there is a linear extension operator from $L^p_{(k)}(aI^n + x)$ into $L^p_{(k)}(\Omega)$ (cf. [12], Chapt. VI). Hence $L^p_{(k)}(aI^n + x)|L^p_{(k)}(\Omega)$. Obviously $L^p_{(k)}(I^n) \sim L^p_{(k)}(aI^n + x)$. Therefore from what was established earlier it follows that $L^p_{(k)}(\Omega)|F$ and $F|L^p_{(k)}(\Omega)$. Hence by the decomposition method $L^p_k(\mathbb{R}^n) \sim F \sim L^p_k(\Omega)$. □

Since $L^p_{(k)}(\mathbb{R}^n)$ is isomorphic to $L^p(0,1)$ (cf. [9] and [10]), Theorem 1 yields

Corollary 4 *If $\Omega \subset \mathbb{R}^n$ is a $L^p_{(k)}$-linear extension domain for $1 < p < \infty$ then the space $L^p_{(k)}(\Omega)$ is isomorphic to $L^p(0,1)$.*

Note (cf. e.g. [10]) that if $n > 1$ and $k = 1, 2, \ldots$ then the spaces $L^1_{(k)}(\Omega)$ and $C^{(k)}_u(\Omega)$ for $\emptyset \neq \Omega \subset \mathbb{R}^n$ are not isomorphic to the corresponding classical Banach spaces.

Theorem 1 extends to differentiable manifolds (cf. [11], p. 21 for definition). We restrict ourselves to a particular case only.

Proposition 5 *Let M be a compact metric n-dimensional Euclidean C^k-manifold without boundary. Then $L^p_{(k)}(M)$ is isomorphic to $L^p_{(k)}(I^n)$ for $1 \leq p < \infty$.*

We recall the definition of $L^p_{(k)}(\Omega)$ for an open subset Ω of M. Let $\mathcal{A} = (A_j, a_j)_{j \in J}$ be a finite atlas (= a system of differentiable coordinates of class k) for M compatible with the differentiable structure of M. Thus the cardinality $\#J < \infty$, $(A_j)_{j\in J}$ is an open covering of M, $a_j : E_j \to A_j$ is a homeomorphism where E_j is an open subset of $\mathbb{R}^n$, and $a_j^{-1}a_i : a_i^{-1}(A_j \cap A_i) \to a_j^{-1}(A_j \cap A_i)$ is a k-times continuously differentiable diffeomorphism for $i, j \in J$. The space $L^p_{(k)}(\Omega)$ consists of all scalar-valued functions f on Ω such that $f|A_j \circ a_j^{-1} \in L^p_{(k)}(E_j)$ for $j \in J$ with $A_j \cap \Omega \neq \emptyset$; we admit $||f||_{\mathcal{A},L^p_{(k)}(\Omega)} = \sum_{\{j\in J: A_j\cap\Omega\neq\emptyset\}} ||F|A_j \circ a_j^{-1}||_{L^p_{(k)}(E_j)}$. Obviously the norm depends on the particular choice of the atlas. However for every two atlases the corresponding norms are equivalent.

Outline of the proof of Proposition 5. By considering open coverings by sets with sufficiently small diameters one constructs the atlases $\mathcal{A}$ and $\mathcal{B}$ so that

(i) $B_j \subset \overline{\bigcap_{\{i\in J: A_i\cap A_j\neq\emptyset\}} A_i}$ for $j \in J$;

(ii) the covering $(A_j)_{j\in J}$ is minimal and $a_j = b_j|A_j$ for $j \in J$;

(iii) for $J_1, J_2 \supset J$ with $J_1 \cap J_2 = \emptyset$ the images of the intersections $\bigcap_{j\in J_1} A_j \cap \bigcap_{j\in J_2} B_j$ under the homeomorphisms b_j are domains which satisfy minimal conditions of smoothness in the sense of [12], Chapt. VI, §3.3 (in particular they are $L^p_{(k)}$-linear extension domains); moreover these domains are homeomorphic to regular parallelepipeds.

Next "slightly increasing" the sets A_j we construct for every m with $1 \leq m < 2^{\#J}$ a sequence of atlases $(\mathcal{A}^r = (A_j^r, a_j^r)_{j\in J})$ indexed by the same set J so that $\mathcal{A} = \mathcal{A}^0$ and

(iv) $\overline{A_j^{r-1}} \subset A_j^r$ for $j \in J$;

(v) the atlases $\mathcal{A}^r$ and $\mathcal{B}$ satisfy (i), (ii), (iii) with $\mathcal{A}$ replaced by $\mathcal{A}^r$.

For $\emptyset \neq J' \subset J$ put $Z^r_{J'} = \bigcap_{j\in J'} A_j^r$, $W_l^r = \bigcup_{\{J':\#J'=l\}} Z^l_{J'}$, $m(\mathcal{A}^r) = \max\{l : W_l^r \neq \emptyset\}$. If we construct the A_j^r's "sufficiently close" to A_j for $r = 1, 2, \ldots, m$ then $Z^0_{J'} \neq \emptyset$ iff $Z^r_{J'} \neq \emptyset$ for $J' \subset J$, hence $m(\mathcal{A}) := m(\mathcal{A}^0) = m(\mathcal{A}^r)$ for $r = 1, 2, \ldots, m(\mathcal{A})$. Next consider for $s = 1, 2, \ldots, m$ the following families of non-empty open sets:

$$U_s = \{Z^{m-s}_{J'} \setminus \overline{\Sigma_s} : \#J' = m(\mathcal{A}) - s + 1 \text{ and } Z^0_{J'} \neq \emptyset\},$$

where $\Sigma_1 = \emptyset$ and $\Sigma_s = \bigcup_{l<s} \bigcup U_l$ and $\bigcup U_l$ denotes the union of sets of the family U_l $(1 \leq l \leq s;\ s = 1, 2, \ldots, m(\mathcal{A}) + 1)$.

One can show that each of the families U_s consists of finitely many open sets which have mutually disjoint closures; each of these sets is homeomorphic to a regular parallelepiped and it is transformed by an appropriate map b_j on a domain satisfying minimal conditions of smoothness. Moreover the intersection of the boundaries of these sets with the closures of the members of Σ_s are finite unions of regular parallelepipeds in $\mathbb{R}^{n-1}$. This allows to establish

Lemma 6 *There is a linear operator of extension* $\Lambda_s : L^p_{(k)}(\bigcup U_s : \Sigma_s) \to L^p_{(k)}(M)$ $(s = 1, 2, \ldots, m(\mathcal{A}))$.

Here $L^p_{(k)}(\Omega : \Omega')$ denotes the subspace of $L^p_{(k)}(\Omega)$ consisting of functions f such that f extends to a function in $L^p_{(k)}(M)$ whose restriction to Ω' is the 0 function (Ω, Ω' open subsets of M).

For $f \in L^p_{(k)}(M)$ we have

$$\begin{aligned} f = {} & \Lambda_1(f|\bigcup U_1) + \Lambda_2((f - \Lambda_1(f|\bigcup U_1))|\bigcup U_2) + \dots \\ & + \Lambda_s(\Lambda_{s-2}(f|\bigcup U_{s-2}) - \Lambda_{s-1}(f|\bigcup U_{s-1}))|\bigcup U_s) + \dots \\ & + \Lambda_{m(\mathcal{A})}(\Lambda_{m(\mathcal{A})-2}(f|\bigcup U_{m(\mathcal{A})-2}) - \Lambda_{m(\mathcal{A})-1}(f|\bigcup U_{m(\mathcal{A})-1}))|\bigcup U_{m(\mathcal{A})}). \end{aligned}$$

The latter identity shows that the space $L^p_{(k)}(M)$ represents as a direct sum of its subspaces isomorphic to $L^p_{(k)}(\bigcup U_s : \Sigma_s)$ $(s = 1, 2, \dots, m(\mathcal{A}))$. On the other hand each of the spaces $L^p_{(k)}(\bigcup U_s : \Sigma_s)$ is an l_p-sum of finitely many spaces each of which is an $L^p_{(k)}$-spaces of functions on regular parallelepiped extending by zero across part of the boundary of the parallelepiped which is the union of finitely many regular parallelepipeds in $\mathbb{R}^{n-1}$. The latter spaces are by the decomposition method isomorphic to $L^p(I^n)$. This shows that $L^p_{(k)}(M)$ is isomorphic to a finite Cartesian power of $L^p_{(k)}(I^n)$ which is isomorphic to $L^p_{(k)}(I^n)$. $\square$

REFERENCES

1. G. M. Fichtenholz, *A Course of Differential and Integral Calculus* (in Russian), Gostekhizdat, Leningrad, 1949.
2. M. R. Hestenes, *Extension of a range of a differentiable function*, Duke Math. J. 8 (1941), 183–192.
3. L. Hörmander, *The Analysis of Partial Differential Operators*, Springer Verlag, Berlin, 1983.
4. P. Jones, *Quasiconformal mapping and extendability of functions in Sobolev spaces*, Acta Math. 147 (1981), 71–88.
5. J. Lindenstrauss and L. Tzafriri, *Banach Spaces*, vol. I, Springer Verlag, Berlin, 1977.
6. V. G. Mazya, *Sobolev Spaces*, Springer Verlag, Berlin, 1985.
7. B. S. Mityagin, *Homotopic structure of the linear group of the Banach space* (in Russian), Usp. Mat. Nauk 25 (1970), 63–106.
8. Z. Ogrodzka, *On simultaneous extension of infinitely differentiable functions*, Studia Math. 28 (1967), 193–207.
9. A. Pełczyński and K. Senator, *On isomorphisms of anisotropic Sobolev spaces with "classical Banach spaces" and a Sobolev type embedding theorem*, Studia Math. 84 (1986), 169–215.
10. A. Pełczyński and M. Wojciechowski, *Sobolev Spaces*, Handbook of the Banach Space Theory, ed. W. B. Johnson and J. Lindenstrauss, Elsevier, to appear.
11. N. Steenrod, *The Topology of Fibre Bundles*, Princeton University Press, Princeton, New Jersey 1951.
12. E. M. Stein, *Singular Integrals and Differential Properties of Functions*, Princeton Univ. Press, Princeton, 1970.
13. H. Whitney, *Analytic extensions of differentiable functions defined in closed sets*, Trans. Amer. Math. Soc. 36 (1934), 63–89.

Recent Progress in Functional Analysis
K.D. Bierstedt, J. Bonet, M. Maestre and J. Schmets

Decomposability and the cyclic behavior of parabolic composition operators

Joel H. Shapiro *

Department of Mathematics, Michigan State University
East Lansing, Michigan 48824, USA
email: shapiro@math.msu.edu

To Professor Manuel Valdivia on the occasion of his seventieth birthday.

Abstract *This paper establishes decomposability for composition operators induced on the Hardy spaces H^p $(1 \leq p < \infty)$ by parabolic linear fractional self-maps of the unit disc that are* not *automorphisms. This result, along with a recent theorem of Miller and Miller [15], shows that no such composition operator is supercyclic. The work here completes part of a previous investigation [1] where the author and Paul Bourdon showed that among linear fractional maps of the disc with no interior fixed point, only the parabolic non-automorphisms induce non-hypercyclic composition operators. Additionally it complements results of Robert Smith [19], who proved decomposability in the case of parabolic* automorphisms, *and it extends recent work of Gallardo and Montes [6] who used different methods to establish the desired non-supercyclicity for the case $p = 2$.*
MCS 2000 Primary 47B33; Secondary 30D55, 47A11, 47A16

Introduction

This paper deals with parabolic linear fractional mappings φ that take the open unit disc U into itself, and the composition operators C_φ that they induce on the Hardy spaces H^p $(1 \leq p < \infty)$ by means of the formula $C_\varphi f = f \circ \varphi$ $(f \in H^p)$. The goal is to show that if φ is *not* an automorphism of U (i.e., if $\varphi(U) \neq U$) then C_φ is: (a) decomposable, and (b) not supercyclic.

To say that an operator T on a Banach space X is *decomposable* means that for every covering of the complex plane $\mathbb{C}$ by a pair $\{V, W\}$ of open sets there is a corresponding pair $\{Y, Z\}$ of closed, T-invariant subspaces such that $X = Y + Z$, the spectrum of $T|_Y$ lies in V, and that of $T|_Z$ lies in W. Decomposability was originally introduced into operator theory in 1963 by Foiaş, but it was not until much later that his definition was shown, by several authors independently, to be equivalent the one given here (see [14, Defn. 1.1.1] and the paragraph that precedes it for the appropriate references).

*Research supported in part by the National Science Foundation

To say that T is *supercyclic* means that there is a vector $x \in X$ such that the *projective orbit* $\{cT^n x : n = 0, 1, 2, \ldots \text{ and } c \in \mathbb{C}\}$ is dense in X. Supercyclicity stands midway between the weaker concept of *cyclicity* (some orbit has dense linear span) and *hypercyclicity* (some orbit is dense). The concept was originally introduced by Hilden and Wallen in [10], who showed that it is possessed by every weighted backward shift on ℓ^2 (in particular, even by some quasinilpotent operators!).

The connection between decomposability and supercyclicity was recently established by Miller and Miller, who proved in [15, Theorem 2] (see also [5, Cor. 6.5] and [14, Prop. 3.3.18]) a result that implies:

Theorem M. *Each supercyclic decomposable operator has its spectrum on some (possibly degenerate) circle centered at the origin.*

Now the composition operators treated in this paper have as spectrum either the interval $[0, 1]$ or a spiral that starts at the point 1 and converges to the origin by winding infinitely often around it (see [2, Theorem 6.1, page 102] or §3.10 below). In any case, their spectra do not lie on any circle, hence once these operators are be shown to be decomposable, their non-supercyclicity will follow from Theorem M.

This paper arises from [1], where Paul Bourdon and I classified the cyclic behavior of linear-fractionally induced composition operators on H^2. We showed that among the linear fractional selfmaps of U fixing no point of U (no others have any chance of being hypercyclic [1, Prop. 0.1, page 3]), the only ones *failing* to induce hypercyclic composition operators are the parabolic maps that are not automorphisms. We showed that, nonetheless, such maps induce *cyclic* operators, and wondered if this cyclicity could be improved to *supercyclicity*. In this regard I was able to prove [18] that for such maps φ, the operator C_φ on H^2 had no hypercyclic scalar multiples (clearly any operator with a hypercyclic scalar multiple is supercyclic, but such operators do not exhaust the supercyclic class [12, page 3.4]). Just recently Gallardo and Montes [6] significantly refined the method of [18] to obtain a proof that C_φ is, indeed, not supercyclic on H^2.

The results from [1] discussed above, while phrased only for H^2, hold as well—with almost the same proofs—for any space H^p with $1 \le p < \infty$, so it makes sense to raise in this more general context the supercyclicity question for composition operators induced by parabolic non-automorphisms. The method of [18], although strongly oriented toward Hilbert space, relied in part on Fourier analysis on the real line, and hinted strongly that decomposability might lie at the heart of the supercyclicity issue for the operators in question—a suspicion strongly supported by Theorem M.

Here is an outline of what follows. After a brief survey of prerequisites (Section 1) the study of parabolically induced composition operators will evolve, in Section 2, into a study of translation operators acting on Hardy spaces of the upper half-plane. This will make it possible, in Section 3, to embed each of our parabolic, non-automorphically induced composition operators into a C^2 functional calculus of Fourier integral operators, and from this will follow the desired decomposability and non-supercyclicity.

Even with the appearance of Laursen and Neumann's long-awaited monograph [14], the subject of decomposable operators is still technically formidable. Thus, in the interests of broadening the reach of this paper, I conclude with a couple of purely expository final sections: one devoted to proving that decomposability follows from the existence of a C^∞

functional calculus, and the other to a direct proof that the decomposability and spectral properties of the operators considered here render them non-supercyclic.

Acknowledgments. I wish to express my gratitude to Eva Gallardo and Alfonso Montes for making their preprint [6] available to me, and to thank Paul Bourdon, Nathan Feldman, and Luis Saldivia for pointing out some errors and inconsistencies in an earlier version of this paper.

1. Prerequisites

1.1. Notation Throughout this paper p denotes an index which, unless otherwise noted, lies in the interval $[1, \infty)$, and:

- U denotes the open unit disc of the complex plane $\mathbb{C}$,
- ∂U is the unit circle,
- m is Lebesgue arc length measure on ∂U, normalized to have unit mass,
- $L^p(\partial U)$ is the L^p space associated with the measure m, and
- Π_+ denotes the open upper half-plane $\{z \in \mathbb{C} : \operatorname{Im} z > 0\}$.

1.2. Hardy spaces The *Hardy space* $H^p = H^p(U)$ is the collection of functions f holomorphic on U with

$$\|f\|_p^p := \sup_{0 \le r < 1} \int_{\partial U} |f(r\zeta)|^p \, dm(\zeta) < \infty.$$

The functional $\|\cdot\|_p$ so defined makes H^p into a Banach space. H^∞ is the space of bounded holomorphic functions on U—a Banach space in the norm $\|f\|_\infty := \sup\{|f(z)| : z \in U\}$. Each $f \in H^p$ has, for $[m]$ almost every $\zeta \in \partial U$, a finite radial limit $f^*(\zeta) := \lim_{r \to 1-} f(r\zeta)$, and the map that associates $f \in H^p$ with its *boundary function* f^* is an isometry taking H^p onto the subspace of $L^p(\partial U)$ consisting of functions whose Fourier coefficients of negative index are all zero. The holomorphic function f can be recovered from f^* by either a Cauchy or a Poisson integral.

1.3. Composition operators A holomorphic selfmap of U is just a function that is holomorphic on U and has all its values in U. Each such map φ induces a linear composition operator C_φ on the space of all functions holomorphic on U:

$$C_\varphi f := f \circ \varphi \qquad (f \text{ holomorphic on } U).$$

A classical (and by no means obvious) theorem of Littlewood guarantees that C_φ restricts to a bounded operator on each H^p space, and the study of how the properties of these operators reflect the function theory of their inducing maps has evolved during the past few decades into a lively enterprise; see the monographs [3] and [17] for introductions to the subject, and the conference proceedings [11] for some more recent developments.

1.4. Parabolic maps For linear fractional selfmaps of U the boundedness of C_φ on H^p is elementary; in this paper I consider only a subclass of these maps, the parabolic ones. These are linear fractional transformations that map U into itself and fix exactly one point of the Riemann sphere, a point which must necessarily lie on the unit circle. Each such map is conformally conjugate, via rotation of the unit disc, to one that fixes the point $1 \in \partial U$. Because the composition operators induced by rotations of the disc are isometric isomorphisms of H^p, this rotational conjugation from an arbitrary fixed point on ∂U to fixed point at 1 translates, at the operator level, to an isometric similarity between composition operators. Because all of the operator-theoretic phenomena to be considered in this paper are similarity-invariant, nothing will therefore be lost by always placing the fixed point of φ at 1.

Suppose, then, that φ is a parabolic selfmap of U with $\varphi(1) = 1$. The map τ defined by

$$\tau(z) = i\,\frac{1+z}{1-z} \qquad z \in \mathbb{C}\backslash\{1\} \tag{1}$$

maps the unit disc conformally onto the upper half-plane $\Pi_+ = \{z \in \mathbb{C} : \operatorname{Im} z > 0\}$, takes $\partial U\backslash\{1\}$ homeomorphically onto the real line, and sends the point 1 to ∞. The map $\Phi := \tau \circ \varphi \circ \tau^{-1}$ is therefore a linear fractional map that takes Π_+ into itself and fixes ∞, hence it must be translation by some $a \in \mathbb{C}$ with $\operatorname{Im} a \geq 0$, that is, $\Phi(w) = w + a$ for $w \in \mathbb{C}$. Let us call a the *translation parameter* of both the translation Φ of Π_+ and the original parabolic mapping φ of U. Note that φ is an automorphism of U precisely when Φ has the same property on Π_+, and this happens if and only if the translation parameter is real.

This characterization of parabolic composition operators suggests that they may be studied most effectively by shifting attention from the unit disc to the upper half-plane; I develop this point of view in the next section.

2. Migrating to the Upper Half-Plane

2.1. Hardy spaces on the upper half-plane There are two ways to define a Hardy space H^p for the upper half-plane:

(a) $H^p(\Pi_+)$ is the space of functions F holomorphic on Π_+ with $F \circ \tau \in H^p(U)$. The norm $\|\cdot\|_p$ defined on $H^p(\Pi_+)$ by $\|F\|_p := \|F \circ \tau\|_p$ (where the norm on the right is the one for $H^p(U)$) makes $H^p(\Pi_+)$ into a Banach space, and insures that the map $C_\tau : H^p(\Pi_+) \to H^p(U)$ is an isometry taking $H^p(\Pi_+)$ onto $H^p(U)$. In particular, for each $F \in H^p(\Pi_+)$ the "radial limit" $F^*(x) = \lim_{y\to 0} F(x+iy)$ exists for a.e. $x \in \mathbb{R}$, and a change of variable involving the map τ shows that the norm of F can be computed by integrating over $\mathbb{R}$:

$$\|F\|_p^p = \frac{1}{\pi}\int_{-\infty}^{\infty} |F^*(x)|^p \,\frac{dx}{1+x^2}.$$

(b) $\mathcal{H}^p(\Pi_+)$ is the space of functions F holomorphic on Π_+ for which

$$\|F\|_p^p := \sup_{y>0} \int_{-\infty}^{\infty} |F(x+iy)|^p \,dx < \infty.$$

Once again the norm defined on the space (which, although denoted by the same symbol as the previous norms, is different from them) makes it into a Banach space.

These two spaces are not the same; the map C_τ takes $\mathcal{H}^p(\Pi_+)$ onto the dense subspace $(1-z)^{2/p}H^p(U)$ of $H^p(U)$, hence $\mathcal{H}^p(\Pi_+)$ is a dense subspace of $H^p(\Pi_+)$. Finally, the norm in $\mathcal{H}^p(\Pi_+)$ can be computed on the boundary: $\|F\|_p^p = \int_{-\infty}^{\infty} |F^*(x)|^p\,dx$, so that $\mathcal{H}^p(\Pi_+)$ can be regarded as a closed subspace of $L^p(\mathbb{R})$. For $p = 1$ it is the subspace consisting of functions whose Fourier transforms vanish on $(-\infty, 0]$, and a similar interpretation can be made for $1 < p \leq 2$. For a detailed exposition of these and other basic facts about Hardy spaces in half-planes I refer the reader to [7, Chapter II], [8, Chapter 8], or [13, Chapter VI].

2.2. Eigenvalues of $\mathbf{C}_\varphi$ Suppose φ is a parabolic selfmap of U with fixed point at 1, and let $a \in \mathbb{C}$ with $\operatorname{Im} a \geq 0$ be its translation parameter, so that $\Phi = \tau \circ \varphi \circ \tau^{-1}$ is just "translation by a" in Π_+. For $t \geq 0$ let $E_t(w) = e^{itw}$ for $w \in \Pi_+$. E_t is a bounded holomorphic function on Π_+, hence

$$e_t(z) = E_t(\tau(z)) = \exp\left\{-t\frac{1+z}{1-z}\right\} \qquad (z \in U),$$

defines bounded holomorphic function on U (the t-th power of the *unit singular function*). Because of this boundedness $e_t \in H^p(U)$, or equivalently, $E_t \in H^p(\Pi_+)$ for each $1 \leq p \leq \infty$. Furthermore $C_\Phi E_t = e^{iat}E_t$ hence also $C_\varphi e_t = e^{iat}e_t$ for each $t \geq 0$. Thus for each such t the function e_t is an eigenvector of $C_\varphi : H^p(U) \to H^p(U)$ with corresponding eigenvalue e^{iat}. Thus $\Gamma_a := \{e^{iat} : t \geq 0\}$ is a subset of the spectrum of C_φ. If a is real, so that φ is an automorphism, then Γ_a covers the unit circle infinitely often, and it turns out that ∂U is precisely the spectrum of C_φ, a result proved over thirty years ago by Nordgren [16]. If $\operatorname{Im} a > 0$ then φ is not an automorphism, and Γ_a is a curve that starts at 1 when $t = 0$ and converges to 0 as $t \to \infty$. If a is pure imaginary then $\Gamma_a = (0, 1]$, otherwise Γ_a spirals infinitely often around the origin, converging to the origin with strictly decreasing modulus. Thus in these non-automorphic cases the spectrum of C_φ contains $\Gamma_a \cup \{0\}$, and it is a (special case of a) result of Cowen [2, Theorem 6.1] that $\Gamma_a \cup \{0\}$ is indeed the whole spectrum. I will give an alternate proof of this fact in Section 3.

2.3. $\mathbf{C}_\Phi$ as a convolution operator We saw in §1.4 that each parabolic selfmap φ of U that fixes the point 1 has the representation $\varphi = \tau^{-1} \circ \Phi \circ \tau$, where τ is the linear fractional mapping of U onto Π_+ given by (1), and Φ is the mapping of translation by some fixed vector a in the closed upper half-plane: $\Phi(w) = w + a$ for $w \in \Pi_+$. At the operator level this conjugacy turns into the similarity $C_\varphi = C_\tau C_\Phi C_\tau^{-1}$, where C_τ is an isometry mapping $H^p(\Pi_+)$ into $H^p(U)$, and C_Φ is a bounded operator on $H^p(\Pi_+)$.

Since all the operator theoretic phenomena being investigated here are preserved by similarity, nothing will be lost (in fact much will be gained) by shifting attention from C_φ on $H^p(U)$ to C_Φ on $H^p(\Pi_+)$. The advantage here is that when the original parabolic mapping φ of U is not an automorphism, the operator C_Φ on $H^p(\Pi_+)$ can be represented as a convolution operator. The key is that each $F \in H^p(\Pi_+)$ is the Poisson integral of its boundary function:

$$F(x+iy) = \frac{1}{\pi}\int_{-\infty}^{\infty} \frac{y}{(x-t)^2+y^2} F^*(t)\,dt \qquad (x+iy \in \Pi_+).$$

Since φ is not an automorphism, its translation parameter $a = \alpha + i\beta$ lies in the (open) upper half-plane, and $C_\Phi F(w) = F(w+a)$ for $w \in \Pi_+$. Thus for each $F \in H^p(\Pi_+)$ and $x \in \mathbb{R}$:

$$(C_\Phi F)^*(x) = F(x+\alpha+i\beta) = \int_{-\infty}^{\infty} P_a(x-t)F^*(t)\,dt \tag{2}$$

where

$$P_a(x) = \frac{1}{\pi}\frac{\beta}{(x-\alpha)^2+\beta^2} \tag{3}$$

is the (upper half-plane) Poisson kernel for the point $a \in \Pi_+$.

From now on I will drop the superscript "$*$" that distinguished holomorphic functions from their radial limit functions, and simply regard each function $F \in H^p(\Pi_+)$ to be either a holomorphic function on the upper half-plane, or the associated radial limit function—an element of the space $L^p(\mu)$, where μ is the *Cauchy measure*

$$d\mu(x) := \frac{1}{\pi}\frac{dx}{1+x^2}, \tag{4}$$

a Borel probability measure on $\mathbb{R}$. In each case, either the context or an explicit statement will make clear which interpretation of F is intended.

Correspondingly, the operator C_Φ can now be given two different interpretations: either as the original composition operator on holomorphic functions, or—by (2) and (3) above—as the restriction to $H^p(\Pi_+)$-boundary functions of the convolution operator

$$C_\Phi F = F * P_a \qquad (F \in L^p(\mu)). \tag{5}$$

From this convolution representation arises the functional calculus which lies at the heart of this paper.

3. A functional calculus

The goal of this section is to prove:

3.1. Theorem *If φ is a parabolic linear fractional selfmap of U that is not an automorphism, then $C_\varphi : H^p \to H^p$ has a C^2 functional calculus.*

For our purposes the conclusion means that there is an algebra homomorphism $\gamma \to \gamma(C_\varphi)$ from $C^2(\mathbb{C})$ into $\mathcal{L}(H^p)$, the algebra of bounded linear operators on H^p, such that if β and γ belong to C^2, then (letting $\sigma(C_\varphi)$ denote the spectrum of C_φ):

(FC1) If $\beta \equiv \gamma$ on $\sigma(C_\varphi)$ then $\beta(C_\varphi) = \gamma(C_\varphi)$,
(FC2) If $\gamma(z) \equiv z$ on $\sigma(C_\varphi)$ then $\gamma(C_\varphi) = C_\varphi$, and
(FC3) If $\gamma \equiv 1$ on $\sigma(C_\varphi)$ then $\gamma(C_\varphi)$ is the identity operator on H^p.

From this functional calculus will follow the decomposability and, therefore the non-supercyclicity, of C_φ. It turns out that for any operator on a Banach space, properties (FC1)–(FC3) follow from the weaker assumption that (FC2) and (FC3) hold with the spectrum replaced by the whole complex plane (see [14, Theorem 1.4.10]). However for the functional calculus constructed here, the full strength of (FC1)–(FC3) will be immediately apparent.

Because the existence of a functional calculus is similarity invariant, it will be enough to carry out the construction in $H^p(\Pi_+)$, with the translation operator C_Φ on that space standing in for C_φ. The work of this section takes place exclusively on the boundary, so that $H^p(\Pi_+)$ will be interpreted as a subspace of $L^p(\mu)$, where μ is the Cauchy probability measure on $\mathbb{R}$ given by (4).

We construct our functional calculus by using (5) to view C_Φ as a convolution operator on $L^p(\mu)$, and then restricting to the invariant subspace $H^p(\Pi_+)$. The following well known sufficient condition is the key to proving boundedness for the operators in question. Even though it is stated here only for the Cauchy measure μ on the Borel sets of $\mathbb{R}$, it is valid for any positive measure on any measure space.

3.2. The Schur Test ([4, Page 518, Problem 54]). *Suppose K is a non-negative Borel measurable function on $\mathbb{R}^2$, and that there exists a positive, finite constant C such that:*

(a) $\displaystyle\int K(x,y)\,d\mu(y) \le C \quad$ *for a.e.* $x \in \mathbb{R}$,

(b) $\displaystyle\int K(x,y)\,d\mu(x) \le C \quad$ *for a.e.* $y \in \mathbb{R}$.

For each non-negative Borel function f on $\mathbb{R}$ define

$$T_K f(x) := \int K(x,y) f(y)\,d\mu(y) \qquad (x \in \mathbb{R}). \tag{6}$$

Then $\|T_K f\|_p \le C\|f\|_p$ for each $1 \le p \le \infty$; in particular T_K can now be defined by (6) on all of $L^p(\mu)$, where it acts as a bounded linear operator.

The Schur Test yields the following criterion for a convolution operator to be bounded on $L^p(\mu)$.

3.3. Proposition *Suppose $k : \mathbb{R} \to \mathbb{C}$ is a bounded Borel measurable function such that*

$$|k(x)| = O(|x|^{-2}) \quad \textit{as } |x| \to \infty. \tag{7}$$

Then the mapping $f \to k * f$ is a bounded linear operator on $L^p(\mu)$ for each $1 \le p \le \infty$.

Proof. The decay condition (7) insures that there is no difficulty in convolving k with any function in $L^p(\mu)$. To apply the Schur test let

$$K(x,y) = \pi(1+y^2)k(x-y) \qquad (x, y \in \mathbb{R}),$$

so that for $x \in \mathbb{R}$:

$$k * f(x) := \int k(x-y)f(y)\,dy = \int K(x,y)f(y)\,d\mu(y)$$

(unadorned integral signs now refer to integration over the entire real line). So to prove the boundedness of the convolution operator it suffices to show that $|K|$ satisfies the hypotheses of Schur's Test. Hypothesis (a) is easy; for each $x \in \mathbb{R}$:

$$\int |K(x,y)|\,d\mu(y) = \int |k(x-y)|\,dy = \int |k(y)|\,dy < \infty$$

where the integrability of k follows from the decay condition (7).

For hypothesis (b) note that for every $y \in \mathbb{R}$:

$$\begin{aligned}\int |K(x,y)|\,d\mu(x) &= (1+y^2)\int |k(x-y)|\frac{dx}{1+x^2}\\ &\le C(1+y^2)\int \frac{1}{1+(x-y)^2}\cdot\frac{1}{1+x^2}\,dx\ ,\end{aligned}$$

where the inequality arises from (7), with C independent of y. The last integral in this display is, by (3) above, a constant multiple of $P_i * P_i$, the convolution of the Poisson kernel for the point $i \in \Pi_+$ with itself. Now this convolution square is just the Poisson kernel for the point $2i$, namely $(2/\pi)(4+y^2)^{-1}$ (see the next paragraph for details). This establishes the boundedness of $\int |K(x,y)|\,d\mu(y)$, and with it, that of the operator of convolution by k on $L^p(\mu)$. □

3.4. Remark on the Poisson kernel The identity $P_i * P_i = P_{2i}$ used in the proof of Proposition 3.3 is a special case of the semigroup identity

$$P_a * P_b = P_{a+b} \qquad (a, b \in \Pi_+), \tag{8}$$

which one can prove using either the Fourier transform or the Poisson integral representation of harmonic functions. Since the Fourier transform of the Poisson kernel will play a crucial role in the sequel, I'd like to take a moment to show how it leads to (8).

The Fourier transform of P_i is well known; it is

$$\widehat{P_i}(\lambda) := \frac{1}{\pi}\int \frac{e^{-i\lambda t}}{1+t^2}\,dt = e^{-|\lambda|} \qquad (\lambda \in \mathbb{R}). \tag{9}$$

Now $P_a(t) = \beta^{-1}P_i((t-\alpha)/\beta)$ for each $a = \alpha + i\beta \in \Pi_+$, so it follows from (9) and a change of variable that:

$$\widehat{P_a}(\lambda) = e^{-i\alpha\lambda}e^{-|\lambda|\beta} = \begin{cases} e^{-i\bar{a}\lambda} & (\lambda \ge 0)\\ e^{-ia\lambda} & (\lambda \le 0)\end{cases} \tag{10}$$

from which it follows easily that for $a, b \in \Pi_+$, $\widehat{P_a * P_b} = \widehat{P_a}\,\widehat{P_b} = \widehat{P_{a+b}}$ at every point of $\mathbb{R}$. This, by uniqueness of Fourier transforms, implies the desired semigroup property. □

Proposition 3.3 provides the foundation for the next result, which is the major building block in the construction of our functional calculus. For all that follows we fix $a = \alpha + i\beta \in \Pi_+$.

3.5. Proposition *Suppose* $\gamma \in C^2(\mathbb{C})$ *with* $\gamma(0) = \gamma(1) = 0$. *Let* k_γ *be the inverse Fourier transform of* $\gamma \circ \widehat{P_a}$. *Then the convolution operator* $f \to k_\gamma * f$ *is bounded on* $L^p(\mu)$ *and maps* $H^p(\Pi_+)$ *into itself.*

Proof. It follows from (10) that $|\widehat{P_a}(\lambda)| = e^{-\beta|\lambda|}$ for each $\lambda \in \mathbb{R}$. Because γ is differentiable and vanishes at 0, the composition $\gamma \circ \widehat{P_a}$ inherits the exponential decay of $\widehat{P_a}$ at $\pm\infty$, and is therefore integrable, hence there is no problem in defining k_γ, its inverse Fourier transform. In fact the first and second derivatives of $\gamma \circ \widehat{P_a}$ on both half-intervals $(0, \infty)$ and $(-\infty, 0)$ have the same exponential decay, and thus k_γ can be estimated by splitting its defining Fourier integral into two pieces—one over each half-interval—and integrating the results twice by parts, using the condition $\gamma(0) = \gamma(1) = 0$ to get rid of the boundary terms at the first stage. The result is that the asymptotic estimate (7) is valid for k_γ, hence by Proposition 3.3 the associated convolution operator is bounded.

As for H^p-preservation, observe first that $\mathcal{H}^p(\Pi_+)$ is a dense subspace of $H^p(\Pi_+)$. One way to see this is to note that C_τ takes $\mathcal{H}^p(\Pi_+)$ to $(1-z)^{2/p}H^p(U)$ (see [7, Lemma 1.2, page 51] or [8, page 130]), and (by definition) $H^p(\Pi_+)$ to $H^p(U)$. An application of Beurling's theorem then seals the argument. Now the functions in $L^1(\mathbb{R}) \cap L^p(\mathbb{R})$ whose Fourier transforms vanish on the negative real axis form a dense subspace of $\mathcal{H}^p(\Pi_+)$, and therefore of $H^p(\Pi_+)$, so it is enough to prove that $k_\gamma * f \in \mathcal{H}^p(\Pi_+)$ for each such function f. Clearly this convolution lies in $L^p(\mathbb{R}) \cap L^1(\mathbb{R})$, and its Fourier transform, which is $\widehat{k_\gamma} \cdot \hat{f}$, vanishes where $\hat{f}$ does—on the negative real axis. Thus $k_\gamma * f \in \mathcal{H}^p(\Pi_+) \subset H^p(\Pi_+)$ and the proof is complete. □

3.6. The functional calculus for $\mathbf{C_\Phi}$ on $\mathbf{L^p}(\mu)$ As usual, we denote by Φ the mapping of "translation by $a \in \Pi_+$" on $\mathbb{C}$. Let $\mathcal{G}$ denote the class of functions γ that satisfy the hypotheses of Proposition 3.5—twice continuously differentiable on $\mathbb{C}$ and vanishing at both 0 and 1. For $\gamma \in \mathcal{G}$ define $\gamma(C_\Phi)$ to be the operator of convolution with k_γ, acting on $L^p(\mu)$. According to the work just completed, $\gamma(C_\Phi)$ is a bounded operator on $L^p(\mu)$ that leaves invariant the closed subspace $H^p(\Pi_+)$ (still being viewed as a space of functions on the real line).

If γ_1 and γ_2 belong to $\mathcal{G}$ and coincide on $\widehat{P_a}(\mathbb{R})$, then so do their left compositions with $\widehat{P_a}$, and hence so do the inverse Fourier transforms of these compositions. Since these inverse Fourier transforms are just the convolution kernels k_{γ_1} and k_{γ_2}, it follows that $\gamma_1(C_\Phi) = \gamma_2(C_\Phi)$.

The map $\gamma \to \gamma(C_\Phi)$ is clearly additive and homogeneous with respect to scalar multiplication. To see that it is also multiplicative, let γ_1 and γ_2 belong to $\mathcal{G}$ and observe that

$$\widehat{k_{\gamma_1 \cdot \gamma_2}} = (\gamma_1 \cdot \gamma_2) \circ \widehat{P_a} = (\gamma_1 \circ \widehat{P_a}) \cdot (\gamma_2 \circ \widehat{P_a}) = \widehat{k_{\gamma_1}} \cdot \widehat{k_{\gamma_2}},$$

hence $k_{\gamma_1 \cdot \gamma_2} = k_{\gamma_1} * k_{\gamma_2}$. It follows that for each $f \in L^p(\mu) \cap L^1(\mathbb{R})$ (a dense subspace of $L^p(\mu)$):

$$(\gamma_1 \cdot \gamma_2)(C_\Phi)f := k_{\gamma_1 \cdot \gamma_2} * f = k_{\gamma_1} * (k_{\gamma_2} * f) = \gamma_1(C_\Phi)[\gamma_2(C_\Phi)f],$$

which establishes the desired multiplicative property. □

The arguments so far have shown that the map $\gamma \to \gamma(C_\Phi)$ is an algebra homomorphism of $\mathcal{G}$ into $\mathcal{L}(L^p(\mu))$. It remains to extend this map appropriately to all of $C^2(\mathbb{C})$. For this it suffices to note that each $\gamma \in C^2(\mathbb{C})$ can be written uniquely as

$$\gamma(z) = a + bz + \gamma_0(z) \qquad (z \in \mathbb{C}),$$

where $a = \gamma(0)$, $b = \gamma(1) - \gamma(0)$, and $\gamma_0 \in \mathcal{G}$. Thus

$$\gamma(C_\Phi) := aI + bC_\Phi + \gamma_0(C_\Phi)$$

defines a bounded linear operator on $L^p(\mu)$ that takes $H^p(\Pi_+)$ into itself. One checks easily that the homomorphic property previously noted on $\mathcal{G}$ for the mapping $\gamma \to \gamma(C_\Phi)$ carries over to the extension just defined on $C^2(\mathbb{C})$, and that this extension has all the properties needed to be a functional calculus for C_Φ on $L^p(\mu)$, except that the uniqueness conditions (FC1)–(FC3), which are supposed to hold for $\sigma(C_\Phi)$, have been proven instead for $\widehat{P_a}(\mathbb{R})$. The next result shows that (FC1)–(FC3) hold just as advertised.

3.7. Proposition $\sigma(C_\Phi : L^p(\mu) \to L^p(\mu)) = \widehat{P_a}(\mathbb{R}) \cup \{0\}$.

Proof. For $t \in \mathbb{R}$ let $E_t(x) = e^{itx}$ $(x \in \mathbb{R})$. Since these functions are continuous and bounded (in fact, unimodular) on $\mathbb{R}$, they all belong to $L^p(\mu)$. For $t \geq 0$ these functions turned out to be eigenvectors of $C_\Phi : H^p(\Pi_+) \to H^p(\Pi_+)$. The first order of business is to show that the full collection serves as eigenvectors for C_Φ on $L^p(\mu)$. For this, fix x and t in $\mathbb{R}$ and note that:

$$C_\Phi(E_t(x)) := P_a * E_t(x) = \int e^{it(x-\xi)} P_a(\xi)\, d\xi = e^{itx}\widehat{P_a}(t) = \widehat{P_a}(t)E_t(x)$$

so $\widehat{P_a}(t)$ is an eigenvalue of $C_\Phi : L^p(\mu) \to L^p(\mu)$ corresponding to the eigenvector E_t. Thus $\widehat{P_a}(\mathbb{R})$ is contained in the $L^p(\mu)$ spectrum of C_Φ, hence so is its closure $\widehat{P_a}(\mathbb{R}) \cup \{0\}$.

To complete the proof it suffices to show that $\lambda \notin \widehat{P_a}(\mathbb{R}) \cup \{0\}$ implies $\lambda \notin \sigma(C_\Phi)$. For each such λ there exists a function $\gamma \in C^2(\mathbb{C})$ with $\gamma(z) = (z - \lambda)^{-1}$ for $z \in \widehat{P_a}(\mathbb{R})$. Now $\psi(z) = z - \lambda$ is also a C^2 function on $\mathbb{C}$, and $\psi \cdot \gamma \equiv 1$ on $\widehat{P_a}(\mathbb{R})$. Thus by the properties derived so far for our functional calculus:

$$\gamma(C_\Phi)(C_\Phi - \lambda I) = \gamma(C_\Phi)\psi(C_\Phi) = (\gamma \cdot \psi)(C_\Phi) = I,$$

where I is the identity map on $L^p(\mu)$. This display shows, because all the operator factors therein commute, that $C_\Phi - \lambda I$ is invertible on $L^p(\mu)$, hence $\lambda \notin \sigma(C_\Phi)$. □

3.8. Remark Recall from §2.2 our observation that the set

$$\Gamma_a := \widehat{P_a}([0, \infty)) = \{e^{iat} : t \geq 0\},$$

is a curve that spirals from the point 1 asymptotically into the origin. By formula (10), $\widehat{P_a}(\mathbb{R})$ is the union of Γ_a and its reflection in the x-axis, a double-spiral joining the point 1 to the origin.

It remains only to check that the functional calculus constructed above for C_Φ on $L^p(\mu)$ restricts properly to the subspace $H^p(\Pi_+)$, which we have already seen is invariant for all the operators involved (Proposition 3.5). This is the content of the next two results.

3.9. Proposition *Suppose* $\gamma_1, \gamma_2 \in C^2(\mathbb{C})$ *with* $\gamma_1 \equiv \gamma_2$ *on* Γ_a. *Then* $\gamma_1(C_\Phi) = \gamma_2(C_\Phi)$ *on* $H^p(\Pi_+)$.

Proof. It is enough to prove that the two operators coincide on the dense subspace $H^p(\Pi_+) \cap L^1(\mathbb{R})$ of $H^p(\Pi_+)$. For f in this subspace the Fourier transform $\hat{f}$ vanishes on the negative real axis. Our hypothesis guarantees that $\gamma_1 \circ \widehat{P_a} = \gamma_2 \circ \widehat{P_a}$ on $[0, \infty)$, so at each point of $\mathbb{R}$ we have:

$$[\gamma_1(C_\Phi)f]\hat{} = [k_{\gamma_1} * f]\hat{} = \widehat{k_{\gamma_1}} \cdot \hat{f} = (\gamma_1 \circ \widehat{P_a}) \cdot \hat{f} = (\gamma_2 \circ \widehat{P_a}) \cdot \hat{f} = [\gamma_2(C_\Phi)f]\hat{}$$

hence $\gamma_1(C_\Phi)f = \gamma_2(C_\Phi)f$. □

3.10. Corollary $\sigma(C_\Phi : H^p(\Pi_+) \to H^p(\Pi_+)) = \Gamma_a \cup \{0\}$.

Proof. We have already seen that each $\lambda \in \Gamma_a$ is an eigenvalue of $C_\Phi : H^p(\Pi_+) \to H^p(\Pi_+)$, so the spectrum of this operator contains $\Gamma_a \cup \{0\}$. To go the other way it is enough to show that if $\lambda \notin \Gamma_a \cup \{0\}$ then λ is not in the spectrum, i.e. that $C_\Phi - \lambda I$ is invertible on $H^p(\Pi_+)$. Now the hypothesis on λ is that $z - \lambda$ is bounded away from zero on Γ_a, hence there exists $\gamma \in C^2(\mathbb{C})$ with $\gamma(z) = (z - \lambda)^{-1}$ on Γ_a. Since $(z - \lambda)\gamma(z) \equiv 1$ on Γ_a it follows from Proposition 3.9 that, just as in the proof of Proposition 3.7, the operator $\gamma(C_\Phi)$ is the inverse, on $H^p(\Pi_+)$, of $C_\Phi - \lambda I$. □

4. Decomposability

As promised in the Introduction, I include these final two sections entirely for the convenience of the reader. While there may be some originality in the organization of Section 5, the material in this section comes right out of [14, Theorem 1.4.10].

In the last section we constructed, for each composition operator induced on H^p by a parabolic non-automorphism, a C^2-functional calculus. The point of this section is that every Banach space operator with even a C^∞ functional calculus is decomposable.

So assume that X is a Banach space and T a bounded linear operator on X, and that T has a C^∞ functional calculus in the sense of the discussion following Theorem 3.1.

To each compact subset K of $\mathbb{C}$ let us attach the subspace $E(K)$ of X formed by intersecting the null spaces of all the operators $\eta(T)$ where $\eta \in C^\infty(\mathbb{C})$ and $K \cap \operatorname{spt} \eta = \emptyset$. Everything depends on the following result.

4.1. Lemma *For each compact subset* K *of* $\mathbb{C}$, *the subspace* $E(K)$ *is closed and* $T-$*invariant, with* $\sigma(T|_{E(K)}) \subset K$. *Moreover, if* λ *is an eigenvalue of* T *then the following are equivalent:*

(a) $\lambda \in K$.

(b) Every λ*-eigenvector of* T *lies in* $E(K)$.

(c) Some λ*-eigenvector of* T *lies in* $E(K)$.

Proof. That $E(K)$ is closed and T-invariant is routine, so I omit the argument. For the spectral inclusion, suppose $\lambda \in \mathbb{C}\backslash K$. We wish to show that $T - \lambda I$ is invertible on $E(K)$. Choose an open set V that contains K but whose closure does not contain λ, and observe that there is a C^∞ function η on the plane with $\eta(z) = (z-\lambda)^{-1}$ on V. Thus $\gamma(z) := (z-\lambda)\eta(z)$ is C^∞ on the plane, and $\equiv 1$ on V, and so $1-\gamma$ has support disjoint from K. Therefore if $x \in E(K)$ we have (by the definition of $E(K)$) $(1-\gamma)(T)x = 0$, hence: $I = \gamma(T) = (T-\lambda I)\eta(T) = \eta(T)(T-\lambda I)$ on $E(K)$, which establishes the desired invertibility.

As for eigenvalues and eigenfunctions, note first that if λ is an eigenvalue and x an eigenvector for λ then it is easy to check that x is a $\gamma(\lambda)$-eigenvector for $\gamma(T)$ for any $\gamma \in C^\infty(\mathbb{C})$. The equivalence of (a), (b), and (c) follows easily from this and the fact that $\lambda \in K$ if and only if $\gamma(\lambda) = 0$ for every $\gamma \in C^\infty(\mathbb{C})$ with support disjoint from K. □

4.2. Theorem *Suppose X is a Banach space and $T \in \mathcal{L}(X)$ has a C^∞ functional calculus in the sense of §3.1. Then T is decomposable.*

Proof. Suppose V and W are nonvoid open subsets that cover the plane. Recall from the Introduction that the goal is to find closed T-invariant subspaces Y and Z whose sum is X such that the restrictions of T to Y and Z have spectra that lie, respectively, in U and V.

To make the decomposition, let $\{\beta, \gamma\}$ be a C^∞ partition of unity on $\sigma(T)$ subordinate to the open covering $\{U, V\}$. Because $\beta + \gamma \equiv 1$ on $\sigma(T)$, the operator $\beta(T) + \gamma(T)$ is the identity on X. Thus $\beta(T)X + \gamma(T)X = X$.

Let $Y = E(\operatorname{spt}\beta)$ and $Z = E(\operatorname{spt}\gamma)$. Then the spectral inclusions follow immediately from Lemma 4.1. To see that $X = Y + Z$ just note that if x is in the range of $\beta(T)$, say $x = \beta(T)x'$ for some $x' \in X$, and if $\eta \in C^\infty(\mathbb{C})$ has support disjoint from that of β, then $\eta \cdot \beta \equiv 0$, so $0 = (\eta \cdot \beta)(T)x' = \eta(T)\beta(T)x' = \eta(T)x$, hence $x \in Y$. In other words $\operatorname{ran}\beta(T) \subset Y$, and similarly $\operatorname{ran}\gamma(T) \subset Z$. Since, as noted above, X is the sum of the smaller subspaces, it is also the sum of the larger ones. □

4.3. Remarks. (a) Operators with a C^∞ functional calculus are called *generalized scalar operators*. These form a proper subclass of the decomposable operators (see the discussion following Theorem 1.4.10 of [14] for references).

(b) If T is a Banach space operator whose spectrum lies in the unit circle, and for which there is a positive integer N such that

$$\|T^k\| = \mathrm{O}(|k|^N) \qquad (|k| \to \infty), \tag{11}$$

then a C^∞ functional calculus can be constructed for T by setting $\gamma(T) := \sum_{-\infty}^{\infty} \hat{\gamma}(k)T^k$, where for $\gamma \in C^\infty(\mathbb{C})$ and $\hat{\gamma}(k)$ is the k-th Fourier coefficient of the restriction of γ to the unit circle. If φ is a parabolic *automorphism* of U then it is well known that $T = C_\varphi$ obeys (11) (see [16], for example), and is therefore—as was first noted by Robert Smith in [19]—decomposable. In contrast to the operators we have been considering here, these automorphically-induced composition operators C_φ *are* supercyclic; in fact *hypercyclic* [1, Thm. 2.2, page 25].

5. (Non)Supercyclicity

So far we have seen that for $1 \leq p < \infty$, composition operators induced on H^p by parabolic non-automorphisms of U are decomposable, and that no such map has its spectrum lying on a circle. As previously mentioned, Theorem M of the Introduction then asserts that no such operator can be supercyclic.

Because the proof of Theorem M requires considerable background, I include for the reader's convenience this final section, which provides a mostly self-contained proof of non-supercyclicity for the class of composition operators we are considering here. The key to the argument is the following result, which occurs in [5, Theorem 6.1] and [6, Prop. 2.1].

5.1. Lemma *A bounded linear operator T on a Banach space X is* not *supercyclic on X whenever its spectrum can be split into a disjoint union of nonvoid compact sets K_1 and K_2, where $K_1 \subset \{|z| < r\}$ and $K_2 \subset \{|z| > r\}$ for some positive r.*

Proof. An operator is supercyclic if and only if every one of its non-zero scalar multiples is supercyclic, so we may, without loss of generality, assume that $r = 1$. The Riesz functional calculus provides a direct sum decomposition $X = X_1 \oplus X_2$ where X_i is a closed, T-invariant subspace of X and $\sigma(T|_{X_i}) \subset K_i$ $(i = 1, 2)$. Because the spectrum of $T|_{X_1}$ lies in the open unit disc, the spectral radius formula implies that the positive powers of this operator converge to zero in the operator norm. Similarly, the spectrum of the restriction of T to X_2 lies outside the closed unit disc, hence by an easy argument, $||T^n x|| \to \infty$ for every $0 \neq x \in X_2$ (see, e.g., [5, Lemma 6.3] for details).

Now suppose $0 \neq x \in X$. The goal is to show that x is not a supercyclic vector. In the decomposition $x = x_1 + x_2$ with $x_i \in X_i$ $(i = 1, 2)$ this will be trivial if either x_1 or x_2 is the zero vector. So suppose otherwise, in which case the Hahn-Banach theorem provides a bounded linear functional Λ on X that vanishes identically on X_2, but has $\Lambda(x_1) \neq 0$. Let y be a non-zero vector that is a limit point of the projective T-orbit of x, so there exist a sequence $\{c_j\}$ of scalars and a strictly increasing sequence $\{n_j\}$ of non-negative integers such that $c_j T^{n_j} \to y$. Therefore:

$$\frac{|\Lambda(y)|}{||y||} = \lim_j \frac{|\Lambda(c_j T^{n_j} x)|}{||c_j T^{n_j} x||} = \lim_j \frac{|\Lambda(T^{n_j} x_1)|}{||T^{n_j}(x_1 + x_2)||} .$$

In the last fraction the numerator is bounded above by $||\Lambda||\, ||T^{n_j} x_1||$, which converges to zero, while the denominator is bounded below by $||T^{n_j} x_2|| - ||T^{n_j} x_1||$, which converges to ∞. Thus the fraction itself converges to zero, and so $\Lambda(y) = 0$.

This shows that for any $0 \neq x \in X$ there is a nontrivial bounded linear functional Λ on X such that each limit point of the projective T-orbit of x lies in the null space of Λ. This projective orbit is therefore not dense in X so, as desired, x is not a supercyclic vector for T. □

5.2. Remark This result lies at the heart of the proof that for any supercyclic operator T there must exist a (possibly degenerate) circle centered at the origin that intersects every component of the spectrum of T (see [9, Proposition 3.1] [5, Theorem 6.1] or [6,

Prop. 2.1]). This "circle theorem" can, in turn, be considered an extension of a result of Kitai, who proved in [12, Theorem 2.8] that every component of the spectrum of a *hypercyclic* operator must intersect the unit circle. I thank Alfonso Montes for pointing out the reference to Herrero's paper.

5.3. Restriction and quotient maps If T is a bounded linear operator on X, then any closed, T-invariant subspace Y of X gives rise to two further operators: the usual *restriction operator* $T|_Y : Y \to Y$ and the perhaps less familiar *quotient operator* $T/Y : X/Y \to X/Y$, defined by: $(T/Y)(x+Y) := Tx+Y$ $(x \in X)$. Let $\sigma_f(T)$ denote the union of $\sigma(T)$ with all the bounded components of its complement, the so-called *full spectrum* of T. It is well known that $\sigma(T|_Y) \subset \sigma_f(T)$, but less familiar is the following result for quotient maps:

5.4. Lemma *If* $X = Y + Z$ *where* Y *and* Z *are closed,* T*-invariant subspaces of* X*, then* $\sigma(T/Z) \subset \sigma_f(T|_Y)$.

The result follows immediately from the one about restriction operators when X is the *direct* sum of Y and Z, for then the quotient map T/Z is similar to the restriction of T to Y. The general case follows from the restriction theorem and the (easily checked) fact that the map $y + (Y \cap Z) \to y + Z$ is an isomorphism of $Y/(Y \cap Z)$ onto X/Z that establishes a similarity between T/Z and $(T|_Y)/(Y \cap Z)$ (see [14, Proposition 1.2.4] for the details).

With these preliminaries out of the way we can now prove the main result of this section.

5.5. Theorem *If* φ *is a parabolic linear fractional selfmap of* U *that is not an automorphism, then* C_φ *is not supercyclic on* H^p *for* $1 \le p < \infty$.

Proof. Recall that $\sigma(C_\varphi)$ is either the closed interval $[0,1]$ or a curve that starts at 1 and converges to the origin by spiralling infinitely often around it, with distance to the origin decreasing monotonically. Choose any numbers $0 < r_1 < \rho_1 < \rho_2 < r_2 < 1$ and note that, because of this monotonicity, $\sigma(C_\varphi)$ intersects $\{|z| > r_1\}$ in an arc that contains the point 1. Let $V = \{|z| < \rho_1\} \cup \{|z| > \rho_2\}$ and $W = \{r_1 < |z| < r_2\}$, so that $\{V, W\}$ is an open covering of the plane. Because C_φ is decomposable on H^p there exist C_φ-invariant subspaces Y and Z such that $H^p = Y + Z$, $\sigma(C_\varphi|_Y) \subset V$, and $\sigma(C_\varphi|_Z) \subset W$.

Because supercyclicity (indeed any form of cyclicity) is inherited by quotient maps, the proof will be finished if we can show that the quotient map C_φ/Z is not supercyclic on H^p/Z. To this end observe that $\sigma(C_\varphi/Z) \subset \sigma_f(C_\varphi|_Y) \subset \sigma_f(C_\varphi) = \sigma(C_\varphi)$, where the first containment follows from Lemma 5.4, the second was pointed out in §5.3, and the final equality is a consequence of the spiral shape of the H^p-spectrum of C_φ (Corollary 3.10). Thus the spectrum of C_φ/Z lies in V, and therefore decomposes into a disjoint union of two compact sets, $K_1 \subset \{|z| < \rho_1\}$ and $K_2 \subset \{|z| > \rho_2\}$. Lemma 5.1 will then complete the job once we establish that neither K_1 nor K_2 is empty.

For this, recall from §4 that the decomposing subspaces Y and Z were constructed by choosing a C^∞ partition of unity $\{\beta, \gamma\}$ on $\sigma(C_\varphi)$ with $\operatorname{spt}\beta \subset V$ and $\operatorname{spt}\gamma \subset W$, and then setting $Y = E(\operatorname{spt}\beta)$ and $Z = E(\operatorname{spt}\gamma)$. From Lemma 4.1 we know that each point

e^{iat} of spt β is an eigenvalue of C_φ for which the corresponding eigenvector e_t lies in Y (here $a \in \Pi_+$ is the translation parameter of φ). Moreover, if $e^{iat} \notin$ spt γ then Lemma 4.1 insures that $e_t \notin Z$, so that the coset $e_t + Z$ is not the zero-element of the quotient space H^p/Z. Thus every point $e^{iat} \in$ spt $\beta \backslash$spt γ, is a C_φ/Z eigenvalue. Since spt $\beta\backslash$spt γ has points in both components of V, and $\sigma(C_\varphi/Z) \subset V$, we see that $\sigma(C_\varphi/Z)$ is split by an origin-centered circle. Thus Lemma 5.1 insures that C_φ/Z is not supercyclic, and therefore neither is C_φ. □

REFERENCES

1. P. S. Bourdon and J. H. Shapiro, *Cyclic phenomena for composition operators,* Memoirs Amer. Math. Soc. #596, AMS, Providence, R.I., 1997.
2. C. C. Cowen, *Composition operators on H^2*, J. Operator Th. 9 (1983), 77–106.
3. C. C. Cowen and B. D. MacCluer, *Composition Operators on Spaces of Analytic Functions*, CRC Press, Boca Raton, 1995.
4. N. Dunford and J. T. Schwartz, *Linear Operators, Vol. 1*, Wiley, 1957.
5. N. S. Feldman, T. L. Miller, and V. G. Miller, *Hypercyclic and supercyclic cohyponormal operators*, preprint 1999.
6. E. Gallardo and A. Montes, *The role of the angle in supercyclic behavior*, preprint, 1999.
7. J. B. Garnett, *Bounded Analytic Functions*, Academic Press, 1981
8. K. Hoffman, *Banach Spaces of Analytic Functions*, Prentice Hall, 1962.
9. D. A. Herrero, *Limits of hypercyclic and supercyclic operators*, J. Functional Analysis 99 (1991) 179–190.
10. H. M. Hilden and L. J. Wallen, *Some cyclic and non-cyclic vectors of certain operators*, Indiana Univ. Math. J. 23 (1974), 557–565.
11. F. Jafari et al., editors, *Studies on Composition Operators,* Contemp. Math. Vol. 213, American Math. Soc. 1998.
12. C. Kitai, *Invariant closed sets for linear operators*, Thesis, U. Toronto, 1982.
13. P. Koosis, *Introduction to H_p Spaces*, Cambridge University Press, 1980.
14. K. Laursen and M. Neumann, *Introduction to Local Spectral Theory*, Oxford University Press, 2000.
15. T. L. Miller and V. G. Miller, *Local spectral theory and orbits of operators*, Proc. Amer. Math. Soc., 127 (1999), 1029–1037.
16. E. A. Nordgren, *Composition operators*, Canad. J. Math. 20 (1968), 442–449.
17. J. H. Shapiro, *Composition Operators and Classical Function Theory*, Springer-Verlag, New York, 1993.
18. J. H. Shapiro, unpublished lectures, Michigan State University, 1997.
19. R. C. Smith, *Local spectral theory for invertible composition operators,* Integral Equations and Operator Theory 25 (1996), 329–335.

Recent Progress in Functional Analysis
K.D. Bierstedt, J. Bonet, M. Maestre and J. Schmets

Algebras of subnormal operators on the unit polydisc

J. Eschmeier

Fachbereich Mathematik, Universität des Saarlandes, D-66041 Saarbrücken, Germany

To Professor Manuel Valdivia on the occasion of his seventieth birthday.

Abstract

In this note it is shown that the dual algebra generated by a completely non-unitary subnormal tuple S with an isometric w^*-continuous H^∞-functional calculus over the unit polydisc satisfies the factorization property $(\mathbb{A}_{1,\aleph_0})$. This observation is used to deduce that S is reflexive and possesses a dense set of vectors generating an analytic invariant subspace.

MCS 2000 Primary 47B20; Secondary 47A13, 47L45, 47A15

1. Introduction

A result of Scott Brown [4] from 1978 shows that each subnormal operator $S \in L(H)$ on a complex Hilbert space H has a non-trivial invariant subspace Using Scott Brown's methods Olin and Thomson [20] proved that each subnormal operator S on H is reflexive. Thus Olin and Thomson extended earlier results of Sarason on the reflexivity of normal operators and analytic Toeplitz operators and of Deddens on the reflexivity of isometries.

A result of Yan [22] shows that each subnormal n-tuple $S \in L(H)^n$, that is, each system $S \in L(H)^n$ that extends to a system $N = (N_1, \ldots, N_n) \in L(K)^n$ of commuting normal operators on a larger Hilbert space K, possesses a non-trivial joint invariant subspace. It is an open question whether each subnormal n-tuple $S \in L(H)^n$ is reflexive. It was shown by Bercovici [3] that each commuting system of isometries on a Hilbert space is reflexive. In [2] Azoff and Ptak extended this result to the case of jointly quasinormal systems.

It is the aim of the present note to show that each completely non-unitary subnormal tuple S with an isometric H^∞-functional calculus over the open unit polydisc in $\mathbb{C}^n$ is reflexive. At the same time we prove that the dual operator algebra generated by S satisfies the factorization property $(\mathbb{A}_{1,\aleph_0})$.

More precisely, let $S \in L(H)^n$ be a subnormal tuple, and let $\mathfrak{A}_S$ be the smallest w^*-closed unital subalgebra of $L(H)$ containing $S_1, \ldots, S_n$. Our proof is based on the observation that each element in the natural predual $Q(S) = C^1(H)/{}^\perp\mathfrak{A}_S$ of $\mathfrak{A}_S$ can be approximated by the equivalence classes of rank–one operators. If in addition, the tuple S possesses an isometric w^*–continuous H^∞–functional calculus $\Phi : H^\infty(\mathbb{D}^n) \to L(H)$, then this density result means precisely that Φ possesses the almost factorization property in the sense of [12]. If S is completely non–unitary, that is, possesses no non–zero reducing subspace M such that the restriction of S onto M is a commuting tuple of unitary operators, then S satisfies the weak $C_{\cdot 0}$–condition

$$\lim_{k\to\infty} S^{*k}x = 0 \quad (x \in H),$$

where $S^{*k} = (S_1^* \cdot \ldots \cdot S_n^*)^k$ for each natural number k.

The above observations together with results proved in [12] allow us to deduce that the dual algebra $\mathfrak{A}_S$ generated by a completely non–unitary subnormal tuple $S \in L(H)^n$ with an isometric w^*–continuous H^∞–functional calculus over the unit polydisc possesses the factorization property $(\mathbb{A}_{1,\aleph_0})$. As consequences we obtain that the vectors generating an analytic invariant subspace for S form a dense subset of H and that the tuple S is reflexive. Analogous results for subnormal tuples over the unit ball in $\mathbb{C}^n$ were obtained in [11].

In the one–variable case, Sarason's decomposition theorem for compactly supported measures and corresponding decomposition theorems for subnormal operators due to Conway and Olin [9] allow the reduction of the reflexivity problem for single subnormal operators to the case of subnormal operators with isometric w^*–continuous H^∞–functional calculus over the unit disc. Since a reduction of this type is missing in the multivariable case, it is not clear whether a general reflexivity proof for subnormal tuples is possible along these lines.

2. Preliminaries

Let H be a complex Hilbert space, and let $L(H)$ be the Banach algebra of all continuous linear operators on H. We regard $L(H)$ as the norm–dual of the space $\mathcal{C}^1(H)$ of all trace–class operators on H. Let $T = (T_1, \ldots, T_n) \in L(H)^n$ be a commuting tuple. The smallest w^*–closed unital subalgebra $\mathfrak{A}_T$ of $L(H)$ containing $T_1, \ldots, T_n$ is isometrically isomorphic to the norm–dual of the quotient space $Q(T) = \mathcal{C}^1(H)/{}^\perp\mathfrak{A}_T$. In this way $\mathfrak{A}_T$ becomes a dual algebra, that is, $\mathfrak{A}_T$ is the norm–dual of a suitable Banach space such that the multiplication in $\mathfrak{A}_T$ is separately w^*–continuous. If A and B are dual algebras, then a dual algebra isomorphism $\varphi : A \to B$ is by definition an algebra homomorphism between A and B that is an isometric isomorphism and a w^*–homeomorphism.

For $x, y \in H$, we denote by $[x \otimes y] \in Q(T)$ the equivalence class of the rank–one operator $H \to H$, $\xi \mapsto \langle \xi, y\rangle x$. Let p, q be any cardinal numbers with $1 \le p, q \le \aleph_0$. The dual algebra $\mathfrak{A}_T$ possesses property $(\mathbb{A}_{p,q})$ if, for each matrix (L_{ij}) of functionals $L_{ij} \in Q(T)$ $(0 \le i < p,\ 0 \le j < q)$, there are vectors $(x_i)_{0\le i<p}$, $(y_j)_{0\le j<q}$ in H with

$$L_{ij} = [x_i \otimes y_j] \quad (0 \le i < p,\ 0 \le j < q).$$

If $p = q$, then we write $(\mathbb{A}_p)$ instead of $(\mathbb{A}_{p,q})$.

Let K be a compact set in $\mathbb{C}^n$, and let $M(K)$ be the space of all complex regular Borel measures on K. We write $M_1^+(K)$ for the subset consisting of all probability measures on K. Let $\mu \in M(K)$ be a positive measure. We denote by $P^\infty(\mu)$ the w^*–closure of the set of all polynomials in $L^\infty(\mu)$ with respect to the duality $\langle L^1(\mu), L^\infty(\mu)\rangle$. The space $P^\infty(\mu)$ is a dual algebra with predual $Q(\mu) = L^1(\mu)/^\perp P^\infty(\mu)$. For $1 \leq q < \infty$, we define $P^q(\mu)$ as the norm–closure of the polynomials in $L^q(\mu)$.

Let $\mathbb{D}^n$ be the open unit polydisc in $\mathbb{C}^n$. The Banach algebra $H^\infty(\mathbb{D}^n)$ of all bounded analytic functions on $\mathbb{D}^n$ is regarded as the norm–dual of the quotient $Q = L^1(\mathbb{D}^n)/^\perp H^\infty(\mathbb{D}^n)$. Here we use that $H^\infty(\mathbb{D}^n)$ is a w^*–closed subspace of $L^\infty(\mathbb{D}^n)$ relative to the duality $\langle L^1(\mathbb{D}^n), L^\infty(\mathbb{D}^n)\rangle$ (formed with respect to the $(2n)$–dimensional Lebesgue measure). A subset σ of $\mathbb{C}^n$ is called dominating in $\mathbb{D}^n$ if $\|f\| = \sup\{|f(z)|;\, z \in \mathbb{D}^n \cap \sigma\}$ for all $f \in H^\infty(\mathbb{D}^n)$.For $\lambda \in \mathbb{D}^n$ and $k \in \mathbb{N}^n$, the w^*–continuous linear functionals

$$\mathcal{E}_\lambda : H^\infty(\mathbb{D}^n) \to \mathbb{C}, \quad f \mapsto f(\lambda),$$

$$\mathcal{E}^{(k)} : H^\infty(\mathbb{D}^n) \to \mathbb{C}, \quad f \mapsto f^{(k)}(0)/k!$$

are regarded as elements in Q. If $\Phi : H^\infty(\mathbb{D}^n) \to L(H)$ is a w^*–continuous algebra homomorphism, then for $x, y \in H$, we regard the w^*–continuous linear functional

$$x \otimes y : H^\infty(\mathbb{D}^n) \to \mathbb{C}, \quad x \otimes y(f) = \langle \Phi(f)x, y\rangle$$

as an element in Q.

For a commuting tuple $T = (T_1, \ldots, T_n) \in L(H)^n$ and $k \in \mathbb{N}$, $\alpha = (\alpha_1, \ldots, \alpha_n) \in \mathbb{N}^n$, we use the standard abbreviations $T^k = (T_1 \cdot \ldots \cdot T_n)^k$ and $T^\alpha = T_1^{\alpha_1} \cdot \ldots \cdot T_n^{\alpha_n}$. We write $\operatorname{Lat}(T)$ for the lattice of all closed linear subspaces of H which are invariant under $T_1, \ldots, T_n$. We denote by $\sigma(T)$ the Taylor spectrum of the commuting tuple T. For the definition and properties of the Taylor spectrum the reader is referred to [14].

3. Factorization results

Let $S \in L(H)^n$ be a subnormal tuple on a Hilbert space H. Our first aim is to approximate the elements $L \in Q(S)$ by the equivalence classes of suitable rank–one operators. Via the spectral theorem for commuting normal tuples, this problem will be reduced to the following measure–theoretic result proved in [13] (Lemma 1.3).

Lemma 3.1 *Let $K \subset \mathbb{C}^n$ be a compact set, and let $\mu \in M_1^+(K)$ be a probability measure on K. For any given element $L \in Q(\mu) = L^1(\mu)/^\perp P^\infty(\mu)$ and any real number $\varepsilon > 0$, there are functions $f, g \in P^2(\mu)$ with $\|f\|, \|g\| \leq \|L\|^{1/2}$ and*

$$\|L - [f\overline{g}]\| < \varepsilon.$$

Let $N \in L(K)^n$ be the minimal normal extension of the subnormal tuple $S \in L(H)^n$, where K is a Hilbert space containing H. The multidimensional analogue of Proposition V.17.4 from [8] (proved in exactly the same way as in the one–variable case) allows us to choose a separating vector h for N in H. Let

$$\mu : B(K) \to [0, \infty), \ \mu(A) = \langle E(A)h, h \rangle,$$

where $K = \sigma(N)$ and E is the operator–valued spectral measure for N, be the scalar spectral measure of N given by the vector h. We denote by

$$H_h = \vee(S^k h; k \in \mathbb{N}^n) \in \mathrm{Lat}(S)$$

the smallest invariant subspace for S containing the vector h, and we write

$$U : P^2(\mu) \to H_h$$

for the unique unitary operator satisfying $Up = p(S)h$ for all polynomials p in n variables.

The w^*–continuous isomorphism of von Neumann algebras

$$\Psi : L^\infty(\mu) \to W^*(N)$$

associated with N induces a dual algebra isomorphism $\gamma : P^\infty(\mu) \to \mathfrak{A}_S$ (cf. [7] or Section 1 in [11]). We denote by $\gamma_* : Q(S) \to Q(\mu)$ the predual of this map.

Lemma 3.2 *Let $L \in Q(S)$ and let $\varepsilon > 0$ be a given real number. Then there are vectors $x, y \in H$ with $||L - [x \otimes y]|| < \varepsilon$ and $||x||, ||y|| \leq ||L||^{1/2}$.*

Proof. By Lemma 3.1 there are functions $f, g \in P^2(\mu)$ with $||f||, ||g|| \leq ||L||^{1/2}$ and

$$||\gamma_*(L) - [f\overline{g}]||_{Q(\mu)} < \varepsilon.$$

Since

$$\langle L - [U(f) \otimes U(g)], \gamma(\varphi) \rangle = \langle \gamma_*(L) - [f\overline{g}], \varphi \rangle$$

holds for all functions $\varphi \in P^\infty(\mu)$, the vectors $x = U(f)$ and $y = U(g)$ satisfy all the assertions of the lemma. □

Let us suppose that, in addition, the subnormal tuple S possesses an isometric w^*–continuous H^∞–functional calculus

$$\Phi : H^\infty(\mathbb{D}^n) \to L(H).$$

The induced dual algebra isomorphism $H^\infty(\mathbb{D}^n) \xrightarrow{\Phi} \mathfrak{A}_S$ is the adjoint of an isometric isomorphism

$$Q = L^1(\mathbb{D}^n)/{}^\perp H^\infty(\mathbb{D}^n) \xleftarrow{\Phi_*} Q(S).$$

For x and y in H, we regard the w^*–continuous linear functional

$$x \otimes y : H^\infty(\mathbb{D}^n) \to \mathbb{C}, \quad f \mapsto \langle \Phi(f)x, y\rangle$$

as an element in the predual Q of $H^\infty(\mathbb{D}^n)$. Obviously we have

$$\Phi_*([x \otimes y]) = x \otimes y \qquad (x, y \in H).$$

It is well known that each subnormal tuple $S \in L(H)^n$ with $\sigma(S) \subset \overline{\mathbb{D}}^n$ possesses a unitary dilation (even a regular unitary dilation in the sense of [21]). Indeed, let as before $N \in L(K)^n$ be the minimal normal extension of S with associated functional calculus

$$\Psi : L^\infty(\mu) \to L(K).$$

For each multiindex $j = (j_1, \dots, j_n) \in \{0,1\}^n$ and each vector $x \in H$, we obtain

$$\sum_{0 \le \alpha \le j} (-1)^{|\alpha|}\langle S^{*\alpha} S^\alpha x, x\rangle = \langle \sum_{0 \le \alpha \le j} (-1)^{|\alpha|} N^{*\alpha} N^\alpha x, x\rangle = \langle \Psi(f_j)x, x\rangle \ge 0,$$

where $f_j \in L^\infty(\mu)$ is the non–negative function defined by

$$f_j(z) = (1 - |z_1|^2)^{j_1} \cdot \ldots \cdot (1 - |z_n|^2)^{j_n}.$$

The validity of the above positivity conditions implies the existence of a regular unitary dilation for S (see Theorem I.9.1 in [21]).

Lemma 3.2 implies that each subnormal tuple $S \in L(H)^n$ with w^*–continuous isometric H^∞–functional calculus $\Phi : H^\infty(\mathbb{D}^n) \to L(H)$ possesses the almost factorization property as defined in [12] (Definition 2.5). More precisely, we have the following result.

Corollary 3.3 *For each element $L \in Q$ and each real number $\varepsilon > 0$, there are vectors $x, y \in H$ with $\|x\|, \|y\| \le \|L\|^{1/2}$ and*

$$\|L - x \otimes y\| < \varepsilon.$$

Let $S \in L(H)^n$ be a subnormal tuple with $\sigma(S) \subset \overline{\mathbb{D}}^n$. Suppose that S is completely non–unitary, that is, there is no non–zero reducing subspace M for S such that $S|M$ is a commuting tuple of unitary operators. An elementary argument (Lemma 2.1 in [12]) shows that in this case the product $S_1 \cdot \ldots \cdot S_n \in L(H)$ is a completely non–unitary subnormal contraction on H. Hence (cf. for instance Corollary 2.4 in [10]) it follows that

$$\lim_{k \to \infty} S^{*k} x = 0 \quad (x \in H).$$

For $\lambda = (\lambda_1, \dots, \lambda_n) \in \mathbb{D}^n$, let $\varphi_\lambda : \mathbb{D}^n \to \mathbb{D}^n$ be the biholomorphic map defined by

$$\varphi_\lambda(z_1, \dots, z_n) = \left(\frac{z_i - \lambda_i}{1 - \overline{\lambda}_i z_i}\right)_{i=1}^n.$$

This mapping extends to a holomorphic $\mathbb{C}^n$–valued function on a neighbourhood of $\overline{\mathbb{D}}^n$ which induces a homeomorphism $\overline{\mathbb{D}}^n \to \overline{\mathbb{D}}^n$. For each $\lambda \in \mathbb{D}^n$, the tuple $S_\lambda = \varphi_\lambda(S)$ is again a completely non–unitary subnormal system with $\sigma(S_\lambda) \subset \overline{\mathbb{D}}^n$. Therefore

$$\lim_{k\to\infty} (S_\lambda)^{*k} x = 0$$

for all $x \in H$ and $\lambda \in \mathbb{D}^n$.

As an application of the results from [12] it follows that the dual algebra generated by a completely non–unitary subnormal tuple $S \in L(H)^n$ with an isometric w^*–continuous $H^\infty(\mathbb{D}^n)$–functional calculus possesses the factorization property $(\mathbb{A}_{1,\aleph_0})$. More precisely, we obtain the following result.

Theorem 3.4 *Let $S \in L(H)^n$ be a completely non–unitary subnormal tuple with an isometric w^*–continuous H^∞–functional calculus $\Phi : H^\infty(\mathbb{D}^n) \to L(H)$. For each $\varepsilon > 0$, there is a constant $C = C(\varepsilon) > 0$ such that, for each sequence $(L_k)_{k\geq 1}$ in Q and each vector $a \in H$, there are elements $x, y_k \in H$ $(k \geq 1)$ with $\|x - a\| < \varepsilon$ and*

$$L_k = x \otimes y_k\,, \quad \|y_k\| \leq Ck^2 \|L_k\| \quad (k \geq 1).$$

Proof. By Corollary 3.3 the tuple S possesses the almost factorization property in the sense of [12]. Since S is completely non–unitary, we have

$$\lim_{k\to\infty} (S_\lambda)^{*k} x = 0 \quad (x \in H,\ \lambda \in \mathbb{D}^n).$$

It follows from Proposition 2.6 and Corollary 3.5 from [12] (with $\theta = 0$ and $\gamma = 1$) that the assertion holds with a constant of the form $C(\varepsilon) = C/\varepsilon$, where $C > 0$ is a suitable universal constant. □

For our reflexivity proof we need the following $\mathbb{A}_1$–factorization result with additional control on the factors.

Theorem 3.5 *Let $S \in L(H)^n$ be a completely non–unitary subnormal tuple with an isometric w^*–continuous H^∞–functional calculus $\Phi : H^\infty(\mathbb{D}^n) \to L(H)$. Let $L \in Q$ be a given functional. Then, for any vectors $u, v \in H$, there are vectors $u', v' \in H$ with $L = u' \otimes v'$ and*

$$\|u' - u\| \leq 4\|L - u \otimes v\|^{1/2}, \quad \|v'\| \leq 4(\|v\| + \|L - u \otimes v\|^{1/2}).$$

Proof. Let us fix a co–isometric extension $C \in L(\mathcal{S} \oplus \mathcal{R})^n$ of S as explained in Section 2 or Section 3 of [12].

Suppose that $L \in Q$ and $u, v \in H$ are given with $v = w \oplus b$ $(w \in \mathcal{S}, b \in \mathcal{R})$. Since S possesses the almost factorization property, Proposition 2.6 and Corollary 3.3 from [12] (with $\theta = 0$, $\gamma = 1$, and $N = 1$) allow us to choose vectors $u' \in H$ and $w' \in \mathcal{S}, b' \in \mathcal{R}$ such that $L = u' \otimes (w' + b')$ and

$$\|u' - u\| \leq 4\|L - u \otimes v\|^{1/2},$$
$$\|w' - w\| \leq 2\|L - u \otimes v\|^{1/2},$$
$$\|b'\| \leq 2(\|b\| + \|L - u \otimes v\|^{1/2}).$$

To conclude the proof it suffices to define $v' = P(w' \oplus b')$, where P is the orthogonal projection from $\mathcal{S} \oplus \mathcal{R}$ onto H, and to observe that the estimate

$$\begin{aligned} \|v'\| &\leq \|w'\| + \|b'\| \leq \|w\| + 2\|b\| + 4\|L - u \otimes v\|^{1/2} \\ &\leq 2\sqrt{2}\|v\| + 4\|L - u \otimes v\|^{1/2} \end{aligned}$$

holds. □

4. Reflexivity

Let $S \in L(H)^n$ be a subnormal tuple on a Hilbert space H. We denote by Alg Lat(S) the subalgebra of $L(H)$ consisting of all operators $C \in L(H)$ with Lat$(C) \supset$ Lat(S). We show that S is reflexive, that is, Alg Lat(S) coincides with the WOT–closed unital subalgebra of $L(H)$ generated by S, whenever S is completely non–unitary and possesses an isometric w^*–continuous H^∞–functional calculus over $\mathbb{D}^n$.

Proposition 4.1 *Let $S \in L(H)^n$ be a completely non–unitary subnormal tuple with an isometric H^∞-functional calculus $\Phi : H^\infty(\mathbb{D}^n) \to L(H)$. Then, for any $\varepsilon > 0$ and any vector $a \in H$, there are vectors $x, y^{(k)} \in H$ $(k \in \mathbb{N}^n)$ with $\|x - a\| < \varepsilon$ and*

$$x \otimes y^{(k)} = \mathcal{E}^{(k)} \quad (k \in \mathbb{N}^n)$$

and such that the power series

$$f(\lambda) = \sum_{k \in \mathbb{N}^n} y^{(k)} \lambda^k$$

converges on $\mathbb{D}^n$.

Proof. Choose an enumeration $(L_k)_{k \geq 1}$ of the set

$$\{\mathcal{E}^{(j)};\ j \in \mathbb{N}^n\}$$

such that, for $j, i \in \mathbb{N}^n$, with $|j| \leq |i|$, the functional $\mathcal{E}^{(j)}$ occurs in the sequence $(L_k)_{k\geq 1}$ before the functional $\mathcal{E}^{(i)}$. A very rough estimate shows that, for $i \geq 0$, each functional

$$\mathcal{E}^{(j)} \quad (|j| = i)$$

occurs in the sequence $(L_k)_{k\geq 1}$ with an index $k \leq (i+1)^n$. Since these functionals belong to the closed unit ball of Q (Theorem 2.2.7 in [16]), Theorem 3.4 allows us to choose vectors $x, y^{(j)} \in H$ $(j \in \mathbb{N}^n)$ with $\|x - a\| < \varepsilon$ and

$$\mathcal{E}^{(j)} = x \otimes y^{(j)} \quad (j \in \mathbb{N}^n)$$

and such that with a suitable constant $C > 0$ (independent of j)

$$\|y_j\| \leq C(|j| + 1)^{2n} \quad (j \in \mathbb{N}^n).$$

But then the power series $f(\lambda) = \sum_{j\in\mathbb{N}^n} y_j \lambda^j$ converges on the unit polydisc $\mathbb{D}^n$. □

With the notations from Proposition 4.1 (and with $S, x, y^{(k)}$ as explained there) define

$$M = M_x = \bigvee_{k\in\mathbb{N}^n} S^k x \in \mathrm{Lat}(S)$$

and $\tilde{y}^{(k)} = P_x y^{(k)}$, where P_x is the orthogonal projection from H onto M_x. Then

$$e : \mathbb{D}^n \to M, \quad e(\lambda) = \sum_{k\in\mathbb{N}^n} \tilde{y}^{(k)} \overline{\lambda}^k$$

is a conjugate analytic function such that, for $\lambda \in \mathbb{D}^n$ and $f \in H^\infty(\mathbb{D}^n)$,

$$\langle x \otimes e(\lambda), f \rangle = \sum_{k\in\mathbb{N}^n} \langle x \otimes y^{(k)}, f \rangle \lambda^k = f(\lambda).$$

We conclude that

$$x \otimes e(\lambda) = \mathcal{E}_\lambda \quad (\lambda \in \mathbb{D}^n).$$

Since, for all $f \in H^\infty(\mathbb{D}^n)$, $\lambda \in \mathbb{D}^n$, and $i = 1, \ldots, n$,

$$\langle \Phi(f)x, (\lambda_i - S_i|M)^* e(\lambda) \rangle = \langle \Phi((\lambda_i - z_i)f)x, e(\lambda) \rangle = 0,$$

it follows that

$$(\lambda_i - S_i|M)^* e(\lambda) = 0 \quad (\lambda \in \mathbb{D}^n,\ i = 1, \ldots, n).$$

As in [11], or in corresponding one–dimensional situations, the reflexivity proof is based on the notion of analytic invariant subspaces.

Definition 4.2 *Let $T \in L(H)^n$ be a commuting tuple of contractions. A space M in* $\mathrm{Lat}(T)$ *is an analytic invariant subspace for T if there is a non-zero conjugate analytic function $e : \mathbb{D}^n \to M$ such that $(\lambda_i - T_i|M)^* e(\lambda) = 0$ for $\lambda \in \mathbb{D}^n$ and $i = 1, \ldots, n$.*

Let $T \in L(H)^n$ be a commuting tuple of contractions. For $x \in H$, we denote by M_x the smallest closed invariant subspace for T containing x and by P_x the orthogonal projection from H onto M_x. Following ideas of Chevreau [6] it was shown in [11] that, if M_x is an analytic invariant subspace for T via the conjugate analytic function $e : \mathbb{D}^n \to M_x$, then the zero set of the function e coincides with the set

$$Z(x) = \{\lambda \in \mathbb{D}^n; \langle x, e(\lambda)\rangle = 0\}$$

and the function

$$\mathbb{D}^n \setminus Z(x) \to M_x, \quad \lambda \mapsto e(\lambda)/\langle e(\lambda), x\rangle$$

extends to a conjugate analytic function $k : \mathbb{D}^n \to M_x$. Furthermore, $k : \mathbb{D}^n \to M_x$ is the unique conjugate analytic function with $\langle x, k(\lambda)\rangle = 1$ for all $\lambda \in \mathbb{D}^n$ and

$$k(\lambda) \in \mathrm{Ker}(\lambda_i - T_i|M_x)^* \quad (\lambda \in \mathbb{D}^n,\ i = 1, \ldots, n).$$

If T possesses a w^*-continuous H^∞-functional calculus $\Phi : H^\infty(\mathbb{D}^n) \to L(H)$, then $k : \mathbb{D}^n \to M_x$ is the unique conjugate analytic function with $x \otimes k(\lambda) = \mathcal{E}_\lambda$ for all $\lambda \in \mathbb{D}^n$.

Let us return to the case that $T = S \in L(H)^n$ is a completely non–unitary subnormal tuple with an isometric w^*-continuous functional calculus $\Phi : H^\infty(\mathbb{D}^n) \to L(H)$. It follows from Poposition 4.1 and the remarks following its proof that

$$\mathcal{C} = \{x \in H;\ M_x \text{ is an analytic invariant subspace for } T\}$$

is a dense subset of H.

Let $x \in \mathcal{C}$ and let $k : \mathbb{D}^n \to M_x$ be the unique conjugate analytic function with

$$x \otimes k(\lambda) = \mathcal{E}_\lambda \quad (\lambda \in \mathbb{D}^n).$$

If $C \in \mathrm{Alg\,Lat}(S)$, then $(C|M_x)^* \in \mathrm{Alg\,Lat}\big((S|M_x)^*\big)$, and hence there are unique complex numbers $g(\lambda)$ $(\lambda \in \mathbb{D}^n)$ such that

$$(C|M_x)^* k(\lambda) = \overline{g(\lambda)} k(\lambda) \quad (\lambda \in \mathbb{D}^n).$$

For $u \in M_x$ and $\lambda \in \mathbb{D}^n$,

$$\langle Cu, k(\lambda)\rangle = \langle u, (C|M_x)^* k(\lambda)\rangle = g(\lambda)\langle u, k(\lambda)\rangle.$$

The induced map

$$\Psi = \Psi_x : \mathrm{Alg\,Lat}(S) \to H^\infty(\mathbb{D}^n), \quad C \mapsto g \text{ (defined as above)}$$

is a contractive unital algebra homomorphism with $\Psi(S_i) = z_i$ for $i = 1, \ldots, n$.

Now the reflexivity of S can be shown exactly as in the case of the unit ball studied in [11].

Proposition 4.3 *Let $S \in L(H)^n$ be a completely non–unitary subnormal tuple with an isometric w^*–continuous H^∞–functional calculus $\Phi : H^\infty(\mathbb{D}^n) \to L(H)$. Let $x \in \mathcal{C}$ and let $C \in$ AlgLat(S). Then $g = \Psi_x(C)$ is the unique function in $H^\infty(\mathbb{D}^n)$ with*

$$\Phi(g)|M_x = C|M_x.$$

Proof. Let us define $M = M_x$. Since $\sigma(S|M) = \overline{\mathbb{D}}^n$, the restriction of Φ to M gives an isometric w^*–continuous H^∞–functional calculus for $S|M$. Hence the uniqueness part of the assertion is obvious.

Denote by $k : \mathbb{D}^n \to M$ the unique conjugate analytic function with

$$x \otimes k(\lambda) = \mathcal{E}_\lambda \quad (\lambda \in \mathbb{D}^n)$$

and define $N = \{u \in M; \langle u, k(\lambda)\rangle = 0 \text{ for all } \lambda \in \mathbb{D}^n\}$. To complete the proof it suffices to show that

$$\langle \Phi(g)u, v\rangle = \langle Cu, v\rangle \quad (u \in M \setminus N,\ v \in M).$$

For $u \in M \setminus N$ and $v \in M$ with $u \otimes v \in \mathcal{E} = LH\{\mathcal{E}_\lambda;\ \lambda \in \mathbb{D}^n\}$, the proof follows exactly as in the case of the ball (see the proof of Proposition 3.6 from [11]). Fix u in $M \setminus N$ and v in M with $L = u \otimes v \notin \mathcal{E}$. Choose a sequence $(L_k)_{k\geq 1}$ in $\mathcal{E}$ with $(L_k) \xrightarrow{k} L$. Since $S|M$ is again a completely non–unitary subnormal tuple with isometric w^*–continuous $H^\infty(\mathbb{D}^n)$–functional calculus, Theorem 3.5 allows us to choose sequences $(u_k)_{k\geq 1}$ and $(v_k)_{k\geq 1}$ in M with $L_k = u_k \otimes v_k$ and

$$\|u_k - u\| \leq 4d_k^{1/2}, \quad \|v_k\| \leq 4(\|v\| + d_k^{1/2})$$

for all $k \geq 1$. Here $d_k = \|L_k - u \otimes v\|$. After passing to suitable subsequences we may suppose that $(v_k)_{k\geq 1}$ converges weakly to a vector $w \in M$ and that $u_k \notin N$ for all $k \geq 1$. But then

$$\langle Cu, w\rangle = \lim_{k\to\infty} \langle Cu_k, v_k\rangle = \lim_{k\to\infty} \langle \Phi(g)u_k, v_k\rangle = \langle \Phi(g)u, v\rangle.$$

Since, for all $f \in H^\infty(\mathbb{D}^n)$,

$$u \otimes w(f) = \lim_{k\to\infty} \langle \Phi(f)u_k, v_k\rangle = u \otimes v(f),$$

it follows that $\langle Cu, v\rangle = \langle Cu, w\rangle$, and the proof is complete. □

Let $S \in L(H)^n$ be a completely non–unitary subnormal tuple with an isometric w^*–continuous H^∞–functional calculus over $\mathbb{D}^n$. Let $C \in \text{Alg Lat}(S)$. By Proposition 4.3, for each vector $x \in \mathcal{C}$, there is a unique function g_x in $H^\infty(\mathbb{D}^n)$ with $C|M_x = \Phi(g_x)|M_x$. Since $\mathcal{C}$ is dense in H, the reflexivity of S is proved if we can show that $g_x = g_y$ for all $x, y \in \mathcal{C}$. This can be shown as in the case of the ball (see the proof of Theorem 3.7 from [11]).

Corollary 4.4 *Each completely non–unitary subnormal tuple $S \in L(H)^n$ with an isometric w^*–continuous H^∞–functional calculus $\Phi : H^\infty(\mathbb{D}^n) \to L(H)$ is reflexive.*

Proof. For the convenience of the reader we sketch the main ideas.

Fix an operator $C \in \text{Alg Lat}(S)$. For each $x \in \mathcal{C}$, let $g_x \in H^\infty(\mathbb{D}^n)$ be the unique function in $H^\infty(\mathbb{D}^n)$ with $\Phi(g_x)|M_x = C|M_x$. Since $\Phi|M_x$ is isometric, it follows that $\|g_x\| \leq \|C\|$.

Let $x \in H$ be arbitrary. Choose a sequence $(x_k)_{k\geq 1}$ in $\mathcal{C}$ with $x = \lim_{k\to\infty} x_k$ such that the associated sequence $(g_{x_k})_{k\geq 1}$ has a w^*–limit g in $H^\infty(\mathbb{D}^n)$. Since, for all $y \in H$,

$$\langle \Phi(g)x, y\rangle = \lim_{k\to\infty} \langle \Phi(g_{x_k})x_k, y\rangle = \langle Cx, y\rangle,$$

it follows that $\Phi(g)x = Cx$.

Let $x, y \in \mathcal{C}$ be arbitrary. To show that $g_x = g_y$, choose a function $h \in H^\infty(\mathbb{D}^n)$ with $\Phi(h)(x + y) = C(x + y)$ and observe that $\Phi(g_x - h)x = \Phi(h - g_y)y$. Since the representations $\Phi|M_x$ and $\Phi|M_y$ are isometric, we are allowed to assume that $g_x \neq h$ and that $g_y \neq h$. By Lemma 3.5 in [11] the vector $u = \Phi(g_x - h)x$ belongs to $\mathcal{C}$. Because $M_u \subset M_x$ the uniqueness part of Proposition 4.3 implies that $g_x = g_u$. In exactly the same way one obtains that $g_y = g_u$. Thus the proof is complete. □

Recall that a relexive subalgebra $L \subset L(H)$ is called super–reflexive if each WOT–closed subalgebra L_0 of L containing the identity operator is reflexive ([15]). Obviously, each super–reflexive subalgebra of $L(H)$ consists entirely of reflexive operators.

Corollary 4.5 *Let $S \in L(H)^n$ be a completely non–unitary subnormal tuple with an isometric w^*–continuous H^∞–functional calculus $\Phi : H^\infty(\mathbb{D}^n) \to L(H)$. Then the algebra $\mathfrak{A}_S = \Phi(H^\infty(\mathbb{D}^n))$ is super–reflexive.*

Proof. By Corollary 4.4 the algebra $\mathfrak{A}_S$ is reflexive. According to Theorem 2.3 of [19] it suffices to show that, for each WOT–continuous linear functional u on $\mathfrak{A}_S$, there are vectors $x, y \in H$ with $u(T) = \langle Tx, y\rangle$ for all $T \in \mathfrak{A}_S$. But this is obvious, since $\mathfrak{A}_S$ has the factorization property $(\mathbb{A}_1)$. □

Since each completely non–unitary subnormal contraction on a Hilbert space is of class $C_{\cdot 0}$, results of Apostol ([1]) on H^∞–functional calculi of polynomially bounded n–tuples imply that each commuting tuple $S \in L(H)^n$ of completely non–unitary subnormal contractions possesses a w^*–continuous H^∞–functional calculus $\Phi : H^\infty(\mathbb{D}^n) \to L(H)$. Thus the preceding corollaries apply to this case.

Corollary 4.6 *Let $S \in L(H)^n$ be a commuting tuple of completely non–unitary subnormal operators with $\sigma(S) \subset \overline{\mathbb{D}}^n$ and such that $\sigma(S)$ is dominating in $\mathbb{D}^n$. Then S is reflexive and, moreover, the algebra $\mathfrak{A}_S$ is super–reflexive.*

Proof. The remarks preceding the corollary imply that S has a w^*–continuous H^∞–functional calculus $\Phi : H^\infty(\mathbb{D}^n) \to L(H)$. Since $\sigma(S)$ is assumed to be dominating in $\mathbb{D}^n$ and since $f\big(\sigma(S) \cap \mathbb{D}^n\big) \subset \sigma\big(\Phi(f)\big)$ for each function $f \in H^\infty(\mathbb{D}^n)$, the representation Φ is isometric. Thus the assertion follows as an application of Corollary 4.5. □

The analogue of Corollary 4.4 on the unit ball is valid without the hypothesis that S is completely non–unitary (see Theorem 3.7 in [11]). We expect that the same improvement is possible in the case of the unit polydisc. But an additional idea seems to be necessary to answer this question. A similar remark applies to Corollary 4.6.

REFERENCES

1. C. Apostol, Functional calculus and invariant subspaces, J.Operator Theory 4 (1980), 159–190.
2. E.A. Azoff and M. Ptak, Jointly quasinormal families are reflexive, Acta Sci.Math. 61 (1995), 545–547 .
3. H. Bercovici, A factorization theorem with applications to invariant subspaces and the reflexivity of isometries, Mathematical Research Letters 1 (1994), 511–518.
4. S. Brown, Some invariant subspaces for subnormal operators, Integral Equations Operator Theory 1 (1978), 310–333.
5. S. Brown and B. Chevreau, Toute contraction à calcul fonctionnel isométrique est réflexive, C.R. Acad. Sci. Paris 307 (1988), 185-188.
6. B. Chevreau, A survey of recent reflexivity results, Operator algebras and operator theory (Craiova, 1989), 46–61, Pitman Res. Notes Math.Ser., 271, Longman Sci.Tech., Harlow 1992.
7. J.B. Conway, Towards a functional calculus for subnormal tuples: the minimal normal extension and approximation in several variables, Proc.Symp. Pure Math., Vol.51 (1990), part 1, Amer.Math.Soc., Providence, R.I., pp. 105–112.
8. J.B. Conway The theory of subnormal operators, Math.Surveys and Monographs, Vol.38, AMS, Providence, Rhode Island, 1991.
9. J.B. Conway and R. Olin, A functional calculus for subnormal operators II, Mem.Amer.Math.Soc. 184 (1977).

10. J. Eschmeier, Invariant subspaces for spherical contractions, Proc.London Math.Soc. 75 (1997), 157–176.
11. J. Eschmeier, Algebras of subnormal operators on the unit ball, J.Operator Theory 42 (1999), 37–76.
12. J. Eschmeier, Invariant subspaces for commuting contractions J.Operator Theory, to appear.
13. J. Eschmeier, On the structure of spherical contractions, To appear.
14. J. Eschmeier and M. Putinar, Spectral decompositions and analytic sheaves, LMS Monograph Series, Clarendon Press, Oxford, 1996.
15. D. Hadwin and E. Nordgren, Subalgebras of reflexive algebras, J.Operator Theory 7 (1982), 3–23.
16. L. Hörmander, An introduction to complex analysis in several variables, Van Nostrand, Princeton, New Jersey, 1966.
17. M. Kosiek and M. Ptak, Reflexivity of N–tuples of contractions with rich joint left essential spectrum, Integral Equations Operator Theory 13 (1990), 395–420.
18. M. Kosiek, A. Octavio, and M. Ptak, On the reflexivity of pairs of contractions, Proc.Amer.Math.Soc. 123 (1995), 1229–1236.
19. A. Loginov and V. Sulman, Hereditary and intermediate reflexivity of W^*–algebras, Izv.Akad. SSSR, Ser.Mat. 39 (1975), 1260–1273; Math. USSR, Izv. 9 (1975), 1189–1201.
20. R. Olin and J. Thomson, Algebras of subnormal operators, J.Funct.Anal. 37 (1980), 271–301.
21. B. Sz.-Nagy and C. Foias, Harmonic analysis of operators on Hilbert space, North–Holland, Amsterdam, 1970.
22. K. Yan, Invariant subspaces for joint subnormal systems, Preprint.

Recent Progress in Functional Analysis
K.D. Bierstedt, J. Bonet, M. Maestre and J. Schmets

An example concerning the local radial Phragmén-Lindelöf condition

Rüdiger W. Braun[a], Reinhold Meise[a], and B. A. Taylor[b*]

[a]Mathematisches Institut, Heinrich-Heine-Universität, Universitätsstraße 1, 40225 Düsseldorf, Germany

[b]Department of Mathematics, University of Michigan, Ann Arbor, MI 48109, U.S.A.

To Professor Manuel Valdivia on the occasion of his seventieth birthday.

Abstract

An example of an algebraic surface in $\mathbb{C}^3$ *is given which satisfies the local radial Phragmén-Lindelöf condition* $\mathrm{RPL}_{\mathrm{loc}}(0)$ *but which fails a certain hyperbolicity condition. This provides a counterexample to the converse of a result by the present authors.*

Introduction. Conditions of Phragmén-Lindelöf type for plurisubharmonic functions on algebraic varieties play an important role in the theory of linear partial differential operators with constant coefficients. The first result which demonstrated this fact was Hörmander's characterization of the operators $P(D)$ which are surjective on the space $\mathcal{A}(\Omega)$ of all real-analytic functions on an open convex subset Ω of $\mathbb{R}^n$ in [10]. Since then, it was shown in a number of papers that similar Phragmén-Lindelöf conditions on algebraic varieties can be used to characterize other properties of (systems of) such operators (see, e.g., Boiti and Nacinovich [2], Braun, Meise, and Vogt [7], Franken and Meise [9], Meise, Taylor, and Vogt [11], Momm [13], Palamodov [14], Zampieri [16]).

In most cases where Phragmén-Lindelöf conditions can be proved, this is accomplished by proving radial estimates first. Moreover, such radial estimates are often used as a priori estimates which are then improved by different arguments. Recently the present authors showed in [5] that a geometric condition on an analytic variety V near a real point ξ implies that V satisfies the condition $\mathrm{RPL}_{\mathrm{loc}}(\xi)$ which means that any plurisubharmonic function u on the variety that vanishes on its real points can grow only linearly, $u(z) = O(|z-\xi|)$, near ξ. The precise formulation of this result is stated in Theorem 1. It is applied in [6] to characterize those surfaces in $\mathbb{C}^3$ that satisfy the local Phragmén-Lindelöf condition and leads to a new geometric characterization of those operators $P(D)$ that are surjective on $\mathcal{A}(\mathbb{R}^4)$.

*The authors gratefully acknowledge support of DAAD under the program "Projektbezogene Förderung des Wissenschaftleraustauschs mit den USA in Zusammenarbeit mit der National Science Foundation"

In the present paper we show that this geometric condition in Theorem 1 is only sufficient but not necessary. To do this we prove that the variety

$$V := \{(s, w_1, w_2) \in \mathbb{C}^3 : (s^2 - w_1^2)^2 = w_2(w_1^4 - w_2^4)\}$$

satisfies the condition $\mathrm{RPL}_{\mathrm{loc}}(\xi)$ at $\xi = (0, 0, 0)$ but does not satisfy the geometric condition stated in Theorem 1. This example is close in spirit to the work of Bainbridge [1] concerning the global radial Phragmén-Lindelöf condition (SRPL) introduced in [3]. In [3], Theorem 5.1, a sufficient condition for (SRPL) was derived which is similar in style to the characterization in Theorem 1. Bainbridge presented an example which showed that the sufficient condition in that Theorem is not necessary. The general idea in both cases is to replace the given variety by another one with which it shares essential properties when seen as an analytic cover.

To formulate the results clearly, we need some preparation:

Notation. Throughout the paper, $|\cdot|$ denotes the Euclidean norm on $\mathbb{C}^n$ while $\|\cdot\|$ denotes the l_∞-norm, i.e., $\|z\| := \max_{1 \le j \le n} |z_j|$. For $a \in \mathbb{C}^n$ and $r > 0$ we let

$$B(a, r) := \{z \in \mathbb{C}^n : \|z - a\| < r\},$$

and we denote by $\mathbb{D}$ the open unit disk in $\mathbb{C}$. Also, for a real direction $\zeta \in \mathbb{R}^n, \zeta \neq 0, \varepsilon > 0$, and a zero neighborhood $D \subset B(0, \|\zeta\|)$ we define the truncated cone $\Gamma(\zeta, D, \epsilon)$ with profile D by

$$\Gamma(\zeta, D, \epsilon) := \bigcup_{0<t<\epsilon} t(\zeta + D).$$

Definition. (a) An analytic variety V in an open set G in $\mathbb{C}^n$ is defined to be a closed analytic subset of G (see Chirka [8], 2.1).

(b) Let V be an analytic variety in some open set in $\mathbb{C}^n$ and let Ω be an open subset of V. A function $u: \Omega \to [-\infty, \infty[$ is called plurisubharmonic if it is locally bounded above, plurisubharmonic in the usual sense on Ω_{reg}, the set of all regular points of V in Ω, and satisfies

$$u(z) = \limsup_{\zeta \in \Omega_{\mathrm{reg}}, \zeta \to z} u(\zeta)$$

at the singular points of V in Ω. By PSH(Ω) we denote the set of all plurisubharmonic functions on Ω.

It is easy to check that the following definition is equivalent to the one given in Meise, Taylor, and Vogt [12], 2.3 (see Lemma 7 in [5]).

Definition. Let V be an analytic variety in some ball $B(\xi, r)$ for $\xi \in V \cap \mathbb{R}^n$ and $r > 0$. We say that V satisfies the condition $\mathrm{RPL}_{\mathrm{loc}}(\xi)$ if the following holds:

There exist $A > 0$ and $0 < r_2 \le r_1 \le r$ such that each plurisubharmonic function u on $V \cap B(\xi, r_1)$ which satisfies

(α) $u(z) \le 1, \quad z \in V \cap B(\xi, r_1)$

(β) $u(z) \leq 0, \quad z \in V \cap B(\xi, r_2) \cap \mathbb{R}^n$

already satisfies

(γ) $u(z) \leq A|z - \xi|, \quad z \in V \cap B(\xi, r_1)$.

Definition. Let $V \subset \mathbb{C}^n$ be an analytic variety in some ball $B(p, r)$, $p \in V$, $r > 0$. Let T_pV denote the tangent cone to V at p in the sense of Whitney [15], 7.1G. To describe T_pV in an equivalent way, let f be analytic in some neighborhood of a point p. Then the localization f_p of f at the point p is defined as the lowest degree homogeneous polynomial in the Taylor series expansion of f at p which does not vanish. With this notation we have

$$T_pV = \{z \in \mathbb{C}^n : f_p(z) = 0 \text{ for all } f, \text{ analytic near } p \text{ and vanishing on } V\}, \qquad (1)$$

by Whitney [15], 7.4D.

Definition. (a) Let V be an analytic variety in $\mathbb{C}^n$ of pure dimension $k \geq 1$, $p \in V$, and $\pi : \mathbb{C}^n \to \mathbb{C}^n$ a projection map. We say that π is a noncharacteristic projection for V at p if $\operatorname{im} \pi$ and $\ker \pi$ are spanned by real vectors, rank $\pi = k$, and $T_pV \cap \ker \pi = \{0\}$.

(b) Let V be an analytic variety in some open set in $\mathbb{C}^n$ and $p \in V \cap \mathbb{R}^n$. V is said to be 1-hyperbolic at p with respect to $\xi \in T_pV \cap \mathbb{R}^n$, $\xi \neq 0$, if there exist a cone $\Gamma = \Gamma(\xi, D, \epsilon)$ and a noncharacteristic projection π for V at p such that $\pi : (V - p) \cap \Gamma \to \pi((V - p) \cap \Gamma)$ is proper and $z \in (V - p) \cap \Gamma$ is real whenever $\pi(z)$ is real.

The expression "1-hyperbolic" stems from our paper [6], where the more general concept of "d-hyperbolicity" is used.

In [5], the following theorem was proved, which was used as an essential tool in [6], to characterize when an analytic surface V in $\mathbb{C}^3$ satisfies the condition $\mathrm{PL}_{\mathrm{loc}}(\xi)$ for $\xi \in V \cap \mathbb{R}^3$.

Theorem 1 *Let V be an analytic variety of pure dimension $k \geq 1$ in some ball $B(\xi, r)$ in $\mathbb{C}^n$, where $\xi \in V \cap \mathbb{R}^n$. Assume that for each irreducible component W of $T_\xi V$ there is $\eta \in W \cap \mathbb{R}^n$ such that V is 1-hyperbolic at ξ with respect to η and to $-\eta$. Then V satisfies* $\mathrm{RPL}_{\mathrm{loc}}(\xi)$.

The aim of the present paper is to present an example which shows that the sufficient condition in Theorem 1 is not necessary. This example is a modification of the one which was constructed by Bainbridge [1], to show that the sufficient condition for the property (SRPL) in [3] is not necessary.

Example 2 *Define the algebraic variety V as*

$$V := \{(s, w_1, w_2) \in \mathbb{C}^3 : (s^2 - w_1^2)^2 = w_2(w_1^4 - w_2^4)\}.$$

Then the following assertions hold:

(a) V satisfies $\mathrm{RPL}_{\mathrm{loc}}(0)$.

(b) *There is no $\xi \in T_0V \cap \mathbb{R}^3$, $\xi \neq 0$, such that V is 1-hyperbolic at the origin with respect to both ξ and to $-\xi$.*

The statements in Example 2 will be a consequence of several lemmas. We begin with the following one:

Lemma 3 *For the variety V defined in Example 2, let $\pi : V \to \mathbb{C}^2$, $\pi(s,w) := w$, and*

$$E := \{x \in [-1,1]^2 : \pi^{-1}(x) \subset \mathbb{R}^3\}.$$

Then $E = E_1 \cup E_2$, where

$$E_1 := \{x \in \mathbb{R}^2 : x_2 \geq 0 \text{ and } |x_2| \leq |x_1| \leq 1\}$$

$$E_2 := \{x \in \mathbb{R}^2 : x_2 \leq 0 \text{ and } \left(\tfrac{|x_2|^5}{1+|x_2|}\right)^{1/4} \leq |x_1| \leq |x_2| \leq 1\}.$$

Proof. $E \subset E_1 \cup E_2$: If $x \in E$ then $x \in [-1,1]^2$ and for each $s \in \mathbb{C}$ satisfying $(s,x) \in V$ we have $s \in \mathbb{R}$ and consequently

$$x_2(x_1^4 - x_2^4) = (s^2 - x_1^2)^2 \geq 0.$$

If $x_2 \geq 0$ this implies $|x_2| \leq |x_1|$, hence $x \in E_1$.

If $x_2 \leq 0$ we get $|x_1| \leq |x_2|$ and $s^2 = x_1^2 \pm \sqrt{x_2(x_1^4 - x_2^4)}$. Since the latter equation has only real roots s, we must have $x_1^2 \geq \sqrt{x_2(x_1^4 - x_2^4)}$ and consequently

$$|x_1|^4 = x_1^4 \geq x_2(x_1^4 - x_2^4) = -|x_2|(|x_1|^4 - |x_2|^4).$$

This implies

$$|x_1|^4(1 + |x_2|) \geq |x_2|^5,$$

hence $x \in E_2$.

$E_1 \cup E_2 \subset E$: If $x \in E_1$ then $x_2 \geq 0$ and $x_2^4 \leq x_1^4$. This implies $x_2(x_1^4 - x_2^4) \geq 0$. Since $0 \leq x_2 \leq |x_1| \leq 1$, we have for $x_2 > 0$

$$1 - \left(\frac{x_2}{x_1}\right)^4 \leq 1 \leq \frac{1}{x_2} \text{ and hence } x_2(x_1^4 - x_2^4) \leq x_1^4,$$

even for $x_2 = 0$. Thus $x_1^2 \pm \sqrt{x_2(x_1^4 - x_2^4)} \geq 0$. Therefore, all roots of the equation $(s^2 - x_1^2)^2 = x_2(x_1^4 - x_2^4)$ are real, i.e., $x \in E$.

If $x \in E_2$ then

$$-x_2^5 = |x_2|^5 \leq |x_1|^4(1 + |x_2|) = x_1^4(1 - x_2)$$

and consequently $0 \leq x_2(x_1^4 - x_2^4) \leq x_1^4$. As before, this implies $x \in E$. □

The following lemma is an easy exercise in calculus.

Lemma 4 *If the real numbers a, b satisfy $|a| > (\sqrt{3}/\sqrt[3]{2})|b|^{2/3}$ then the equation $t^3 - ta^2 + b^2 = 0$ has three different real solutions.*

Lemma 5 *For each $\alpha \in]0,1[$ there exists $C \geq 1$ such that each function $\varphi : \mathbb{D} \to [-\infty, \infty[$ which is subharmonic on $\mathbb{D}$ and satisfies*

(a) $\varphi(z) \leq 1$ *for all* $z \in \mathbb{D}$.

(b) $\varphi(x) \leq |x|^\alpha$, $x \in]-1,1[$.

already satisfies

$$\varphi(z) \leq C|z|^\alpha, \ z \in \mathbb{D}.$$

Proof. Fix $\alpha \in]0,1[$ and denote by $z \mapsto z^\alpha$ the function which is holomorphic on $\mathbb{C} \setminus]-\infty, 0]$ and positive when $z > 0$. This means $(re^{it})^\alpha = r^\alpha e^{it\alpha}$ for $r > 0$ and $-\pi < t < \pi$. Since $0 < \alpha < 1$ we have

$$\min\{\cos((t - \frac{\pi}{2})\alpha) : 0 \leq t \leq \pi\} = \cos\left(\frac{\alpha\pi}{2}\right) > 0.$$

Now let $A := 1/\cos\left(\frac{\alpha\pi}{2}\right)$ and define $h(z) := A \operatorname{Re}((-iz)^\alpha)$ for $z \in \mathbb{C}$, $\operatorname{Im} z > 0$. Then h is harmonic in $\mathbb{D}_+ := \{z \in \mathbb{D} : \operatorname{Im} z > 0\}$ and extends continuously to $\overline{\mathbb{D}}_+$. The definition of h and the hypotheses on φ imply

$$h(e^{it}) \geq 1 \text{ for } t \in [0,\pi] \text{ and } h(x) \geq A|x|^\alpha \cos\left(\frac{\alpha\pi}{2}\right) = |x|^\alpha \geq \varphi(x), \ x \in]-1,1[.$$

Since φ is subharmonic on $\mathbb{D}_+$, these estimates imply

$$\varphi(z) \leq h(z) \leq A|z|^\alpha, \ z \in \mathbb{D}_+,$$

from which the lemma follows. □

Definition. Let E be a compact subset of $\mathbb{R}^2$ and let $r > 0$ be given. Then the local extremal function $\Lambda_E(\cdot; r)$ of E relative to $B(0,r) \subset \mathbb{C}^2$ is defined as

$$\Lambda_E(w;r) := \sup\{u(w) : u \in \operatorname{PSH}(B(0,r)), u \leq 1, u|_{E \cap B(0,r)} \leq 0\}, \ w \in B(0,r).$$

Now we are prepared for the proof of the following lemma which we will use to show the first assertion in Example 2.

Lemma 6 *For the set E defined in Lemma 3 and $0 < r < \frac{1}{2}$ there exist $C > 0$ and $0 < \delta \leq r$ such that*

$$\Lambda_E(w;r) \leq C\|w\| \text{ for all } w \in B(0,\delta).$$

Proof. Fix $0 < r < \frac{1}{2}$ and $u \in \mathrm{PSH}(B(0,r))$ satisfying

$$u \leq 1 \text{ on } B(0,r) \text{ and } u|_{E \cap B(0,r)} \leq 0.$$

Then define $P \in \mathbb{C}[s, w_1, w_2]$ by

$$P(s, w_1, w_2) := w_2(w_1^2 - w_2^2) - s^2,$$

$$V_0 := \{(s, w_1, w_2) \in \mathbb{C}^3 : P(s, w_1, w_2) = 0\},$$

and

$$\varphi : V_0 \cap (\mathbb{C} \times B(0,r)) \to [-\infty, \infty[\text{ by } \varphi(s, w_1, w_2) := u(w_1, w_2).$$

Obviously, φ is plurisubharmonic on $V_0 \cap (\mathbb{C} \times B(0,r))$ and bounded by 1 from above. Next fix $t > 0$ satisfying $2t^{2/3} < r$, and $(s, w_1) \in B(0,t)$. Then recall that a given polynomial $q(z) = \sum_{j=0}^{m} c_{m-j} z^j$ has all zeros in the disk with radius at most $2 \max_{1 \leq j \leq m} |\frac{c_j}{c_0}|^{1/j}$. Hence all roots of the polynomial

$$z \mapsto z^3 - w_1^2 z + s^2$$

lie in the disk of radius $R = R(s, w_1)$, where R can be estimated by

$$R \leq 2 \max(|w_1|, |s^2|^{1/3}) \leq 2 \max(t, t^{2/3}) \leq 2t^{2/3} < r.$$

Hence it follows from Hörmander [10], Lemma 4.4, that the function v, defined on $B(0,t)$ by

$$v(w_1, s) := \max\{\varphi(s, w_1, w_2) : (s, w_1, w_2) \in V_0 \cap B(0,r)\}$$

is plurisubharmonic. Obviously, v is bounded by 1 from above. Next we claim that for $d := (\sqrt{3}/\sqrt[3]{2})$ the following assertion holds:

$$\text{if } (s, w_1) \in B(0,t) \cap \mathbb{R}^2 \text{ satisfies } |w_1| > d|s|^{2/3} \text{ then } v(w_1, s) \leq 0. \tag{2}$$

To prove (2) fix $(s, w_1) \in B(0,t) \cap \mathbb{R}^2$ satisfying the condition in (2). Then it follows from Lemma 4 that the polynomial

$$z \mapsto z^3 - w_1^2 z + s^2$$

has three real roots. Let ξ be one of these and assume first $\xi \geq 0$. Then

$$\xi(w_1^2 - \xi^2) = s^2 \geq 0 \text{ and hence } w_1^2 - \xi^2 \geq 0.$$

From this we get $|\xi| \leq |w_1| < t < 1$. Since $\xi \geq 0$, Lemma 3 implies that (w_1, ξ) belongs to the set E and consequently $\varphi(s, w_1, \xi) = u(w_1, \xi) \leq 0$.

If $\xi \leq 0$ then the same arguments show $|w_1| \leq |\xi|$. To show $(|\xi|^5/(1+|\xi|))^{1/4} \leq |w_1|$ for t sufficiently small, assume that $|\xi|^5/(1+|\xi|) > w_1^4$. Then $|\xi|^5 > |w_1|^4$ and hence $|\xi| > |w_1|^{4/5}$. This implies

$$s^2 = \xi(w_1^2 - \xi^2) \geq |w_1|^{4/5}(|w_1|^{8/5} - |w_1|^2) \geq \frac{1}{2}|w_1|^{12/5}.$$

provided that $0 < t^{2/5} \leq \frac{1}{2}$. The latter inequality together with the hypothesis in (2) then gives

$$|w_1| > d|s|^{2/3} > d \cdot 2^{-1/3}|w_1|^{4/5} \text{ and hence } |w_1| > d^5\, 2^{-5/3}.$$

If we assume that t is so small that $t < d^5\, 2^{-5/3}$ then the assumption on $|\xi|$ leads to a contradiction. Hence

$$\left(\frac{|\xi|^5}{1+|\xi|}\right)^{1/4} \leq |w_1| \leq |\xi| < 1.$$

provided that $t < t_0 := \min((1/2)^{5/2}, d^5 2^{-5/3}, (2r)^{3/2})$. By Lemma 3, this shows that (w_1, ξ) belongs to E. As before, this implies $\varphi(s, w_1, \xi) \leq 0$ and consequently $v(w_1, s) \leq 0$. Hence (2) holds.

Next fix $0 < t < t_0$ and choose $\rho > 0$ so that $4d\rho^{2/3} \leq t$. Then fix $s \in B(0,\rho) \cap \mathbb{R}$ and define

$$\psi : \mathbb{D} \to [-\infty, \infty[, \quad \psi(w) := \max(0, v(tw, s)).$$

Obviously, ψ is subharmonic on $\mathbb{D}$ and bounded by 1 from above. The choice of ρ implies that $\varepsilon := \frac{2}{t}d|s|^{2/3}$ satisfies

$$0 \leq \varepsilon \leq \frac{2}{t}d\rho^{2/3} \leq \frac{1}{2}.$$

Moreover, for $w \in \mathbb{R}$ and $\varepsilon \leq |w| < 1$ we have

$$t > |tw| \geq t\varepsilon = 2d|s|^{2/3} > d|s|^{2/3}.$$

Hence (2) implies $v(tw, s) \leq 0$ and consequently

$$\psi(w) = 0 \text{ for } w \in \mathbb{R}, \quad \varepsilon \leq |w| < 1.$$

Therefore, [4], Lemma 5.8, implies the existence of $C_1 > 0$ such that

$$\psi(w) \leq C_1|\operatorname{Im}\sqrt{w^2 - \varepsilon^2}| = C_1|\operatorname{Im}\sqrt{w^2 + (i\varepsilon)^2}|, \quad |w| \leq \frac{1}{2}.$$

From this and [4], Lemma 5.7 (i), it follows that

$$v(w_1, s) \leq \frac{C_1}{t}\left|\operatorname{Im}\sqrt{w_1^2 + (i2d|s|^{2/3})^2}\right| \leq \frac{C_1}{t}(|\operatorname{Im} w_1| + 2d|s|^{2/3})$$

for $|w_1| \leq \frac{t}{2}$, $s \in B(0,\rho) \cap \mathbb{R}$. If we let $\delta := \min(\rho, \frac{t}{2})$ then there exists $C_2 > 0$ such that

$$v(w_1, s) \leq C_2(|\operatorname{Im} w_1| + |s|^{2/3}), \quad (w_1, s) \in B(0,\delta), \ s \in \mathbb{R}.$$

Next fix $w_1 \in B(0,\delta)$ and define

$$\gamma : \mathbb{D} \to [-\infty, \infty[, \quad \gamma(z) := \frac{1}{C_2}v(w_1, \delta z) - |\operatorname{Im} w_1|.$$

Then γ is subharmonic on $\mathbb{D}$, bounded by 1 from above and satisfies

$$\gamma(x) \leq \delta^{2/3}|x|^{2/3}, \quad x \in \mathbb{R}, \quad |x| < 1.$$

By Lemma 5, this implies the existence of $C_3 > 0$ such that

$$\gamma(z) \leq C_3|\delta z|^{2/3}, \quad z \in \mathbb{D}$$

and consequently

$$v(w_1, s) \leq C_2|\operatorname{Im} w_1| + C_2C_3|s|^{2/3} \leq C_4(|\operatorname{Im} w_1| + |s|^{2/3}),$$

for $(w_1, s) \in B(0, \delta)$.

Next fix $(w_1, w_2) \in B(0, \delta)$ and choose $s \in \mathbb{C}$ with $s^2 = w_2(w_1^2 - w_2^2)$. Then there exists $D > 0$ such that $|s|^2 \leq D\|w\|^3$ and$(s, w_1, w_2) \in V_0$. By the definition of φ and v we have

$$\begin{aligned} u(w_1, w_2) = \varphi(s, w_1, w_2) &\leq v(w_1, s) \leq C_4(|\operatorname{Im} w_1| + |s|^{2/3}) \\ &\leq C_4(|\operatorname{Im} w_1| + D\|w\|) \leq C_5\|w\|. \end{aligned}$$

Since the proof shows that the number δ only depends on r and not on the particular function u, it follows that

$$\Lambda_E(w, r) \leq C_5\|w\| \text{ for } w \in B(0, \delta).$$

□

Proof of assertion (a) in Example 2. Fix $0 < \rho < 1$, let $r := \rho/2$ and let $u \in \operatorname{PSH}(V \cap B(0, \rho))$ satisfy

$$u(z) \leq 1, \ z \in V \cap B(0, \rho) \text{ and } u(x) \leq 0, \ x \in V \cap B(0, \rho) \cap \mathbb{R}^3.$$

Then note that for $w \in B(0, r)$ and $(s, w) \in V$, we have

$$|s^2| = |w_1^2 \pm \sqrt{w_1(w_1^4 - w_2^4)}| < r^2 + \sqrt{2r^5} \leq 4r^2.$$

Hence each point $(s, w) \in V$ with $w \in B(0, r)$ satisfies $|s| < \rho$. Therefore, it follows from Hörmander [10], Lemma 4.4, that the function

$$v : B(0, r) \to [-\infty, \infty[, \ v(w) := \max\{u(s, w) : (s, w) \in V \cap B(0, \rho)\}$$

is plurisubharmonic on $B(0, r)$ and bounded by 1 from above. Moreover, $v(w) \leq 0$ whenever $w \in E$. By the definition of the local extremal function $\Lambda_E(\cdot, r)$, these properties imply

$$v(w) \leq \Lambda_E(w; r), \quad w \in B(0, r).$$

Now note that by Lemma 6 there exist $0 < \delta \leq r$ and $C > 0$ such that

$$\Lambda_E(w; r) \leq C\|w\|, \quad w \in B(0, \delta).$$

Hence we have $v(w) \leq C\|w\|$ for $w \in B(0, \delta)$ and consequently

$$u(s, w) \leq C\|(s, w)\|, \quad (s, w) \in B(0, \delta) \cap V.$$

Since u is bounded by 1, this estimate implies

$$u(s, w) \leq A\|(s, w)\|, \quad (s, w) \in B(0, \rho) \cap V,$$

if we let $A := \max(C, \frac{1}{\delta})$. Hence V satisfies $\mathrm{RPL}_{\mathrm{loc}}(\xi)$. □

To prove also assertion (b) of Example 2 we will use the following lemma.

Lemma 7 *Define* $P(s, w_1, w_2) := (s^2 - w_1^2)^2 - w_1(w_1^4 - w_2^4)$. *Then for each* $a \in \mathbb{R} \setminus \{-1, 0, 1\}$ *and* $\xi = (1, 1, a)$ *or* $\xi = (1, -1, a)$, *there exists* $0 < \delta < 1$ *such that* P *does not vanish on one of the real cones* $\Gamma(\xi, B(0, \delta), 1) \cap \mathbb{R}^3$ *or* $\Gamma(-\xi, B(0, \delta), 1) \cap \mathbb{R}^3$.

Proof. Consider first $\xi = (1, 1, a)$ for fixed $a \in \mathbb{R} \setminus \{-1, 0, 1\}$. Choose $0 < \delta < 1$ so small that $\delta < \min(|a|, \frac{1}{2}|a - 1|, \frac{1}{2}|a + 1|)$ and fix $\zeta = (x, y_1, y_2) \in B(0, \delta)$. Then some computation shows that

$$P(t(\xi + \zeta)) = At^5 + Bt^4, \quad t \in \mathbb{R},$$

where

$$A = (a + y_2)(a - 1 - y_1 + y_2)(a + 1 + y_1 + y_2)((1 + y_1)^2 + (a + y_2)^2),$$
$$B = (x - y_1)^2(2 + x + y_1)^2.$$

Next fix $a \in]-\infty, -1[$ and note that by the choice of δ we have

$$a + y_2 < 0, \ a - 1 - y_1 + y_2 < 0, \text{ and } a + 1 + y_1 + y_2 < 0$$

and hence $A = A(y_1, y_2) < 0$ for all $\zeta = (x, y_1, y_2) \in B(0, \delta) \cap \mathbb{R}^3$. Obviously, this implies

$$P(t(\xi + \zeta)) = At^5 + Bt^4 > 0 \text{ if } t < 0 \text{ and } \zeta \in B(0, \delta) \cap \mathbb{R}^3$$

and consequently

$$P(z) > 0 \text{ for all } z \in \Gamma(-\xi, B(0, \delta), 1) \cap \mathbb{R}^3.$$

The same arguments show that this holds for $a \in]0, 1[$, while for $a \in]-1, 0[$ and $a \in]1, \infty[$ we have

$$P(z) > 0 \text{ for all } z \in \Gamma(\xi, B(0, \delta), 1) \cap \mathbb{R}^3.$$

Next consider $\xi = (1, -1, a)$ for fixed $a \in \mathbb{R} \setminus \{-1, 0, 1\}$. Then we have for $\zeta = (x, y_1, y_2)$

$$P(t(\xi + \zeta)) = A't^5 + B't^4,$$

where

$$A' = (a + y_2)(a - 1 + y_1 + y_2)(a + 1 - y_1 + y_2)((1 - y_1)^2 + (a + y_2)^2),$$
$$B' = (x + y_1)^2(2 + x - y_1)^2.$$

Hence we can argue as in the previous case to complete the proof of the lemma. □

In Lemma 7 a finite set of directions is excluded. The following lemma allows us to treat them by perturbation.

Lemma 8 *Let $P \in \mathbb{C}[z_1, \ldots, z_n]$ and $W := \{z \in \mathbb{C}^n : P(z) = 0\}$. Assume that the origin belongs to W and that W is 1-hyperbolic at 0 with respect to $\xi \in T_0W \cap \mathbb{R}^n$. Then there exists $\delta > 0$ such that for each $\eta \in T_0W \cap \mathbb{R}^n$ satisfying $\|\eta - \xi\| < \delta$ and for each $0 < \rho < \delta$ we have*

$$W \cap \mathbb{R}^n \cap \Gamma(\eta, B(0, \rho), \rho) \neq \emptyset.$$

Proof. Since W is 1-hyperbolic at 0 with respect to $\xi \in T_0W \cap \mathbb{R}^n$, modulo a real linear change of variables, we may assume that $\xi = e_1 := (1, 0, \ldots, 0)$ and that the noncharacteristic projection $\pi : \mathbb{C}^n \to \mathbb{C}^n$ is given by $\pi(z', z_n) = z'$ for $(z', z_n) \in \mathbb{C}^{n-1} \times \mathbb{C}$. In these coordinates we decompose P as $P = \sum_{k=\nu}^{m} P_k$, where P_k is either identically zero or homogeneous of degree k and $\nu \in \mathbb{N}$ is chosen so that $P_\nu \not\equiv 0$. It follows from (1) that

$$T_0W = \{z \in \mathbb{C}^n : P_\nu(z) = 0\}.$$

By hypothesis there exist $\varepsilon > 0$ and an open zero neighborhood $D \subset B(0, \|\xi\|)$ such that for $\Gamma := \Gamma(\xi, D, \varepsilon)$ the map $\pi : W \cap \Gamma \to \pi(W \cap \Gamma)$ is proper and $z \in W \cap \Gamma$ is real whenever $\pi(z)$ is real. Choose $\delta > 0$ so that $B(0, 2\delta) \subset D$. Then fix $\eta \in T_0W \cap \mathbb{R}^n$ satisfying $\|\eta - \xi\| < \delta$. Since T_0W is a homogeneous variety, $t\eta \in T_0W$ for each $t \in \mathbb{C}$.

Next let $e_n := (0, \ldots, 0, 1)$ and consider for fixed $t > 0$ the polynomial

$$q(\cdot; t) : \lambda \mapsto P(t(\eta + \lambda e_n)), \quad \lambda \in \mathbb{C}.$$

Then we have

$$q(\lambda; t) = \sum_{k=\nu}^{m} t^k P_k(\eta + \lambda e_n) = t^\nu \left(P_\nu(\eta + \lambda e_n) + \sum_{k=\nu+1}^{m} t^{k-\nu} P_k(\eta + \lambda e_n) \right).$$

Next note that $P_\nu(e_n) \neq 0$ since the projection π is noncharacteristic for T_0W. Hence we can choose $0 < \rho_0 < \delta$ so small that

$$P_\nu(\eta + \lambda e_n) \neq 0 \text{ for } 0 < |\lambda| \leq \rho_0.$$

Then fix $0 < \rho \leq \rho_0$ and note that

$$c := \inf\{|P_\nu(\eta + \lambda e_n)| : \lambda \in \mathbb{C},\ |\lambda| = \rho\} > 0.$$

Since the polynomials $\lambda \mapsto P_k(\eta + \lambda e_n)$ are bounded on $\overline{B(0, \rho)}$, it follows that we can find $0 < \varepsilon_1 < \varepsilon$ such that

$$\sup_{0 < t \leq \varepsilon_1} \sup \left\{ \sum_{k=\nu+1}^{m} t^{k-\nu} |P_k(\eta + \lambda e_n)| : \lambda \in \overline{B(0, \rho)} \right\} < c.$$

These estimates show that we can apply the Theorem of Rouché to obtain that for each $0 < t \leq \min(\varepsilon_1, \rho)$ the polynomial $q(\lambda; t)$ and the polynomial $r(\lambda; t) := t^\nu P_\nu(\eta + \lambda e_n)$ have the same number of zeros in the disk $B(0, \rho)$. Since

$$r(0; t) = t^\nu P_\nu(\eta) = P_\nu(t\eta) = 0,$$

it follows that there is $\lambda_t \in B(0, \rho)$ such that

$$0 = q(\lambda_t; t) = P(t(\eta + \lambda_t e_n)).$$

Hence $\zeta_t := t(\eta + \lambda_t e_n)$ belongs to W and $\pi(\zeta_t) = t\eta' \in \mathbb{R}^{n-1}$.
Next note that by the previous choices

$$\zeta_t = t(\eta + \lambda_t e_n) \in t(\eta + B(0, \rho)) \subset t(\xi + B(0, 2\delta)) \subset t(\xi + D) \subset \Gamma.$$

Now the hypothesis implies $\zeta_t \in \mathbb{R}^n$ and hence

$$W \cap \mathbb{R}^n \cap \Gamma(\eta, B(0, \rho), \rho) \neq \emptyset.$$

Obviously, this proves the lemma. □

Proof of assertion (b) in Example 2. From (1) it follows that

$$\begin{aligned} T_0V &= \{(s, w) \in \mathbb{C}^3 : (s^2 - w_1^2)^2 = 0\} \\ &= \{(\lambda, \lambda, \mu) : \lambda, \mu \in \mathbb{C}\} \cup \{(\lambda, -\lambda, \mu) : \lambda, \mu \in \mathbb{C}\}. \end{aligned}$$

To show that V is not 1-hyperbolic at zero with respect to both vectors ξ and $-\xi$ for each $\xi \in T_0V \cap \mathbb{R}^3$, $\xi \neq 0$, we distinguish the following two cases:
case 1: ξ is of the form $(1, 1, a)$ or $(1, -1, a)$, $a \in \mathbb{R} \setminus \{-1, 0, 1\}$. In this case Lemma 7 implies the existence of $\delta > 0$ and $\varepsilon > 0$ such that

$$V \cap \mathbb{R}^3 \cap \Gamma(\xi, B(0, \delta), \varepsilon) = \emptyset \text{ or } V \cap \mathbb{R}^3 \cap \Gamma(-\xi, B(0, \delta), \varepsilon) = \emptyset.$$

From this and Lemma 8 (or the definition of 1-hyperbolicity) it follows that V fails to be 1-hyperbolic at zero with respect to at least one of the vectors ξ or $-\xi$.
case 2: ξ is of the form $(0, 0, 1)$, $(1, 1, b)$ or $(1, -1, b)$, $b \in \{-1, 0, 1\}$. To treat this case, note that each vector ξ of this form is a limit of vectors $\eta \in T_0V \cap \mathbb{R}^3$ which are of the form discussed in case 1. Therefore, Lemma 8 together with Lemma 7 implies that V fails to be 1-hyperbolic at zero with respect to at least one of the vectors ξ or $-\xi$.
This completes the proof since each $\zeta \in T_0V \cap \mathbb{R}^3$ is a multiple of a vector ξ treated in case 1 or case 2. □

REFERENCES

1. Bainbridge, D.: *Phragmén-Lindelöf estimates for plurisubharmonic functions of linear growth*, Thesis, Ann Arbor 1998.
2. Boiti, C., Nacinovich, M.: *The overdetermined Cauchy problem*, Ann. Inst. Fourier (Grenoble) **47** (1997), 155–199.
3. Braun, R.W., Meise, R., Taylor, B.A.: *A radial Phragmén-Lindelöf estimate for plurisubharmonic functions on algebraic varieties*, Ann. Polon. Math. **LXXII** (1999), 159–179.
4. Braun, R.W., Meise, R., Taylor, B.A.: *Algebraic varieties on which the classical Phragmén-Lindelöf estimates hold for plurisubharmonic functions*, Math. Z. **232** (1999), 103–135.

5. Braun, R.W., Meise, R., Taylor, B.A.: *Local radial Phragmén-Lindelöf estimates for plurisubharmonic functions on analytic varieties*, Proc. Amer. Math. Soc., to appear.
6. Braun, R.W., Meise, R., Taylor, B.A.: *The geometry of analytic varieties satisfying the local Phragmén-Lindelöf condition and a geometric characterization of partial differential operators that are surjective on $\mathcal{A}(\mathbb{R}^4)$*, manuscript.
7. Braun, R.W., Meise, R., Vogt, D.: *Characterization of the linear partial differential operators with constant coefficients which are surjective on non-quasianalytic classes of Roumieu type on $\mathbb{R}^N$*, Math. Nachr. **168** (1994), 19–54.
8. Chirka, E. M.: *Complex Analytic Sets*, Kluwer, Dordrecht, 1989.
9. Franken, U., Meise, R.: *Extension and lacunas of solutions of partial differential equations*, Ann. Inst. Fourier (Grenoble) **46** (1996), 154–161.
10. Hörmander, L.: *On the existence of real analytic solutions of partial differential equations with constant coefficients*, Invent. Math. **21** (1973), 151–183.
11. Meise, R., Taylor, B.A., Vogt, D.: *Characterization of the linear partial differential operators with constant coefficients that admit a continuous linear right inverse*, Ann. Inst. Fourier (Grenoble) **40** (1990), 619–655.
12. Meise, R., Taylor, B.A., Vogt, D.: *Extremal plurisubharmonic functions of linear growth on algebraic varieties*, Math. Z. **219** (1995), 515–537.
13. Momm, S.: *On the dependence of analytic solutions of partial differential equations on the right hand side*, Trans. Amer. Math. Soc. **345** (1994), 729–752.
14. Palamodov, V.I.: *A criterion for the splitness of differential complexes with constant coefficients*, in *Geometrical and Algebraical Aspects in Several Complex Variables*, C.A. Berenstein and D.C. Struppa (Eds.), Edit El (1991), pp. 265–290.
15. Whitney, H.: *Complex Analytic Varieties*, Addison-Wesley, Reading, Mass., 1972.
16. Zampieri, G.: *An application of the fundamental priciple of Ehrenpreis to the existence of global solutions of linear partial differential equations*, Boll. Un. Mat. Ital. **6** (1986), 361–392.

Recent Progress in Functional Analysis
K.D. Bierstedt, J. Bonet, M. Maestre and J. Schmets

Continuity of monotone functions with values in Banach lattices

Lech Drewnowski*

Faculty of Mathematics and Computer Science, A. Mickiewicz University, ul. Matejki 48/49, 60-769 Poznań, POLAND

To Professor Manuel Valdivia on the occasion of his seventieth birthday.

Abstract

Inspired by some results of Lavrič (1992), *we investigate those Banach lattices E that have the following property* (λ)*: Every increasing function $f : [0,1] \to E$ has only countably many points of discontinuity. We show that the space of regular functions on* $[0,1]$ *contains no copy of ℓ_∞ and yet it lacks property* (λ)*, thus answering in the negative a question of Lavrič. On the positive part, our main results show that if a Banach lattice E has property* (λ)*, then so do the Banach lattices $C(K,E)$ and $L_p(\mu,E)$* ($1 \leq p < \infty$) *for a wide class of compact spaces K and every measure μ.*

MCS 2000 Primary 46B42, 46B03, 46E15, 46E30, 46E40, 54D30.

Introduction

Let us say that a Banach lattice E has *property* (λ) if every increasing (= nondecreasing) function $f : [0,1] \to E$ has at most countably many points of discontinuity. In a 1992 paper [16], B. Lavrič showed that

(L1) *There exists an increasing function $f : [0,1] \to \ell_\infty$ without any points of continuity.*

He then combined this with the classical Lozanovskii – Meyer-Nieberg result (see [2]) to conclude that

(L2) *For a σ-Dedekind complete Banach lattice E the following are equivalent.*
(a) *E has order continuous norm.*
(b) *E has property* (λ).
(c) *E contains no lattice copy of ℓ_∞.*

He also proved that

(L3) *Every separable Banach lattice has property* (λ).

*This research was started while the author held a visiting position in the Department of Mathematics, University of Mississippi, in the Spring Semester of 1998. It was later supported by the State Committee for Scientific Research (Poland), Grant no. 2 P03A 05115.

Finally, since the assumption of σ-Dedekind completeness in (L2) was needed only in the proof that (c) implies (a), and also because of (L3), he raised the following question.

(LQ) *If a Banach lattice contains no lattice copy of ℓ_∞, must it have property (λ)?*

In this paper, after introducing or recalling some notions and collecting a few general facts (some of which were also used in [16]), we give a very short construction of a function required in (L1), and extend (L2) and (L3) to F-normed lattices (which is fairly straightforward). Next, we show that the space of regular functions on $[0,1]$ provides a negative answer to (LQ). Then, in the main part of the paper, we investigate property (λ) for Banach lattices of continuous or measurable vector-valued functions. We prove, for instance, that if E is a Banach lattice with property (λ), then: 1) the Banach lattice $C(K,E)$ has property (λ) whenever K is a compact space with all its separable subspaces metrizable (e.g., an Eberlein compact), or a product of any family of such compact spaces; 2) the Banach lattice $L_p(\mu, E)$ has property (λ) for every measure μ and $1 \leq p < \infty$.

To some extent, the title of this paper is somewhat inadequate. This is particularly visible in Section 5, where we deal with property (λ) for spaces of continuous functions taking values in a Banach lattice. However, an attentive reader will certainly notice that almost all of our results make sense and, in fact, remain valid (with the same proofs) in the case of functions with values in an F-normed lattice, or even an arbitrary ordered metric space with a monotone metric (in the sense of Fact 1.7 below).

In what follows, we denote by I the interval $[0,1]$ (but any interval in $\mathbf{R}$ could be used as well), and all topological spaces occuring below are assumed Hausdorff. The terms *increasing* and *decreasing* are used in the weak sense, meaning the same as *nondecreasing* and *nonincreasing*, respectively. We refer the reader to [1] for locally solid topological Riesz spaces (or vector lattices) and relevant notions like the (σ-) Dedekind completeness or the (σ-) Lebesgue and pre-Lebesgue properties. In general, however, our functional analysis terminology and notation are fairly standard.

Acknowledgments. First of all, I would like to thank Professor Gerard Buskes (University of Mississippi) who, in January 1998, called my attention to Lavrič's paper and the question posed in it—that was a starting point for the research whose results are presented here. I am also very grateful to Dr. Artur Michalak (A. Mickiewicz University, Poznań) for many stimulating discussions and valuable comments on this paper.

1. A few general facts

A function $f : I \to E$, where $E = (E, \tau)$ is a topological space, is said to be *regular* if the right-hand limit $f(t+)$ exists for each $t \in [0,1)$, and the left-hand limit $f(t-)$ exists for each $t \in (0,1]$. In general, for any function f as above, we denote by $D(f)$ the set of points of discontinuity of f. Clearly, if $|D(f)| \leq \aleph_0$, then f has a separable range. If E is a metric space, then $D(f)$ is always an F_σ subset of I (see [14, § 21.III]); in consequence, $|D(f)| \leq \aleph_0$ or $|D(f)| = 2^{\aleph_0}$ (by [14, § 37.I, Thm. 3]).

As is well known, every monotone function $f : I \to \mathbf{R}$ is regular, and $|D(f)| \leq \aleph_0$. If, for instance, f is increasing, the latter follows immediately from the fact that the intervals $(f(t-), f(t+))$ are pairwise disjoint. More generally (see [18, III.2, Thm. 3]):

Fact 1.1 *For a regular function $f : I \to E$, where E is a metric space, $|D(f)| \leq \aleph_0$.*

For, otherwise, we would find $\varepsilon > 0$ and an uncountable set D of points $t \in I$ where f has a "jump" $\geq \varepsilon$, and f could not be regular at any of the accumulation points of D.

Let us agree to say that a topological ordered space $E = (E, \tau, \leq)$ or its topology τ has

- *property* (σ-DCL) if every monotone order bounded sequence in E is τ-convergent;
- *property* (λ) if every increasing (equivalently, every decreasing) function $f : I \to E$ has at most countably many points of discontinuity;
- *property* (λ_0) if every increasing (equivalently, every decreasing) function $f : I \to E$ has at least one point of continuity in the interior of I.

Remark 1.2 If E has property (λ_0), then every increasing function $f : I \to E$ has a dense set of points of continuity. (Consider compositions of f with increasing affine mappings of I onto its closed subintervals.) Of course, (λ) implies (λ_0), but as of this writing it is open whether (or when) also the converse is true; see Problem 1 below.

It is easy to verify the following facts (using Fact 1.1 to get Fact 1.5).

Fact 1.3 *A topological ordered space E has property* (σ-DCL) *iff every monotone (equivalently, every increasing) function $f : I \to E$ is regular.*

Fact 1.4 *Let $(E, \tau, \leq)$ be a topological ordered space, and ρ a weaker Hausdorff topology on E. Suppose τ has property* (σ-DCL). *Then also ρ has property* (σ-DCL), *and τ has property* (λ) *iff ρ has property* (λ).

Fact 1.5 *If a topological ordered space $(E, \tau, \leq)$ has property* (σ-DCL), *and each of its order intervals is τ-closed and metrizable in a weaker topology, then E has property* (λ).

Fact 1.6 *An arbitrary product of topological ordered spaces with property* (σ-DCL) *has property* (σ-DCL), *and an at most countable product of topological ordered spaces with property* (λ) *has property* (λ).

Fact 1.7 *Let E be an ordered metric space with a monotone metric d (that is, such that $x \leq u \leq v \leq y$ implies $d(u, v) \leq d(x, y)$). Then a monotone function $f : I \to E$ has at most countably many discontinuities iff its range $f(I)$ is separable.*

In consequence, if all order intervals in E are separable, then E has property (λ).

Proof (comp. Proof of Lemma 3.1 below). Let $f : I \to E$ be a monotone function with $|D(f)| > \aleph_0$. Then for some $\varepsilon > 0$ one can find an uncountable subset D of $D(f)$ such that $d(f(t), f(u)) \geq \varepsilon$ for all distinct $t, u \in D$. It follows that $f(I)$ is nonseparable. □

Now, let E be a Hausdorff locally solid topological Riesz space. Then property (σ-DCL) translates into the following condition: *Every increasing order bounded positive sequence in E is convergent*, and this in turn is equivalent to: *E is σ-Dedekind complete and has the σ-Lebesgue property.* Moreover, if E is topologically complete and has the Lebesgue property, then it is Dedekind complete (see [1, Thm. 10.3]) and, in consequence, it has property (σ-DCL). As a corollary to Fact 1.5 we thus have the following.

Fact 1.8 *Every F-lattice with the Lebesgue property (or order continuous F-norm) has property* (λ).

2. An increasing function without points of continuity

The construction given in [16] of a function $f : I \to \ell_\infty$ required in (L1) is rather technical and almost two pages long. It actually produces a function f with the stronger property that $\|f(t) - f(s)\| = 1$ whenever $s \neq t$. Here is a much simpler construction.

Proposition 2.1 *There exists an increasing function $f : I \to \ell_\infty$ with $\|f(t) - f(u)\| = 1$ whenever $t \neq u$, and thus without any points of continuity.*

Proof. Let (I_n) be a sequence of closed subintervals in I such that every subinterval of I contains an I_n. (For instance, (I_n) can be an arrangement of the dyadic subintervals $[j2^{-k}, (j+1)2^{-k}]$, where $k \in \mathbf{N}$ and $j = 0, 1, \ldots, 2^k - 1$.) For each n let $f_n : I \to I$ be the increasing continuous picewise linear function such that $f_n(t) = 0$ for $t \leq \min I_n$ and $f_n(t) = 1$ for $t \geq \max I_n$. Then the function $f = (f_n) : I \to \ell_\infty$ is clearly increasing. Moreover, for any $t, u \in I$ with $u < t$ one obviously has $\|f(t) - f(u)\| \leq 1$, and if n is chosen so that $I_n \subset (u, t)$, then $\|f(t) - f(u)\| \geq |f_n(t) - f_n(u)| = 1 - 0 = 1$ □

Remarks 2.2 (a) Evidently, the function f constructed above is continuous when ℓ_∞ is considered with the coordinatewise convergence topology. In fact, f is even weak*-continuous; that is, for each $a^* = (a_n) \in \ell_1$ the scalar function $t \to \langle a^*, f(t)\rangle = \sum_n a_n f_n(t)$ is continuous. To see this assume (as we may) that $\|a^*\|_1 = 1$, take any $\varepsilon > 0$, and then choose $k \in \mathbf{N}$ so that $\sum_{n>k} |a_n| < \varepsilon/2$, and next $\delta > 0$ such that $|f_n(t) - f_n(u)| < \varepsilon/2$ whenever $1 \leq n \leq k$ and $|t - u| < \delta$. Then, as $\sum_{1 \leq n \leq k} |a_n| \leq 1$ and $|f_n(t) - f_n(u)| \leq 1$ for all n, it is easily seen that $|\langle a^*, f(t) - f(u)\rangle| < \varepsilon$ whenever $|t - u| < \delta$.

(b) In general, *a monotone function $f : I \to E$, where E is a Banach lattice, is (left-, right-) continuous at a point $t \in I$ iff f has this property when E is considered with its weak topology.*

This can be easily seen by viewing E as a sublattice in $C(U)$, where U stands for the positive part of the closed unit ball in E^*, and applying the Dini's theorem.

Thus the function f constructed in the proof above has no points of continuity when ℓ_∞ is considered with its weak topology. That is, for every $t \in I$ there is $u^* \in \ell_\infty^*$ such that $u^* \circ f$ is not continuous at t. This can also be verified directly:

Assume that $0 < t \leq 1$ and fix a strictly increasing sequence (k_n) in $\mathbf{N}$ such that $\alpha_n := \min I_{k_n} < \max I_{k_n} =: \beta_n < t$ and $\alpha_n \to t$. Let $\mathcal{U}$ be an ultrafilter on $\mathbf{N}$ containing all the sets $K_m := \{k_n : n \geq m\}$, $m = 1, 2, \ldots$. Then $u^*(a) := \lim_{\mathcal{U}} a_n$, $a = (a_n) \in \ell_\infty$, defines a continuous (and positive) linear functional on ℓ_∞. Let $\bar{t} = u^*(f(t))$. Thus, given $\varepsilon > 0$, there is $U \in \mathcal{U}$ such that $|\bar{t} - f_i(t)| < \varepsilon$ for all $i \in U$. In particular, $|\bar{t} - 1| < \varepsilon$ for all $i \in U \cap K_1 \neq \emptyset$ (because $f_i(t) = 1$ for $i \in K_1$). In consequence, $\bar{t} = 1$.

Now take any $0 \leq s < t$ and let $\bar{s} = u^*(f(s))$. As above, for every $\varepsilon > 0$ there is $U \in \mathcal{U}$ such that $|\bar{s} - f_i(s)| < \varepsilon$ for all $i \in U$. Choose m so that $s < \alpha_m < \beta_m < t$. Then $|\bar{s} - 0| < \varepsilon$ for all $i \in U \cap K_m \neq \emptyset$ (because $f_i(s) = 0$ for $i \in K_m$). In consequence, $\bar{s} = 0$.

Thus $u^*(f(t)) = 1$ and $u^*(f(s)) = 0$ for $0 \leq s < t$, hence $u^* \circ f$ is not left-continuous at t. Similarly, for each $t \in [0, 1)$ there is $u^* \in \ell_\infty^*$ with $u^* \circ f$ not right-continuous at t.

(c) If one does not insist on having additional properties like those in the first part of the preceding remark, then also the (simpler) function $f = (f_n)$ with $f_n =$ the characteristic function of the interval $[\max I_n, 1]$ would do the job required in Proposition 2.1. □

3. Extensions of the results of Lavrič

We start with a simple (and rather obvious) lemma that will be of constant use throughout the rest of this paper.

Lemma 3.1 *For an increasing function $f : I \to E$, where $E = (E, \|\cdot\|)$ is an F-normed lattice, the following are equivalent.*

(a) *f has uncountably many points of discontinuity.*

(b) *There exist an uncountable set $D \subset I$ and $\varepsilon > 0$ such that $\|f(u) - f(t)\| \geq \varepsilon$ for all $t \in D$ and $u > t$, or $\|f(u) - f(t)\| \geq \varepsilon$ for all $t \in D$ and $u < t$.*

(c) *There exist an uncountable set $D \subset I$ and $\varepsilon > 0$ such that $\|f(u) - f(t)\| \geq \varepsilon$ for all distinct $u, t \in D$.*

Moreover, the set D in (b) *and* (c) *can be chosen so that each $t \in D$ is a two-sided condensation point of D. That is, $D \subset (0,1)$ and both $(u,t) \cap D$ and $(t,v) \cap D$ are uncountable whenever $u < t < v$. In addition, one may also require that $|D| = 2^{\aleph_0}$.*

Proof (cf. [16, Proof of Prop. 2]. (a)⇒(b): For each $t \in D(f)$ there is $\varepsilon(t) > 0$ with $\|f(u) - f(t)\| \geq \varepsilon(t)$ for all $u > t$ or all $u < t$. Since $|D(f)| > \aleph_0$, it follows that there exists an uncountable subset D of $D(f)$ and $\varepsilon > 0$ as required in (b).

(b)⇒(c): Obvious.

(c)⇒(a): By removing a countable subset if necessary, we may assume that each $t \in D$ is a condensation point of D, i.e., every neighborhood of t has an uncountable intersection with D (see [12, 1.7.11]). Then, obviously, f is discontinuous at each point $t \in D$.

The first assertion in the last part follows from the fact (which is analogous to that just used above) that if a set $D \subset I$ is uncountable, then there exist only countably many t's in D that fail to be two-sided accumulation points of D. As for the second assertion, it should be enough to recall that if $|D(f)| > \aleph_0$, then we actually have $|D(f)| = 2^{\aleph_0}$ (see the first paragraph of Section 1). □

The result below is an extension of Lavrič's main theorem (L2).

Theorem 3.2 *For a σ-Dedekind complete F-normed lattice $E = (E, \|\cdot\|)$ the following statements are equivalent:*

(a) *E is pre-Lebesgue.*

(b) *E has property (λ).*

(c) *E has property (λ_0).*

(d) *E contains no lattice copy of ℓ_∞.*

Proof. (a)⇒(b): Suppose (b) fails and let $f : I \to E$ be an increasing function with $|D(f)| > \aleph_0$. Apply Lemma 3.1 to find D and ε as required in condition (c) of the lemma. Next, choose a strictly increasing sequence (u_n) in D, and denote $x_n = f(u_{n+1}) - f(u_n)$. Then $x_n \geq 0$, $\|x_n\| > \varepsilon$, and $x_1 + \ldots + x_n = f(u_{n+1}) - f(u_1) \leq f(1) - f(0)$ for each n. We have thus constructed a positive sequence (x_n) in E such that $x_n \not\to 0$ and the sequence $x_1 + \ldots + x_n$ $(n \in \mathbf{N})$ is order bounded. However, this contradicts (a), by [1, Thm. 10.1].

(b)⇒(c) is trivial, and (c)⇒(d) follows from Proposition 2.1.

(d)⇒(a): If (a) were false, then E would contain a lattice copy of ℓ_∞ (see [1, Thm. 10.7] or [10, Thm. 2.7]). (It is here where the σ-Dedekind completeness of E is needed.) □

Problem 1: *Is it true for every F- (or Banach) lattice E that* (λ_0) *implies* (λ)*?*

The next result is an extension of Lavrič's result (L3) and, at the same time, a particular case of Fact 1.7.

Proposition 3.3 *Every separable F-normed lattice has property* (λ).

4. A negative answer to Lavrič's question (LQ)

For a Banach space E, we denote by $R(I, E)$ the Banach space consisting of all regular functions $f : I \to E$, equipped with the sup norm. Of course, if E is a Banach lattice, so is $R(I, E)$ under the pointwise order induced from E.

We first prove the following.

Theorem 4.1 *If a Banach space* E *contains no isomorphic copy of* ℓ_∞, *neither does the space* $R(I, E)$.

Proof. Suppose there exists an isomorphic embedding $T : \ell_\infty \to R(I, E)$. We may assume that $\|Ta\| > 1$ whenever $\|a\| = 1$. Let $m : \mathcal{P}(\mathbf{N}) \to R(I, E)$ be the associated (bounded, finitely additive) measure defined by $m(A) = T(1_A)$, and denote $f_n = Te_n$. For each n choose $t_n \in I$ so that $\|f_n(t_n)\| > 1$, and define a bounded finitely additive measure $m_n : \mathcal{P}(\mathbf{N}) \to E$ by $m_n(A) = m(A)(t_n)$. Since E contains no copy of ℓ_∞, each of the measures m_n is exhaustive, that is, $m_n(A_k) \to 0$ as $k \to \infty$ for any disjoint sequence (A_k) in $\mathcal{P}(\mathbf{N})$ (see [6, Thm. I.4.2]). From this it follows that the set $\{t_n : n \in \mathbf{N}\}$ is infinite. Hence, by passing to a subsequence (t_{n_k}), replacing $\mathbf{N}$ with $\{n_1, n_2, \ldots\}$, and relabeling, we may assume that the sequence (t_n) is strictly monotone, say increasing, and converges to a point $t \in I$. Now, as each $m(A) : I \to E$ is a regular function, the limit

$$\lim_{n\to\infty} m_n(A) \equiv \lim_{n\to\infty} m(A)(t_n) = m(A)(t-)$$

exists for every $A \in \mathcal{P}(\mathbf{N})$. Hence, by the Brooks-Jewett theorem (see [4], [7], [15]), the measures m_n are uniformly exhaustive. That is, whenever (A_k) is a disjoint sequence in $\mathcal{P}(\mathbf{N})$, then $\lim_k m_n(A_k) = 0$ uniformly for $n \in \mathbf{N}$. In particular, $m_n(A_n) \to 0$ as $n \to \infty$. However, for $A_n = \{n\}$ we have $\|m_n(A_n)\| = \|m(A_n)(t_n)\| = \|f_n(t_n)\| > 1$ for every n which is a required contradiction. □

For I and E as above, let $R_{lc}(I, E)$ stand for the closed subspace of $R(I, E)$ consisting of those functions $f \in R(I, E)$ that are continuous from the left at each point $t \in (0, 1]$, and continuous from the right at the point 0. Obviously, if E is a Banach lattice, then $R_{lc}(I, E)$ is a closed sublattice of the Banach lattice $R(I, E)$. Let's simply write $R(I)$ and $R_{lc}(I)$ when $E = \mathbf{R}$.

And now, here's the promised negative answer to Lavrič's question:

Proposition 4.2 *The Banach lattice* $R_{lc}(I)$ *contains no isomorphic copy of* ℓ_∞, *and yet there exists an increasing function* $f : I \to R_{lc}(I)$ *with* $\|f(t) - f(u)\| = 1$ *whenever* $t \neq u$.

Proof. The first part follows directly from Theorem 4.1. A required function f can be defined by setting $f(0) = 0$ (the zero function), and $f(t) = 1_{[0,t]}$ for $t \in (0, 1]$. □

Remarks 4.3 (a) In Proposition 4.2, we preferred using $R_{lc}(I)$ instead of $R(I)$ because of the worth noting fact that the closed linear span of the range of the function f defined in the proof is precisely $R_{lc}(I)$.

(b) Note that since $R_{lc}(I)$ is an AM-space with a strong unit $e = 1_I$ and such that the order interval $[-e, e]$ coincides with the closed unit ball, $R_{lc}(I)$ is lattice isometric to $C(K)$ for some compact space K (see [1, Thm. 10.16]). Likewise for $R(I)$. Thus there exist Banach lattices of type $C(K)$ that are counterexamples to (LQ). For some more concrete counterexamples of this type, see Remark 5.12 below.

(c) Let E be any topological vector space. Then every regular function $f : I \to E$ is bounded; in fact, every sequence in $f(I)$ has a subsequence convergent in E. Define the space $R(I, E)$ as above, and equip it with the topology of uniform convergence on I. Then the same proof as above yields the following generalization of Theorem 4.1:

If E has the property that every bounded finitely additive measure $m : \mathcal{P}(\mathbf{N}) \to E$ is exhaustive, then so does the space $R(I, E)$, hence $R(I, E)$ contains no copy of ℓ_∞.

(d) The preceding observation can be considerably generalized: Let S be a set in which a class $\mathcal{C}$ of 'convergent' sequences is distinguished so that every sequence (t_n) in S has a subsequence $(s_n) \in \mathcal{C}$. Also, let E be a topological vector space. Define $R(S, \mathcal{C}; E)$ to be the vector space of functions $f : S \to E$ such that the limit $\lim_n f(s_n)$ exists in E for every sequence $(s_n) \in \mathcal{C}$. (All such functions are bounded.) Equip the space $R(S, \mathcal{C}; E)$ with the topology of uniform convergence on S. Then an exact analogue of the result stated above holds for the space $R(S, \mathcal{C}, E)$. One of the consequences of this is the following:

If K is a sequentially compact space and E is a Banach (or F-) space that contains no isomorphic copy of ℓ_∞, then neither does $C(K, E)$.

(e) A slight variant of the space $R_{lc}(I)$ appears in an example in Corson [5] and, by the arguments used therein, $R_{lc}(I)/C(I) \approx c_0(I)$. This can be applied to give an alternative proof that $R_{lc}(I)$ contains no copy of ℓ_∞. In fact, if a Banach space X contains a copy of ℓ_∞, and Y is a closed subspace of X, then either Y or X/Y contains a copy of ℓ_∞, see [8].

Corson's result also shows that property (λ) is not a three space property. In fact, both $C(I)$ and $c_0(I) \approx R_{lc}(I)/C(I)$ have (λ), but $R_{lc}(I)$ does not.

5. Spaces of continuous vector functions with property (λ)

For a compact space K and a Banach lattice E, we denote by $C(K, E)$ the Banach lattice of all continuous functions from K to E, with the pointwise order and supremum norm. As usual, we write simply $C(K)$ when $E = \mathbf{R}$.

Given a function $f : I \to C(K, E)$, we will often write $f(t, s)$ instead of $f(t)(s)$.

We are interested in the question which Banach lattices of the type $C(K, E)$ have property (λ). Clearly, if a space $C(K, E)$ has property (λ), so do both $C(K)$ and E; the converse is an open question:

Problem 2: *Let K be a compact space and E a Banach lattice. Is it true that $C(K, E)$ has property (λ) if both $C(K)$ and E have property (λ)?*

In view of Lavrič's result (L3) (or Fact 1.7, or Proposition 3.3), if K is a compact metric space, then $C(K)$ has property (λ). However, it is not so for $K = \beta\mathbf{N}$ by (L1) (or Proposition 2.1) because $\ell_\infty \cong C(\beta\mathbf{N})$, or the spaces K mentioned in Remark 4.3 (b), or the spaces H and H_0 in Remark 5.12 below. Thus it is only natural to raise the following.

Problem 3: *Find an intrinsic characterization of those compact spaces K for which $C(K)$ has property (λ).*

The results proved in this section show that, for any Banach lattice with property (λ), the class of compact spaces K such that $C(K, E)$ has property (λ) is quite large: It contains all metrizable compact spaces (Theorem 5.1), more generally—all Eberlein compacts (Theorem 5.11; see also Theorem 5.13), and arbitrary products of such spaces (Corollary 5.14). Moreover, it is closed under continuous images (Remark 5.15 (a)). However, if $C(K)$ has property (λ) and L is a closed subspace of K, then $C(L)$ need not have property (λ) (see Remark 5.15 (b)).

We start with the metric case to which, as will be seen, the proofs of the other results just mentioned will be ultimately reduced.

Theorem 5.1 *If K is a metrizable compact space and E is a Banach lattice having property (λ), then also the Banach lattice $C(K, E)$ has property (λ).*

Proof. Denote by ρ a metric defining the topology of K, and suppose an increasing function $f : I \to C(K, E)$ has uncountably many discontinuities. Then, by Lemma 3.1 (b), for some uncountable set $D \subset (0, 1]$ and some $\varepsilon > 0$, we have $\|f(u) - f(t)\| \geq \varepsilon$ whenever $u \in D$ and $0 \leq t < u$.

Fix any $u \in D$. Then, for $0 \leq t < u$, the sets $\{s \in K : \|f(u, s) - f(t, s)\| \geq \varepsilon\}$ are compact and nonempty, and they decrease as t increases. Hence the intersection $K(u)$ of all these sets is nonempty. Choose a point $s_u \in K(u)$. Clearly, $\|f(u, s_u) - f(t, s_u)\| \geq \varepsilon$ whenever $u \in D$ and $0 \leq t < u$.

Denote $L = \{s_u : u \in D\}$, and for each $s \in L$ let $D_s = \{u \in D : s_u = s\}$. Clearly, these sets are pairwise disjoint. Moreover, they are nonempty and (at most) countable. For if some D_s were uncountable, then since $\|f(u, s) - f(t, s)\| \geq \varepsilon$ whenever $u \in D_s$ and $0 \leq t < u$, the increasing function $f(\cdot, s) : t \to f(t, s)$ from I to E would have uncountably many discontinuities, contradicting property (λ) of E.

Since $D = \bigcup_{s \in L} D_s$ is uncountable, we conclude that also L is uncountable. For each $s \in L$ select a point $u_s \in D_s$ and a $\delta_s > 0$ such that whenever $s', s'' \in K$ and $\rho(s', s'') < \delta_s$, then $\|f(u_s, s') - f(u_s, s'')\| < \varepsilon/4$. Since L is uncountable, there are an uncountable subset M of L and a number $\delta > 0$ such that $\delta_s > \delta$ for all $s \in M$. Thus whenever $s \in M$, $s', s'' \in K$, and $\rho(s', s'') < \delta$, then $\|f(u_s, s') - f(u_s, s'')\| < \varepsilon/4$.

Let $s_0 \in M$ be a condensation point of M. Denote $M_0 = \{s \in M : \rho(s, s_0) < \delta/2\}$ and $D_0 = \{u_s : s \in M_0\}$. Now, take any distinct points $s, s' \in M_0$; we may assume of course that $u_{s'} < u_s$. Then

$$\begin{aligned} \|f(u_s, s_0) - f(u_{s'}, s_0)\| &\geq \|f(u_s, s) - f(u_{s'}, s)\| - \|f(u_s, s_0) - f(u_s, s)\| \\ &\quad -\|f(u_{s'}, s) - f(u_{s'}, s')\| - \|f(u_{s'}, s') - f(u_{s'}, s_0)\| \\ &> \varepsilon - \frac{3}{4}\varepsilon = \frac{1}{4}\varepsilon. \end{aligned}$$

By Lemma 3.1, the increasing function $f(\cdot, s_0) : t \to f(t, s_0)$ from I to E has uncountably many discontinuities, contradicting property (λ) of E. □

Theorem 5.2 *Let S be a noncompact locally compact space, $\omega S = S \cup \{\infty\}$ the one-point compactification of S, and E a Banach lattice. Assume that for every compact subset K of S the Banach lattice $C(K, E)$ has property (λ). Then also the Banach lattice $C(\omega S, E)$ has property (λ).*

Proof. Suppose $f : I \to C(\omega S, E)$ is an increasing function with uncountably many disconitnuities. Then, by Lemma 3.1 (c), we can find an uncountable set $D \subset I$ and some $\varepsilon > 0$ such that $\|f(t) - f(u)\| > \varepsilon$ for all distinct pairs $t, u \in D$. Moreover, we may assume that each $t \in D$ is a condensation point of D. Note that for each $s \in \omega S$ the function $f(\cdot, s) : I \to E$ is increasing. Since E has property (λ), $f(\cdot, \infty)$ has at most countably many discontinuities. Therefore, we may also assume that the set D is chosen so that $f(\cdot, \infty)$ is continuous at each point of D. Fix any $t_0 \in D \cap (0, 1)$ and assume, as we may, that $D \cap (t, t_0]$ is uncountable for all $0 < t < t_0$. Choose $t_1 \in D \cap (0, t_0)$ so that $\|f(t_0, \infty) - f(t_1, \infty)\| < \varepsilon/3$. Let K be a compact subset of S such that $\|f(t_1, s) - f(t_1, \infty)\| < \varepsilon/3$ and $\|f(t_0, s) - f(t_0, \infty)\| < \varepsilon/3$ for all $s \in \omega S \setminus K$. Now, if $t, u \in D \cap [t_1, t_0]$ and $t < u$, then for all $s \in \omega S \setminus K$,

$$\begin{aligned}\|f(u, s) - f(t, s)\| &\leq \|f(t_1, s) - f(t_0, s)\| \\ &\leq \|f(t_1, s) - f(t_1, \infty)\| + \|f(t_1, \infty) - f(t_0, \infty)\| \\ &\quad + \|f(t_0, \infty) - f(t_0, s)\| < \varepsilon;\end{aligned}$$

thus $\|f(u, s) - f(t, s)\| < \varepsilon$. But $\|f(u) - f(t)\| > \varepsilon$, so $\sup_{s \in K} \|f(u, s) - f(t, s)\| > \varepsilon$. Finally, define an increasing function $g : I \to C(K)$ by $g(t) = f(t)|K$. Then, by the above, $\|g(u) - g(t)\| > \varepsilon$ for all distinct t, u from the uncountable set $D \cap [t_1, t_0]$. Hence, by Lemma 3.1, $|D(g)| > \aleph_0$, contradicting the assumption on S. □

From the preceding two results we derive the following.

Corollary 5.3 *Let S be a locally compact space such that every compact subspace of S is metrizable, and E a Banach lattice with property (λ). Then also the Banach lattice $C(\omega S, E)$ has property (λ).*

In particular, the following holds.

Corollary 5.4 *For any set Γ, if E is a Banach lattice with property (λ), then also the Banach lattice $c(\Gamma, E)$ has property (λ).*

Recall that $c(\Gamma, E)$ consists of all functions $x = (x_\gamma) : \Gamma \to E$ that have a limit "at infinity" when Γ is regarded as a discrete topological space. That is, there exists an element $x_\infty \in E$ such that for every $\varepsilon > 0$ the inequality $\|x_\infty - x_\gamma\| > \varepsilon$ may hold only for finitely many points $\gamma \in \Gamma$. Note that then $x_\gamma = x_\infty$ on the complement of a countable subset of Γ. The norm in this space is the sup norm. Of course, $c(\Gamma, E) \cong C(\omega\Gamma, E)$.

Remark 5.5 Corollary 5.4 can be deduced directly from Theorem 5.2. Indeed, compact subspaces K of the discrete space Γ are finite, hence the assumption that the spaces $C(K, E)$ have property (λ) is satisfied because finite products $E \times \ldots \times E$ have property (λ) whenever E has it (see Fact 1.6).

For the compact space denoted below by Ω, see [12, Examples 3.1.27 and 4.4.11].

Corollary 5.6 *Let Ω denote the compact space of all ordinals $\gamma \le \omega_1$, where ω_1 is the first uncountable ordinal. If E is a Banach lattice with property (λ), then also the Banach lattice $C(\Omega, E)$ has property (λ).*

In some of the proofs below we will make use of the following two lemmas. The first is just a simple (but useful) observation.

Lemma 5.7 *Let E and F be a Banach lattices, and let $f : I \to E$ and $g : I \to F$ be increasing functions. Suppose $\|g(r') - g(r)\| \ge \|f(r') - f(r)\|$ for all r, r' from a dense subset Q of I. Then $\|g(u) - g(t)\| \ge \|f(w) - f(v)\|$ whenever $0 \le t < v < w < u \le 1$.*

In consequence, if f has uncountably many discontinuities, so does g.

Proof. Choose $r, r' \in Q$ so that $t < r < v$ and $w < r' < u$. Then

$$\|g(u) - g(t)\| \ge \|g(r') - g(r)\| \ge \|f(r') - f(r)\| \ge \|f(w) - f(v)\|.$$

To finish apply Lemma 3.1. □

Our second lemma is a well known fact due to Y. Mibu (1944); see [11, p. 221] and [12, 3.2.H] for more information. It is usually proved by applying the Stone-Weierstrass theorem (see, for instance, reference [7] in [11]). However, it also admits a completely elementary proof which is worth incluiding here.

Lemma 5.8 *Let K be a compact subset of the product $\prod_{j \in J} K_j$ of compact spaces. Then every continuous map f from K to any metric space $Z = (Z, d)$ depends on a countable number of coordinates $j \in J$. That is, there exists a countable subset J_0 of J such that whenever $s', s'' \in K$ and $s'|J_0 = s''|J_0$, then $f(s') = f(s'')$.*

Proof. For each $n \in \mathbf{N}$ there is a finite cover $\mathcal{U}_n$ of K consisting of sets of the form $U(s) = \prod_{j \in J} U_j(s_j)$ such that $s = (s_j) \in K$, $U_j(s_j)$ is a neighborhood of s_j in K_j, the set $J(s) := \{j \in J : U_j(s_j) \ne K_j\}$ is finite, and $d(f(s), f(t)) \le 1/n$ whenever $t \in K \cap U(s)$. Let J_n denote the union of the sets $J(s)$ associated with the members $U(s)$ of $\mathcal{U}_n$. Then the union J_0 of all these J_n's is as required. In fact, let $s', s'' \in K$ and $s'|J_0 = s''|J_0$. For each n there is a $U(s)$ in $\mathcal{U}_n$ such that $s' \in U(s)$; observe that then also $s'' \in U(s)$. Therefore, $d(f(s'), f(s)) \le 1/n$ and $d(f(s''), f(s)) \le 1/n$. In consequence, $d(f(s'), f(s'')) \le 2/n$. It follows that $f(s') = f(s'')$. □

Theorem 5.9 *Let E be a Banach lattice and let $\{K_j : j \in J\}$ be a family of compact spaces. For $\varphi \subset J$ let $K_\varphi = \prod_{j \in \varphi} K_j$. Assume that*

(∗) *for every finite set $\varphi \subset J$ the space $C(K_\varphi, E)$ has property (λ).*

Then $C(K_J, E)$ has property (λ).

Proof. Suppose there is an increasing function $f : I \to C(K_J, E)$ with $|D(f)| > \aleph_0$.

Denote by Q the set of rational numbers in I. Applying Lemma 5.8 we find a countable subset J_0 of J such that $f(r, s) = f(r, s')$ whenever $r \in Q$, $s, s' \in K_J$, and $s|J_0 = s'|J_0$.

Fix an element $\bar{s} \in K_{J\setminus J_0}$ and define a function $g : I \to C(K_{J_0}, E)$ by $g(t,s) = f(t,s')$, where $s \in K_{J_0}$, and $s' \in K_J$ is such that $s'|J_0 = s$ and $s'|(J \setminus J_0) = \bar{s}$.

Note that the function g is increasing and that $f(r,s) = g(r, s|J_0)$ for $r \in Q$ and $s \in K_\varphi$. Also, observe that $C(K_{J_0}, E)$ can be viewed as a sublattice of $C(K_J, E)$ (via the embedding that assigns to each $h \in C(K_{J_0}, E)$ the function $\bar{h} \in C(K_J, E)$ defined by $\bar{h}(s) = h(s|J_0)$). Hence, by Lemma 5.7, g has uncountably many discontinuities.

We thus may assume that the index set J is countable. In view of Lemma 3.1, there are an uncountable set $D \subset I$ and a number $\varepsilon > 0$ such that $\|f(u) - f(t)\| \geq \varepsilon$ whenever $u \in D$ and $0 \leq t < u$. Then, as in the proof of Theorem 5.1, assign to each $u \in D$ a point $s_u \in K_J$ so that $\varepsilon \leq \|f(u, s_u) - f(t, s_u)\| \geq \varepsilon$ whenever $u \in D$ and $0 \leq t < u$.

Denote $L = \{s_u : u \in D\}$ and, for each $s \in L$, let $D_s = \{u \in D : s_u = s\}$. Clearly, these sets are pairwise disjoint. Moreover, they are nonempty and (at most) countable. For if some D_s is uncountable, then $\|f(u,s) - f(t,s)\| \geq \varepsilon$ whenever $u \in D_s$ and $0 \leq t < u$. In consequence, the increasing function $f(\cdot, s) : t \to f(t,s)$ from I to E has uncountably many discontinuities, contradicting property (λ) of E.

Since $D = \bigcup_{s \in L} D_s$ is uncountable, we conclude that also L is uncountable. For each $s \in L$ select a point $u_s \in D_s$. Next, by the (uniform) continuity of the function $f(u_s)$, choose a finite set $\varphi_s \subset J$ such that whenever $s', s'' \in K_J$ and $s'|\varphi_s = s''|\varphi_s$, then $\|f(u_s, s') - f(u_s, s'')\| < \varepsilon/3$. Since L is uncountable, there are an uncountable subset M of L and a finite set $\varphi \subset J$ such that $\varphi_s = \varphi$ for all $s \in M$. Thus whenever $s \in M$, $s', s'' \in K_J$, and $s'|\varphi = s''|\varphi$, then $\|f(u_s, s') - f(u_s, s'')\| < \varepsilon/3$.

Fix a point $\bar{s} \in K_{J\setminus\varphi}$, and for every $s \in K_\varphi$ denote by s' the point in K_J such that $s'|\varphi = s$ and $s'|(J \setminus \varphi) = \bar{s}$. Define an increasing function $h : I \to C(K_\varphi, E)$ by $h(t,s) = f(t,s')$. Take any distinct $s_1, s_2 \in M$, and assume that $u_{s_1} < u_{s_2}$. Then, writing σ_k for $s_k|\varphi$ $(k = 1,2)$, we have

$$\begin{aligned} h(u_{s_2}, \sigma_2) - h(u_{s_1}, \sigma_2) &= f(u_{s_2}, \sigma_2') - f(u_{s_1}, \sigma_2') \\ &= [f(u_{s_2}, \sigma_2') - f(u_{s_2}, s_2)] + [f(u_{s_2}, s_2) - f(u_{s_1}, s_2)] \\ &\quad + [f(u_{s_1}, s_2) - f(u_{s_1}, \sigma_2')], \end{aligned}$$

hence

$$\begin{aligned} \|h(u_{s_2}, \sigma_2) - h(u_{s_1}, \sigma_2)\| &\geq \|f(u_{s_2}, s_2) - f(u_{s_1}, s_2)\| - \|f(u_{s_2}, \sigma_2') - f(u_{s_2}, s_2)\| \\ &\quad - \|f(u_{s_1}, s_2) - f(u_{s_1}, \sigma_2')\| > \varepsilon/3. \end{aligned}$$

Thus $\|h(u_{s_2}) - h(u_{s_1})\| > \varepsilon/3$ for all distinct $s_1, s_2 \in M$. In consequence, the function h has uncountably many points of discontinuity, which contradictis assumption $(*)$. □

The assumption $(*)$ in the preceding theorem suggests the following problem (which is of course a special case of Problem 2).

Problem 4: *Is it true that if K_1, K_2 are compact spaces such that both $C(K_1)$ and $C(K_2)$ have property (λ), then so does $C(K_1 \times K_2)$? The same for the spaces of continuous E-valued functions, where E is a Banach lattice with property (λ).*

Remark 5.10 Let E be a Banach lattice. If K_1 is a metrizable compact space and K_2 is a compact space for which $C(K_2, E)$ has property (λ), then $C(K_1 \times K_2, E)$ has property (λ). To see this, note that $C(K_1 \times K_2, E) \cong C(K_1, C(K_2, E))$ and apply Theorem 5.1.

Recall that a compact space that is homeomorphic to a weakly compact set in some Banach space is called an *Eberlein compact*; see [3] and [17] for more information. Every metrizable compact space K is an Eberlein compact; in fact, it is even isometric to a norm compact subset of $C(I)$.

Theorem 5.11 *Let K be an Eberlein compact. If a Banach lattice E has property (λ), so does the space $C(K, E)$.*

Proof. Suppose there is an increasing function $f : I \to C(K, E)$ with $|D(f)| > \aleph_0$.

By a theorem of Amir and Lindenstrauss [3], we may assume that K is a weakly compact subset of the Banach space $c_0(\Gamma)$ for some Γ. Then, in particular, K is a compact subspace of the product space $[a, b]^\Gamma$ for some $[a, b] \subset \mathbf{R}$.

Denote by Q the set of rational numbers in I. By Lemma 5.8, there exists a countable subset Δ of Γ such that $f(r, s) = f(r, s')$ whenever $r \in Q$, $s, s' \in K$, and $s|\Delta = s'|\Delta$. Let $K_\Delta = \{s|\Delta : s \in K\}$. It is easily verified that K_Δ is a (sequentially) weakly compact subset of the Banach space $c_0(\Delta) \cong c_0$. It follows that K_Δ is a metrizable compact space. Hence, by Theorem 5.1, $C(K_\Delta, E)$ has property (λ).

To get a contradiction, select for each $s \in K_\Delta$ a point $s' \in K$ so that $s'|\Delta = s$, and then define a function $g : I \to C(K_\Delta, E)$ by $g(t, s) = f(t, s')$. Note that if $t < u$ and $s \in K_\Delta$, then $g(t, s) = f(t, s') \leq f(u, s') = g(u, s)$. Thus g is increasing and $f(r, s) = g(r, s|\Delta)$ for $r \in Q$ and $s \in K$. By Lemma 5.7, g has uncountably many discontinuities, which is a desired contradiction. □

Remark 5.12 As is well known, Eberlein compacts are sequentially compact. However, in general it is not true that if K is a compact and sequentially compact space, then $C(K)$ has property (λ). In fact, even additional assumptions on K like separability and the first axiom of countability will not force $C(K)$ to have property (λ).

We will see that this is the case for $K = H$, the *Helly space*, that is, the compact space consisting of all nondecreasing functions $x : I \to I$ equipped with the pointwise convergence topology. It is known (see e.g. [13, Exerc. to Ch. V]) that H is nonmetrizable, separable, and sequentially compact; in fact, each of its points has a countable base of neighborhoods. Now, define a function $f : I \to C(H)$ by $f(t)(x) = x(t)$ $(t \in I,\ x \in H)$. Clearly, f is nondecreasing. Take any $t, u \in I$ with $u < t$, any $v \in (u, t)$, and let x be the characteristic function of the interval $[v, 1]$. Then $f(t)(x) = 1$ while $f(u)(x) = 0$ and it follows that $\|f(t) - f(u)\| = 1$. Thus f is discontinuous at each point of I.

Observe that that also for the closed subspace H_0 of H, consisting of functions with values in $\{0, 1\}$, the function $t \to f(t)|H_0$ shows that the Banach lattice $C(H_0)$ fails to have property (λ).

It has been recently shown by A. Michalak that for *each* nonmetrizable closed subset K of H the space $C(K)$ lacks property (λ). Note that, by the result stated at the end of Remark 4.3 (d), none of these spaces $C(K)$ can contain a copy of ℓ_∞. Thus these spaces provide additional counterexamples to (LQ). □

The following result has been inspired by some comments about an earlier version of this paper made by A. Michalak.

Theorem 5.13 *Let K be a compact space in which every separable subspace is metrizable. Then, if a Banach lattice E has property (λ), so does the Banach lattice $C(K, E)$.*

Proof. Suppose there is an increasing function $f : I \to C(K, E)$ with $|D(f)| > \aleph_0$.

Let Q denote the set of rationals in I. For each pair r, r' of distinct points in Q choose a point $s_{r,r'} \in K$ so that $\|f(r) - f(r')\| = \|f(r, s_{r,r'}) - f(r', s_{r,r'})\|$, and let K_0 denote the closure of the set of all these points $s_{r,r'}$. By the assumption on K, the subspace K_0 is metrizable. Therefore, by Theorem 5.1, the space $C(K_0, E)$ has property (λ).

On the other hand, consider the function $g : I \to C(K_0, E)$ defined by $g(t) = f(t)|K_0$. Clearly, Lemma 5.7 is applicable and shows that $|D(g)| > \aleph_0$. A contradiction. □

Note that Eberlein compacts have the property imposed on K in the above result. Hence Theorem 5.11 is a direct consequence of Theorems 5.1 and 5.13.

Since finite products of Eberlein compacts are Eberlein compacts, from Theorems 5.9 and 5.11 the following is immediate.

Corollary 5.14 *If a compact space K is the product of a family of Eberlein compacts, then for every Banach lattice E with property (λ) the space $C(K, E)$ has property (λ).*

In particular, it is so for the products of metrizable compact spaces (which is also a direct consequence of Theorems 5.1 and 5.9). Moreover, in view of Theorem 5.13, an analogous result holds for the products where each factor is a compact space all of whose separable subspaces are metrizable.

Remarks 5.15 (a) If $C(K, E)$ has property (λ) and a compact space K' is a continuous image of K by a map φ, then also $C(K', E)$ has property (λ). It is so because then $C(K', E)$ embeds in $C(K, E)$ via the composition operator associated with φ. From this and Corollary 5.14 it follows, for example, that if K is a dyadic space (see [12]) and E is a Banach lattice with property (λ), then also $C(K, E)$ has (λ).

(b) It is not true that if a Banach lattice E has property (λ) and F is a closed ideal in E, then the Banach lattice E/F has property (λ). In fact, let $K = [0, 1]^I$. Then, by Corollary 5.14, $C(K)$ has (λ), while $C(H)$, where $H \subset K$ is the Helly space, does not (see Remark 5.12). To finish note that, thanks to the Urysohn-Tietze extension theorem, $C(H) \cong C(K)/F$, where $F = \{x \in C(K) : x|H = 0\}$. □

6. Spaces of measurable vector functions with property (λ)

Theorem 6.1 *Let (S, Σ, μ) be a measure space and E a Banach lattice with property (λ). Then also $L_p(S, \Sigma, \mu, E)$, $1 \le p < \infty$, has property (λ).*

Proof. We may assume that μ is complete so that Bochner μ-measurable functions from S to E coincide with those that are Borel measurable and μ-almost separable-valued.

Let $f : I \to L_p(\mu, E)$ be an increasing function; we may assume that $f(0) = 0$. Since $0 \le f(t) \le f(1)$ and $f(1)$ has a support of σ-finite μ-measure, we may also assume that μ is σ-finite. Furthermore, it is easy to reduce the proof that f may have only countably many discontinuities to the case where μ is finite. Thus in what follows we shall assume that μ is a probability measure.

Suppose f has uncountably many discontinuities, and denote by Q the set of rational numbers in I. Then there exists a countably generated sub-σ-algebra $\mathcal{A}$ of Σ such that $f(r)$ is $\mathcal{A}$-measurable for each $r \in Q$. Define a function $g : I \to L_p(S, \mathcal{A}, \mu; E)$ by

$$g(t) = \mathbf{E}(f(t) \mid \mathcal{A}), \qquad t \in I.$$

Note that g is increasing and that $g(r) = f(r)$ for $r \in Q$. Hence, according to Lemma 5.7, g must have uncountably many discontinuities. Hence, by Lemma 3.1, there exist an uncountable set $D \subset I$ and $\varepsilon > 0$ such that $\|g(u) - g(t)\|_p \geq \varepsilon$ for all distinct $t, u \in D$.

Fix an increasing sequence $(\mathcal{A}_n)$ of finite subalgebras of $\mathcal{A}$ such that $\mathcal{A}$ is the smallest σ-algebra containing all the $\mathcal{A}_n$'s. Note that for all $h \in L_p(S, \mathcal{A}, \mu; E)$,

$$\lim_{n\to\infty} \|h - \mathbf{E}(h \mid \mathcal{A}_n)\|_p = 0;$$

see [6, Cor. V.2.2]. For $u \in D$ and $n \in \mathbf{N}$ let $g_n(u) = \mathbf{E}(g(u) \mid \mathcal{A}_n)$. By the preceding observation, for each $u \in D$ there is $n(u) \in \mathbf{N}$ such that $\|g(u) - g_{n(u)}(u)\|_p < \varepsilon/3$. Since D is uncountable, it has an uncountable subset D_0 such that $n(u) = m$ for all $u \in D_0$ and some $m \in \mathbf{N}$. Then it is easily seen that for all distinct $t, u \in D_0$,

$$\|g_m(u) - g_m(t)\|_p > \varepsilon/3.$$

Hence the increasing function $g_m : I \to L_p(S, \mathcal{A}_m, \mu; E)$ has uncountably many discontinuities. However, as the algebra $\mathcal{A}_m$ is finite, the space $L_p(S, \mathcal{A}_m, \mu; E)$ is isomorphic to a finite product $E \times \ldots \times E$ which obviously has property (λ). A contradiction. □

Remarks 6.2 (a) In particular, *if a Banach lattice E has property (λ), then so does the Banach lattice $l_p(\Gamma, E)$ for $1 \leq p < \infty$ and any set Γ.* A direct proof of this is much simpler than that of Theorem 6.1:

Suppose $f = (f_\gamma) : I \to l_p(\Gamma, E)$ is an increasing function with $|D(f)| > \aleph_0$. Then there are an uncountable set $D \subset I$ and $\varepsilon > 0$ such that $\|f(u) - f(t)\|_p > \varepsilon$ for all distinct $t, u \in D$. Assume $f(0) = 0$, and choose a finite set $A \subset \Gamma$ such that $\|f(1)1_{\Gamma\setminus A}\|_p < \varepsilon/3$. Then also $\|f(t)1_{\Gamma\setminus A}\|_p < \varepsilon/3$ for all $t \in I$. Consider the function $g : I \to l_p(A, E)$ defined by $g(t) = f(t)1_A$. It is obviously increasing and, as easily verified, $\|g(u) - g(t)\|_p > \varepsilon/3$ for all distinct $t, u \in D$. Hence g has uncountably many discontinuities which is impossible because $l_p(A, E)$ is isomorphic to a finite product $E \times \ldots \times E$ and thus has property (λ).

The result stated above can be generalized replacing $l_p(\Gamma)$ by any solid F-space $L \subset \mathbf{R}^\Gamma$ with an order continuous F-norm $\|\cdot\|_L$. That is, whenever $a = (\alpha_\gamma) \in L$, then for every $\varepsilon > 0$ there is a finite set $A \subset \Gamma$ with $\|a1_{\Gamma\setminus A}\|_L < \varepsilon$.

(b) Theorem 6.1 can be extended to Banach lattices F consisting of Bochner measurable functions from S to a Banach lattice E and such that F satisfies the conditions listed in [9, Thm. 3.2]; in particular, to Banach lattices $F = L(E)$, where L is a Köthe function space with order continuous norm. We give an otline of the proof:

Assume E satisfies (λ), and let $f : I \to F$ be an increasing function with $|D(f)| > \aleph_0$ and $f(0) = 0$. Since $f(1)$ has a support of σ-finite measure and the norm of F is absolutely continuous, there is $A \in \Sigma$ with $0 < \mu(A) < \infty$ such that $k = f(1)1_A \in L_1(\mu; E)$ and the increasing function $f_A : I \to F_A = \{h1_A : h \in F\}$, defined by $f_A(t) = f(t)1_A$, has uncountably many discontinuities. Now, on the order interval $[0, k] \subset F_A$ the L_1-norm is stronger than that induced from F. In consequence, $f_A : I \to L_1(A, \mu, E)$ has uncountably many discontinuities, contradicting case $p = 1$ of Theorem 6.1.

REFERENCES

1. C. Aliprantis and O. Burkinshaw, Locally solid Riesz spaces, Academic Press, 1978.
2. C. Aliprantis and O. Burkinshaw, Positive operators, Academic Press, 1985.
3. D. Amir and J. Lindenstrauss, The structure of weakly compact sets in Banach spaces, Ann. of Math. 88 (1968), 35–46.
4. J. K. Brooks and R. S. Jewett, On finitely additive vector measures, Proc. Nat. Acad. Sci. USA 60 (1970), 1294–1298.
5. H. H. Corson, The weak topology of Banach spaces, Trans. Amer. Math. Soc. 101 (1961), 1–15.
6. J. Diestel and J. J. Uhl, Jr., Vector Measures, Math. Surveys 15, Amer. Math. Soc., Providence, RI, 1977.
7. L. Drewnowski, Equivalence of Brooks-Jewett, Vitali-Hahn-Saks and Nikodym theorems, Bull. Acad. Polon. Sci. Sér. Sci. Math. Astronom. Phys. 20 (1972), 725–731.
8. L. Drewnowski, On Banach spaces containing a copy of ℓ_∞, and some three-space properties, to appear.
9. L. Drewnowski and G. Emmanuele, On Banach spaces with the Gelfand-Phillips property, Rend. Circ. Mat. Palermo 38 (1989), 377–391.
10. L. Drewnowski and I. Labuda, Copies of c_0 and ℓ_∞ in topological Riesz spaces, Trans. Amer. Math. Soc. 350 (1998), 3555–3570.
11. R. Engelking, On functions defined on Cartesian products, Fund. Math. 59 (1966), 221–231.
12. R. Engelking, General Topology, PWN – Polish Scientific Publishers, Warszawa, 1977.
13. J. L. Kelley, General Topology, D. van Nostrand, New York, 1955.
14. K. Kuratowski, Topology, vol. I, Academic Press, New York and London; PWN – Polish Scientific Publishers, Warszawa, 1966.
15. I. Labuda, Sur quelques généralisations des théorèmes de Nikodym et de Vitali-Hahn-Saks, Bull. Acad. Polon. Sci., Sér. Sci. Math. Astronom. Phys. 20 (1972), 447–456.
16. B. Lavrič, A characterization of Banach lattices with order continuous norm, Radovi Matematicki 8 (1992), 37–41.
17. J. Lindenstrauss, Weakly compact sets—their topological properties and the Banach spaces they generate, Symposium on Infinite-Dimensional Topology (Louisiana State Univ., Baton Rouge, La., 1967), Ann. of Math. Studies 69, Princeton Univ. Press, Princeton, N. J., 1972, 235–273.
18. L. Schwartz, Analyse mathématique, vol. I, Hermann, Paris, 1967.

Recent Progress in Functional Analysis
K.D. Bierstedt, J. Bonet, M. Maestre and J. Schmets

Remarks on Gowers' dichotomy

Anna Maria Pelczar

Institute of Mathematics, Jagiellonian University, Reymonta 4, 30-059 Kraków, Poland

Abstract

In this paper we present some general method of reasoning, which provides a proof of Gowers' dichotomy, as well as direct proofs of particular cases of the dichotomy for different types of unconditional-like basic sequences. This method generalizes the proof of the particular case of Gowers' dichotomy given by Maurey.

MCS 2000 Primary 46B20; Secondary 46B15, 03E05

1. INTRODUCTION

W.T.Gowers proved in [3] a dichotomy for sets of finite block sequences in Banach spaces, a Ramsey-type theorem which has important applications in the theory of Banach spaces.

The special case of Gowers' dichotomy, given in [3], claims that every Banach space contains either an unconditional basic sequence or a HI space (in which no closed infinitely dimensional subspace is a direct sum of two closed infinitely dimensional subspaces). Let us recall that Gowers and Maurey gave an example of a space containing no unconditional sequences (in fact a HI space), solving the unconditional basic sequence problem [5]. The special case of Gowers' dichotomy, mentioned above, combined with the result of Komorowski and Tomczak-Jaegermann [6] brings a positive solution to the homogeneous Banach problem: a Banach space isomorphic to all its closed infinitely dimensional subspaces is isomorphic to a Hilbert space. Later on Gowers generalized the dichotomy to analytic sets of infinite block sequences and gave further applications to the classification of Banach spaces [4].

In this paper we present dichotomies concerning different types of unconditional-like sequences and geometric properties of convex sets in Banach spaces. We provide also another proof of the abstract Gowers' dichotomy for finite block sequences.

Let E be a Banach space. Denote by $\mathcal{G}(E)$ the family of all infinitely dimensional and closed subspaces of E, by $\mathcal{F}(E)$ - the family of finitely dimensional subspaces of E.

Denote by B_E the closed unit ball, by S_E - the unit sphere of E. Given a set $A \subset E$ by $\mathrm{span}(A)$ (resp. $\overline{\mathrm{span}}(A)$) denote the vector subspace (resp. the closed vector subspace) spanned by A. We will denote by Θ the origin in the space E in order to distinct it from the number zero.

Assume now that E is a Banach space with a basis $\{e_n\}_{n=1}^{\infty}$. A support of a vector $x = \sum_{n=1}^{\infty} x_n e_n$ is the set $\operatorname{supp} x = \{n \in \mathbf{N} : \ x_n \neq 0\}$. We use notation $x < y$ for $x, y \in E$, if every element of $\operatorname{supp} x$ is smaller than every element of $\operatorname{supp} y$, and $n < x$ (resp. $n > x$) for $x \in E$, $n \in \mathbf{N}$, if every element of $\operatorname{supp} x$ is greater (resp. is less) than n. A block sequence with respect to $\{e_n\}$ is any sequence of non-zero finitely supported vectors $x_1 < x_2 < \ldots$, a block subspace - a closed subspace spanned by a block sequence.

We will work on a special class of block subspaces spanned by a dense subset of E. By $\mathcal{Q}$ denote the vector space over $\mathbf{Q}$, the set of rational numbers, if E is a real Banach space, or over $\mathbf{Q} + i\mathbf{Q}$, if E is a complex Banach space, spanned by the basis $\{e_n\}$. Obviously $\mathcal{Q}$ is a countable dense set in E.

By $\mathcal{G}_\bullet(E)$ denote the family of all infinitely dimensional block subspaces spanned by block sequences of vectors from the set $\mathcal{Q}$. By $\mathcal{F}_\bullet(E)$ denote the family of all finitely dimensional subspaces spanned by vectors from $\mathcal{Q}$. Given a subspace $M \in \mathcal{G}_\bullet(E)$ put

$$\mathcal{G}_\bullet(M) = \mathcal{G}_\bullet(E) \cap \mathcal{G}(M), \quad \mathcal{F}_\bullet(M) = \mathcal{F}_\bullet(E) \cap \mathcal{F}(M)$$

2. THE "STABILIZING" LEMMA

Let E be a Banach space. Define a quasi-ordering relation on the family $\mathcal{G}(E)$: for subspaces $L, M \in \mathcal{G}(E)$ write $L \leq M$ iff there exists a finitely dimensional subspace F of E such that $L \subset M + F$. In our consideration we use a simple observation: for any subspaces $L, M \in \mathcal{G}(E)$ satisfying $L \leq M$ we have $L \cap M \in \mathcal{G}(M)$.

We present now the Lemma which will form a useful tool in our argumentation. In the proof we generalize the argumentation given in the proof of some properties of "zawada" (Lemma 1.21) in [15], which uses a standard now diagonalization.

Lemma 2.1 *Let E be a Banach space. Let τ be a mapping defined on the family $\mathcal{G}(E)$ with values in the family 2^{Σ} of subsets of some countable set Σ.*

If the mapping τ is monotone with regard to the relation $\leq$ in $\mathcal{G}(E)$ and the inclusion $\subset$ in 2^{Σ}, ie.

$$\forall N, M \in \mathcal{G}(E): \quad N \leq M \Longrightarrow \tau(N) \subset \tau(M),$$

then there exists a subspace $M \in \mathcal{G}(E)$ which is stabilizing for τ, ie.

$$\forall L \in \mathcal{G}(M): \quad \tau(L) = \tau(M)$$

Proof. Suppose that for any subspace $N \in \mathcal{G}(E)$ there is a subspace $L \leq N$ such that $\tau(L) \subset \tau(N)$, $\tau(L) \neq \tau(N)$. We will construct a transfinite sequence $\{L_\xi\}_{\xi:\,\xi<\omega_1} \subset \mathcal{G}(E)$, where ω_1 is the first uncountable ordinal, such that

$$\xi < \eta \implies L_\eta \leq L_\xi, \ \tau(L_\eta) \subset \tau(L_\xi), \ \tau(L_\eta) \neq \tau(L_\xi)$$

For $\xi = 0$ put $L_0 = E$. Take an ordinal number $\xi < \omega_1$ and assume that we have defined subspaces L_η for $\eta < \xi$. We consider two cases:

1. ξ is of the form $\eta + 1$. Then by our hypothesis there exists a subspace $L_\xi \subset L_\eta$ such that $\tau(L_\xi) \subset \tau(L_\eta)$, $\tau(L_\xi) \neq \tau(L_\eta)$.

2. ξ is a limit ordinal number. Since $\xi < \omega_1$, ξ is a limit of some increasing sequence $\{\xi_n\}$ of ordinal numbers.

By the induction hypothesis we have that the sequence $\{L_{\xi_n}\}$ is decreasing with respect to the relation $\leq$ and the sequence $\{\tau(L_{\xi_n})\}$ is strictly decreasing with respect to the inclusion.

By the monotonicity of the sequence $\{L_{\xi_n}\}$ we have $L_{\xi_1} \cap \ldots \cap L_{\xi_n} \in \mathcal{G}(E)$, $n \in \mathbf{N}$. Choose by induction a basic sequence $\{a_n\}$ such that $a_n \in L_{\xi_1} \cap \ldots \cap L_{\xi_n}$, $n \in \mathbf{N}$, and put $L_\xi = \overline{\text{span}}\{a_n\}_{n\in\mathbf{N}}$. Then obviously $L_\xi \leq L_{\xi_n}$ for $n \in \mathbf{N}$, thus $\tau(L_\xi) \subset \tau(L_{\xi_n}) \subset \tau(L_{\xi_{n-1}})$, therefore also $\tau(L_\xi) \neq \tau(L_{\xi_n})$ for $n \in \mathbf{N}$, which ends the construction.

Hence we have constructed an uncountable family $\{\tau(L_\xi)\}_{\xi<\omega_1}$ of strongly decreasing (with respect to the inclusion) subsets of the set Σ, which contradicts the countability of Σ. □

Remark 2.2 Let E be a Banach space with a basis. Notice that Lemma 2.1 holds also for the family of all block subspaces or the family $\mathcal{G}_\bullet(E)$. Indeed, one can repeat the reasoning from the proof above picking, where appropriate, vectors with finite support or from the set $\mathcal{Q}$ which form a block sequence.

3. GOWERS' DICHOTOMY FOR FINITE SEQUENCES

We describe now Gowers game [3,4]. Let E be a Banach space with a basis $\{g_n\}$. Given a subset $A \subset E$ let $\Sigma(A)$ be the set of all finite block sequences contained in $B_E \cap A$.

Given a set $\sigma \subset \Sigma(E)$ we define Gowers game of two players, S and P, in the following way: in the first step player S picks a block subspace $L_1 \in \mathcal{G}(E)$, then player P picks a vector $x_1 \in B_{L_1}$. In the second step player S picks a block subspace $L_2 \in \mathcal{G}(E)$, then player P a vector $x_2 \in B_{L_2}$. They continue in this way choosing alternately block subspaces (player S) and vectors (player P). We say that player P wins the game, if at some stage the sequence of vectors chosen by P belongs to σ. If it never happens player S wins.

By a strategy of player S we mean a method of picking subspaces, defined for all possible choices of player P, ie. a fixed strategy **S** of player S is an inductively defined function which indicates the subspace to be chosen by player S in the first move, and associates with any sequence of vectors $x_1, \ldots, x_n$ chosen by player P in the first n moves, during which player S applied the strategy **S**, a subspace $L_{n+1} = \mathbf{S}(x_1, \ldots, x_n)$ to be chosen by player S in the $(n+1)-$th move. A strategy of player P is defined analogously; a strategy **P** of player P is a function which, for any $n \in \mathbf{N}$, associates with every sequence of vectors $x_1, \ldots, x_n$ chosen by player P according to **P** in the first n moves and subspace L_{n+1} chosen by S in the $(n+1)-$th move, a vector $x_{n+1} = \mathbf{P}(x_1, \ldots, x_n, L_n)$ to be chosen by P in the $(n+1)-$th move. We say that a strategy of player S (resp. P) is winning, if applying it player S (resp. P) wins every game.

Given a set $\sigma \subset \Sigma(E)$ and a sequence of positive scalars $\Delta = \{\delta_i\}_i$ by σ_Δ denote the "$\Delta-$envelope" of the set σ in the set $\Sigma(E)$:

$$\sigma_\Delta = \{\ \{x_1, \ldots, x_n\} \in \Sigma(E) : \exists\{y_1, \ldots, y_n\} \in \sigma : \|x_i - y_i\| < \delta_i,\ i = 1, \ldots, n\ \}$$

Given a set $\sigma \subset \Sigma(E)$ and vectors $x_1, \ldots, x_n \in E$ put

$$\sigma(x_1, \ldots, x_n) = \{\{y_1, \ldots, y_m\} \in \Sigma(E) : \ \{x_1, \ldots, x_n, y_1, \ldots, y_m\} \in \sigma\}$$

Theorem 3.1 Gowers' dichotomy, [3]
Fix a set $\sigma \subset \Sigma(E)$ and a sequence Δ of positive scalars. Then there exists a block subspace $E_1 \in \mathcal{G}(E)$ such that either $\sigma \cap \Sigma(E_1) = \emptyset$ or player P has a winning strategy for σ_Δ in Gowers game restricted to E_1.

A set $\sigma \subset \Sigma(E)$ satisfying for any sequence Δ the assertion of Gowers' dichotomy is called weakly-Ramsey. An open question concerns the class of weakly-Ramsey sets of infinite block sequences. Gowers [4] proved that analytic sets (in the product norm topology) of infinite block sequences are weakly-Ramsey.

When dealing with this property it could be useful to concern a certain weaker property, ie. the determinacy of the game. A set σ of block sequences is called determining if for any sequence Δ of positive scalars there exists a block subspace of E in which either player S has a winning strategy for the set σ or player P has a winning strategy for the set σ_Δ. This question can be reduced to the one concerning the game, which we will call Mycielski-Steinhaus game in order to distinguish it from Gowers game, which is easier to deal with.

Fix a set X and a set $\tilde{\sigma} \subset X^{\mathbf{N}}$ of infinite sequences of elements of X. By Mycielski-Steinhaus game for X and $\tilde{\sigma}$ we mean an infinite game of two players $\tilde{\mathrm{S}}$ and $\tilde{\mathrm{P}}$, picking alternately elements of the set X. The result of a game is a sequence $\{x_1, x_2, \ldots\}$, whose odd elements were chosen by player $\tilde{\mathrm{P}}$, and even elements - by player $\tilde{\mathrm{S}}$. Each of players in every move knows the set X, $\tilde{\sigma}$ and previously chosen elements. Player $\tilde{\mathrm{P}}$ wins if $\{x_1, x_2, \ldots\} \in \tilde{\sigma}$, player $\tilde{\mathrm{S}}$ wins if $\{x_1, x_2, \ldots\} \notin \tilde{\sigma}$.

Proposition 3.2 ([8], Thm 1.1.1, Thm 1.1.2, Prop. 6.3.1) *For any set σ of block sequences there exists a set $\tilde{\sigma} \subset \mathcal{Q}^{\mathbf{N}}$ such that if player $\tilde{S}$ (resp. $\tilde{P}$) has a winning strategy in Mycielski-Steinhaus game for sets $X = \mathcal{Q}$ and $\tilde{\sigma}$ then player S (resp. P) has a winning strategy in Gowers game for the set σ (resp. σ_Δ).*

If, in addition, the set σ contains only finite sequences then the set $\tilde{\sigma}$ can be chosen to be open in the space $\mathcal{Q}^{\mathbf{N}}$ endowed with the product topology of the discrete topology.

The problem of determinacy of Mycielski-Steinhaus game has been extensively studied. By standard results Mycielski-Steinhaus game is determined for open subsets [2] and borel subsets [9] of the set $X^{\mathbf{N}}$ endowed with the product topology of the discrete topology, where X is a countable set. Mycielski and Steinhaus introduced the following

Axiom of determinacy (AD) [11] *Mycielski-Steinhaus game for any countable set X and any set $\tilde{\sigma} \subset X^{\mathbf{N}}$ is determined, ie. one of players has a winning strategy.*

This axiom contradicts the axiom of choice, but implies the countable axiom of choice which is sufficient for the large number of applications. Gowers, using the axiom of choice, proved that for any Banach space E there exists a set σ of infinite block sequences that is not determining (Theorem 8.1, [4]), hence with the axiom of choice assumed, not every set is weakly-Ramsey. By Proposition 3.2, with AD assumed, every set of block sequences in a Banach space is determining. However, after dealing with determinacy of a given set σ to obtain the weakly-Ramsey property it remains to show that if for a fixed sequence Δ player S has a winning strategy in every block subspace for the set σ_Δ, then the set σ misses completely some block subspace.

Generalizing the method of Maurey's proof of Gowers' dichotomy for unconditional sequences [10] we prove the following

Theorem 3.3 *For any set $\sigma \subset \Sigma(E)$ and any sequence Δ of positive scalars there exists a block subspace $E_1 \in \mathcal{G}(E)$ such that either $\sigma \cap \Sigma(E_1) = \emptyset$ or player S does not have a winning strategy for σ_Δ in Gowers game restricted to E_1.*

Notice that by the result of determinacy of Mycielski-Steinhaus game for open sets and Proposition 3.2 this reasoning provides a version of the proof of Gowers' dichotomy for finite block sequences.

Proof of Theorem 3.3. Fix a set $\sigma \subset \Sigma(E)$ and a sequence $\Delta = \{\delta_i\}_i$ of positive scalars. Put $\Delta_m = \{\delta_i/2^m\}_i$ for $m \in \mathbf{N}$. Given a block subspace $M \in \mathcal{G}(E)$ define the set $\tau(M) \subset \Sigma(E) \times \mathbf{N}$ in the following way: the system $(x_1, \dots, x_n; m) \in \Sigma(E) \times \mathbf{N}$ belongs to $\tau(M)$ iff player S has a winning strategy for $\sigma_{\Delta_m}(x_1, \dots, x_n)$ in Gowers game restricted to M. Obviously, if $k < s$ and $(x_1, \dots, x_n; k) \in \tau(M)$, then $(x_1, \dots, x_n; s) \in \tau(M)$.

In order to apply Lemma 2.1 we consider a countable set. Put $\tau_\bullet(M) = \tau(M) \cap \Sigma(\mathcal{Q})$. Obviously if $N \le L$ then $\tau_\bullet(N) \subset \tau_\bullet(L)$, thus by Lemma 2.1 applied to the set $\Sigma(\mathcal{Q})$ there exists a block subspace $E_0 \in \mathcal{G}(E)$ which is stabilizing for the mapping $\tau_\bullet$.

We restrict now our consideration to the subspace E_0.

Given a system $(x_1, \dots, x_n; m) \in \tau(E_0)$ by $\mathcal{S}_m(x_1, \dots, x_n)$ denote the set of all block subspaces that can be chosen by player S according to some winning strategy in the first move of Gowers game restricted to E_0 for the set $\sigma_{\Delta_m}(x_1, \dots, x_n)$.

Notice that any sequences $\{u_1, \dots, u_n\}, \{v_1, \dots, v_n\} \in \Sigma(E)$ satisfy for any $m \in \mathbf{N}$ the following: if $\{u_1, \dots, u_n\} \notin \sigma_{\Delta_m}$ and $\|u_i - v_i\| < \delta_i/2^{m+1}$ for $i = 1, \dots, n$, then $\{v_1, \dots, v_n\} \notin \sigma_{\Delta_{m+1}}$.

Hence the following Lemma is true:

Lemma 3.4 *For any sequences $\{x_1, \dots, x_n\}, \{y_1, \dots, y_n\} \in \Sigma(E_0)$ and $m \in \mathbf{N}$ satisfying*

$$(x_1, \dots, x_n; m) \in \tau(E_0), \quad \|x_i - y_i\| < \delta_i/2^{m+1} \quad \textit{for} \quad i = 1, \dots, n$$

we have $(y_1, \dots, y_n; m+1) \in \tau(E_0)$ *and* $\mathcal{S}_m(x_1, \dots, x_n) \subset \mathcal{S}_{m+1}(y_1, \dots, y_n)$.

This Lemma means that a subspace "winning" for player S in Gowers game for the envelope of a set given by a sequence $\{x_1, \dots, x_n\}$ is also "winning" in the game for a somewhat smaller envelopes of sets given by sequences which are close to $\{x_1, \dots, x_n\}$.

Lemma 3.4 and the denseness of $\mathcal{Q}$ in E imply some sort of stabilization of the mapping τ on the subspace E_0:

Lemma 3.5 *If $(x_1, \dots, x_n; m) \in \tau(E_0)$ then for any block subspace $L \in \mathcal{G}(E_0)$ holds $(x_1, \dots, x_n; m+2) \in \tau(L)$.*

Notice that if player S has a winning strategy in E_0 then by the above Lemma he has large freedom in choosing "winning" subspaces, ie. given by some winning strategies of player S.

Now we consider the dichotomy: either $(\Theta; 1) \in \tau(E_0)$ (ie. player S has a winning strategy for σ_Δ in E_0) or $(\Theta; 1) \notin \tau(E_0)$. If the second statement holds put $E_1 = E_0$.

Assume the first case occurs. We construct a block subspace E_1 satisfying $\sigma \cap \Sigma(E_1) = \emptyset$, which will finish the proof of the Theorem. In order to achieve it we will choose a block sequence spanning a suitable subspace, applying Lemma 3.5, Lemma 3.4 and the compactness of sets of the form $\Sigma(F)$, where F is a finitely dimensional subspace, in its natural topology. However the idea of the construction is simple, it requires some calculation.

We construct by induction a decreasing (with respect to the inclusion) sequence of block subspaces $\{L_n\}$ and a block sequence $\{e_n\}$ satisfying the following conditions:

1. $\Sigma_n \times \{3n\} \subset \tau(E_0)$, where $\Sigma_n = \Sigma(\text{span}\{e_1, \dots, e_n\})$ for $n \in \mathbf{N}$,

2. $L_1 \in \mathcal{G}(E_0) \cap \mathcal{S}_1(\Theta)$ and $L_{n+1} \in \mathcal{G}(L_n) \cap \mathcal{S}_{3n+3}(x_1, \dots, x_k)$ for $n \in \mathbf{N}$ and for any sequence $\{x_1, \dots, x_k\} \in \Sigma_n$,

3. $e_{n+1} \in L_{n+1}$ and $e_{n+1} > e_n$ for $n \in \mathbf{N}$.

By 2. and 3. the subspace $E_1 = \overline{\text{span}}\{e_n\}_{n=1}^{\infty}$ satisfies $\sigma \cap \Sigma(E_1) = \emptyset$.

The construction. Let L_1 be an arbitrary subspace from the family $\mathcal{G}(E_0) \cap \mathcal{S}_1(\Theta)$ and let e_1 be an arbitrary vector with a finite support from L_1.

Assume now we have chosen subspaces $L_1, \dots, L_n$ and vectors $e_1, \dots, e_n$ satisfying conditions 1., 2. and 3.

The family Σ_n is a closed subset of the compact space

$$\bigcup_{k=1}^{n} (\text{span}\{e_1, \dots, e_n\} \cap B_E)^k$$

endowed with its natural topology (ie. the union of product norm topologies).

Given a sequence $\{x_1, \dots, x_k\} \in \Sigma_n$ put

$$U(x_1, \dots, x_k) = \Big\{\{y_1, \dots, y_k\} \in \Sigma_n : \ \|x_i - y_i\| < \delta_i/2^{3n+3}, \ i = 1, \dots, k\Big\}$$

The family $\{U(x_1, \dots, x_k)\}_{\{x_1,\dots,x_k\}\in\Sigma_n}$ forms an open covering of the space Σ_n, hence by compactness we can choose a finite subcovering

$$\left\{U\left(x_1^j, \dots, x_{k_j}^j\right)\right\}_{j=1}^{j_n}$$

By conditions 1., 2. and 3. $(x_1^j, \dots, x_{k_j}^j; 3n) \in \tau(E_0)$ for $j = 1, \dots, j_n$, hence by Lemma 3.5 there exists a decreasing (with respect to the inclusion) sequence of block subspaces $\{N_j\}_{j=1}^{j_n} \subset L_n$ satisfying $N_j \in \mathcal{S}_{3n+2}(x_1^j, \dots, x_{k_j}^j)$ for $j = 1, \dots, j_n$. By Lemma 3.4 we have for $j = 1, \dots, j_n$

$$\mathcal{S}_{3n+2}(x_1^j, \dots, x_{k_j}^j) \subset \mathcal{S}_{3n+3}(y_1, \dots, y_{k_j}) \quad \text{for} \quad \{y_1, \dots, y_{k_j}\} \in U(x_1^j, \dots, x_{k_j}^j)$$

Therefore the subspace $L_{n+1} = N_{j_n}$ belongs to the family $\mathcal{S}_{3n+3}(y_1, \dots, y_k)$ for any sequence $\{y_1, \dots, y_k\} \in \Sigma_n$ (condition 2.).

Let e_{n+1} be an arbitrary vector with a finite support from the subspace L_{n+1} such that $e_{n+1} > e_n$ (condition 3.). We verify now condition 1. for $(n+1)$. Take a sequence $\{x_1, \dots, x_k\} \in \Sigma_{n+1}$. For $k = 1$ we have $x_1 \in L_1 \in \mathcal{S}_1(\Theta)$, hence $(x_1, 3n+3) \in \tau(E_0)$.

For $k > 1$ we have $\{x_1, \ldots, x_{k-1}\} \in \Sigma_s$ and $x_k \in L_s$ for some number $1 < s \leq n+1$. Since $L_s \in \mathcal{S}_{3s+3}(x_1, \ldots, x_{k-1})$ (by conditions 2. and 3. for $1 \leq s \leq n+1$), therefore $(x_1, \ldots, x_k, 3n+3) \in \tau(E_0)$, which ends the inductive construction and therefore the proof of the Theorem. □

Remark 3.6 Notice that the whole reasoning presented above remains true if σ is a set of infinite block sequences such that in the family of all strategies of player S, regarded as functions on subsets of $\Sigma(E)$ with values in the family $\mathcal{G}(E)$ endowed with the discrete topology, the set of winning strategies for σ is closed in the topology of pointwise convergence. Hence with AD assumed, any set satisfying the condition given above is weakly-Ramsey.

Let us have a closer look on the method of proof of Theorem 3.3. First applying Lemma 2.1 we restrict our attention to a "stabilizing" subspace E_0, ie. for which some specific property, say $(*)$, of finite block sequences, guaranteeing their extension inside previously defined set is hereditary. Afterwards we consider the following dichotomy: either the origin of the space E has in the stabilizing subspace E_0 the property $(*)$ or it does not. If the first case occurs, using the heredity of the property $(*)$ in E_0 we extend by induction the origin to a basic sequence spanning an infinitely dimensional subspace E_1, in which each finite block sequence has property $(*)$, and in particular belongs to the previously defined set.

4. GEOMETRIC ASPECTS OF DICHOTOMIES FOR UNCONDITIONAL SEQUENCES AND HI SPACES

Now we examine dichotomies concerning unconditional sequences. We start with some notation. Given subspaces $L, M \subset E$, $L \cap M = \{\Theta\}$, denote by $P_{L,M}$ the projection $P_{L,M} : L + M \ni x + y \mapsto x \in L$.

A sequence $\{e_n\}$ is called $C-$unconditional for some constant $C > 0$, if for any sequence of scalars $\{a_n\}$ and any sequence $\{\varepsilon_n\}$ of scalars with modulus 1 we have

$$\left\|\sum \varepsilon_n a_n e_n\right\| \leq C \left\|\sum a_n e_n\right\|$$

A Banach space E is called decomposable, if there exist subspaces $L, M \in \mathcal{G}(E)$ such that $L \cap M = \{\Theta\}$ and $L + M = E$. A Banach space is called hereditarily indecomposable, if none of its subspaces is decomposable.

Definition 4.1 *A Banach space E is called a HI(C) space, for $C \geq 1$, if for any subspaces $L, M \in \mathcal{G}(E)$, $L \cap M = \{\Theta\}$, holds $\|P_{L,M}\| \geq C$.*

Gowers' obtained as a corollary the following dichotomy:

Theorem 4.2 [3] *Let E be a Banach space. Fix a scalar $C \geq 1$. Then E contains either a $2C-$unconditional sequence or a HI(C) subspace.*

Maurey [10] and Zbierski independently [20] provided a direct proof of Theorem 4.2. The method of reasoning presented in the previous section is a generalization of his argumentation slightly modified by use of Lemma 2.1. ¿From Theorem 4.2 by using the standard diagonalization one can derive the isomorphic version of the dichotomy:

Theorem 4.3 [3,10] *Every Banach space E contains either an unconditional sequence or a HI subspace.*

Obviously every Banach space is of type HI(1), hence Dichotomy 4.2 for $C = 1$ is trivial. Now we present a dichotomy concerning geometric structure of the unit ball in a Banach space, which covers this extremal case. First we state the Lemma showing the relation between absolutely convex bodies and bounded projections in Banach spaces.

Definition 4.4 *Let C be an absolutely convex subset of a Banach space E. Then C is called almost bounded if for some finitely codimensional subspace M of E the set $C \cap M$ is bounded. The set C is called essentially unbounded if it is not almost bounded.*

Lemma 4.5 *If for a subspace $L \in \mathcal{G}(E)$ there exists an essentially unbounded absolutely convex body C satisfying $C \cap L \subset cB_L$ for some scalar $c \geq 1$, then for any $\varepsilon > 0$ there is a subspace $M \in \mathcal{G}(E)$ such that $M \cap L = \{\Theta\}$ and $\|P_{L,M}\| \leq c + \varepsilon$.*

This Lemma follows from Corollary 2.3, [12], of Kato Theorem (Prop. 2.c.4, [7]). We will consider absolutely convex bodies in Banach spaces of a specific form presented below.

Definition 4.6 *For a subspace L of a Banach space E put*

$$W_L = \bigcap_{f \in J(S_L)} f^{-1}([-1,1]),$$

where J denotes the duality mapping $J: S_E \ni x \mapsto J(x) = \{f \in S_{E^*}:\ f(x) = 1\} \in 2^{S_{E^*}}$.

Notice that for any subspace L we have $W_L = (J(S_L))^\circ$, consequently the set W_L is bounded iff the set $J(S_L)$ is norming for E.

Theorem 4.7 *Every Banach space contains either a subspace E_1 such that for any infinitely dimensional subspace L of E_1 the set $W_L \cap E_1$ is almost bounded or an unconditional basic sequence $\{e_n\}_{n \in \mathbf{N}}$ satisfying the following:*

$$(*) \qquad \|P_i : \overline{\operatorname{span}}\{e_n\}_{n\in\mathbf{N}} \to \overline{\operatorname{span}}\{e_n\}_{n\neq i}\| \leq 1 + \varepsilon_i, \quad i \in \mathbf{N}$$

for some (any) sequence $\{\varepsilon_n\}_{n\in\mathbf{N}}$ of positive scalars.

In order to prove this dichotomy we recall a theorem guaranteeing the existence of an unconditional basis with property $(*)$ and show a technical Lemma, using the technique described in the previous section.

Theorem 4.8 [15] *If a Banach space E satisfies the condition*

$(\circ)$ for any subspaces $M \in \mathcal{G}(E)$, $F \in \mathcal{F}(M)$, $H \in \mathcal{G}(M)$, $F + H = M$ and any $\varepsilon > 0$ there exist a subspace $L \in \mathcal{G}(M)$, $L \supset F$, and a vector $a \in H$ such that $\|P_{L,\mathbf{R}a}\| \leq 1 + \varepsilon$,

then there exists an unconditional basic sequence $\{e_n\}$ satisfying $(*)$.

Lemma 4.9 *If any subspace $M \in \mathcal{G}(E)$ contains a further subspace $L \in \mathcal{G}(E)$ such that $\inf\{\|P_{L,N}\| :\ N \in \mathcal{G}(M)\} = 1$, then there exists a subspace $E_1 \in \mathcal{G}(E)$ satisfying*

$(\circ\circ)$ for any subspaces $M \in \mathcal{G}(E_1)$, $F \in \mathcal{F}(M)$ and any $\varepsilon > 0$ there exist subspaces $L, N \in \mathcal{G}(M)$, $L \supset F$, such that $\|P_{L,N}\| \leq 1 + \varepsilon$.

Proof of Theorem 4.7. Notice that for any subspaces $M \in \mathcal{G}(E)$, $L \in \mathcal{G}(M)$, if the set $W_L \cap M$ is essentially unbounded, then, by Lemma 4.5, $\inf\{\|P_{L,N}\| : N \in \mathcal{G}(M)\} = 1$.

Assume that no subspace of E satisfies the first condition of the dichotomy in Theorem 4.7. By the previous remark the space E satisfies the assumptions of Lemma 4.9. Obviously the condition ($\circ\circ$) implies the condition ($\circ$), which by Theorem 4.8 ends the proof of Theorem 4.7. □

Notice that the above reasoning considering projections of norms close to 1 exhibits an extremal case of the isometric dichotomy given in Theorem 4.2, namely statement that every Banach space contains either an unconditional sequence satisfying ($*$) or a space whose no subspace admits projections on it of norm arbitrarily close to 1.

Proof of Lemma 4.9. First we introduce some notation. Given a subspace $G \subset E$ and a scalar $\delta > 0$ put

$$\mathcal{Z}(G,\delta) = \{H \in \mathcal{G}(E) \cup \mathcal{F}(E) : \ S_G \subset S_H + \delta B_E\}$$

Given subspaces $M \in \mathcal{G}(E)$, $L \in \mathcal{G}(M)$ put

$$p(L,M) = \inf\{\|P_{L,N}\| : \ N \in \mathcal{G}(M)\}$$

Assume that E is a Banach space with a basis and satisfies the assumptions of the Lemma. Given a subspace $M \in \mathcal{G}(E)$ put

$$\tau(M) = \{(F,\delta) \in (\mathcal{F}_\bullet(E) \times \mathbf{Q}) : \exists L \leq M, \ L \in \mathcal{Z}(F,\delta), \ p(L,M) = 1\}$$

If $N \leq M$ then $\tau(N) \subset \tau(M)$, hence by Lemma 2.1 and Remark 2.2 there exists a subspace E_0 that is stabilizing for the mapping τ. Let L be a subspace of E_0 satisfying $p(L, E_0) = 1$. We will show that L satisfies ($\circ\circ$).

CLAIM. For any subspace $F \in \mathcal{F}_\bullet(L)$ and a scalar $\rho > 0$ there exists a scalar $\delta = \delta(F,\rho)$ such that if $L_1 \in \mathcal{Z}(F,\delta)$, $L_1 \leq L$ then $L_1 \in \mathcal{Z}(L_2,\rho)$ for some subspace $L_2 \in \mathcal{G}(L)$ containing F.

Proof of the Claim. Let $H \in \mathcal{G}(L)$ be a complement of F in L (ie. $H \cap F = \{\Theta\}$, $H + F = L$). For a scalar $\delta > 0$ take a subspace $L_1 \leq L$ satisfying $L_1 \in \mathcal{Z}(F,\delta)$ and put $L_2 = (L_1 \cap H) + F \in \mathcal{G}(L)$. Take a vector of the form $x + y \in S_{L_2}$, where $x \in F$, $y \in L_1 \cap H$. By the choice of L_1 there exists a vector $z \in L_1$ such that $\|x - z\| \leq 2\delta\|x\|$. Therefore $\|(x+y) - (x+z)\| \leq 2\delta\|P_{F,H}\|$. Hence for sufficiently small δ (but dependent only on ρ, F and the choice of H) $L_1 \in \mathcal{Z}(L_2,\rho)$. □

Fix a subspace $F \in \mathcal{F}(L)$ and $0 < \varepsilon < 1$.

Let $\delta > 0$ be the scalar associated on the basis of the Claim with F and $\rho = \varepsilon/6$. Pick a subspace $G \in \mathcal{F}_\bullet(E)$ satisfying $G \in \mathcal{Z}(F,\delta/2)$ and $L \in \mathcal{Z}(G,\delta/2)$. Then we have $(G,\delta) \in \tau(E_0) = \tau(L)$. Hence there exists a subspace $L_1 \leq L$ satisfying $L_1 \in \mathcal{Z}(G,\delta/2)$ and $p(L_1,L) = 1$. By the choice of L_1 there exists a subspace $N \in \mathcal{G}(L)$ such that $\|P_{L_1,N}\| \leq 1 + \varepsilon/2$. Since $L_1 \in \mathcal{Z}(F,\delta)$, by the choice of δ there exists a subspace $L_2 \in \mathcal{G}(L)$ containing F and satisfying $L_1 \in \mathcal{Z}(L_2,\rho)$ for $\rho = \varepsilon/6$. We will show that $\|P_{L_2,N}\| \leq 1 + \varepsilon$. Indeed, by the definition of the norm of projection we have

$$\|P_{L_2,N}\| \leq (1+\rho)\|P_{L_1,N}\| + \rho$$

Hence $$\|P_{L_2,N}\| \leq 1 + \varepsilon/2 + 3\varepsilon/6 = 1 + \varepsilon \qquad \square$$

Let us recall that for any $C > 1$ the space l_2 can be renormed so as to contain no $C-$unconditional basic sequence [1], which is closely related to the distortion problem.

If E is a HI space, then by Theorem 4.7 for any renorming there is a subspace of E satisfying the first condition stated in Theorem 4.7. However, by a direct application of Lemma 4.5 we obtain the following characterization of hereditarily indecomposable spaces:

Proposition 4.10 [13] *For a Banach space E the following conditions are equivalent:*

1. *the space E is hereditarily indecomposable,*
2. *for any equivalent norm on E and any infinitely dimensional subspace L of E the set W_L is almost bounded.*

This characterization is a particular case of the following general theorem, which can be proved similarly:

Theorem 4.11 [12,13,16] *For a Banach space E the following conditions are equivalent:*

1. *the space E is hereditarily indecomposable,*
2. *the intersection of any two unbounded absolutely convex bodies in E, containing no line, is unbounded.*
3. *the intersection of any unbounded absolutely convex body in E, containing no line, and any infinitely dimensional subspace of E is unbounded.*

5. CONES AND BASIC SEQUENCES

We will examine now what will happen if we consider cones instead of vector subspaces. We call a set $K \subset E$ a cone if it is closed, convex and satisfies $\mathbf{R}^+x \subset K$ for any vector $x \in K$. Given a set $A \subset E$ by $\overline{\text{cone}}(A)$ denote the cone spanned by A, ie. the smallest cone containing A. In particular for a basic sequence $\{e_n\}$ we have

$$\overline{\text{cone}}\{e_n\} = \overline{\left\{\sum_{n=1}^{N} a_n e_n,\ a_n \geq 0,\ N \in \mathbf{N}\right\}}$$

The notion of the basis of a cone is analogous to the case of vector subspaces; we assume always that a basis of a cone of a Banach space is a basic sequence in this space. The sphere of a cone K is the set $S_K = S_E \cap K$. The distance between two cones $K, H \subset E$ is given by

$$\rho(K, H) = \inf\{\|x - y\| : \ x \in S_K,\ y \in S_H\}$$

Let E be a Banach space with a basis. Given a cone $K \subset E$ by $\mathcal{C}_b(K)$ denote the family of all block cones (ie. spanned by block bases) in K.

We recall now some notation and definition introduced in [17,18]. Given a basic sequence $\{e_n\}$ and a vector $x = \sum_{n=1}^{N} a_n e_n$ by $|x|$ denote the vector $\sum_{n=1}^{N} |a_n| e_n$.

Definition 5.1 *We say that a basis* $\{e_n\}$ *of the space* E *is of type*

1. *UL, if there exists a constant* $C \geq 1$ *such that* $|||x||| \leq C||x||$ *for any vector* $x \in E$ *with a finite support,*

2. *UL*, if there exists a constant* $C \geq 1$ *such that* $||x|| \leq C|||x|||$ *for any vector* $x \in E$ *with a finite support.*

Obviously a basic sequence is unconditional iff it is both UL and UL*. Let us recall some examples of UL bases: unit vectors in James space J and vectors in l_1 of the form $e_1, e_2 - e_1, e_3 - e_2, \ldots$, where $\{e_n\}$ are the standard unit vectors in l_1. On the other hand, Schauder basis in the space of continuous functions C[0,1], the summing basis in the space c of convergent sequences, unit vectors in dual J^* to James space form UL* bases. The universal Schauder basis of Pełczyński is neither of type UL nor UL*.

We will focus now on the property UL*. Standard argument (as in the case of unconditional sequences) shows that a basic sequence $\{e_n\}$ is UL* iff there exists a constant $C \geq 1$ such that

$$\left\|\sum_{n\in A} a_n e_n\right\| \leq C \left\|\sum_{n\in B} a_n e_n\right\|$$

for any sets $A \subset B \subset \mathbf{N}$ and any sequence of positive scalars $\{a_n\}$.

In the case of unconditional sequences, ie. without assumption that scalars are positive, the condition above means that projections on suitable subspaces are uniformly bounded, ie. spheres of suitable subspaces are uniformly separated. In the case of cones the condition means that spheres of suitable cones are uniformly separated.

Lemma 5.2 *A basic sequence* $\{e_n\}$ *is of type UL* iff there exists a constant* $c > 0$ *such that for any disjoint finite subsets* $I, J \subset \mathbf{N}$ *we have* $\rho(K_I, -K_J) \geq c$*, where* K_I *denotes* $\overline{\text{cone}}\{e_n,\ n \in I\}$.

Proof. Assume that $\{e_n\}$ is a UL* sequence. Notice that for $I, J \subset \mathbf{N}$ we have

$$\rho(K_I, -K_J) = \inf\left\{\left\|\sum_{n\in I} a_n e_n + \sum_{n\in J} a_n e_n\right\| : \left\|\sum_{n\in I} a_n e_n\right\| = 1, \left\|\sum_{n\in J} a_n e_n\right\| = 1, a_n \geq 0\right\}$$

and thus $\rho(K_I, -K_J)$ is bounded from below by some constant $C \geq 1$ by the property UL*.

Assume now that spheres of suitable cones are uniformly separated. Fix finite disjoint sets $I, J \subset \mathbf{N}$. Let $x = \sum_{n\in I} a_n e_n$, $y = \sum_{n\in J} a_n e_n$. Let $||x|| \leq ||y||$. Simple calculus shows that

$$\left\|\frac{x}{||x||} + \frac{y}{||x||}\right\| \geq \frac{1}{2}\left\|\frac{x}{||x||} + \frac{y}{||y||}\right\|$$

whereas the right-hand side by the assumption is bounded from below by some constant $c > 0$. Hence $\{e_n\}$ is an UL* sequence. □

Notice that by Lemma 5.2 a Banach space E containing a cone K with a basic sequence satisfying $\rho(K, -K) > 0$ contains also a UL* basic sequence. Actually, a stronger statement is true. In order to prove it one can repeat the reasoning from the proof of Theorem 4.2 considering block cones instead of vector subspaces and distances $\rho(K,-H)$ for $K,H \in \mathcal{C}_b(E)$ instead of norms $\|P_{L,M}\|$ for $L,M \in \mathcal{G}(E)$ ([10,14]). We get the following

Theorem 5.3 *Any cone K with a basis of a Banach space E contains either an UL* sequence or a block subcone K_1 such that for any subcones $H_1, H_2 \in \mathcal{C}_b(K_1)$ we have $\rho(H_1,-H_2)=0$.*

The HI space constructed in [5] satisfies in fact also the second condition from Dichotomy 5.3, any two block cones in this space are arbitrarily close. Hence having an UL* basis is not a hereditary property, since the Schauder system is an UL* basis of the space of all continuous functions on the interval [0,1] ([17]).

Now we will consider the property UL.

Lemma 5.4 *Let $\{e_n\}$ be a basic sequence in E. If there exists a constant $c>0$ such that for any disjoint finite sets $I,J\subset \mathbf{N}$ we have $\rho(K_I,K_J)\geq c$, where $K_I=\overline{\text{cone}}\{e_n,\ n\in I\}$, then $\{e_n\}$ is a UL sequence.*

Proof. Take a vector $x=\sum_{n=1}^N a_n e_n$. For a set $I\subset\{1,\cdot,N\}$ put $x_I=\sum_{n\in I}a_n e_n$. Put $I=\{n\in\mathbf{N}:\ a_n>0\}$, $J=\{n\in\mathbf{N}:\ a_n<0\}$. As before we have

$$2\inf\left\{\left\|\frac{x_J}{\|x_J\|}+\frac{x_I}{\|x_J\|}\right\|,\left\|\frac{x_J}{\|x_I\|}+\frac{x_I}{\|x_I\|}\right\|\right\}\geq\left\|\frac{x_J}{\|x_J\|}+\frac{x_I}{\|x_I\|}\right\|,$$

thus by the assumption we get $4\|x_I+x_J\|\geq c(\|x_I\|+\|x_J\|)\geq c\|x_I-x_J\|$. Hence $\{e_n\}$ is a UL sequence. □

As before, one can repeat the reasoning from the proof of Theorem 4.2 ([10,14]), considering block cones instead of subspaces and distances $\rho(K,H)$ for $K,H\in\mathcal{C}_b(E)$ instead of norms $\|P_{L,M}\|$ for $L,M\in\mathcal{G}(E)$, and obtain the following

Theorem 5.5 *Every cone K with a basis in a Banach space E contains either an UL sequence or a block subcone K_1 such that for any two subcones $H_1,H_2\in\mathcal{C}_b(K_1)$ we have $\rho(H_1,H_2)=0$.*

This result reveals another geometric property of a HI space. Recall that any subspace of a space with a UL basis contains an unconditional sequence ([18]). Hence we get the following

Corollary 5.6 *Every cone with a basis in a HI space contains a block subcone K such that for any two subcones $H_1,H_2\in\mathcal{C}_b(K)$ we have $\rho(H_1,H_2)=0$.*

Using the "stabilizing" Lemma one can also provide a direct proof of the dichotomy for asymptotic unconditional sequences, given in [19].

The presented paper is a part of the author's PhD thesis written under the supervision of Professor Edward Tutaj.

REFERENCES

1. P.G.Casazza, N.J.Kalton, D.Kutzarova, M.Mastyło, Complex interpolation and complementably minimal spaces, Lecture Notes in Pure and Appl. Math. vol. 175 (1996), 135-143.
2. D.Gale, F.Stewart, Infinite games with perfect information, *Ann. Math. Studies* **28** (1953), 245-266.
3. W.T.Gowers, A new dichotomy for Banach spaces, *Geom. Funct. Anal.* **6** (1996), 1083-1093.
4. W.T.Gowers, Infinite Ramsey theorem and some Banach-space dichotomies, submitted.
5. W.T.Gowers, B.Maurey, The unconditional basic sequence problem, *Journal of AMS* **6** (1993), 851-874.
6. R.Komorowski, N.Tomczak-Jaegermann. Banach spaces without local unconditional structure. *Israel J. Math.* **89** (1995), no. 1-3, 205-226.
7. J.Lindenstrauss, L.Tzafriri, Classical Banach spaces, vol. I (Springer Verlag, 1977).
8. J. Lopez Abad, Weakly-Ramsey sets in Banach spaces, PhD Thesis. Universitat de Barcelona, 2000.
9. D.Martin, Borel determinacy, *Ann. of Math.* **102** (1975), 363-371.
10. B.Maurey, A note on Gowers' dichotomy theorem, *Conv. Geom. Anal.* **34** (1998), 149-157.
11. J.Mycielski, H.Steinhaus, A mathematical axiom contradicting the axiom of choice, *Bull. Acad. Polon. Sci., Série Math., Astr. et Phys.* **10** (1962), 1-3.
12. A.M.Pelczar, On a certain property of hereditarily indecomposable Banach spaces, to be published in Uniw. Iagel. Acta. Math.
13. A.M.Pelczar, O Dychotomii Gowersa (On Gowers' Dichotomy), PhD Thesis. Uniwersytet Jagielloński (Kraków, 2000).
14. A.M.Pelczar, Remarks on Gowers' dichotomy, preprint IMUJ 2000/01.
15. E.Tutaj, O pewnych warunkach wystarczających do istnienia w przestrzeni Banacha ciągu bazowego bezwarunkowego (On some conditions sufficient for the existence of an unconditional basic sequence in a Banach space). PhD Thesis, Uniwersytet Jagielloński (Kraków, 1974).
16. E.Tutaj, On some restatement of the problem of existence of a closed direct sum in Banach spaces I,II,III, *Bull. Pol. Ac. Sci.: Math* **34** (1986), 41-45, 441-449, 450-456.
17. E.Tutaj, Some observations concerning the classes of unconditional-like basic sequences, *Bull. Pol. Ac. Sci.: Math.* **35** (1987), 35-42.
18. E.Tutaj, A remark on unconditional basic sequences, *Bull. Pol. Ac. Sci.: Math.* **37** (1988), 1-3.
19. R.Wagner, Gowers' dichotomy for asymptotic structure, *Proc. AMS* (10) **124** (1996), 3089-3095.
20. P.Zbierski, lecture, Uniwersytet Warszawski, May 1999.

Norm attaining operators and James' Theorem

M. D. Acosta [a] *, J. Becerra Guerrero [b] † and M. Ruiz Galán [c] *

[a]Departamento de Análisis Matemático, Universidad de Granada,
18071 Granada (SPAIN)

[b]Departamento de Matemática Aplicada, Universidad de Granada,
18071 Granada (SPAIN)

[c]Departamento de Matemática Aplicada, Universidad de Granada,
18071 Granada (SPAIN)

To Professor Manuel Valdivia on the occasion of his seventieth birthday.

Abstract

There are several results relating isomorphic properties of a Banach space and the set of norm attaining functionals. Here, we show versions for operators of some of these results. For instance, a Banach space X has to be reflexive if it does not contain ℓ_1 and for some non trivial Banach space Y and positive r, the unit ball of the space of operators from X into Y is the closure (weak operator topology) of the convex hull of the norm one operators satisfying that balls centered at any of them with radius r are contained in the set of norm attaining operators. We also prove a similar result by using a very weak isometric condition on the space instead of non containing ℓ_1.

MCS 2000 Primary 46B10, 46B28; Secondary 47A30

Given a Banach space X, B_X and S_X will be the closed unit ball and the unit sphere, respectively. We will write X^* for the topological dual of X and $NA(X)$ will be the subset of norm attaining functionals, that is,

$$NA(X) = \{x^* \in X^* : \exists x \in X, \|x\| = 1, x^*(x) = \|x^*\|\}.$$

Bourgain and Stegall showed that a separable Banach space whose unit ball is not dentable satisfies that $NA(X)$ is first Baire category in the dual space X^* [4, Theorem 3.5.5 and Problem 3.5.6]. Up to now, it remains unknown whether or not the previous result holds also in the non separable case. However for spaces of type $C(K)$ (K Hausdorff and

*Research partially supported by DGES, project no. BFM 2000-1467.
†Research partially supported by Junta de Andalucía, Grant FQM0199

compact), Kenderov, Moors and Sciffer, proved that $NA(C(K))$ is first Baire category if K is infinite [16]. Jiménez Sevilla and Moreno [15] showed that a Banach space X satisfying the Mazur intersection property is reflexive provided that $NA(X)$ has non empty interior. Acosta and Ruiz Galán [2,3] proved the parallel result for spaces satisfying some smoothness condition (Hahn-Banach smooth or very smooth) instead of the Mazur intersection property.

Let us note that if $X = Y^*$ is a dual space, then $Y \subseteq Y^{**} = X^*$ is a closed subspace w^*-dense in X^*, which is contained in $NA(X)$. Petunin and Plichko proved that a separable Banach space X is isometric to a dual space as soon as there is a norm closed and w^*-dense subspace of X^* contained in the set $NA(X)$ [18]. Debs, Godefroy and Saint-Raymond [5] showed that a separable Banach space X whose dual unit ball satisfies that $NA(X) \cap B_{X^*}$ contains a w^*-open set of the unit ball, has to be reflexive. This result was extended to the general case by Jiménez Sevilla and Moreno [15, Proposition 3.2].

James' Theorem [14] states that a Banach space X satisfying that $NA(X)$ contains a ball centered at zero, is reflexive. However, one cannot expect that a Banach space is reflexive as soon as the set of norm attaining functionals has non empty interior. For instance, this assumption is satisfied by ℓ_1. In this case, it is easy to check that under the usual identification $\ell_1^* = \ell_\infty$, for any finite subset F of $\mathbb{N}$, the open set

$$P(F) := \{z \in \ell_\infty : \max_{n \in F} |z(n)| > \sup_{k \notin F} |z(k)|\}$$

is contained in $NA(\ell_1)$. Then, any $z \in S_{\ell_\infty}$ with finite support satisfies

$$z + \frac{1}{2} B_{\ell_\infty} \subseteq P(\text{supp } z).$$

In fact, this behaviour is quite general, as the following result shows:

Proposition 1([2, Lemma 1]). *Every Banach space can be equivalently renormed so that the set of norm attaining functionals has non empty interior (norm topology).*

Inspired by the space ℓ_1 we looked for an isomorphic condition weaker that reflexivity, so that this condition and some extra assumption on the set $NA(X)$ implies reflexivity.

Coming back to the concrete example ℓ_1, since the extreme points of the dual unit ball in this case is the subset of sequences of scalars satisfying $|z(n)| = 1$ for every n, then, these points can be approximated (w^*-topology) by elements which are in the interior of $NA(\ell_1)$ by using the sequence $\{P_n(z)\}$, where P_n is given by

$$P_n(z)(k) = z(k) \quad (k \leq n), \; P_n(z)(k) = 0 \quad (k > n).$$

It is clear that $P_n(z) + \frac{1}{2} B_{\ell_\infty} \subseteq NA(\ell_1)$ $(n \in \mathbb{N})$. Since the convex hull of the extreme points of B_{ℓ_∞} is w^*-dense in the ball (Krein-Milman Theorem), then it is satisfied

$$B_{\ell_\infty} = \overline{\text{co}}^{w^*} \{z \in S_{\ell_\infty} : z + \frac{1}{2} B_{\ell_\infty} \subseteq NA(\ell_1)\}$$

and ℓ_1 is obviously not reflexive. By assuming the previous condition on the dual unit ball it was proved that ℓ_1 is "essentially" the only non reflexive example of such a space.

In fact, if a Banach space does not contain an isomorphic copy of ℓ_1 and for some $r > 0$ it holds

$$B_{X^*} = \overline{co}^{w^*}\{x^* \in S_{X^*} : (x^* + rB_{X^*}) \subset NA(X)\},$$

then X is reflexive ($\overline{co}^{w^*}$ is the w^*-closure of the convex hull) [1, Theorem 2].

Now we will extend the previous result to operators. For any Banach spaces X and Y, we will denote by $L(X,Y)$ the set of all *bounded and linear operators* from X into Y and $NA(X,Y)$ will be the subset of *norm attaining operators* and for a positive number r we will write

$$NA_r(X,Y) = \{T \in L(X,Y) : T + rB_{L(X,Y)} \subset NA(X,Y)\}.$$

Also w_{op} will denote the *weak operator topology* of $L(X,Y)$. We will assume that all the spaces considered are **real**.

Theorem 2. *Let X be a Banach space not containing an isomorphic copy of ℓ_1 and assume that for some $r > 0$ and some Banach space Y it holds*

$$B_{L(X,Y)} = \overline{co}^{w_{op}}\{S_{L(X,Y)} \cap NA_r(X,Y)\}.$$

Then X is reflexive.

Proof. We will argue by contradiction. Hence, assume that X is not reflexive. Since X does not contain ℓ_1, X does not have the Grothendieck property (see [19, Theorem 1] or [12, Proposition 1]) and so, by the proof of [3, Lemma 2] there is $0 \neq \Phi \in X^{***}$ so that $\Phi(X) = 0$ and satisfies for any $T \in S_{L(X,Y)} \cap NA_r(X,Y)$

$$\|T^{**}(x^{**})\| + r\Phi(x^{**}) \leq \|x^{**}\|, \qquad \forall x^{**} \in X^{**}. \tag{1}$$

Now, let us fix $x_0^{**} \in S_{X^{**}}$ and $\varepsilon > 0$. By using again that X does not contain ℓ_1 and [10, Theorem 1] there is $x_0 \in S_X, \alpha > 0$ so that

$$\mathrm{Osc}\ \mathrm{x}_0^{**}(\mathrm{S}(\mathrm{B}_{\mathrm{X}^*}, \mathrm{x}_0, \alpha)) < \varepsilon, \tag{2}$$

where $S(B_{X^*}, x_0, \alpha) = \{x^* \in B_{X^*} : x^*(x_0) > 1 - \alpha\}$ and Osc denote the oscillation (i.e. $\sup - \inf$).

We fix $y_0 \in S_Y, y_0^* \in S_{Y^*}$ and $x_0^* \in S_{X^*}$ so that

$$1 = x_0^*(x_0) = y_0^*(y_0).$$

Since the operator $T = x_0^* \otimes y_0$ satisfies $\|T\| = 1$, by using the assumption, there is a sequence of elements $\{T_n\} \in S_{L(X,Y)}$ so that

$$y_0^* T_n x_0 \to y_0^* T x_0 = 1$$

and such that each operator T_n can be written as

$$T_n = \sum_{i=1}^{n} t_i S_i \quad (t_i \geq 0, \sum t_i = 1), \quad \|S_i\| = 1, \quad S_i + rB_{L(X,Y)} \subset NA(X,Y).$$

By chosing an appropriate element in the convex combination, we can assume that in fact

$$T_n + rB_{L(X,Y)} \subset NA(X,Y) \quad \text{and} \quad y_0^* T_n x_0 \to y_0^* T x_0 = 1. \tag{3}$$

Let us choose a w^*-cluster point $x^{***} \in X^{***}$ of the sequence $\{T_n^* y_0^*\}$. By (3) we can apply inequality (1) to each operator T_n and the element $x_0 + tx_0^{**}$ $(t > 0)$, so

$$\|T_n^{**}(x_0 + tx_0^{**})\| + rt\Phi(x_0^{**}) \leq \|x_0 + tx_0^{**}\|.$$

As a consequence,

$$T_n^{**}(x_0 + tx_0^{**})(y_0^*) + rt\Phi(x_0^{**}) \leq \|x_0 + tx_0^{**}\|.$$

Since x^{***} is a w^*-cluster point of $\{T_n^* y_0^*\}$, by varying n and using (3) we get

$$1 + tx^{***}(x_0^{**}) + rt\Phi(x_0^{**}) \leq \|x_0 + tx_0^{**}\|,$$

that is,

$$x^{***}(x_0^{**}) + r\Phi(x_0^{**}) \leq \frac{\|x_0 + tx_0^{**}\| - 1}{t}, \quad \forall t > 0. \tag{4}$$

By using [8, Theorem V.9.5], $\lim_{t \to 0^+} \frac{\|x_0 + tx_0^{**}\| - 1}{t} = \max V(x_0, x_0^{**})$ where

$$V(x_0, x_0^{**}) = \{y^{***}(x_0^{**}) : y^{***} \in S_{X^{***}}, y^{***}(x_0) = 1\}.$$

Hence, it follows from inequality (4) that

$$x^{***}(x_0^{**}) + r\Phi(x_0^{**}) \leq \max V(x_0, x_0^{**}) \tag{5}$$

By Goldstine's Theorem the slice $S(B_{X^*}, x_0, \alpha)$ is w^*-dense in $S(B_{X^{***}}, x_0, \alpha)$, so it holds

$$\text{Osc } x_0^{**}(S(B_{X^{***}}, x_0, \alpha)) = \text{Osc } x_0^{**}(S(B_{X^*}, x_0, \alpha)).$$

Since $x^{***}(x_0) = 1$, by using the estimation (2) in inequality (5) and the previous observation we get

$$x^{***}(x_0^{**}) + r\Phi(x_0^{**}) \leq x^{***}(x_0^{**}) + \varepsilon.$$

The inequality $r\Phi(x_0^{**}) \leq \varepsilon$, valid for any $\varepsilon > 0$ and $x_0^{**} \in S_{X^{**}}$, gives $\Phi = 0$, a contradiction.

□

Now, we will check that the assumptions posed in Theorem 2 are sharp.

Remark 3. *For any Banach space Y, $B_{L(\ell_1,Y)}$ is the closure in the strong operator topology of the convex hull of the set*

$$S_{L(\ell_1,Y)} \cap NA_{\frac{1}{2}}(\ell_1, Y).$$

Proof. We will use without comment the fact that $\|T\| = \sup\{\|Te_n\| : n \in \mathbb{N}\}$.

The assertion is trivial if $Y = \{0\}$. Therefore, let us fix an operator $0 \neq T \in B_{L(\ell_1,Y)}$ and n large enough so that $T(e_{i_0}) \neq 0$ for some $i_0 \leq n$ ($\{e_n\}$ is the canonical basis of ℓ_1). We define the operators $T_1, T_2 \in L(\ell_1, Y)$ given by

$$T_1(x) = \sum_{k=1, k\neq i_0}^{n} x(k)T(e_k) + x(i_0)\frac{T(e_{i_0})}{\|T(e_{i_0})\|},$$

$$T_2(x) = \sum_{k=1, k\neq i_0}^{n} x(k)T(e_k) - x(i_0)\frac{T(e_{i_0})}{\|T(e_{i_0})\|} \qquad (x \in \ell_1).$$

Since it is clearly satisfied

$$T_1(e_i) = T_2(e_i) = T(e_i) \quad (i \leq n, i \neq i_0), \qquad T_1(e_i) = T_2(e_i) = 0 \quad (i > n)$$

and

$$T(e_{i_0}) = tT_1(e_{i_0}) + (1-t)T_2(e_{i_0}) \qquad \left(t = \frac{1 + \|T(e_{i_0})\|}{2}\right),$$

then

$$TP_n = tT_1 + (1-t)T_2,$$

where P_n is the natural projection on the subspace $[e_i : i \leq n]$. Since $\{e_n\}$ is a basis of ℓ_1, by varying n, T can be approximated in the strong operator topology by a convex combination of operators as above. It is clear that the operators T_1, T_2 are in the unit sphere of $L(\ell_1, Y)$. We will check that the operator T_1 satisfies

$$T_1 + \frac{1}{2}B_{L(\ell_1,Y)} \subset NA(\ell_1, Y).$$

Since $\|T_1(e_{i_0})\| = 1$, for any $S \in T_1 + \frac{1}{2}B_{L(\ell_1,Y)}$, then

$$\|S(e_j)\| \leq \|T_1(e_j)\| + \|S - T_1\| \leq \|T_1(e_j)\| + \frac{1}{2} = \frac{1}{2} \qquad (j > n),$$

meanwhile $\|S\| \geq \|T_1(e_{i_0})\| - \|(S - T_1)(e_{i_0})\| \geq \frac{1}{2}$. But $\|S\| = \max\{\|SP_n\|, \|S(I - P_n)\|\}$, so in this case $\|S\| = \|SP_n\|$ and S attains its norm in the subspace $[e_i : i \leq n]$. Of course T_2 satisfies similar conditions and so $T_2 + \frac{1}{2}B_{L(\ell_1,Y)} \subset NA(\ell_1, Y)$. □

Also the second assumption posed in Theorem 2 is sharp in the following sense:

Proposition 4. *For any Banach space Z, there is a Banach space X isomorphic to Z so that the unit ball of $L(X, Y)$ is the norm closure of the set*

$$\mathrm{co}\left(S_{L(X,Y)} \cap \mathrm{int}\, NA(X, Y)\right)$$

for any Banach space $Y \neq \{0\}$.

Proof. Of course, we can assume that Z satisfies dim $Z \geq 2$. Therefore, let M be a closed linear subspace of Z and $0 \neq z_0 \in Z$ so that $Z = \mathbb{R}z_0 \oplus M$ and consider

$$X = \mathbb{R}z_0 \oplus_1 M,$$

that is, the norm $|\ |$ of X is given by

$$|\lambda z_0 + m| = |\lambda| + \|m\| \qquad (\lambda \in \mathbb{R}, m \in M).$$

Now, let us fix $T \in B_{L(X,Y)}$ and define the operators

$$T_1(x) = z_0^*(x)y_0 + TP_M(x), \qquad T_2(x) = -z_0^*(x)y_0 + TP_M(x) \qquad (x \in X),$$

where $y_0 = \frac{Tz_0}{\|Tz_0\|}$ if $Tz_0 \neq 0$ (or some fixed vector in S_Y in some other case), P_M is the natural projection from X onto M and z_0^* is the norm one functional in X given by $z_0^*(\lambda z_0 + m) = \lambda$. It is clear that $1 = \|T_i\| = \|T_i(z_0)\|,\ \ i = 1,2$ and

$$T = \frac{1 + \|Tz_0\|}{2} T_1 + \frac{1 - \|Tz_0\|}{2} T_2,$$

so,

$$T \in \text{co}\ \{S \in S_{L(X,Y)} : \|Sz_0\| = 1\}.$$

To finish with, let us note that any operator S in the unit sphere of $L(X,Y)$ satisfying $\|Sz_0\| = 1$ can be approximated in the norm topology by the sequence of operators S_n given by

$$S_n(x) = z_0^*(x)S(z_0) + \left(1 - \frac{1}{n}\right) SP_M(x).$$

Now these operators are in the interior of the set of norm attaining functionals since

$$\|S_n\| = \max\{\|S_n P_M\|, \|S_n z_0\|\} = 1,$$

and $1 = \|S_n z_0\| > \|S_n P_M\| = 1 - \frac{1}{n}$, so, in some ball centered at S_n, the same inequality happens and all the operators close enough attain the norm at z_0, that is, finally

$$T \in \overline{\text{co}} \left(S_{L(X,Y)} \cap \text{int}\ NA(X,Y) \right)$$

□

Now we will use an assumption not comparable with the first condition posed in Theorem 2. We will assume that the Banach space is non rough (instead of not containing ℓ_1). First let us recall that a Banach space is rough if for some $\varepsilon > 0$ it holds

$$\limsup_{h \to 0} \frac{\|x + h\| + \|x - h\| - 2\|x\|}{\|h\|} \geq \varepsilon, \qquad \forall x \in X.$$

By [7, Proposition I.1.11], X is non rough if, and only if, for any $\varepsilon > 0$ there is $x \in S_X, \alpha > 0$ so that

$$\text{diam}\ S(B_{X^*}, x, \alpha) < \varepsilon,$$

where $S(B_{X^*}, x, \alpha)$ is the w^*-slice given by

$$S(B_{X^*}, x, \alpha) = \{x^* \in B_{X^*} : x^*(x) > 1 - \alpha\}.$$

Note that the lack of roughness is weaker than the Mazur intersection property, since the last property is equivalent to the norm denseness in the dual unit sphere of points

x^* contained in w^*-slices of arbitrarily small diameter (see [11, Theorem 2.1]). Therefore, there are non rough spaces containing ℓ_1 (even spaces with the Mazur intersection property).

It is clear that any Asplund space has non rough norm. In fact, Leach and Whitfield proved that a Banach space so that every equivalent norm is non rough, is an Asplund space (see [17] or [7, Theorem I.5.3]). By virtue of Proposition 1 there are non reflexive Asplund spaces so that the set of norm attaining functionals has non empty interior. As a consequence, the same assertion holds for non rough spaces. However, by assuming additional conditions one gets reflexivity.

Proposition 5 ([1, Proposition 5]). *If X is non rough and for some $r > 0$*

$$B_{X^*} = \overline{\text{co}}^{w^*}(NA_r(X, \mathbb{R}) \cap S_{X^*}),$$

then X is reflexive. As a consequence, an Asplund space whose dual unit ball satisfies the previous condition, has to be reflexive.

Our purpose now is to give a characterization of reflexivity valid for spaces of operators and assuming non roughness. First we will need the following result, whose proof follows the argument used to check that the product of two strongly exposed points is also strongly exposed in the projective tensor product (see [9, p. 46], for instance). By $K(X,Y)$ we will denote the space of *compact operators* from X into Y.

Lemma 6. *Let X and Y be Banach spaces such that X^* and Y are non rough. Then $K(X,Y)$ and $L(X,Y)$ are also non rough.*

Proof. We will first check the statement for $L(X,Y)$. Let us consider the set

$$D = \{x \otimes y^* : x \in B_X, y^* \in B_{Y^*}\} \subseteq E^*,$$

where we denoted by $E = L(X,Y)$ and we consider any element $x \otimes y^*$ acting on E by

$$(x \otimes y^*)(T) = y^*(T(x)) \quad \forall T \in E.$$

It is clear that under this identification $D \subseteq E^*$, and since D is a 1-norming set of B_{E^*}, then its convex hull is w^*-dense in B_{E^*}. We have to prove that E^* has w^*-slices of small diameter (see [7, Proposition I.1.11]). If we fix $\varepsilon > 0$, since X^* and Y are non rough, there are $x_0^* \in S_{X^*}$, $y_0 \in S_Y$ and $0 < \delta < \varepsilon$ satisfying

$$\text{diam } S(B_X, x_0^*, \delta) < \varepsilon, \quad \text{diam } S(B_{Y^*}, y_0, \delta) < \varepsilon. \tag{6}$$

We fix elements $x_0 \in S(B_X, x_0^*, \delta)$, $y_0^* \in S(B_{Y^*}, y_0, \delta)$ and we will prove that

$$\text{diam } S(B_{E^*}, T_0, \delta^2) \leq 8\varepsilon$$

for the operator $T_0 = x_0^* \otimes y_0$.

Let us choose $e^* \in S(B_{E^*}, T_0, \delta^2)$. We can clearly assume that $e^* \in \text{co } (D)$, that is,

$$e^* = \sum_{j=1}^n t_j x_j \otimes y_j^*, \ t_j \geq 0, \ \sum_{j=1}^n t_j = 1, \ x_j \in B_X, \ y_j^* \in B_{Y^*}.$$

We denote by $A = \{j \in \{1, ..., n\} : x_0^*(x_j)y_j^*(y_0) \leq 1-\delta\}$. By the choice of e^* it is satisfied

$$1 - \delta^2 < e^*(T_0) = \Big(\sum_{j=1}^{n} t_j x_j \otimes y_j^*\Big)(T_0) =$$

$$= \sum_{j=1}^{n} t_j x_0^*(x_j) y_j^*(y_0) \leq \sum_{j\in A} t_j(1-\delta) + \sum_{j\notin A} t_j = 1 - \delta \sum_{j\in A} t_j \ ,$$

and so $\sum_{j\in A} t_j < \delta$.

For any $j \notin A$, it holds $x_0^*(x_j)y_j^*(y_0) \geq 1 - \delta$, and by changing the sign, if necessary, we can assume $x_0^*(x_j)$, $y_j^*(y_0) \geq 1 - \delta$, so $x_j \in S(B_X, x_0^*, \delta)$, $y_j^* \in S(B_{Y^*}, y_0, \delta)$ and from (5) it follows

$$\|x_j - x_0\| < \varepsilon \ , \quad \|y_j^* - y_0^*\| < \varepsilon \ .$$

As a consequence,

$$\|e^* - x_0 \otimes y_0^*\| = \|\sum_{j=1}^{n} t_j(x_j \otimes y_j^* - x_0 \otimes y_0^*)\| \leq$$

$$\leq \sum_{j\in A} t_j \|x_j \otimes y_j^* - x_0 \otimes y_0^*\| + \sum_{j\notin A} t_j (\|(x_j - x_0) \otimes y_j^*\| + \|x_0 \otimes (y_j^* - y_0^*)\|) <$$

$$< 2\delta + 2\varepsilon < 4\varepsilon.$$

Since e^* is any element of $S(B_{E^*}, T_0, \delta^2)$, we proved that

$$\operatorname{diam} S(B_{E^*}, T_0, \delta^2) \leq 8\varepsilon.$$

The operator T_0 is, in fact, compact and any element in $K(X,Y)^*$ is the restriction to $K(X,Y)$ of a functional on $L(X,Y)$ with the same norm, so

$$\operatorname{diam} S(B_{K(X,Y)^*}, T_0, \delta^2) \leq \operatorname{diam} S(B_{E^*}, T_0, \delta^2) \leq 8\varepsilon,$$

and also $K(X,Y)$ is non rough, as we wanted to show. □

Theorem 7. *Let X and Y be Banach spaces and let E be either the space $K(X,Y)$ or $L(X,Y)$. The following assertions are equivalent:*

i) E is reflexive.

ii) $B_{E^*} = \overline{\text{co}}^{w^*}(S_{E^*} \cap NA_r(E, \mathbb{R}))$ (some $r > 0$) and X^*, Y are non rough.

Proof. i) ⇒ii) If E is reflexive, then $NA(E) = E^*$ and X^* and Y are also reflexive, and so, they are non rough.

ii) ⇒i) If X^* and Y are non rough, by Lemma 6, then E is non rough and we can apply Proposition 5 to deduce that E is reflexive. □

Let us note that in the case that X or Y has the approximation property, then if $L(X,Y)$ is reflexive, it holds that $K(X,Y)^{**} = L(X,Y)$ [6, Proposition 16.7], and so $K(X,Y) = L(X,Y)$. Also, if $K(X,Y) = L(X,Y)$ and X, Y are reflexive, then $L(X,Y)$

is reflexive [13]. Therefore, if either X or Y has the approximation property, then all statements posed in Theorem 7 are equivalent and they hold if, and only if, $K(X,Y) = L(X,Y)$ and the spaces X, Y are reflexive.

There are Banach spaces X, Y such that X^* and Y are non rough, $NA(X,Y)$ has non empty interior but $L(X,Y)$ is not reflexive. It is enough to take $Y = \mathbb{R}$ and X a space isomorphic to ℓ_1 such that its dual X^* satisfies that the interior of $NA(X^*)$ is not empty (see [2, Proposition 1]).

Also the assumption of lack of roughness is necessary in Theorem 7. For instance, for $X = c_0$ and $Y = \mathbb{R}$, it holds that $B_{X^{**}} = B_{\ell_\infty} = \overline{\text{co}}^{w^*}(S_{\ell_\infty} \cap NA_{\frac{1}{2}}(\ell_1, \mathbb{R}))$ and X is not reflexive.

Before finishing, we will state some open questions related to the results:

Open problems.

1) Suppose that for all (equivalent) norms on a Banach space the set of norm attaining functionals has non empty interior. Is the space reflexive?

2) Assume that the unit ball of a Banach space is non dentable, is the set of norm attaining functionals of first Baire category?

3) Suppose that X has the Mazur intersection property and $NA(X,Y)$ has non empty interior ($Y \neq \{0\}$), is X reflexive?

4) If X is separable and the unit ball is non dentable, is $NA(X,Y)$ first Baire category, for any Banach space Y?

5) Assume that the dual unit ball contains a weak-open set (related to the ball) of norm attaining functionals, is X reflexive?

It is known that a separable Banach space which is not weakly sequentially complete admits an equivalent norm for which the set of norm attaining functionals has empty interior (see [2, Corollary 7]). This provides a partial answer to Problem 1. Jiménez Sevilla and Moreno proved that the answer to Question 5 is positive for separable spaces. Also they gave some positive answers in the general case by assuming extra conditions on the space [15].

REFERENCES

1. M. D. Acosta, J. Becerra Guerrero and M. Ruiz Galán, Dual spaces generated by the set of norm attaining functionals, preprint.
2. M.D. Acosta and M. Ruiz Galán, New characterizations of the reflexivity in terms of the set of norm attaining functionals, Canad. Math. Bull. 41 (1998), 279–289.
3. M.D. Acosta and M. Ruiz Galán, Norm attaining operators and reflexivity, Rend. Circ. Mat. Palermo 56 (1998), 171–177.
4. R.D. Bourgin, Geometric aspects of convex sets with the Radon-Nikodym property, Lecture Notes in Math. 993, Springer-Verlag, Berlin, 1983.

5. G. Debs, G. Godefroy and J. Saint Raymond, Topological properties of the set of norm-attaining linear functionals, Canad. J. Math. 47 (1995), 318–329.
6. A. Defant and K. Floret, Tensor norms and operators ideals, Noth-Holland Math. Studies 176, North-Holland, Amsterdam, 1993.
7. R. Deville, G. Godefroy and V. Zizler, Smoothness and renormings in Banach spaces, Pitman Monographs Surveys Pure Appl. Math. 64, Longman Sci. Tech., New York, 1993.
8. N. Dunford and J.T. Schwarz, Linear Operators Part I: General theory, Interscience Publishers, New York, 1958.
9. M. Feder and P. Saphar, Spaces of compact operators and their dual spaces, Israel J. Math. 21 (1975), 38–49.
10. J.R. Giles, Comparable differentiability characterisations of two classes of Banach spaces, Bull. Austral. Math. Soc. 56 (1997), 263–272.
11. J.R. Giles, D.A. Gregory and B. Sims, Characterisation of normed linear spaces with Mazur's intersection property, Bull. Austral. Math. Soc. 18 (1978), 105–123.
12. M. González and J.M. Gutiérrez, Polynomial Grothendieck properties, Glasgow Math. J. 37 (1995), 211-219.
13. J.R. Holub, Reflexivity of $L(E,F)$, Proc. Amer. Math. Soc. 39 (1973), 175–177.
14. R.C. James, Weak compactness and reflexivity, Israel J. Math. 2 (1964), 101–119.
15. M. Jiménez Sevilla and J.P. Moreno, A note on norm attaining functionals, Proc. Amer. Math. Soc. 126 (1998), 1989–1997.
16. P.S. Kenderov, W.B. Moors and S. Sciffer, Norm attaining functionals on $C(T)$, Proc. Amer. Math. Soc. 126 (1998), 153–157.
17. E.B. Leach and J.H.M. Whitfield, Differentiable functions and rough norms on Banach spaces, Proc. Amer. Math. Soc. 33 (1972), 120–126.
18. J.I. Petunin and A.N. Plichko, Some properties of the set of functionals that attain a supremum on the unit sphere, Ukrain. Math. Zh. 26 (1974), 102–106, MR 49:1075.
19. M. Valdivia, Fréchet spaces with no subspaces isomorphic to ℓ_1, Math. Japon. 38 (1993), 397–411.

Recent Progress in Functional Analysis
K.D. Bierstedt, J. Bonet, M. Maestre and J. Schmets

The extension theorem for norms on symmetric tensor products of normed spaces

Klaus Floret

Department of Mathematics, University of Oldenburg, D-26111 Oldenburg, Germany, e-mail: `floret@mathematik.uni-oldenburg.de`

To Professor Manuel Valdivia on the occasion of his seventieth birthday.

Abstract

It is shown that every s-tensor norm on n-th symmetric tensor products of normed spaces (n fixed) is equivalent to the restriction on symmetric tensor products of a tensor norm (in the sense of Grothendieck) on "full" n-fold tensor products of normed spaces. As a consequence a large part of the isomorphic *theory of norms on symmetric tensor products can be deduced from the theory of "full" tensor norms, which usually is easier to handle. Dually, the* isomorphic *theory of maximal normed ideals of n-homogeneous polynomials can be treated, to a certain extent, through the theory of maximal normed ideals of n-linear functions or mappings.*
MCS 2000 Primary 46B28; Secondary 46M05, 46G25

1. Introduction and definitions

1.1. Symmetric and full tensor products of vector spaces (over $\mathbb{K} = \mathbb{R}$ or $\mathbb{C}$) are defined by universal properties (linearizing all symmetric n-linear or all n-linear mappings respectively) where $n \in \mathbb{N}$ is fixed (the case $n = 1$ is trivial). The n-th symmetric tensor product $\otimes^{n,s} E$ of a vector space E can be obtained as the range $\operatorname{im} \sigma_E^n$ of the symmetrization map $\sigma_E^n : \otimes^n E \longrightarrow \otimes^n E$ (full n-fold tensor product) which linearizes

$$E^n \ni (x_1, \ldots, x_n) \rightsquigarrow x_1 \vee \cdots \vee x_n := \frac{1}{n!} \sum_{\eta \in S_n} x_{\eta(1)} \otimes \cdots \otimes x_{\eta(n)} =$$

$$= \frac{1}{n!} \int_\Omega \varepsilon_1(w) \cdots \varepsilon_n(w) \otimes^n \left[\sum_{k=1}^n \varepsilon_k(w) x_k \right] P(dw) \in \otimes^n E$$

where S_n denotes the group of permutations of $\{1, \ldots, n\}$ and (Ω, P) is a probability space with $\varepsilon_k : \Omega \longrightarrow \mathbb{K}$ being stochastically independent, normalized ($\int_\Omega |\varepsilon_k|^2 \, dP = 1$) and centralized ($\int_\Omega \varepsilon_k \, dP = 0$) variables. The injection

$$\otimes^{n,s} E := \operatorname{im} \sigma_E^n = \operatorname{span} \{ \otimes^n x \mid x \in E \} \hookrightarrow \otimes^n E$$

will be denoted by ι_E^n and $\sigma_E^n : \otimes^n E \longrightarrow \otimes^{n,s} E$ is the canonical projection (from now on σ_E^n is considered as a map onto $\otimes^{n,s} E$); clearly $\sigma_E^n \circ \iota_E^n = \mathrm{id}_{\otimes^{n,s} E}$. If $\langle E, F \rangle$ is a separating dual system, then the following diagrams of natural maps

$$
\begin{array}{ccc}
\otimes^{n,s}E & \hookrightarrow & (\otimes^{n,s}F)^* \\
\downarrow \iota_E^n & & \downarrow (\sigma_F^n)^* \\
\otimes^n E & \hookrightarrow & (\otimes^n F)^*
\end{array}
\qquad\qquad
\begin{array}{ccc}
\otimes^{n}E & \hookrightarrow & (\otimes^{n}F)^* \\
\downarrow \sigma_E^n & & \downarrow (\iota_F^n)^* \\
\otimes^{n,s} E & \hookrightarrow & (\otimes^{n,s} F)^*
\end{array}
$$

commute. See [5] for details on the algebraic theory of symmetric tensor products; they were introduced by R. Ryan [10] to Functional Analysis.

1.2. A *tensor norm* β *of order* n assigns to each n-tuple $(E_1, \ldots, E_n)$ of normed spaces a norm $\beta(\cdot; E_1, \ldots, E_n)$ on $\otimes(E_1, \ldots, E_n)$ (notation $\otimes_\beta(E_1, \ldots, E_n)$ or $\otimes_{\beta,j=1}^n E_j$ or $\otimes_\beta^n E$ if all $E_j = E$) such that

(1) $\varepsilon \leq \beta \leq \pi$ *where* ε *and* π *are the natural injective and projective norms.*

(2) *The metric mapping property: If* $T_j \in \mathcal{L}(E_j; F_j)$, *then*

$$\| \otimes_{j=1}^n T_j : \otimes_{\beta,j=1}^n E_j \longrightarrow \otimes_{\beta,j=1}^n F_j \| = \|T_1\| \cdots \|T_n\| \, .$$

Equivalently it is enough that

(0′) $\beta(\cdot; E_1, \ldots, E_n)$ *is a seminorm for all normed spaces* $E_1, \ldots, E_n$.

(1′) $\beta(\otimes^n 1; \mathbb{K}, \ldots, \mathbb{K}) = 1$

(2′) *Same as* (2) *with* $\leq \|T_1\| \cdots \|T_n\|$.

It is clear that the same definitions can be made if the class NORM of all normed spaces is replaced by the class FIN of all finite dimensional normed spaces. To distinguish these norms from the s-tensor norms (on symmetric tensor products tensor products to be defined in a moment) it might be helpful to label them as "full tensor norms". For a given "full" tensor norm β of order n one can define the tensor norms

$$
\begin{aligned}
\overrightarrow{\beta}(z; E_1, \ldots, E_n) &:= \inf\{\beta(z; M_1, \ldots, M_n) \mid M_j \in \mathrm{FIN}\,(E_j), z \in \otimes_{j=1}^n M_j\} \\
\overleftarrow{\beta}(z; E_1, \ldots, E_n) &:= \sup\Big\{\beta((\otimes_{j=1}^n Q_{L_j}^{E_j})(z); E_1/L_1, \ldots, E_n/L_n) \Big| L_j \in \mathrm{COFIN}\,(E_j)\Big\}
\end{aligned}
$$

(where FIN (E) is the set of finite dimensional subspaces of E, COFIN (E) the set of finite codimensional subspaces of E and $Q_F^E : E \longrightarrow E/F$ the natural quotient mapping); note that for these constructions it is enough to know β on finite dimensional spaces. The mapping property implies $\overleftarrow{\beta} \leq \beta \leq \overrightarrow{\beta}$ and all three coincide if $E_1, \ldots, E_n$ have the metric approximation property (proof as in [3, 13.2.] for $n = 2$). β is called *finitely generated* if $\beta = \overrightarrow{\beta}$ and *cofinitely generated* if $\beta = \overleftarrow{\beta}$. It is easy to see that $\varepsilon = \overleftarrow{\varepsilon} = \overrightarrow{\varepsilon}$ and $\pi = \overrightarrow{\pi} \neq \overleftarrow{\pi}$ – the latter using spaces without m.a.p. (see [3, 16.2.] for $n = 2$). If β is a tensor norm of order n, then the *dual tensor norm* β' is defined on FIN by

$$\otimes_{\beta'}(M_1, \ldots, M_n) :\overset{1}{=} [\otimes_\beta(M_1', \ldots, M_n')]'$$

and on NORM by the finite hull $\overrightarrow{\beta'}$. A tensor norm β is called *injective* (resp. *semi-injective*) if

$$\otimes_\beta(F_1, \ldots, F_n) \longrightarrow \otimes_\beta(E_1, \ldots, E_n)$$

is a metric (resp. isomorphic) injection whenever $F_j \subset E_j$ are subspaces. In the case of semi-injectivity it is easy to see that there are universal constants for the equivalence of the norms on $\otimes(F_1, \ldots, F_n)$. The tensor norm β is called *projective* (resp. *semi-projective*) if the natural map

$$\otimes_{j=1}^n Q_{F_j}^{E_j} : \otimes_\beta^n(E_1, \ldots, E_n) \longrightarrow \otimes_\beta^n(E_1/F_1, \ldots, E_n/F_n)$$

is a metric surjection (resp. open) whenever $F_j \subset E_j$ are closed subspaces; again there are universal constants in the semi-projective case.

1.3. The theory of "full" tensor norms is well-developed in the case $n = 2$ (see e.g. [3]), it is due to Grothendieck [8] and, up to a certain extent, also to Schatten [11]. Many results are easily extended from 2 to n.

1.4. Again being $n \in \mathbb{N}$ fixed, *an s-tensor norm α of order n* assigns to each normed space E a norm $\alpha(\cdot; \otimes^{n,s}E)$ on $\otimes^{n,s}E$ such that

(1) $\varepsilon_s \leq \alpha \leq \pi_s$

(2) *The metric mapping property: If $T \in \mathcal{L}(E; F)$, then*

$$\| \otimes^{n,s} T : \otimes_\alpha^{n,s} E \longrightarrow \otimes^{n,s} F \| = \|T\|^n .$$

The theory of the natural projective s-tensor norm π_s and natural injective s-tensor norm ε_s is presented e.g. in [5]. As in the case of "full" tensor norms there is a useful test: α is an s-tensor norm of order n if

(0′) *$\alpha(\cdot; \otimes^{n,s}E)$ is a seminorm on $\otimes^{n,s}E$ (for all normed spaces E)*

(1′) $\alpha(\otimes^n 1; \otimes^{n,s}\mathbb{K}) = 1$

(2′) *Like* (2) *with* $\leq \|T\|^n$.

Restricting E to be in FIN (or on Hilbert spaces) one obtains the definition of an s-tensor norm of order n on FIN (or on the class of all Hilbert spaces). Having the definitions for full tensor norms in mind it is clear how to define the *finite hull* $\overrightarrow{\alpha}$, the *cofinite hull* $\overleftarrow{\alpha}$ and the dual norm α'; note that for $M \in$ FIN

$$(\otimes_\alpha^{n,s} M)' \stackrel{1}{=} \otimes_{\alpha'}^{n,s} M'$$

(by definition) and hence

$$|\langle z, z' \rangle| \leq \alpha(z; \otimes^{n,s} M)\alpha'(z'; \otimes^{n,s} M')$$

for all $z \otimes^{n,s} M$ and $z' \in \otimes^{n,s}M'$; this sort of trace-duality is often useful. It follows from the definition of ε_s (see e.g. [5, 3.1.]) that $\pi_s' = \varepsilon_s$. Moreover, it is obvious how to define α to be *finitely generated, cofinitely generated, injective, semi-injective, projective and semi-projective.*

An introduction to the theory of s-tensor norms will be presented in [6]; in this paper I shall need only the basic definitions.

2. The norm extension theorem

2.1. Fix $n \in \mathbb{N}$. If β is a "full" tensor norm of order n, then

$$\beta|_s(z; \otimes^{n,s} E) := \beta(\iota_E^n(z); {}^n E)$$

(interprete ${}^n E$ as $E, \ldots, E$ (n-times)) defines an s-tensor norm of order n with $\varepsilon|_s \leq \beta|_s \leq \pi|_s$. Since $\pi|_s \leq \pi_s$ but $\pi|_s \neq \pi_s$ (and $\varepsilon_s \neq \varepsilon|_s$; see [5]) not all s-tensor norms are of this form. However, for every s-tensor norm α there is a full β such that α and $\beta|_s$ are equivalent (notation: $\alpha \sim \beta|_s$): This is the main content of the norm extension theorem which will be proved. β can be chosen to be *symmetric*, i.e. the natural map $R_\eta : \otimes_\beta(E_1, \ldots, E_n) \longrightarrow \otimes_\beta(E_{\eta(1)}, \ldots, E_{\eta(n)})$ is an isometry (onto) for all permutations $\eta \in S_n$. Clearly the projective norm π and the injective norm ε are symmetric. For $n = 2$ the norms w_2 and w_2^* are symmetric.

2.2. It is worthwhile to have good information about the constants in the norm extension theorem. For this define for an s-tensor norm α

$$K_2(\alpha) := \sqrt{n!}\alpha(e_1 \vee \cdots \vee e_n; \otimes^{n,s} \ell_2^n)$$

where the e_j are the unit vectors in ℓ_2^n.

Lemma. $\left(\frac{n!}{n^n}\right)^{1/2} = K_2(\varepsilon_s) \leq K_2(\alpha) \leq K_2(\pi_s) = \left(\frac{n^n}{n!}\right)^{1/2}$.

Proof: It is clear that $K_2(\varepsilon_s) \leq K_2(\alpha) \leq K_2(\pi_s)$ for all s-tensor norms α. For an upper estimate of $K_2(\pi_s)$ use the Rademacher functions $r_k : [0,1] \longrightarrow \{-1, +1\}$ in the polarization formula

$$e_1 \vee \cdots \vee e_n = \frac{1}{n!} \int_0^1 r_1(t) \cdots r_n(t) \otimes^n (r_1(t), \ldots, r_n(t)) dt$$

hence

$$\pi_s(e_1 \vee \cdots \vee e_n; \otimes^{n,s} \ell_2^n) \leq \frac{1}{n!} \int_0^1 \|(r_1(t), \ldots, r_n(t))\|_{\ell_2^n}^n dt \leq \frac{(\sqrt{n})^n}{n!} .$$

For ε_s one obtains:

$$\begin{aligned} \varepsilon_s(e_1 \vee \cdots \vee e_n; \otimes^{n,s} \ell_2^n) &= \sup\{|\langle e_1 \vee \cdots \vee e_n; \otimes^n x' \rangle| \mid x' \in B_{\ell_2^n}\} = \\ &= \sup\left\{|x_1' \cdots x_n'| \;\middle|\; \sum_{k=1}^n |x_k'|^2 = 1\right\} = n^{-n/2} \end{aligned}$$

– where the last equality can be proved using Lagrange multipliers. Now observe

$$\begin{aligned} \langle e_1 \vee \cdots \vee e_n, e_1 \vee \cdots \vee e_n \rangle &= \left(\frac{1}{n!}\right)^2 \sum_{\eta, \sigma \in S_n} \langle e_{\eta(1)} \otimes \cdots \otimes e_{\eta(n)}, e_{\sigma(1)} \otimes \cdots \otimes e_{\sigma(n)} \rangle \\ &= \frac{1}{n!} \end{aligned}$$

hence the "trace-duality" $(\otimes_{\varepsilon_s}^{n,s}\ell_2^n)' \stackrel{1}{=} \otimes_{\pi_s}^{n,s}\ell_2^n$ implies

$$\begin{aligned}\frac{1}{n!} = \langle e_1 \vee \cdots \vee e_n, e_1 \vee \cdots \vee e_n\rangle &\le \pi_s(e_1 \vee \cdots \vee e_n; \otimes^{n,s}\ell_2^n)\varepsilon_s(e_1 \vee \cdots \vee e_n; \otimes^{n,s}\ell_2^n) \\ &\le \frac{n^{n/2}}{n!} \cdot n^{-n/2} = \frac{1}{n!}\end{aligned}$$

and therefore $\pi_s(e_1 \vee \cdots \vee e_n; \otimes^{n,s}\ell_2^n) = \frac{n^{n/2}}{n!}$. □

The trace-duality gives $K_2(\alpha)K_2(\alpha') \ge 1$ for all s-tensor norms α and it would be interesting to check whether equality norms α and it would be interesting to check whether equality holds (as in the case of $\alpha = \varepsilon_s$).

2.3. Everything is prepared to state and prove the

Norm Extension Theorem. *For every s-tensor norm α of order n there is a full symmetric tensor norm β of order n with $\beta|_s$ being equivalent to α. More precisely: there is a construction which gives for every s-tensor norm α of order n a full symmetric tensor norm $\Phi(\alpha)$ of order n such that*

(1) $\|\sigma_E^n : \otimes_{\Phi(\alpha)}^n E \longrightarrow \otimes_\alpha^{n,s} E\| \le \left(\frac{n^n}{n!}\right)^{1/2} K_2(\alpha) \le \frac{n^n}{n!}$ *for all normed spaces E.*

(2) $\|\iota_E^n : \otimes_\alpha^{n,s} E \longrightarrow \otimes_{\Phi(\alpha)}^n E\| \le \left(\frac{n^n}{n!}\right)^{1/2} K_2(\alpha)^{-1} \le \frac{n^n}{n!}$ *for all normed spaces E.*

(3) *In particular:*

$$\frac{n!}{n^n}\Phi(\alpha)|_s \le \left(\frac{n!}{n^n}\right)^{1/2} K_2(\alpha)\,\Phi(\alpha)|_s \le \alpha \le \left(\frac{n^n}{n!}\right)^{1/2} K_2(\alpha)\,\Phi(\alpha)|_s \le \frac{n^n}{n!}\Phi(\alpha)|_s$$

(4) *If $\alpha_1 \le c\,\alpha_2$, then $K_2(\alpha_1)\,\Phi(\alpha_1) \le c\,K_2(\alpha_2)\,\Phi(\alpha_2)$.*

(5) *If α is finitely generated (resp. cofinitely generated), then $\Phi(\alpha)$ is finitely generated (resp. cofinitely generated).*

(6) *If α is injective, then $\Phi(\alpha)$ is injective.*

(7) *If α is semi-projective (resp. semi-injective), then $\Phi(\alpha)$ is semi-projective (resp. semi-injective).*

(8) *For the dual norm α' one has $\Phi(\alpha') \sim \Phi(\alpha)'$, more precisely*

$$\Phi(\alpha)' \le K_2(\alpha)\,K_2(\alpha')\,\Phi(\alpha') \le n^{n/2}\Phi(\alpha)'\,.$$

(9) *If γ is a full symmetric tensor norm of order n, then $\Phi(\gamma|_s) \sim \gamma$; with constants:*

$$\frac{1}{\sqrt{n!}}K_2(\gamma|_s)\,\Phi(\gamma|_s) \le \gamma \le \sqrt{n!}\,K_2(\gamma|_s)\Phi(\gamma|_s)\,.$$

PROOF: (a) If $E_1, \dots, E_n$ are normed spaces, $\ell_2^n(E_j) := \ell_2^n(E_1, \dots, E_n)$ and $P_k : \ell_2^n(E_j) \longrightarrow E_k$ and $I_k : E_k \hookrightarrow \ell_2^n(E_j)$ the natural projections and injections, then it is straightforward to see that $\mathrm{id}_{\otimes_{j=1}^n E_j} = Q_{E_1,\dots,E_n} \circ J_{E_1,\dots,E_n}$ where

$$J_{E_1,\dots,E_n} : \otimes_{j=1}^n E_j \xrightarrow{\sqrt{n!}\, I_1 \otimes \cdots \otimes I_n} \otimes^n \ell_2^n(E_j) \xrightarrow{\sigma^n_{\ell_2^n(E_j)}} \otimes^{n,s} \ell_2^n(E_j)$$

$$Q_{E_1,\dots,E_n} : \otimes^{n,s} \ell_2^n(E_j) \xrightarrow{\iota^n_{\ell_2^n(E_j)}} \otimes^n \ell_2^n(E_j) \xrightarrow{\sqrt{n!}\, P_1 \otimes \cdots \otimes P_n} \otimes_{j=1}^n E_j$$

(see [5, 1.10] for the origin of this factorization). Note that

$$J_{E_1,\dots,E_n}(x_1 \otimes \cdots \otimes x_n) = \sqrt{n!}\,(x_1, 0, \dots, 0) \vee \cdots \vee (0, \dots, 0, x_n)\,.$$

(b) The definition

$$\beta_\alpha(z; E_1, \dots, E_n) := \alpha(J_{E_1,\dots,E_n}(z); \otimes^{n,s} \ell_2^n(E_j))$$

gives a norm on $\otimes_{j=1}^n E_j$ which satisfies the metric mapping property (2') from 1.2.: to see this take $\|T_j : E_j \longrightarrow F_j\| \le 1$ and define $T : \ell_2^n(E_j) \longrightarrow \ell_2^n(F_j)$ by $T(x_1, \dots, x_n) := (T_1 x_1, \dots, T_n x_n)$; then $\|T\| \le 1$ and

$$\begin{array}{ccc}
\otimes^n_{\beta_\alpha, j=1} E_j & \xrightarrow{J_{E_1,\dots,E_n}} & \otimes^{n,s}_\alpha \ell_2^n(E_j) \\
{\scriptstyle T_1 \otimes \cdots \otimes T_n} \downarrow & & \downarrow {\scriptstyle \otimes^{n,s} T} \\
\otimes^n_{\beta_\alpha, j=1} F_j & \xrightarrow{J_{F_1,\dots,F_n}} & \otimes^{n,s}_\alpha \ell_2^n(F_j)
\end{array}$$

commutes. This shows $\|T_1 \otimes \cdots \otimes T_n : \cdots\| \le 1$. To see that β_α is symmetric, note first that for every $\eta \in S_n$ the natural map $S_\eta : \ell_2^n(E_1, \dots, E_n) \longrightarrow \ell_2^n(E_{\eta(1)}, \dots, E_{\eta(n)})$ is an isometry (onto); moreover, the diagram

$$\begin{array}{ccc}
\otimes^n_{j=1} E_j & \xrightarrow{J_{E_1,\dots,E_n}} & \otimes^{n,s} \ell_2^n(E_j) \\
{\scriptstyle R_\eta} \downarrow & & \downarrow {\scriptstyle \otimes^{n,s} S_\eta} \\
\otimes^n_{j=1} E_{\eta(j)} & \xrightarrow{J_{E_{\eta(1)},\dots,E_{\eta(n)}}} & \otimes^{n,s} \ell_2(E_{\eta(j)})
\end{array}$$

commutes, since

$$\begin{aligned}
(\sqrt{n!})^{-1}(\otimes^{n,s} S_\eta) J_{E_1,\dots,E_n}(x_1 \otimes \cdots \otimes x_n) &= (\otimes^{n,s} S_\eta)(x_1, 0, \dots, 0) \vee \cdots = \\
&= S_\eta(x_1, 0, \dots, 0) \vee \cdots \vee S_\eta(0, \dots, 0, x_n) = \\
&= (0, \dots, x_1, \dots, 0) \vee \cdots \vee (0, \dots, x_n, \dots, 0) = \\
&= (x_{\eta(1)}, 0, \dots, 0) \vee \cdots \vee (0, \dots, 0, x_{\eta(n)}) = \\
&= (\sqrt{n!})^{-1} J_{E_{\eta(1)},\dots,E_{\eta(n)}} x_{\eta(1)} \otimes \cdots \otimes x_{\eta(n)}\,.
\end{aligned}$$

The metric mapping property of α implies that β_α is symmetric. Since $\beta_\alpha(\otimes^n 1; \mathbb{K}, \dots, \mathbb{K}) = K_2(\alpha)$ one obtains that

$$\Phi(\alpha) := K_2(\alpha)^{-1}\beta_\alpha$$

defines a full symmetric tensor norm of order n.

(c) To estimate σ_E^n consider $S : \ell_2^n(E) \longrightarrow E$ defined by $S(x_1, \dots, x_n) := \sum_{k=1}^n x_k$; clearly $\|S\| = \sqrt{n}$. For $x_1, \dots, x_n \in E$ one obtains

$$\begin{aligned} \sigma_E^n(x_1 \otimes \cdots \otimes x_n) &= x_1 \vee \cdots \vee x_n = S(x_1, 0, \dots, 0) \vee \cdots \vee S(0, \dots, 0, x_n) = \\ &= (n!)^{-1/2}[\otimes^{n,s} S] \circ J_{^nE}(x_1 \otimes \cdots \otimes x_n) \in \otimes^{n,s} E \end{aligned}$$

hence $\sigma_E^n = (n!)^{-1/2}[\otimes^{n,s} S] \circ J_{^nE}$ which implies

$$\|\sigma_E^n : \otimes_{\beta_\alpha}^n E \xrightarrow{J_{^nE}} \otimes_\alpha^{n,s} \ell_2^n(E) \overset{(n!)^{-1/2}\otimes^{n,s}S}{\longrightarrow} \otimes_\alpha^{n,s} E\| \leq (n!)^{-1/2}\, \|S\|^n = \left(\frac{n^n}{n!}\right)^{1/2}.$$

(d) To see the estimate for ι_E^n consider Rademacher functions $r_k : [0,1] \longrightarrow \{-1, +1\}$ and, for every $t \in [0,1]$, the operators

$$\begin{aligned} D_t :\ & \ell_2^n(E) \longrightarrow \ell_2^n(E), & (x_1, \dots, x_n) &\rightsquigarrow (r_1(t)x_1, \dots, r_n(t)x_n). \\ \Delta :\ & E \longrightarrow \ell_2^n(E), & x &\rightsquigarrow (x, \dots, x). \end{aligned}$$

Thus $\|D_t\| = 1$ and $\|\Delta\| = \sqrt{n}$. For $x \in E$ one gets

$$\begin{aligned} J_{^nE}(\otimes^n x) &= \sqrt{n!}(x, 0, \dots, 0) \vee \cdots \vee (0, \dots, 0, x) = \\ &= \frac{1}{\sqrt{n!}} \int_0^1 r_1(t) \cdots r_n(t) \otimes^n (r_1(t)x, \dots, r_n(t)x)dt = \\ &= \frac{1}{\sqrt{n!}} \int_0^1 r_1(t) \cdots r_n(t)[\otimes^{n,s}(D_t \circ \Delta)] \otimes^n x\, dt \end{aligned}$$

and therefore for all $z \in \otimes^{n,s} E$

$$J_{^nE}(z) = \frac{1}{\sqrt{n!}} \int_0^1 r_1(t) \cdots r_n(t)[\otimes^{n,s}(D_t \circ \Delta)]z\, dt \in \otimes^{n,s} \ell_2^n(E).$$

It follows for all $z \in \otimes^{n,s} E$ that

$$\begin{aligned} \beta_\alpha(\iota_E^n(z); {}^nE) &= \alpha(J_{^nE}(z); \otimes^{n,s} \ell_2^n(E)) \leq \frac{1}{\sqrt{n!}} \|\Delta\|^n \alpha(z; \otimes^{n,s} E) \leq \\ &\leq \left(\frac{n^n}{n!}\right)^{1/2} \alpha(z; \otimes^{n,s} E) \end{aligned}$$

hence (2). Note that this gives in particular

$$\|Q_{E_1,\dots,E_n} : \otimes_\alpha^{n,s} \ell_2^n(E_j) \longrightarrow \otimes_{\beta_\alpha, j=1}^n E_j\| \leq \left(\frac{n^n}{n!}\right)^{1/2} \cdot \sqrt{n!} = n^{n/2}$$

and $Q_{E_1,\dots,E_n}$ is continuous; the fact that $\mathrm{id}_{\otimes^n E_j} = Q_{E_1,\dots,E_n} \circ J_{E_1,\dots,E_n}$ implies that $Q_{E_1,\dots,E_n}$ is even open and hence $\Phi(\alpha)$ is equivalent to the quotient norm of $Q_{E_1,\dots,E_n}$.

(e) If $\alpha_1 \leq c\,\alpha_2$, then $\beta_{\alpha_1} \leq c\,\beta_{\alpha_2}$ which is (4).

(f) Assume that α is a finitely generated s-tensor norm. Since every $M \in \mathrm{FIN}\,(\ell_2^n(E_j))$ is contained in some $\ell_2^n(M_j)$ with $M_j \in \mathrm{FIN}\,(E_j)$ one obtains

$$\begin{aligned}\beta_\alpha(z; E_1, \ldots, E_n) &= \alpha(J_{E_1,\ldots,E_n}(z); \otimes^{n,s}\ell_2^n(E_j)) = \\ &= \inf\{\alpha(J_{E_1,\ldots,E_n}(z); \otimes^{n,s}\ell_2^n(M_j) \mid M_j \in \mathrm{FIN}\,(E_j), J\ldots(z) \in \otimes^{n,s}\ell_2^n(M_j)\} = \\ &= \inf\{\alpha(J_{M_1,\ldots,M_n}(z); \otimes^{n,s}\ell_2^n(M_j) \mid M_j \in \mathrm{FIN}\,(E_j),\ z \in \otimes_{j=1}^n M_j\} = \\ &= \inf\{\beta_\alpha(z; M_1, \ldots, M_n) \mid M_j \in \mathrm{FIN}\,(E_j),\ z \in \otimes_{j=1}^n M_j\}\end{aligned}$$

which shows that $\Phi(\alpha)$ is finitely generated.

(g) If α is cofinitely generated take $L \in \mathrm{COFIN}\,(\ell_2^n(E_j))$, recall $\ell_2^n(E_j)' = \ell_2^n(E_j')$ and choose $L_j \in \mathrm{COFIN}\,(E_j)$ with

$$L \supset \ker[\ell_2^n(E_j) \overset{1}{\twoheadrightarrow} \ell_2^n(E_j/L_j)] =: L_0 \in \mathrm{COFIN}\,(\ell_2^n(E_j))\,.$$

The diagram

$$\begin{array}{ccc} \otimes_{\beta_\alpha, j=1}^n E_j & \xrightarrow{\quad J_{E_1,\ldots,E_n} \quad} & \otimes_\alpha^{n,s}\ell_2^n(E_j) \\ \Big\downarrow \otimes_{j=1}^n Q_{L_j}^{E_j} & & \Big\downarrow \otimes^{n,s} Q_{L_0}^{\ell_2^n(E_j)} \\ \otimes_{\beta_\alpha, j=1}^n E_j/L_j & \xrightarrow{\quad J_{E_1/L_1,\ldots,E_n/L_n} \quad} & \otimes_\alpha^{n,s}\ell_2^n(E_j)/L_0 \overset{1}{=} \otimes_\alpha^{n,s}\ell_2^n(E_j/L_j) \end{array}$$

commutes which easily gives that β_α and hence $\Phi(\alpha)$ is cofinitely generated.

(h) If α is (semi-)injective it is immediate, by the construction, that β_α is (semi-) injective and hence $\Phi(\alpha)$. The fact that $\Phi(\alpha)$ is equivalent to the quotient norm of $Q_{E_1,\ldots,E_n}$ (see part (d) of this proof) implies easily that $\Phi(\alpha)$ is semi-projective if α is semi-projective.

(i) To show the statement (8) about the dual norms, take an s-tensor norm α and observe first, that $\Phi(\alpha')$ is finitely generated by (5); since $\Phi(\alpha)'$ anyhow is finitely generated it is enough to show that $\Phi(\alpha)'$ and $\Phi(\alpha')$ are equivalent on FIN (with constants independent of the space): for this take $M := (M_1, \ldots, M_n) \in \mathrm{FIN}^n$ and define, for convenience, $M' := (M_1', \ldots, M_n')$. Observe that

$$[I_k : M_k \longrightarrow \ell_2^n(M)]' = [P_k : \ell_2^n(M') \longrightarrow M_k'] \quad \text{and} \quad P_k' = I_k\,;$$

moreover, $\left(\sigma_{\ell_2^n(M)}^n\right)' = \iota_{\ell_2^n(M')}^n$ and $\left(\iota_{\ell_2^n(M)}^n\right)' = \sigma_{\ell_2^n(M')}^n$ by the diagrams at the end of 1.1. This gives $J_M' = Q_{M'}$ and $Q_M' = J_{M'}$. Now define β_α' by

$$\otimes_{\beta_\alpha'}^n M := (\otimes_{\beta_\alpha}^n M')'$$

and note that $\beta_\alpha' = K_2(\alpha)^{-1}\Phi(\alpha)'$. Dualizing

$$\otimes_{\beta_\alpha}^n M' \xrightarrow{\quad J_{M'} \quad} \otimes_\alpha^{n,s}\ell_2^n(M') \xrightarrow{\quad Q_{M'} \quad} \otimes_{\beta_\alpha}^n M'$$

to

$$\otimes^n_{\beta'_\alpha} M \xrightarrow{Q'_{M'}=J_M} \otimes^{n,s}_{\alpha'} \ell^n_2(M) \xrightarrow{J'_{M'}=Q_M} \otimes^n_{\beta'_\alpha} M$$

gives (see end of (d))

$$\begin{aligned} \beta_{\alpha'}(z;M) &= \alpha'(J_M(z); \otimes^{n,s}\ell^n_2(M)) \le \|Q_{M'}\|\, \beta'_\alpha(z;M) \le n^{n/2}\, \beta'_\alpha(z;M) \\ \beta'_\alpha(z;M) &= \beta'_\alpha(J'_{M'} \circ J_M(z); M) \le \|J'_{M'}\|\, \alpha'(J_M(z); \otimes^{n,s}\ell^n_2(M)) = \beta_{\alpha'}(z;M) \end{aligned}$$

which implies (8).

(j) Finally let γ be a full symmetric tensor norm, then $c_E := \|\sigma^n_E : \otimes^n_\gamma E \longrightarrow \otimes^{n,s}_{\gamma|_s} E\| \le 1$ by the symmetry. Define

$$\gamma_1 := \beta_{\gamma|_s} = K_2(\gamma|_s)\Phi(\gamma|_s)\,,$$

then

$$\begin{aligned} \gamma_1(z; E_1, \ldots, E_n) &\le c_{\ell^n_2(E_j)} \sqrt{n!}\, \gamma([I_1 \otimes \cdots \otimes I_n](z);\ \ell^n_2(E_j), \ldots, \ell^n_2(E_j)) \le \\ &\le \sqrt{n!}\, \gamma(z; E_1, \ldots, E_n) \end{aligned}$$

hence $\gamma_1 \le \sqrt{n!}\,\gamma$ and $\Phi(\gamma|_s) \le K_2(\gamma|_s)^{-1}\sqrt{n!}\,\gamma$. On the other hand

$$\begin{aligned} \gamma(z; E_1, \ldots, E_n) &= \gamma(Q_{E_1,\ldots,E_n} \circ J_{E_1,\ldots,E_n}(z); E_1, \ldots, E_n) \le \\ &\le \sqrt{n!}\, \gamma|_s(J_{E_1,\ldots,E_n}(z); \otimes^{n,s}\ell^n_2(E_j)) = \sqrt{n!}\, \gamma_1(z; E_1, \ldots, E_n) \end{aligned}$$

hence $\gamma \le \sqrt{n!}\, K_2(\gamma|_s)\Phi(\gamma|_s)$. □

2.4. Some comments on the construction: it is clear that one may take $\ell^n_p(E_j)$ in the construction (or any symmetric norm on $\mathbb{R}^n$ instead of the ℓ_2-norm); I took $p = 2$ to facilitate the dualization.

2.5. Taking $\widetilde{\beta}_\alpha$ to be the quotient norm of $Q_{E_1,\ldots,E_n}$ would give (after normalization as in part (b) of the proof) a full symmetric tensor norm $\Psi(\alpha)$ of order n, equivalent to $\Phi(\alpha)$ (see the end of (d) of the proof in 2.3.), satisfying (1)–(5), (7)–(9), (with other constants, a priori), and: *If α is projective, then $\Psi(\alpha)$ is projective.* Is $\Psi(\alpha) \neq \Phi(\alpha)$?

2.6. Note that (9) (and (4)) imply that two symmetric full tensor norms (of order n) are equivalent if they are equivalent on symmetric tensor products. In other words: There is (up to equivalence) at most one full symmetric tensor norm extending a given s-tensor norm.

2.7. Since

$$\begin{aligned} \tfrac{n^n}{n!} &= \|\sigma^n_{\ell_1} : \otimes^n_\pi \ell_1 \longrightarrow \otimes^{n,s}_{\pi_s}\ell_1\| \le \|\sigma^n_{\ell_1} : \otimes^n_{\Phi(\pi_s)}\ell_1 \longrightarrow \otimes^{n,s}_{\pi_s}\ell_1\| \le \tfrac{n^n}{n!} \\ \tfrac{n^n}{n!} &= \|\iota^n_{c_0} : \otimes^{n,s}_{\varepsilon_s} c_0 \longrightarrow \otimes^n_\varepsilon c_0\| \le \|\iota^n_{c_0} : \otimes^{n,s}_{\varepsilon_s} c_0 \longrightarrow \otimes^n_{\Phi(\varepsilon_s)} c_0\| \le \tfrac{n^n}{n!} \end{aligned}$$

(see e.g. [5, 2.1., 2.3. and 5.3]) it follows that the constant $\frac{n^n}{n!}$ is best possible.

2.8. I do not know whether any s-tensor norm α with $\varepsilon|_s \leq \alpha \leq \pi|_s$ can be extended to a full norm γ with $\gamma|_s = \alpha$.

2.9. For the investigation of individual spaces the constants

$$c(n,\alpha,E) := \|\sigma_E^n : \otimes^n_{\Phi(\alpha)} E \longrightarrow \otimes_\alpha^{n,s} E\| \leq \left(\tfrac{n^n}{n!}\right)^{1/2} K_2(\alpha) \leq \tfrac{n^n}{n!}$$
$$d(n,\alpha,E) := \|\iota_E^n : \otimes_\alpha^{n,s} \longrightarrow \otimes^n_{\Phi(\alpha)} E\| \leq \left(\tfrac{n^n}{n!}\right)^{1/2} K_2(\alpha)^{-1} \leq \tfrac{n^n}{n!}$$

may be of interest.

3. Some applications

3.1. Every continuous n-homogeneous polynomial on a normed space E, notation: $q \in \mathcal{P}^n(E)$, has a canonical extension $\overline{q} \in \mathcal{P}^n(E'')$, usually called the Aron-Berner extension (see e.g. [5, 6.5.]). Let us use the identification $\mathcal{P}^n(E) \stackrel{1}{=} (\otimes_{\pi_s}^{n,s} E)'$, $q \rightsquigarrow q^L$. It would be interesting to know whether

$$\|\overline{q}^L\|_{(\otimes_\alpha^{n,s} E')'} = \|q^L\|_{(\otimes_\alpha^{n,s} E)'} \in [0,\infty]$$

holds. For $\alpha = \pi_s$ (Davie-Gamelin [4]) and $\alpha = \varepsilon_s$ (Carando-Zalduendo [2], see [5] for an alternative proof) this is true, but not at all trivial.

Isomorphic Extension Lemma. *Let α be a finitely generated s-tensor norm of order n, E normed and $q \in \mathcal{P}^n(E)$. Then $q^L \in (\otimes_\alpha^{n,s} E)'$ if and only if $\overline{q}^L \in (\otimes_\alpha^{n,s} E'')'$.*

PROOF: Since $E \stackrel{1}{\hookrightarrow} E''$, the metric mapping property implies one direction. For the other direction take a finitely generated full tensor norm β of order n such that $\beta|_s \sim \alpha$ (it exists due to the norm extension theorem). It can be seen (more or less as in the case $n = 2$, see [3, 13.2.]) that the canonical Arens-extension $\overline{\varphi}$ (see [5, 6.1.]) is in $(\otimes_\beta^n E'')'$ if φ is in $(\otimes_\beta^n E)'$. Setting $\varphi := q^L \circ \sigma_E^n$ gives $\overline{q}^L = \overline{\varphi} \circ \iota_{E''}^n$ hence the result follows from the properties of β. □

3.2. It is worthwhile to mention that a normed ideal $\mathcal{Q}$ of n-homogeneous scalar-valued polynomials is maximal if and only if it is of the form

$$\mathcal{Q}(E) \stackrel{1}{=} (\otimes_\alpha^{n,s} E)'$$

for some finitely generated s-tensor norm α of order n; see [7]. The maximal normed ideals $\mathcal{A}$ of n-linear continuous functionals are of the form

$$\mathcal{A}(E_1,\ldots,E_n) \stackrel{1}{=} (\otimes_\beta(E_1,\ldots,E_n))'$$

(for all Banach spaces E_j) for some finitely generated full tensor norm of order n (see also [7]). The norm extension theorem (use in particular that $\Phi(\alpha)$ is finitely generated if α is) easily gives the

Proposition. *For every maximal normed ideal $\mathcal{Q}$ of n-homogeneous scalar-valued polynomials there is a maximal normed ideal $\mathcal{A}$ of n-linear functionals such that $q \in \mathcal{P}^n(E)$ (with associated symmetric n-linear form $\check{q}$) is in $\mathcal{Q}(E)$ if and only if $\check{q} \in \mathcal{A}(E, \ldots, E)$.*

Checking the constants gives (if α is the "associated" finitely generated s-tensor norm to $\mathcal{Q}$ and $\mathcal{A}$ the ideal "associated" with the full tensor norm $\Phi(\alpha)$)

$$\|\check{q}\|_{\mathcal{A}} \leq c(n, \alpha, E) \|q\|_{\mathcal{Q}}$$

$$\|q\|_{\mathcal{Q}} \leq d(n, \alpha, E) \|\check{q}\|_{\mathcal{A}}$$

3.3. If β is a finitely generated full tensor norm of order 2, then the canonical map $* : \tilde{\otimes}^2_\beta E \longrightarrow \tilde{\otimes}^2_\varepsilon E$ is injective if E has the approximation property ($\tilde{}$ stands for the completion; see e.g. [3, 17.20.] for a proof). The norm extension theorem easily gives: *If α is a finitely generated s-tensor norm of order 2 and E a Banach space with the approximation property, then the canonical map $** : \tilde{\otimes}^{2,s}_\alpha E \longrightarrow \tilde{\otimes}^{2,s}_{\varepsilon_s} E$ is injective.* It is likely but not known whether $*$ (and hence $**$) is injective also for arbitrary n. If E has even the metric approximation property it is an easy consequence of the duality theory of s-tensor norms (to be presented in [6]) that $**$ is injective for all n.

3.4. In the proof of $\Phi(\gamma|_s) \sim \gamma$ in the norm extension theorem it was not used that γ is symmetric, only that $\|\sigma^n_E : \otimes^n_\gamma E \longrightarrow \otimes^{n,s}_{\gamma|_s} E\| \leq 1$.

Proposition. *If γ is a full tensor norm of order n such that $\sigma^n_E : \otimes^n_\gamma E \longrightarrow \otimes^{n,s}_{\gamma|_s} E$ is continuous for all normed spaces E, then there is a full symmetric tensor norm β of order n equivalent to γ.*

PROOF: It is easy to see that there is a universal constant c with $\|\sigma^n_E \ldots\| \leq c$. Now use just the part (j) of the proof of the norm extension theorem. □

Note that β can be chosen finitely generated, cofinitely generated, injective or projective if γ is.

Corollary. *For every full tensor norm γ of order n the following statements are equivalent:*

(1) *For every normed space E the symmetrization map $\sigma^n_E : \otimes^n_\gamma E \longrightarrow \otimes^n_\gamma E$ is continuous.*

(2) *For every permutation $\eta \in S_n$ and all normed spaces $E_1, \ldots, E_n$ the natural map*

$$R_\eta : \otimes_\gamma(E, \ldots, E_n) \longrightarrow \otimes_\gamma(E_{\eta(1)}, \ldots, E_{\eta(n)})$$

is continuous.

Note that, for $n = 2$, Cohen's tensor norm w_1 is not symmetric since w_1 is associated with the operator ideal of 1-factorable operators and $w_1^t = w_\infty$ with the one of ∞-factorable operators (see [3, 17.8. and 17.12] for details). If follows that $\sigma^2_E : \otimes^2_{w_1} E \longrightarrow \otimes^2_{w_1} E$ is, in general, not continuous.

3.5. A direct proof of the non-trivial part (1) $\curvearrowright$ (2) of this latter result runs as follows:

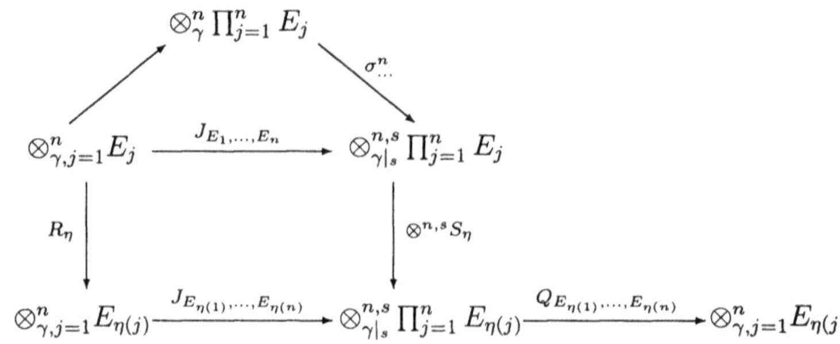

(see parts (a) and (b) of the proof of the norm extension theorem): the $J\dots$'s are continuous by the mapping property and the continuity of the $\sigma_{\dots}^n$, the lower map is the identity, $Q_{\dots}$ and $\otimes^{n,s} S_\eta$ are continuous and hence also R_η. This proof holds also for locally convex spaces and tensor topologies of order n, defined as follows (see [1]): a *tensor topology* τ *of order* n assigns to each n-tuple of locally convex spaces $(E_1, \dots, E_n)$ a locally convex topology $\tau(E_1, \dots, E_n)$ on $\otimes(E_1, \dots, E_n)$ such that

(1) $\otimes : E_1 \times \dots \times E_n \longrightarrow \otimes_\tau(E_1, \dots, E_n)$ is separately continuous.

(2) If $D_j \subset E_j'$ are equicontinuous, then $\{x_1' \otimes \dots \otimes x_n' \mid x_j' \in D_j\} \subset [\otimes(E_1, \dots, E_n)]^*$ is τ-equicontinuous.

(3) If $T_j \in \mathcal{L}(E_j; F_j)$, then $\otimes T_j : \otimes_\tau(E_1, \dots, E_n) \longrightarrow \otimes_\tau(F_1, \dots, F_n)$ is continuous.

It is worthwhile to note, that $\otimes_\tau(E_1, \dots, E_n)$ is separated if all E_j are separated: To see this, just observe that $\langle \otimes_{j=1}^n E_j, \otimes_{j=1}^n E_j \rangle$ is a separating dual system and $\otimes_{j=1}^n E' \subset (\otimes_{\tau,j=1}^n E_j)'$ by property (2).

The above proof (actually one needed only the mapping property (3)) gives the

Proposition. *For every tensor topology τ of order n the following are equivalent:*

(1) *For every locally convex space E the symmetrization map $\sigma_E^n : \otimes_\tau^n E \longrightarrow \otimes_\tau^n E$ is continuous.*

(2) *For every $\eta \in S_n$ and locally convex spaces $E_1, \dots, E_n$ the natural map $\otimes_\tau(E_1, \dots, E_n) \longrightarrow \otimes_\tau(E_{\eta(1)}, \dots, E_{\eta(n)})$ is continuous (hence a homomorphism).*

3.6. If β is a full tensor norm of order n, one can construct a tensor topology of order n (the so-called *tensor norm topology* associated with β) which, for each n-tuple of locally convex spaces $(E_1, \dots, E_n)$, is defined by the seminorms $\otimes_\beta(p_1, \dots, p_n)$

$$\otimes_\beta(p_1 \dots, p_n) := \beta((\otimes_{j=1}^n Q_{p_j})(z)); E_1/\ker p_1, \dots, E_n/\ker p_n)$$

(where p_j runs through a basis of continuous seminorms on E_j and $Q_{p_j} : E_j \longrightarrow E_j/\ker p_j$ are the natural quotient maps); these topologies were introduced by Harksen in 1979 for

$n = 2$ (see [9] and [3, §35]). Notation: $\otimes_\beta(E_1, \ldots, E_n)$ or $\otimes^n_\beta E$ if $E = E_1 = \cdots E_n$. If α is an s-tensor norm of order n the same idea

$$(\otimes^{n,s}_\alpha p)(z) := \alpha((\otimes^{n,s} Q_p)(z); \otimes^{n,s} E/\ker p)$$

defines a locally convex topology on $\otimes^{n,s} E$; notation: $\otimes^{n,s}_\alpha E$.

Proposition. *For each s-tensor norm α of order n there is a full tensor norm β of order n such that for all locally convex spaces E the space $\otimes^{n,s}_\alpha E$ is a complemented topological subspace of $\otimes^n_\beta E$ (via ι^n_E).*

This is an immediate consequence of the norm extension theorem. The properties (4)–(9) have consequences for $\otimes^n_\beta$: For example, $\otimes^n_\beta$ respects subspaces/quotients topologically if $\otimes^{n,s}_\alpha$ does.

REFERENCES

1. J. Ansemil and K. Floret, The symmetric tensor product of a direct sum of locally convex spaces, Stud. Math. 129 (1998) 285–295.
2. D. Carando and L. Zalduendo, A Hahn-Banach theorem for integral polynomials, Proc. Amer. Math. Soc. 127 (1999) 241-250.
3. A. Defant and K. Floret, Tensor Norms and Operator Ideals, North Holland Math. Studies 176, 1993.
4. A. Davie and T. Gamelin, A theorem on polynomial-star approximation, Proc. Amer. Math. Soc. 106 (1989) 351–356.
5. K. Floret, Natural norms on symmetric tensor products of normed spaces, Note di Matematica (2nd Trier-conference 1997) 17 (1997) 153–188.
6. K. Floret, The metric theory of symmetric tensor products of normed spaces, in preparation.
7. K. Floret and S. Hunfeld, Ultrastability of ideals of homogeneous polynomials and multilinear mappings on Banach spaces, to appear in Proc. Amer. Math. Soc.
8. A. Grothendieck, Résumé de la théorie métrique des produits tensoriels topologiques, Bol. Soc. Mat. São Paulo 8 (1956) 83–110.
9. J. Harksen, Charakterisierung lokalkonvexer Räume mit Hilfe von Tensorprodukt-topologien, Math. Nachr. 106 (1982) 347–374.
10. R. Ryan, Application of Topological Tensor Products to Infinite Dimensional Holomorphy, doctoral thesis, Trinity College Dublin, 1980.
11. R. Schatten, A Theory of Cross Norms, Ann. of Math. Studies 26, 1950.

Recent Progress in Functional Analysis
K.D. Bierstedt, J. Bonet, M. Maestre and J. Schmets

Remarks on p-summing multipliers.

Oscar Blasco *

Departamento de Análisis Matemático, Universidad de Valencia,
Burjassot 46100, Valencia (SPAIN)

To Professor Manuel Valdivia on the occasion of his seventieth birthday.

Abstract

Let X and Y be Banach spaces and $1 \leq p < \infty$, a sequence of operators (T_n) from X into Y is called a p-summing multiplier if $(T_n(x_n))$ belongs to $\ell_p(Y)$ whenever (x_n) satisfies that $(\langle x^*, x_n \rangle)$ belongs to ℓ_p for all $x^* \in X^*$. We present several examples of p-summing multipliers and extend known results for p-summing operators to this setting. We get, using almost summing and Rademacher bounded operators, some sufficient conditions for a sequence to be a p-summing multiplier between spaces with some geometric properties.
MCS 2000 Primary 47B10; Secondary 47D50, 42A45

1. Introduction.

Let X and Y be two real or complex Banach spaces and let $E(X)$ and $F(Y)$ be two Banach spaces whose elements are defined by sequences of vectors in X and Y (containing any eventually null sequence in X or Y). A sequence of operators $(T_n) \in \mathcal{L}(X,Y)$ is called a *multiplier sequence* from $E(X)$ to $F(Y)$ if there exists a constant $C > 0$ such that

$$||(T_j x_j)_{j=1}^n||_{F(Y)} \leq C||(x_j)_{j=1}^n||_{E(X)}$$

for all finite families $x_1, \ldots, x_n$ in X. The set of all multiplier sequences is denoted by $(E(X), F(Y))$.

Given a real or complex Banach space X and $1 \leq p \leq \infty$, we denote by $\ell_p(X)$ and $\ell_p^w(X)$ the Banach spaces of sequences in X with norms $||(x_n)||_{\ell_p(X)} = ||(||x_n||)||_{\ell_p}$ and $||(x_n)||_{\ell_p^w(X)} = \sup_{||x^*||=1} ||(\langle x^*, x_n\rangle)||_{\ell_p}$ respectively. $Rad_p(X)$ stands for the space of sequences $(x_n) \in X$ such that $\sup_n (\int_0^1 ||\sum_{j=1}^n r_j(t)x_j||^p dt)^{1/p} < \infty$, where $(r_j)_{j\in\mathbb{N}}$ are the Rademacher functions on $[0,1]$ defined by $r_j(t) = sign(\sin 2^j \pi t)$.

It is easy to see that $Rad_\infty(X) = \ell_1^w(X)$. It follows from Kahane's inequalities (see [11], page 211) that $Rad_p(X) = Rad_q(X)$ with equivalent norms for all $1 \leq p, q < \infty$. This space will then be denoted $Rad(X)$, and we shall use the L^1-norm throughout the paper.

*The research was partially supported by the Spanish grant DGESIC PB98-1426

The reader is referred to [4],[5],[6],[7] for the study of multiplier sequences in the case $E(X) = H^1(\mathbb{T}, X)$, corresponding to vector-valued Hardy spaces, and $F(Y) = \ell_p(Y)$ or $F(Y) = BMOA(\mathbb{T}, Y)$, to [3] ,[8],[17] and [27] for $E(X) = Rad(X)$ and $F(Y) = Rad(Y)$, to [2] for the particular cases $p = q$, $X = Y$ and $T_j = \alpha_j Id_X$ and to [1] for the case $E(X) = \ell_p^w(X)$ and $F(Y) = \ell_q(\mathbb{K})$.

In this article we shall consider the case of the classical sequence spaces $E(X) = \ell_p^w(X)$ and $F(Y) = \ell_p(Y)$. A sequence $(T_j)_{j\in\mathbb{N}}$ of operators in $\mathcal{L}(X, Y)$ is a *p-summing multiplier* if there exists a constant $C > 0$ such that, for any finite collection of vectors $x_1, x_2, \ldots x_n$ in X, it holds that

$$\Big(\sum_{j=1}^n \|T_j x_j\|^p\Big)^{1/p} \leq C \sup_{\|x^*\|=1} \Big(\sum_{j=1}^n |\langle x^*, x_j\rangle|^p\Big)^{1/p}.$$

Note that a constant sequence $T_j = T$ for all $j \in \mathbb{N}$ belongs to $(\ell_p^w(X), \ell_p(Y))$ if and only if T is a p-summing operator, usually denoted $T \in \Pi_p(X, Y)$. This fact suggests the use of the notation $\ell_{\pi_p}(X, Y)$ instead of $(\ell_p^w(X), \ell_p(Y))$.

In the paper [1] J.L. Arregui and the author introduced and considered the notion of (p, q)-summing multipliers and concentrated on the case $Y = \mathbb{K}$. It was shown that some geometric properties on X can be described using $\ell_{\pi_{p,q}}(X, \mathbb{K})$ and also that classical theorems, like Grothendieck theorem and others, can be rephrased into this setting.

Let us now recall the basic notions on Banach space theory and absolutely summing operators to be used later on.

An operator $T \in \mathcal{L}(X, Y)$ is *absolutely summing* if for every unconditionally convergent series $\sum x_j$ in X it holds that $\sum Tx_j$ is absolutely convergent in Y.

For $1 \leq p < \infty$, an operator $T: X \to Y$ is *p-summing* (see [22]) if it maps sequences $(x_j) \in \ell_p^w(X)$ into sequences $(Tx_j) \in \ell_p(Y)$, equivalently, if there exists a constant C such that

$$\Big(\sum_{j=1}^n \|Tx_j\|^p\Big)^{1/p} \leq C \sup_{\|x^*\|=1} \Big(\sum_{j=1}^n |\langle x^*, x_j\rangle|^p\Big)^{1/p}$$

for any finite family $x_1, x_2, \ldots x_n$ of vectors in X.

The least of such constants is the p-summing norm of u, denoted by $\pi_p(T)$. The space $\Pi_p(X, Y)$ of all p-summing operators from X to Y then is a Banach space for $1 \leq p < \infty$. It is well known that the space of absolutely summing operators coincides with the space of 1-summing operators.

For $1 \leq p \leq 2$ (respect. $q \geq 2$), a Banach space X is said to have *(Rademacher) type p* (resp. *(Rademacher) cotype q*) if there exists a constant C such that

$$\int_0^1 \Big\|\sum_{j=1}^n x_j r_j(t)\Big\| dt \leq C\Big(\sum_{j=1}^n \|x_j\|^p\Big)^{1/p}$$

(resp.

$$\Big(\sum_{j=1}^n \|x_j\|^q\Big)^{1/q} \leq C \int_0^1 \Big\|\sum_{j=1}^n x_j r_j(t)\Big\| dt),$$

for any finite family $x_1, x_2, \ldots x_n$ of vectors in X .

A Banach space X is said to have the *Orlicz property* there exists a constant C such that

$$(\sum_{j=1}^{n} \|x_j\|^2)^{1/2} \leq C \sup_{\|x^*\|=1} \sum_{j=1}^{n} |\langle x_j, x^*\rangle|$$

for any finite family $x_1, x_2, \ldots x_n$ of vectors in X.

Let us recall that Grothendieck's theorem establishes, in this setting, that, for any compact set K, any measure space (Ω, Σ, μ) and any Hilbert space H,

$$\mathcal{L}(L_1(\mu), H) = \Pi_1(L_1(\mu), H). \tag{1}$$

or

$$\mathcal{L}(C(K), L^1(\mu)) = \Pi_2(C(K), L^1(\mu)). \tag{2}$$

Because of that a Banach space X is called a GT-space, i.e. X satisfies the Grothendieck theorem if (see [24], page 71)

$$\mathcal{L}(X, \ell_2) = \Pi_1(X, \ell_2). \tag{3}$$

The basic theory of p-summing operators, type and cotype can be found, for example, in the books [11], [9], [16], [26], [23], [24] or [27] .

In this paper we restrict ourselves to the case $p = q$ for simplicity, although some of the results presented here can be easily stated in the general case. The paper is divided into three sections. In the first one we shall give several examples of p-summing multipliers. In the second one we show some general results extending known facts in the study of p-summing operators to p-summing multipliers. In the last section we relate this new notion to the class of almost summing operators or Rademacher bounded sequences and find some sufficient conditions for a sequence to belong to $\ell_{\pi_p}(X, Y)$, at least for certain spaces X and Y.

Throughout the paper (e_j) denotes the canonical basis of the sequence spaces ℓ_p and c_0, $\langle x^*, x\rangle$ the duality pairing between X^* and X, p' the conjugate exponent of p, $\mathbb{K}$ stands for $\mathbb{R}$ or $\mathbb{C}$ and, as usual, C denotes a constant that may vary from line to line.

2. Definition and examples.

It is not difficult to show (see [1] Proposition 2.1) that $(\ell_p(X), \ell_p(Y)) = \ell_\infty(\mathcal{L}(X, Y))$ for any couple of Banach spaces X and Y and $1 \leq p \leq \infty$. Let us give a name to the multipliers corresponding to $(\ell_p^w(X), \ell_p(Y))$.

Definition 2.1 *(see [1]) Let X and Y be Banach spaces, and let $1 \leq p, q < \infty$. A sequence $(T_j)_{j\in\mathbb{N}}$ of operators in $\mathcal{L}(X, Y)$ is a (p, q)-summing multiplier if there exists a constant $C > 0$ such that, for any finite collection of vectors $x_1, x_2, \ldots x_n$ in X, it holds that*

$$\Big(\sum_{j=1}^{n} ||T_j x_j||^p\Big)^{1/p} \leq C \sup_{||x^*||=1} \Big(\sum_{j=1}^{n} |\langle x^*, x_j\rangle|^q\Big)^{1/q}.$$

We use $\ell_{\pi_{p,q}}(X,Y)$ to denote the set of (p,q)-summing multipliers, and $\pi_{p,q}[T_j]$ is the least constant C for which (T_j) verifies the inequality in the definition. In order to avoid ambiguities, sometimes we shall use $\pi_{p,q}[T_j; X, Y]$.

We shall only deal with the case $p = q$. The space $\ell_{\pi_{p,p}}(X,Y)$ will be denoted $\ell_{\pi_p}(X,Y)$, its norm π_p and its elements will be called p-summing multipliers. It is not difficult to show (see [1]) that if X and Y are Banach spaces and $1 \leq p < \infty$ then $(\ell_{\pi_p}(X,Y), \pi_p)$ is a Banach space.

Remark 2.1 *A sequence $(T_j) \in \ell_{\pi_1}(X,Y)$ if and only if it holds that for any unconditionally convergent series $\sum x_j$ in X we have $(T_j(x_j))_j \in \ell_1(Y)$ (see [1]).*

Remark 2.2 *Let $1 \leq p < \infty$. A sequence $(T_j) \in \ell_{\pi_1}(X,Y)$ if and only if the map $(y_j^*) \to (T_j^*(y_j^*))$ is bounded from $\ell_{p'}(Y^*)$ into $\ell_{\pi_{1,p}}(X,\mathbb{K})$.*

Moreover $\pi_p[T_n; X, Y] = \sup_{||y_n||_{\ell_{p'}(Y^*)}=1} \pi_{1,p}[T_n^*(y_n); X, \mathbb{K}]$.

Let us now mention some basic examples of p-summing multipliers in different contexts.

Example 2.1 *Let $1 \leq p < \infty$ and μ be a probality measure on a compact set K. Let (ϕ_n) be a sequence of continuous functions and define $T_n : C(K) \to L^p(\mu)$ by $T_n(\psi) = \phi_n\psi$.*

If $(\sum_{n=1}^{\infty} |\phi_n|^{p'})^{1/p'} \in L^p(\mu)$ then $T_n \in \ell_{\pi_p}(C(K), L^p(\mu))$.

Proof. Assume $p > 1$ (the case p=1 is left to the reader). Let $n \in \mathbb{N}$ and $\psi_1, \psi_2, ..., \psi_n$ in $C(K)$. Recalling that

$$||(\psi_n)||_{\ell_p^w(C(K))} = ||(\sum_{k=1}^{n} |\psi_k|^p)^{1/p}||_\infty \tag{4}$$

then

$$\begin{aligned}\sum_{k=1}^{n} ||T_k(\psi_k)||^p_{L^p(\mu)} &= \int_K \sum_{k=1}^{n} |\phi_k\psi_k|^p d\mu \\ &\leq \int_K (\sum_{k=1}^{n} |\phi_k|^{p'})^{p/p'} (\sum_{k=1}^{n} |\psi_k|^p) d\mu \\ &\leq ||(\sum_{k=1}^{n} |\psi_k|^p)^{1/p}||^p_\infty \int_K (\sum_{k=1}^{n} |\phi_k|^{p'})^{p/p'} d\mu.\end{aligned}$$

This shows that $\pi_p[T_j] \leq (\int_K (\sum_{k=1}^{n} |\phi_k|^{p'})^{p/p'} d\mu)^{1/p}$. ■

Example 2.2 *Let $1 \leq p < \infty$, (Ω, Σ, μ) and (Ω', Σ', μ') be finite measure spaces. Let $(f_n) \subset L^p(\mu, L^1(\mu'))$ and consider the operators $T_n : L^\infty(\mu') \to L^p(\mu)$ given by $T_n(\phi) = \langle \phi, f_n \rangle = \int_{\Omega'} \phi(w') f_n(., w') d\mu(w')$.*

If $\sup_n |f_n(w, w')| \in L^p(\mu, L^1(\mu'))$ then $T_n \in \ell_{\pi_p}(L^\infty(\mu'), L^p(\mu))$.

Proof. Given $n \in \mathbb{N}$ and $\phi_1, \phi_2, ..., \phi_n$ in $L^\infty(\mu')$ then

$$\begin{aligned}
\sum_{k=1}^{n} ||T_k(\phi_k)||^p_{L^p(\mu)} &= \sum_{k=1}^{n} \int_\Omega |\langle \phi_k, f_k(w)\rangle|^p d\mu(w) \\
&= \int_\Omega \sum_{k=1}^{n} |\int_{\Omega'} \phi_k(w') f_k(w,w') d\mu(w')|^p d\mu(w) \\
&\leq \int_\Omega \Big(\int_{\Omega'} (\sum_{k=1}^{n} |\phi_k(w')|^p |f_k(w,w')|^p)^{1/p} d\mu(w') \Big)^p d\mu(w) \\
&\leq \int_\Omega \Big(\int_{\Omega'} \sup_k |f_k(w,w')| (\sum_{k=1}^{n} |\phi_k(w')|^p)^{1/p} d\mu(w') \Big)^p d\mu(w) \\
&\leq ||(\sum_{k=1}^{n} |\phi_k(w')|^p)^{1/p}||^p_{L^\infty(\mu')} \int_\Omega \Big(\int_{\Omega'} \sup_k |f_k(w,w')| d\mu(w') \Big)^p d\mu(w).
\end{aligned}$$

This shows, using (4), that $\pi_p[T_j] \leq ||sup_n|f_n(w,w')|||_{L^p(\mu, L^1(\mu'))}$. ■

Example 2.3 *Let $1 \leq p < \infty$ and (A_n) be a sequence of matrices such that $T_n((\lambda_k)) = (\sum_{k=1}^\infty A_n(k,j)\lambda_k)_j$ defines bounded operators from c_0 to ℓ_p. If*

$$\sum_{k=1}^{\infty} \sup_n (\sum_{j=1}^{\infty} |A_n(k,j)|^p)^{1/p} < \infty$$

then $(T_n) \in \ell_{\pi_1}(c_0, \ell_p)$.

Proof. Note that $T_n = \sum_{k=1}^\infty e_k^* \otimes y_{n,k}$ where $(y_{n,k}) \in \ell_p$ is given by $y_{n,k} = (A_n(k,j))_j$. Hence, if $x_n = (\lambda_{n,k})_k$ then

$$\begin{aligned}
\sum_{n=1}^{\infty} ||T_n(x_n)|| &\leq \sum_{n=1}^{\infty}\sum_{k=1}^{\infty} |\langle e_k^*, x_n\rangle| ||y_{n,k}|| \\
&= \sum_{k=1}^{\infty}\sum_{n=1}^{\infty} |\lambda_{n,k}| (\sum_{j=1}^{\infty} |A_n(k,j)|^p)^{1/p} \\
&\leq \sum_{k=1}^{\infty} \sup_n (\sum_{j=1}^{\infty} |A_n(k,j)|^p)^{1/p} \sum_{n=1}^{\infty} |\lambda_{n,k}| \\
&\leq (\sup_k \sum_{n=1}^{\infty} |\lambda_{n,k}|) \sum_{k=1}^{\infty} \sup_n (\sum_{j=1}^{\infty} |A_n(k,j)|^p)^{1/p} \\
&= ||(x_n)||_{\ell_1^w(c_0)} \sum_{k=1}^{\infty} \sup_n (\sum_{j=1}^{\infty} |A_n(k,j)|^p)^{1/p}.
\end{aligned}$$

■

Example 2.4 *Let $f \in L^1([0,1] \times [0,1])$ and measurable sets $E_n \subset [0,1]$ for $n \in \mathbb{N}$. Let $T_n : L^\infty([0,1]) \to L^1([0,1])$ be defined by $T_n(\phi)(t) = (\int_0^1 f(t,s)\phi(s)ds)\chi_{E_n}(t)$. Then $(T_n) \in \ell_{\pi_2}(L^\infty([0,1]), L^1([0,1]))$.*

Proof. First observe that f can be regarded as a function in $L^1([0,1], L^1([0,1]))$ and then $\phi \to \int_0^1 f(.,s)\phi(s)ds$ defines a bounded operator from $L^\infty([0,1])$ to $L^1([0,1])$ with norm ≤ 1.

Given $n \in \mathbb{N}$ and $\phi_1, \phi_2, ..., \phi_n$ in $L^\infty([0,1])$ we have, using 2

$$\begin{aligned}\sum_{k=1}^n ||T_k(\phi_k)||_{L^1}^2 &= \sum_{k=1}^n ||(\int_0^1 f(.,s)\phi_k(s)ds)\chi_{E_k}||_{L^1}^2 \\ &\leq \sum_{k=1}^n ||(\int_0^1 f(.,s)\phi_k(s)ds)||_{L^1}^2 \\ &\leq K_G^2 ||f||_{L^1}^2 ||(\sum_{k=1}^n |\phi_k|^2)^{1/2}||_\infty^2.\end{aligned}$$

This shows that $\pi_2[T_j] \leq K_G ||f||_{L^1}$. ■

Example 2.5 *Let $u \in h^2(\mathbb{D})$, i.e. a harmonic function on the unit disc $\mathbb{D}$ such that $\sup_{0<r<1} \int_{-\pi}^{\pi} |u_r(e^{it})|^2 \frac{dt}{2\pi} < \infty$ where $u_r(e^{it}) = u(re^{it})$. Let us fix an increasing sequence r_n converging to 1 and define $T_n : L^1(\mathbb{T}) \to L^2(\mathbb{T})$ by $T_n(\psi) = \psi * u_{r_n}$. Then $(T_n) \in \ell_{\pi_1}(L^1(\mathbb{T}), L^2(\mathbb{T}))$.*

Proof. It is well known (see [13]) that $u_r = P_r * \phi$ for some $\phi \in L^2(\mathbb{T})$ where P_r stands for the Poisson kernel. Therefore $T_n(\psi) = \psi * \phi * P_{r_n}$.

Given $n \in \mathbb{N}$ and $\psi_1, \psi_2, ..., \psi_n$ we have, using now (1) for the operator $T : L^1(\mathbb{T}) \to L^2(\mathbb{T})$ given by $T(\psi) = \psi * \phi$,

$$\begin{aligned}\sum_{k=1}^n ||T_k(\psi_k)||_{L^2} &= \sum_{k=1}^n ||\psi_k * \phi * P_{r_n}||_{L^2} \\ &\leq \sum_{k=1}^n ||\psi_k * \phi||_{L^2} \\ &\leq K_G ||(\psi_k)||_{\ell_1^w(L^1)} ||\phi||_{L^2}.\end{aligned}$$

Therefore one gets $\pi_2[T_j] \leq K_G ||\phi||_{L^2} = K_G ||u||_{h^2}$. ■

3. General facts on p-summing multipliers.

Let us start with some simple observations to get examples of p-summing multipliers. Examples 2.4 and 2.5 fall under the following general principle whose proof is left to the reader.

Proposition 3.1 *Let X, Y and Z be Banach spaces and $1 \leq p < \infty$. If $T \in \Pi_p(X, Y)$ and $(S_n) \in \ell_\infty(\mathcal{L}(Y, Z))$ then $(S_n T) \in \ell_{\pi_p}(X, Z)$.*

Moreover $\pi_p[S_n T] \leq \pi_p[T] \sup_n ||S_n||$.

Example 2.3 is also a particular case of the following:

Proposition 3.2 *Let X, Y be Banach spaces and $1 \leq p < \infty$.*
Given $(y_{n,k}) \subset \ell_\infty(\mathbb{N}\times\mathbb{N}, Y)$ and $(x_k^*) \in \ell_1^w(X^*)$ let us consider $T_n = \sum_{k=1}^\infty x_k^* \otimes y_{n,k}$.
If $\sum_{k=1}^\infty ||x_k^*||(\sup_n ||y_{n,k}||) < \infty$ then $T_n \in \ell_{\pi_1}(X, Y)$.

Proof. Notice that

$$\begin{aligned}\sum_{n=1}^\infty ||T_n(x_n)|| &\leq \sum_{n=1}^\infty\sum_{k=1}^\infty |\langle x_k^*, x_n\rangle|||y_{n,k}|| \\ &= \sum_{k=1}^\infty\sum_{n=1}^\infty |\langle \frac{x_k^*}{||x_k^*||}, x_n\rangle|||x_k^*||||y_{n,k}|| \\ &\leq (\sup_{||x^*||=1}\sum_{n=1}^\infty |\langle x^*, x_n\rangle|)\sum_{k=1}^\infty ||x_k^*||(\sup_n ||y_{n,k}||).\end{aligned}$$

■

Lemma 3.3 *Let X be a Banach space, $n \in \mathbb{N}$, $x_1, x_2, ..., x_n \in X$ and $x_1^*, x_2^*, ..., x_n^* \in X^*$. Then*

$$\sum_{k=1}^n |\langle x_k^*, x_k\rangle| \leq (\sup_{||x^*||=1}\sum_{k=1}^n |\langle x_k, x^*\rangle|)\int_0^1 ||\sum_{k=1}^n x_k^* r_k(t)||dt.$$

Proof.

$$\begin{aligned}\sum_{k=1}^n |\langle x_k^*, x_k\rangle| &= \sup_{|\alpha_k|=1} |\sum_{k=1}^n \langle x_k^*, x_k\alpha_k\rangle| \\ &= \sup_{|\alpha_k|=1} |\int_0^1 \langle \sum_{k=1}^n x_k\alpha_k r_k(t), \sum_{k=1}^n x_k^* r_k(t)\rangle dt| \\ &\leq \sup_{|\alpha_k|=1}\sup_{t\in[0,1]} ||\sum_{k=1}^n x_k\alpha_k r_k(t)|| \int_0^1 ||\sum_{k=1}^n x_k^* r_k(t)||dt \\ &\leq (\sup_{||x^*||=1}\sum_{k=1}^n |\langle x_k, x^*\rangle|)\int_0^1 ||\sum_{k=1}^n x_k^* r_k(t)||dt.\end{aligned}$$

■

Proposition 3.4 *Let X and Y be Banach spaces. If $(T_n) \subset \mathcal{L}(X, Y))$ is such that*

$$\sup_{||x||=1}\sum_{k=1}^\infty ||T_k(x)|| < \infty$$

then $T_n \in \ell_{\pi_1}(X, Y)$. Moreover $\pi_1[T_n] \leq \sup_{||x||=1}\sum_{k=1}^\infty ||T_k(x)||$.

Proof. Given $n \in \mathbb{N}$ and $x_1, x_2, ..., x_n \in X$ we have, using Lemma 3.3,

$$\begin{aligned}
\sum_{k=1}^{n} ||T_k(x_k)|| &= \sup_{||y_k^*||=1} \sum_{k=1}^{n} |\langle T_k(x_k), y_k^*\rangle| \\
&= \sup_{||y_k^*||=1} \sum_{k=1}^{n} |\langle x_k, T_k^*(y_k^*)\rangle| \\
&\leq ||(x_n)||_{\ell_1^w(X)} \sup_{||y_k^*||=1} \int_0^1 || \sum_{k=1}^{n} T_k^*(y_k^*) r_k(t)||dt \\
&\leq ||(x_n)||_{\ell_1^w(X)} \sup_{||y_k^*||=1} \sup_{t\in[0,1]} || \sum_{k=1}^{n} T_k^*(y_k^*) r_k(t)|| \\
&\leq ||(x_n)||_{\ell_1^w(X)} \sup_{||y_k^*||=1, ||x||=1} \sum_{k=1}^{n} |\langle T_k^*(y_k^*), x\rangle| \\
&\leq ||(x_n)||_{\ell_1^w(X)} \sup_{||x||=1} \sum_{k=1}^{n} ||T_k(x)||.
\end{aligned}$$

■

Theorem 3.5 *Let X, Y and Z be Banach spaces and $1 \leq p < \infty$.*

i) If $(T_n) \in \ell_{\pi_p}(X,Y)$ and $(S_n) \in \ell_\infty(\mathcal{L}(Y,Z))$ then $(S_nT_n) \in \ell_{\pi_p}(X,Z)$. Moreover $\pi_p[S_nT_n] \leq \pi_p[T_n] \sup_n ||S_n||$.

ii) If $(S_n) \in \ell_1^w(\mathcal{L}(X,Y))$ and $(T_n) \in \ell_{\pi_p}(Y,Z)$ then $(T_nS_n) \in \ell_{\pi_p}(X,Z)$. Moreover $\pi_p[T_nS_n] \leq \pi_p[T_n]||(S_n)||_{\ell_1^w(\mathcal{L}(X,Y))}$.

iii) If $T \in \mathcal{L}(X,Y)$ and $(T_n) \in \ell_{\pi_p}(Y,Z)$ then $T_nT \in \ell_{\pi_p}(X,Z)$. Moreover $\pi_p[T_nT] \leq \pi_p[T_n]||T||$.

iv) If $T \in \Pi_2(X,Y)$ and $(T_n) \in \ell_{\pi_2}(Y,Z)$ then $T_nT \in \ell_{\pi_1}(X,Z)$. Moreover $\pi_1[T_nT] \leq \pi_2[T_n]\pi_2[T]$.

Proof. (i) Take $n \in \mathbb{N}$ and $x_1, x_2, ..., x_n \in X$. Then

$$\begin{aligned}
\sum_{k=1}^{n} ||S_kT_k(x_k)||^p &\leq \sum_{k=1}^{n} ||S_k||^p||T_k(x_k)||^p \\
&\leq \sup_n ||S_n||^p \pi_p^p[T_n]||(x_n)||^p_{\ell_p^w(X)}.
\end{aligned}$$

(ii) Take $n \in \mathbb{N}$ and $x_1, x_2, ..., x_n \in X$. Then

$$\begin{aligned}
(\sum_{k=1}^{n} ||T_kS_k(x_k)||^p)^{1/p} &\leq \pi_p[T_n] \sup_{||y^*||=1} (\sum_{k=1}^{n} |\langle S_k(x_k), y^*\rangle|^p)^{1/p} \\
&= \pi_p[T_n] \sup_{||y^*||=1} (\sum_{k=1}^{n} |\langle x_k, S_k^*(y^*)\rangle|^p)^{1/p}
\end{aligned}$$

$$
\begin{aligned}
&= \pi_p[T_n] \sup_{\|y^*\|=1} \sup_{\|(\alpha_n)\|_{\ell_{p'}}=1} |\sum_{k=1}^{n} \langle x_k\alpha_k, S_k^*(y^*)\rangle| \\
&= \pi_p[T_n] \sup_{\|y^*\|=1} \sup_{\|(\alpha_n)\|_{\ell_{p'}}=1} |\int_0^1 \langle \sum_{k=1}^{n} x_k\alpha_k r_k(t), \sum_{k=1}^{n} S_k^*(y^*) r_k(t)\rangle dt| \\
&\leq \pi_p[T_n] (\sup_{\|(\alpha'_n)\|_{\ell_{p'}}=1} \|\sum_{k=1}^{n} x_k\alpha'_k\|)(\sup_{\|y^*\|=1} \sup_{t\in[0,1]} \|\sum_{k=1}^{n} S_k^*(y^*) r_k(t)\|) \\
&= \pi_p[T_n] \|(x_k)\|_{\ell_p^w(X)} \sup_{\|y^*\|=1, \|x\|=1} \sum_{k=1}^{n} |\langle S_k^*(y^*), x\rangle| \\
&= \pi_p[T_n] \|(x_k)\|_{\ell_p^w(X)} \sup_{\|y^*\|=1, \|x\|=1} \sum_{k=1}^{n} |\langle y^*, S_k(x)\rangle| \\
&= \pi_p[T_n] \|(x_k)\|_{\ell_p^w(X)} \|(S_n)\|_{\ell_1^w(\mathcal{L}(X,Y))}.
\end{aligned}
$$

(iii) Take $n \in \mathbb{N}$ and $x_1, x_2, ..., x_n \in X$. Then

$$
\begin{aligned}
\sum_{k=1}^{n} \|T_k T(x_k)\|^p &\leq \pi_p^p[T_n] \sup_{\|z^*\|=1} \sum_{k=1}^{n} |\langle T(x_k), z^*\rangle|^p \\
&= \pi_p^p[T_n] \sup_{\|z^*\|=1} \sum_{k=1}^{n} |\langle x_k, T^*(z^*)\rangle|^p \\
&\leq \pi_p^p[T_n] \sup_{\|x^*\|=1} \sum_{k=1}^{n} |\langle x_k, x^*\rangle|^p.
\end{aligned}
$$

(iv) Given $(x_n) \in \ell_1^w(X)$ and $T \in \Pi_2(X,Y)$ then $T(x_n) = \alpha_n x'_n$ where $\alpha_n \in \ell_2$ and $x'_n \in \ell_2^w(X)$ and $\|(x'_n)\|_{\ell_2^w(X)} \leq \|(x_n)\|_{\ell_1^w(X)}^{1/2} \pi_2[T]$ and $\|(\alpha_n)\|_{\ell_2} \leq \|(x_n)\|_{\ell_1^w(X)}^{1/2}$ (see [11] page 53). Hence, for each $n \in \mathbb{N}$

$$
\begin{aligned}
\sum_{k=1}^{n} \|T_k(Tx_k)\| &= \sum_{k=1}^{n} \|T_k(x'_k)\| |\alpha_k)| \\
&\leq \pi_2[T_n] \|(x'_n)\|_{\ell_2^w(X)} \|(\alpha_n)\|_{\ell_2} \\
&\leq \pi_2[T_n] \pi_2[T] \|(x_n)\|_{\ell_1^w(X)}.
\end{aligned}
$$

■

Let us now prove the natural generalization of the fact that $T \in \Pi_p(X,Y)$ if and only if $T^{**} \in \Pi_p(X^{**}, Y^{**})$. We need the following lemma.

Lemma 3.6 *(see [1], Proposition 2.9) Let X be a Banach space, $1 \leq p < \infty$ and let (x_j^*) be a sequence in X^*. Then $(x_j^*) \in \ell_{\pi_{1,p}}(X, \mathbb{K})$ if and only if there exists $C > 0$ such that*

$$
\sum_{j=1}^{n} |\langle x_j^{**}, x_j^*\rangle| \leq C \sup_{\|x^*\|=1} \Big(\sum_{j=1}^{n} |\langle x_j^{**}, x^*\rangle|^p \Big)^{1/p}
$$

for every $x_1^{**}, \ldots, x_n^{**}$ in X^{**}.

Theorem 3.7 *Let X and Y be Banach spaces, $1 \leq p < \infty$ and let $(T_n) \in \mathcal{L}(X,Y)$. Then $(T_n) \in \ell_{\pi_p}(X,Y)$ if and only if $(T_n^{**}) \in \ell_{\pi_p}(X^{**},Y^{**})$.*

Proof. The only thing to show is that if $(T_n) \in \ell_{\pi_p}(X,Y)$ then $(T_n^{**}) \in \ell_{\pi_p}(X^{**},Y^{**})$.

We have to show that there exists $C > 0$ for which

$$\Big(\sum_{j=1}^{n} ||T_j^{**}(x_j^{**})||^p\Big)^{1/p} \leq C$$

for any $x_1^{**}, \ldots, x_n^{**}$ in X^{**} such that $\sup_{||x^*||=1}\Big(\sum_{j=1}^{n} | < x_j^{**}, x^* > |^p\Big)^{1/p} = 1$.

Given $(y_j^*) \in \ell_{p'}(Y^*)$, Remark 2.2 shows that $(T_j^*(y_j^*)) \in \ell_{\pi_{1,p}}(X,\mathbb{K})$. Now Lemma 3.6 gives

$$\sum_{j=1}^{n} |\langle x_j^{**}, T_j^*(y_j^*)\rangle| = \sum_{j=1}^{n} |\langle T_j^{**}(x_j^{**}), y_j^*\rangle| \leq C.$$

Therefore the result is achieved from the duality $(\ell_{p'}(Y^*))^* = \ell_p(Y^{**})$. ■

sectionConnections with other classes of operators and geometry of Banach spaces.

Regarding embeddings between the spaces, let us mention that for $1 \leq p \leq q < \infty$ one has $\ell_{\pi_p}(X,Y) \subset \ell_{\pi_q}(X,Y)$. The reader is referred to [1] for general embedding theorems. The next result generalizes the well known fact of the coincidence of the classes $\Pi_1(X,Y) = \Pi_2(X,Y)$ under the assumption of cotype 2 of X (see [11], Corollary 11.16). The following is essentially contained in Corollaries 3.12 and 3.13 in [1], but we include a proof here for completeness.

Theorem 3.8

i) If X has cotype 2 then $\ell_{\pi_1}(X,Y) = \ell_{\pi_2}(X,Y)$.

ii) If X has cotype $q > 2$ then $\ell_{\pi_1}(X,Y) = \ell_{\pi_p}(X,Y)$ for any $p < q'$.

Proof. (i) Let us take $(T_n) \in \ell_{\pi_2}(X,Y)$ and let $(x_n) \in \ell_1^w(X)$. According to the identification with $\mathcal{L}(c_0,X)$ we have that the sequence $x_n = u(e_n)$ for some $u \in \mathcal{L}(c_0,X)$. Using now the cotype 2 assumption we have $\mathcal{L}(c_0,X) = \Pi_2(c_0,X)$ (see [11],Theorem 11.14). Now, since $(e_n) \in \ell_1^w(c_0)$ and $u \in \Pi_2(c_0,X)$ then (see [11], Lemma 2.23) $u(e_n) = \alpha_n x'_n$ where $\alpha_n \in \ell_2$ and $x'_n \in \ell_2^w(X)$ and $||(x'_n)||_{\ell_2^w(X)} \leq \pi_2[u]$ and $||(\alpha_n)||_{\ell_2} \leq 1$ Hence, for each $n \in \mathbb{N}$

$$\begin{aligned}
\sum_{k=1}^{n} ||T_k(x_k)|| &= \sum_{k=1}^{n} ||T_k(x'_k)|||\alpha_k)| \\
&\leq \pi_2[T_n]||(x'_n)||_{\ell_2^w(X)}||(\alpha_n)||_{\ell_2} \\
&\leq \pi_2[T_n]\pi_2[u] \\
&\leq K_G\pi_2[T_n]||(x_n)||_{\ell_1^w(X)}.
\end{aligned}$$

(ii) follows the same lines (using Theorem 11.14 and Lemma 2.23 in [11]) for $q > 2$. ■

Definition 3.9 *(see [11], page 234) Let X and Y be Banach spaces. A linear operator $T: X \to Y$ is said to be almost summing, to be denoted $T \in \Pi_{as}(X,Y)$, if there exists $C > 0$ such that*

$$\int_0^1 \|\sum_{j=1}^n T(x_j)r_j(t)\|dt \leq C \sup_{\|x^*\|=1} (\sum_{j=1}^n |\langle x^*, x_j\rangle|^2)^{1/2}$$

for any finite family $x_1, x_2, \ldots x_n$ of vectors in X.

The least of such constants is the as-summing norm of u, denoted by $\pi_{as}(u)$.

Let us now relate these operators with p-summing multipliers.

Theorem 3.10 *Let X and H be a Banach and a Hilbert space, respectively. If $(T_n) \subset \mathcal{L}(X,H)$ are such that $T_n^* \in \Pi_{as}(H, X^*)$ for all $n \in \mathbb{N}$ and*

$$\sup_n \int_0^1 \pi_{as}[\sum_{k=1}^n T_k^* r_k(s)]ds < \infty$$

then $T_n \in \ell_{\pi_1}(X,H)$.

Moreover $\pi_1[T_n] \leq \sup_n \int_0^1 \pi_{as}[\sum_{k=1}^n T_k^* r_k(s)]ds$.

Proof. Let $(x_n) \in \ell_1^w(X)$. Then

$$\begin{aligned}
\sum_{k=1}^n \|T_k(x_k)\| &= \sum_{k=1}^n (\sum_{j=1}^\infty |\langle T_k(x_k), e_j\rangle|^2)^{1/2} \\
&\leq C \sum_{k=1}^n \int_0^1 |\sum_{j=1}^\infty \langle T_k(x_k), e_j\rangle r_j(t)|dt \\
&\leq C \int_0^1 \sum_{k=1}^n |\langle x_k, \sum_{j=1}^\infty T_k^*(e_j) r_j(t)\rangle|dt.
\end{aligned}$$

First note that, since $(e_n) \in \ell_2^w(H)$, then $S \in \Pi_{as}(H, X^*)$ implies

$$\int_0^1 \|\sum_{j=1}^\infty S(e_j)r_j(t)\|dt \leq \pi_{as}(S).$$

Now using Lemma 3.3 we get

$$\begin{aligned}
\sum_{k=1}^n \|T_k(x_k)\| &\leq C\|(x_n)\|_{\ell_1^w(X)} \int_0^1 (\int_0^1 \|\sum_{k=1}^n (\sum_{j=1}^\infty T_k^*(e_j)r_j(t))r_k(s)\|ds)dt \\
&\leq C\|(x_n)\|_{\ell_1^w(X)} \int_0^1 (\int_0^1 \|(\sum_{k=1}^n T_k^* r_k(s))(\sum_{j=1}^\infty e_j r_j(t))\|dt)ds \\
&= C\|(x_n)\|_{\ell_1^w(X)} \int_0^1 (\int_0^1 \|\sum_{j=1}^\infty (\sum_{k=1}^n T_k^* r_k(s))(e_j)r_j(t)\|dt)ds \\
&\leq C\|(x_n)\|_{\ell_1^w(X)} \int_0^1 \pi_{as}[\sum_{k=1}^n T_k^* r_k(s)]ds.
\end{aligned}$$

■

Theorem 3.11 *Let $2 \leq q < \infty$, H be a Hilbert space and X be a Banach space with the Orlicz property (for $q = 2$) or cotype $q > 2$. If $(T_n) \subset \mathcal{L}(X, H)$ are such that $T_n^* \in \Pi_{as}(H, X^*)$ for all $n \in \mathbb{N}$ and*

$$\sup_n \int_0^1 (\sum_{k=1}^n ||T_k^*(\sum_{j=1}^\infty e_j r_j(t))||^{q'})^{1/q'} dt < \infty$$

then $(T_n) \in \ell_{\pi_1}(X, H)$.

Proof. Let $(x_n) \in \ell_1^w(X)$. Then for each $n \in \mathbb{N}$, the argument in Theorem 3.10 gives

$$\sum_{k=1}^n ||T_k(x_k)|| \leq C \int_0^1 \sum_{k=1}^n |\langle x_k, \sum_{j=1}^\infty T_k^*(e_j) r_j(t)\rangle dt|.$$

Now the assumption on X allows us to write

$$\begin{aligned}\sum_{k=1}^n ||T_k(x_k)|| &\leq C \int_0^1 \sum_{k=1}^n |\langle x_k, \sum_{j=1}^\infty T_k^*(e_j) r_j(t)\rangle dt| \\ &\leq C \int_0^1 (\sum_{k=1}^n ||x_k||^q)^{1/q} (\sum_{k=1}^n || \sum_{j=1}^\infty T_k^*(e_j) r_j(t)||^{q'})^{1/q'} dt \\ &\leq C ||(x_k)||_{\ell_1^w(X)} \int_0^1 (\sum_{k=1}^n || \sum_{j=1}^\infty T_k^*(e_j) r_j(t)||^{q'})^{1/q'} dt.\end{aligned}$$

■

Definition 3.12 *(see [3], [8]) Let X and Y be Banach spaces. A sequence $(T_j)_{j \in \mathbb{N}}$ of operators in $\mathcal{L}(X, Y)$ is called Rademacher bounded if there exists a constant $C > 0$ such that*

$$\int_0^1 || \sum_{k=1}^n T_k(x_k) r_k(t)|| dt \leq C \int_0^1 || \sum_{k=1}^n x_k r_k(t)|| dt$$

for any finite collection of vectors $x_1, x_2, \ldots x_n$ in X.

We use $Rad(X, Y)$ to denote the set of Rademacher bounded sequences, and $rad[T_j]$ is the least constant C for which (T_j) verifies the inequality in the definition.

Remark 3.1

i) If $T_n = T$ for all $n \in \mathbb{N}$ then $(T_n) \in Rad(X, Y)$.

ii) If $(T_n) \in Rad(X, Y)$ and $(x_n) \in \ell_1^w(X)$ then $(T_n(x_n)) \in \ell_2^w(X)$.

Let us mention the following simple observations whose proofs follow easily from the definitions.

Proposition 3.13 *Let X, Y be Banach spaces.*

i) If X has the Orlicz property (resp. cotype $q > 1$) then $\ell_2(\mathcal{L}(X, Y)) \subset \ell_{\pi_1}(X, Y)$ (resp. $\ell_{q'}(\mathcal{L}(X, Y)) \subset \ell_{\pi_1}(X, Y)$).

ii) If Y has type 2 then $\ell_{\pi_2}(X,Y) \subset Rad(X,Y)$.

iii) If X has cotype q, Y has type p and $1/r = (1/p) - (1/q)$ then $\ell_r(\mathcal{L}(X,Y)) \subset Rad(X,Y)$. In particular, if X has cotype 2 and Y has type 2 then $\ell_\infty(\mathcal{L}(X,Y)) = Rad(X,Y)$.

iv) If Z has cotype 2, $T \in \Pi_{as}(X,Y)$ and $(T_n) \in Rad(Y,Z)$ then $(T_nT) \in \ell_{\pi_2}(X,Z)$.

Proof. (i) Let $n \in \mathbb{N}$ and $x_1, x_2, ..., x_n$ in X. Then we have

$$\begin{aligned}\sum_{k=1}^{n} ||T_k(x_k)|| &\leq (\sum_{k=1}^{n} ||T_k||^2)^{1/2}(\sum_{k=1}^{n} ||x_k||^2)^{1/2} \\ &\leq C(\sum_{k=1}^{n} ||T_k||^2)^{1/2}||(x_k)||_{\ell_1^w(X)}.\end{aligned}$$

Obvious modifications give the case $q > 2$.

(ii) Let $n \in \mathbb{N}$ and $x_1, x_2, ..., x_n$ in X. Then we have

$$\begin{aligned}\int_0^1 ||\sum_{k=1}^{n} T_k(x_k)r_k(t)||dt &\leq C(\sum_{k=1}^{n} ||T_k(x_k)||^2)^{1/2} \\ &\leq C\pi_2[T_n]||(x_k)||_{\ell_2^w(X)} \\ &\leq C\pi_2[T_n]\int_0^1 ||\sum_{k=1}^{n} x_k r_k(t)||dt.\end{aligned}$$

(iii) Let $n \in \mathbb{N}$ and $x_1, x_2, ..., x_n$ in X. Then we have

$$\begin{aligned}\int_0^1 ||\sum_{k=1}^{n} T_k(x_k)r_k(t)||dt &\leq C(\sum_{k=1}^{n} ||T_k(x_k)||^p)^{1/p} \\ &\leq C(\sum_{k=1}^{n} ||T_k||^r)^{1/r}(\sum_{k=1}^{n} ||x_k||^q)^{1/q} \\ &\leq C(\sum_{k=1}^{n} ||T_k||^r)^{1/r}\int_0^1 ||\sum_{k=1}^{n} x_k r_k(t)||dt.\end{aligned}$$

(iv) Let $n \in \mathbb{N}$ and $x_1, x_2, ..., x_n$ in X. Then we have

$$\begin{aligned}(\sum_{k=1}^{n} ||T_kT(x_k)||^2)^{1/2} &\leq C\int_0^1 ||\sum_{k=1}^{n} T_kT(x_k)r_k(t)||dt \\ &\leq Crad[T_n]\int_0^1 ||\sum_{k=1}^{n} T(x_k)r_k(t)||dt \\ &\leq Crad[T_n]\pi_{as}[T]\sup_{||x^*||=1}(\sum_{k=1}^{n} |\langle x_k, x^*\rangle|^2)^{1/2}.\end{aligned}$$

■

We are now going to get the main results of this section. We need the following lemma.

Lemma 3.14 *(see [24], Theorem 6.6 and Corollary 6.7) If X is a GT-space of cotype 2 then there exists a constant $C > 0$ such that*

$$\sum_{k=1}^{n} |\langle x_k, x_k^* \rangle| \leq C\left(\int_0^1 ||\sum_{k=1}^{n} x_k r_k(t)||dt\right) \sup_{||x||=1} \left(\sum_{k=1}^{n} |\langle x_k^*, x\rangle|^2\right)^{1/2}. \tag{5}$$

If X^* is a GT-space of cotype 2 then there exists a constant $C > 0$ such that

$$\sum_{k=1}^{n} |\langle x_k^*, x_k \rangle| \leq C\left(\int_0^1 ||\sum_{k=1}^{n} x_k^* r_k(t)||dt\right) \sup_{||x^*||=1} \left(\sum_{k=1}^{n} |\langle x_k, x^*\rangle|^2\right)^{1/2}. \tag{6}$$

Theorem 3.15 *Let (Ω, Σ, μ) be a measure space and X a Banach space.*
If $T_n \subset \mathcal{L}(L^1(\mu), X)$ are such that

$$\sup_{||\phi||=1} \sum_{n=1}^{\infty} ||T_n(\phi)||^2 < \infty$$

then $(T_n) \in \ell_{\pi_1}(L^1(\mu), X)$.

Proof. Let $(\phi_n) \subset L^1(\mu)$. Since $L^1(\mu)$ is a GT-space of cotype 2, Lemma 3.14 gives

$$\begin{aligned}
\sum_{k=1}^{n} ||T_k(\phi_k)|| &= \sup_{||x_n^*||=1} \sum_{k=1}^{n} |\langle T_k(\phi_k), x_k^*\rangle| \\
&= \sup_{||x_n^*||=1} \sum_{k=1}^{n} |\langle \phi_k, T_k^*(x_k^*)\rangle| \\
&\leq C\left(\int_0^1 ||\sum_{k=1}^{n} \phi_k r_k(t)||dt\right) \sup_{||x_n^*||=1} \sup_{||x||=1} \left(\sum_{k=1}^{n} |\langle T_k^*(x_k^*), x\rangle|^2\right)^{1/2} \\
&\leq C\left(\int_0^1 ||\sum_{k=1}^{n} \phi_k r_k(t)||dt\right) \sup_{||x||=1, ||x_n^*||=1} \left(\sum_{k=1}^{n} |\langle x_k^*, T_k(x)\rangle|^2\right)^{1/2} \\
&\leq C\left(\sup_{t\in[0,1]} ||\sum_{k=1}^{n} \phi_k r_k(t)||\right) \sup_{||x||=1} \left(\sum_{k=1}^{n} ||T_k(x)||^2\right)^{1/2} \\
&\leq C||(\phi_n)||_{\ell_1^w(L^1)} \sup_{||x||=1} \left(\sum_{k=1}^{n} ||T_k(x)||^2\right)^{1/2}.
\end{aligned}$$

■

Theorem 3.16 *Let X^* be a GT-space of cotype 2 and let Y^* have type 2. If $T_n \in \mathcal{L}(X, Y)$ and $(T_n^*) \in Rad(Y^*, X^*)$ then $(T_n) \in \ell_{\pi_2}(X, Y)$.*
In particular if $T_n : c_0 \to \ell_q$ for $q \geq 2$ and $(T_n^*) \in Rad(\ell_{q'}, \ell_1)$ then $T_n \in \ell_{\pi_2}(c_0, \ell_q)$.

Proof. Let $(x_n) \in \ell_2^w(X)$. Using Lemma 3.14 for X^*, one gets

$$\begin{aligned}
(\sum_{k=1}^{n} ||T_k(x_k)||^2)^{1/2} &= \sup_{||(y_n^*)||_{\ell_2(Y^*)}=1} \sum_{k=1}^{n} |\langle T_k(x_k), y_k^*\rangle| \\
&= \sup_{||(y_n^*)||_{\ell_2(Y^*)}=1} \sum_{k=1}^{n} |\langle x_k, T_k^*(y_k^*)\rangle| \\
&\leq C \sup_{||(y_n^*)||_{\ell_2(Y^*)}=1} (\int_0^1 ||\sum_{k=1}^{n} T_k^*(y_k^*) r_k(t)||dt) \sup_{||x^*||=1} (\sum_{k=1}^{n} |\langle x_k, x^*\rangle|^2)^{1/2} \\
&\leq C||(x_n)||_{\ell_2^w(X)} rad[T_n^*] \sup_{||(y_n^*)||_{\ell_2(Y^*)}=1} (\int_0^1 ||\sum_{k=1}^{n} y_k^* r_k(t)||dt) \\
&\leq C||(x_n)||_{\ell_2^w(X)} rad[T_n^*],
\end{aligned}$$

where the last inequality follows from the type 2 condition on Y^*. ■

REFERENCES

1. J.L. Arregui and O. Blasco, (p,q)-summing sequences, to appear.
2. S. Aywa and J.H. Fourie, On summing multipliers and applications, to appear.
3. E. Berkson and A. Gillespie, Spectral decomposition and harmonic analysis on UMD spaces, Studia Math. 112 (1994) 13-49.
4. O. Blasco, A characterization of Hilbert spaces in terms of multipliers between spaces of vector valued analytic functions, Michigan Math. J. 42 (1995) 537-543.
5. O. Blasco, Vector valued analytic functions of bounded mean oscillation and geometry of Banach spaces, Illinois J.Math. 41 (1997) 532-557.
6. O. Blasco, Remarks on vector valued BMOA and vector valued multipliers, Positivity (2000).
7. O. Blasco and A. Pełczyński, Theorems of Hardy and Paley for vector valued analytic functions and related classes of Banach spaces, Trans. Amer. Math. Soc. 323 (1991) 335-367.
8. P. Clement, B de Pagter, F.A. Sukochev and H. Witvliet, Schauder decompositions and multiplier theorems, Studia Math. 138 (2000) 135-163.
9. A. Defant and K. Floret, Tensor Norms and Operator Ideals, North-Holland (1993).
10. J. Diestel, Sequences and Series in Banach Spaces, Springer-Verlag (1984).
11. J. Diestel, H. Jarchow, A. Tonge Absolutely Summing Operators, Cambridge University Press (1995).
12. J. Diestel and J.J.Uhl Jr., Vector Measures , Amer. Math. Soc. Series (1977).
13. P. Duren, Theory of H^p-spaces, Academic Press, New York (1970).
14. A. Dvoretzky and C. A. Rogers, Absolute and unconditional convergence in normed linear spaces, Proc. Nat. Acad. Sci. USA 36 (1950) 192–197.
15. A. Grothendieck, Résumé de la théorie métrique des produits tensoriels topologiques, Bol. Soc. Mat. São Paulo 8 (1953/1956) 1–79.
16. G.J.O. Jameson, Summing and Nuclear Norms in Banach Space Theory, Cambridge University Press (1987).
17. N. Kalton and L. Weis, The H^∞ calculus and sums of closed operators , to appear.

18. S. Kwapień, Some remarks on (p, q)-summing operators in ℓ_p-spaces, Studia Math. 29 (1968) 327–337.
19. J. Lindenstrauss and A. Pełczyński, Absolutely summing operators in $\mathcal{L}_p$-spaces and their applications, Studia Math. 29 (1968) 275–326.
20. B. Maurey and G. Pisier, Séries de variables aléatories vectorielles indépendantes et propiétés géométriques des espaces de Banach, Studia Math. 58 (1976) 45–90.
21. W. Orlicz, Über unbedingte konvergenz in funktionenräumen (I), Studia. Math. 4 (1933) 33–37.
22. A. Pietsch, Absolut p-summierende Abbildungen in normierten Räumen, Studia Math. 27 (1967) 333–353.
23. A. Pietsch, Operator Ideals, North-Holland (1980) .
24. G. Pisier, Factorization of Operators and Geometry of Banach spaces. CBM 60, Amer. Math. Soc. Providence R.I. (1985).
25. M. Talagrand, Cotype and $(q, 1)$-summing norm in a Banach space, Inventiones Math. 110 (1992) 545–556.
26. N. Tomczak-Jaegermann, Banach-Mazur Distances and Finite-dimensional Operator Ideals , Longman Scientific and Technical (1989).
27. L. Weis, Operator valued Fourier multiplier theorems and maximal regularity, to appear.
28. P. Wojtaszczyk, Banach Spaces for Analysts, Cambridge University Press (1991).

Recent Progress in Functional Analysis
K.D. Bierstedt, J. Bonet, M. Maestre and J. Schmets

Bergman projection on simply connected domains

J. Taskinen*

Department of Mathematics, University of Joensuu,
P.O. Box 111, FIN–80101 Joensuu, Finland

To Professor Manuel Valdivia on the occasion of his seventieth birthday.

Abstract

We study the problem of finding a substitute for the space $H^\infty(\Omega)$ which is the continuous image of the corresponding L^∞-type space under the Bergman projection. The spaces are defined on quite general simply connected domains.

1. Introduction.

It is a classical fact that the Szegö and Bergman projections and the harmonic conjugation operator are not bounded between the spaces $H^\infty(\mathbb{D})$, $L^\infty(\partial\mathbb{D})$ and $L^\infty(\mathbb{D})$. Here $\mathbb{D}$ is the open unit disc of the complex plane. However, in [8] we constructed substitutes for these spaces which behave in the optimal way with respect to these operators.

In the present work we continue the study by replacing the unit disc by more general simply connected complex domains Ω. As in [8] our function spaces will be endowed with weighted sup–seminorms. If Ω is bounded, the most natural weights on Ω are functions of the boundary distance $d(z) := \inf(|z-w| \mid w \in \partial\Omega)$, where $\partial\Omega$ denotes the boundary.

We consider the weighted spaces $L_V^\infty := L_V^\infty(\Omega)$ and $H_V^\infty := H_V^\infty(\Omega)$. The weights defining their topologies are of the above mentioned type. We prove that the Bergman projection is a continuous operator from L_V^∞ onto H_V^∞, and that these spaces are in a sense smallest possible extensions of $H^\infty(\Omega)$ and $L^\infty(\Omega)$ having this property.

Our results will follow in principle by a quite straightforward use of the Riemann conformal mapping. However, the formulation of the results and applicability to concrete situations is not a priori completely clear, hence, it is worthwhile to present the details. So the nontrivial part of this note consists of the examples in Section 3. Another motivation is that we are presenting a bit unusual combination of "hard analysis" techniques applied to "soft" locally convex spaces.

For unexplained notation and terminology concerning analytic function spaces we refer to [10]. For locally convex space theory, see the books [4], [5] and [7]. The two–dimensional

*Academy of Finland project no. 38954

Lebesgue measure is denoted by dA. By C, C' etc. we denote strictly positive constants, the value of which may vary from equation to equation, but which are independent on variables and indices in the equations (or the dependence on e.g. n is denoted by C_n).

On the disc, the Bergman projection $R_{\mathbb{D}}$ is defined as follows. For, say, $f \in L^1(\mathbb{D}, dA)$, $R_{\mathbb{D}}f$ is an analytic mapping of $z \in \mathbb{D}$ defined by

$$R_{\mathbb{D}}f(z) := \frac{1}{\pi}\int_{\mathbb{D}} \frac{f(\zeta)}{(1-z\bar{\zeta})^2} dA(\zeta). \tag{1}$$

This operator is bounded e.g. from $L^p(\mathbb{D}, dA)$ onto the Bergman space $A^p(\mathbb{D})$, if $1 < p < \infty$. For $p = 1, \infty$ it is not bounded.

For general domains there are several possibilities to define Bergman–type projections (see [9], Section 4). Let $\varphi : \Omega \to \mathbb{D}$ be a conformal mapping. The general formula for the Bergman projection for simply connected domains (see [1]) reads as follows:

$$Rf(z) := \frac{1}{\pi}\int_{\Omega} \frac{\varphi'(z)\bar{\varphi}'(\zeta)}{(1-\varphi(z)\bar{\varphi}(\zeta))^2} f(\zeta) dA(\zeta) \tag{2}$$

However, for our purposes the definition

$$R_\Omega f := (R_{\mathbb{D}}(f \circ \psi)) \circ \varphi \tag{3}$$

is more convenient. Here $\psi := \varphi^{-1} : \mathbb{D} \to \Omega$.

Acknowledgement. The author wishes to thank the referee for some comments and a simplification of the main proof.

2. General result.

In this section Ω denotes a bounded simply connected complex domain, and $\varphi : \Omega \to \mathbb{D}$ is a conformal mapping produced by the Riemann mapping theorem. We denote $\psi := \varphi^{-1}$.

The results of this section will be valid for domains Ω such that the conformal map φ satisfies

$$c(1 - |\varphi(z)|)^a \le d(z) \le C(1 - |\varphi(z)|)^b \tag{4}$$

for some $a, b, c, C > 0$, uniformly for $z \in \Omega$. The validity of (4) depends only on the domain, not on the particular choice of the conformal mapping. This follows basically from the fact that every conformal mapping $\Omega \to \mathbb{D}$ can be obtained from a fixed φ by composing it with a Möbius transformation of $\mathbb{D}$.

Definition 2.1. Let us denote by V the set of functions $v : \Omega \to \mathbb{R}^+$ which are continuous, depend on $d(z)$ only, are decreasing as $d(z) \to 0$, and satisfy for all $n \in \mathbb{N}$

$$(|\log d(z)| + 1)^n v(z) \le C_n \quad \text{for all } z \in \Omega. \tag{5}$$

Of course V depends heavily on Ω but for notational simplicity we do not display it.

Definition 2.2. We define

$$H_V^\infty := H_V^\infty(\Omega) := \Big\{ f : \Omega \to \mathbb{C} \text{ analytic} \Big| ||f||_v := \sup_{z\in\Omega} |f(z)|v(z) < \infty \text{ for every } v \in V \Big\}. \tag{6}$$

The space L_V^∞ is defined in the same way by replacing the word "analytic" by "measurable" and "sup" by "ess sup".

These spaces are Hausdorff locally convex spaces when endowed with the family $\{\|\cdot\|_v \mid v \in V\}$ of seminorms. The spaces are non–metrizable, since V is essentially uncountable.

Theorem 1.6 of [2] implies the following:

Proposition 2.3. *Let us define for all $n \in \mathbb{N}$ the weight $v_n = (1 + |\log d(z)|)^{-n}$ and the Banach space $H_{v_n}^\infty := \left\{ f : \Omega \to \mathbb{C} \text{ analytic} \;\middle|\; \|f\|_{v_n} < \infty \right\}$. We have*

$$H_V^\infty = \mathrm{ind}_{n\to\infty} H_{v_n}^\infty, \tag{7}$$

that is, the space H_V^∞ is the inductive limit of the spaces $H_{v_n}^\infty$ as $n \to \infty$.

The analogous statement holds for L_V^∞ as well.

We can now formulate the main result:

Theorem 2.4. *Assume that the bounded simply connected domain Ω satisfies the condition (4). Then the Bergman projection R_Ω is a continuous mapping from $L_V^\infty(\Omega)$ onto $H_V^\infty(\Omega)$.*

Let us denote by $V_{\mathbb{D}}$ the weight system V of [8]: it consists of radial, continuous, decreasing weights $v(r)$ on $\mathbb{D}$ such that

$$\sup_{r\in[0,1[} |\log(1-r)|^n v(r) \le C_n \tag{8}$$

for every $n \in \mathbb{N}$. We first claim:

Lemma 2.5. *Let Ω satisfy (4).*
(i) Let $v : \mathbb{D} \to \mathbb{R}^+$ be a radial, decreasing continuous function. If $v \in V_{\mathbb{D}}$, then

$$w(z) := \sup_{\zeta\in\Omega, d(\zeta)=d(z)} v \circ \varphi(\zeta) \in V. \tag{9}$$

(ii) Let $v : \Omega \to \mathbb{R}^+$ be a function of $d(z)$, decreasing for $d(z) \to 0$. If $v \in V$, then

$$w(r) := \sup_{z\in\mathbb{D}, |z|=r} v \circ \psi(z) \in V_{\mathbb{D}}. \tag{10}$$

Proof. Let $v \in V_{\mathbb{D}}$ and $n \in \mathbb{N}$ be given. We obtain by (4)

$$\sup_{z\in\Omega}(1 + |\log d(z)|)^n w(z) = \sup_{z\in\Omega}(1 + |\log d(z)|)^n \sup_{d(\zeta)=d(z)} v \circ \varphi(\zeta) \tag{11}$$

$$= \sup_{z\in\Omega} \sup_{d(\zeta)=d(z)} (1 + |\log d(\zeta)|)^n v \circ \varphi(\zeta) = \sup_{z\in\Omega}(1 + |\log d(z)|)^n v \circ \varphi(z) \tag{12}$$

$$\leq C' \sup_{z \in \Omega} |\log(1-|\varphi(z)|)|^n v \circ \varphi(z) \tag{13}$$

$$= C' \sup_{z \in \mathbb{D}} |\log(1-|z|)|^n v(z) \leq C''. \tag{14}$$

The other one is similar. □

The proof of the Theorem 2.4 is now an easy consequence. By Lemma 2.5, the composition operator $C_\varphi : f \mapsto f \circ \varphi$ is an isomorphism from $L^\infty_{V_{\mathbb{D}}}$ onto $L^\infty_V(\Omega)$ and similarly from $H^\infty_{V_{\mathbb{D}}}$ onto $H^\infty_V(\Omega)$. (Here $H^\infty_{V_{\mathbb{D}}}$ is the space (6) with Ω replaced by $\mathbb{D}$ and V replaced by $V_{\mathbb{D}}$ and so on.) Its inverse is the composition operator C_ψ. Hence, the result follows from (3) and [8], Theorem 14.

For the same reasons, one can strengthen Theorem 2.4 using [8], Theorem 14, as follows:

Theorem 2.6. *The space $L^\infty_V(\Omega)$ is the smallest space (consisting of measurable functions $\Omega \to \mathbb{C}$) having the following properties:*

1° *the Bergman projection R_Ω is a continuous operator on $L^\infty_V(\Omega)$ with $R_\Omega(L^\infty_V(\Omega)) = H^\infty_V(\Omega)$,*

2° *$L^\infty(\Omega) \subset L^\infty_V(\Omega)$, and*

3° *the topology of L^∞_V can be given by weighted sup–seminorms with continuous weights v which depend on $d(z)$ only and are decreasing as $d(z) \to 0$.*

3. Examples.

The nontrivial part of our work is to show that a reasonable class of domains satisfies the condition (4). For example, one has

Proposition 3.1. *Every polyhedron satisfies the condition (4).*

This will be a special case of the more general result for bounded *regulated domains*. The main reference for their elementary properties is [6], Chapter 3. To introduce regulated domains briefly, let us start with a simply connected, bounded domain $\Omega \subset \mathbb{C}$ with a locally connected boundary. In this case a Riemann conformal map $\psi : \mathbb{D} \to \Omega$ has a continuous extension to $\overline{\mathbb{D}}$ (still denoted by ψ; see [6], Section 2.1.). We can thus define the curve $w(t) = \psi(e^{it})$, $0 \leq t \leq 2\pi$. According to [6], Section 3.5, Ω is called a regulated domain, if each point of $\partial\Omega$ is attained only finitely often by ψ, and if

$$\beta(t) := \lim_{\tau \to t^+} \arg(w(\tau) - w(t)) \in \mathbb{R} \tag{15}$$

exists for all t and defines a regulated function. (Recall that β is regulated, if it can be approximated uniformly by step functions, i.e. for every $\varepsilon > 0$ there exist $0 = t_0 < t_1 < \ldots < t_n = 2\pi$ and constants $\gamma_1, \ldots, \gamma_n$ such that

$$|\beta(t) - \gamma_j| < \varepsilon \quad \text{for } t_{j-1} < t < t_j \;, \quad j = 1, \ldots, n.) \tag{16}$$

Geometrically, β is the *direction angle of the forward tangent of $\partial\Omega$ at $w(t)$*. So, a polyhedron is a regulated domain: in this case β is piecewise constant with finitely many jumps each of which is strictly between $-\pi$ and π. The polyhedron does not need to be convex.

Regulated domains can be characterized as follows.

Proposition 3.2. *Let $\Omega \subset I\!D$ be a simply connected domain with locally connected boundary. Then Ω is regulated if and only if, for a Riemann conformal map $\psi : I\!D \to \Omega$,*

$$\log \psi'(z) = \log|\psi'(0)| + \frac{i}{2\pi}\int_0^{2\pi} \frac{e^{it}+z}{e^{it}-z}(\beta(t) - t - \frac{\pi}{2})dt, \tag{17}$$

where $\beta : [0, 2\pi] \to I\!R$ is a regulated function.

For a proof, see [6]. In the situation of Proposition 3.2 the function β coincides with the direction angle defined above. Notice that the extra t (after β) is needed to remove the jump at $t = 2\pi$; think about the circle.

The following condition for β will be enough to imply the condition (4):

There exists a $0 < \tau < 1/2$ and a $\delta > 0$ such that

$$\sup_{\substack{|s-\theta|\le\tau \text{ or} \\ ||s-\theta|-2\pi|\le\tau}} |\beta(\theta) - \theta - \beta(s) + s| \le \pi - \delta \tag{18}$$

for every $\theta \in [0, 2\pi]$.

This means a restriction for the jumps in the direction angle. A domain with a cusp does not satisfy (18), since at a cusp the direction angle has a jump of π or $-\pi$. On the other hand, every polyhedron satisfies (18), see the explanation above.

Theorem 3.3. *Assume that Ω is a regulated domain with β in (17) satisfying (18). Then $\varphi := \psi^{-1}$ also satisfies (4), and hence Theorems 2.4 and 2.6 apply.*

Proof. Applying the conformal map ψ once, (4) is equivalent to the existence of constanst $a, b, c, C > 0$ such that

$$c(1-|z|)^a \le d(\psi(z)) \le C(1-|z|)^b \tag{19}$$

for all $z \in I\!D$.

Among other things we will use the Koebe distortion theorem in the form (see [6], Corollary 1.4)

$$c(1-|z|)|\psi'(z)| \le d(\psi(z)) \le C(1-|z|)|\psi'(z)|, \tag{20}$$

where $0 < c < C$ are constants independent of z.

So let $z = re^{i\theta}$. Fixing $\tau > 0$ as in (18), we may assume that r is so large that $1 - r < \tau/2$. In order to avoid inessential notational troubles near the endpoints of $[0, 2\pi]$ we assume that $\theta \in [\pi/2, 3\pi/2]$. This is actually general enough, since if (19) is proven for such values of z, then one obtains the missing values by proving (19) for $\theta \in [\pi/2, 3\pi/2]$ but ψ composed with a disc rotation of magnitude π.

We estimate (17). The real part of its right hand side determines the absolute value of ψ'. Hence, in order to estimate the modulus of the real part of the right hand side, we need to consider the imaginary part of $(e^{it}+z)/(e^{it}-z)$: it equals

$$\operatorname{Im}\frac{e^{it}+z}{e^{it}-z} = \frac{2r\sin(\theta-t)}{1+r^2-2r\cos(\theta-t)}, \tag{21}$$

see [3], Chapter 3. If, say $|\theta - t| \leq 1/2$, then one can write it as

$$\operatorname{Im}\frac{e^{it}+z}{e^{it}-z} = \frac{2r(\theta-t)}{(1-r)^2+r(\theta-t)^2} + K(z,t), \tag{22}$$

where $|K(z,t)| < C$ for all z, t; use $\sin x \cong x$ and $\cos x \cong 1 - x^2/2$ for small x.

Since the derivative with respect to t of the denominator essentially equals the numerator, one can compute that the integral

$$\int_0^{2\pi} \frac{r\sin(\theta-t)}{1+r^2-2r\cos(\theta-t)}dt \tag{23}$$

equals 0. Hence, we do not need to care for the $\pi/2$ in (17).

We divide the integration interval into three parts. First, if $|\theta - t| \geq \tau/2$ (recall also $|\theta - t| \leq 3\pi/2$ by our choice of θ), we have

$$|1+r^2-2r\cos(\theta-t)| \geq |1+r^2-2r(1-\tau^2/3)| = (1-r)^2 + 2r\tau^2/3 \geq \tau^2/3. \tag{24}$$

We thus obtain the bound (β is a bounded function; τ is a constant depending on the domain only)

$$\Big| \int\limits_{|t-\theta|\geq\tau/2} \frac{2r\sin(\theta-t)}{1+r^2-2r\cos(\theta-t)}(\beta(t)-t)dt \Big| \leq \int\limits_0^{2\pi} \frac{C}{\tau^2}dt \leq C'. \tag{25}$$

Moreover, by (22)

$$\Big| \int\limits_{|t-\theta|\leq 1-r} \frac{2r\sin(\theta-t)}{1+r^2-2r\cos(\theta-t)}(\beta(t)-t)dt \Big| \tag{26}$$

$$\leq \int\limits_{|t-\theta|\leq 1-r} \frac{Cr|\theta-t|}{(1-r)^2+r(\theta-t)^2}dt \leq \int\limits_{|t-\theta|\leq 1-r} \frac{C'r(1-r)}{(1-r)^2}dt \leq C''. \tag{27}$$

(The last integral is bounded since the measure of the integration interval is $1-r$.) The remaining integral is the crucial one. Let us denote

$$\beta^- := \inf_{\substack{\theta+\tau/2\geq t\geq \\ \theta+1-r}} \beta(t) - t \ \text{ and } \beta^+ := \sup_{\substack{\theta-\tau/2\leq t\leq \\ \theta-(1-r)}} \beta(t) - t \tag{28}$$

Since the sign of the integrand depends on the sin and $\beta - t$ only, we obtain the upper bound

$$\int\limits_{\substack{\tau/2\geq|t-\theta|\geq \\ 1-r}} \frac{2r\sin(\theta-t)}{1+r^2-2r\cos(\theta-t)}(\beta(t)-t)dt \tag{29}$$

$$\leq \int\limits_{\substack{\theta+\tau/2\geq t\geq\\ \theta+1-r}} \frac{2\beta^- r\sin(\theta-t)}{1+r^2-2r\cos(\theta-t)}dt + \int\limits_{\substack{\theta-\tau/2\leq t\leq\\ \theta-(1-r)}} \frac{2\beta^+ r\sin(\theta-t)}{1+r^2-2r\cos(\theta-t)}dt \tag{30}$$

$$= \int\limits_{\substack{\theta+\tau/2\geq t\geq\\ \theta+1-r}} \frac{2(\beta^+-\beta^-)r\sin(t-\theta)}{1+r^2-2r\cos(\theta-t)}dt. \tag{31}$$

But by (18), $\beta^+ - \beta^- \leq \pi - \delta$. Hence, the expression (31) is bounded by

$$(\pi-\delta)\int\limits_{\substack{\theta+1-r\\ \leq t<\theta+1}} \frac{2r\sin(t-\theta)}{1+r^2-2r\cos(\theta-t)}dt \tag{32}$$

$$\leq (\pi-\delta)\Big[\log(1+r^2-2r\cos(\theta-t))\Big]_{t=\theta+1-r}^{\theta+1}. \tag{33}$$

Applying again $\cos(1-r) \cong 1-(1-r)^2/2$ we obtain the bound $C+2(\pi-\delta)|\log(1-r)|$. (Only the integration limit $\theta+1-r$ is essential; the limit $\theta+1$ only gives an inessential constant.) In the same way one proves the lower estimate $-C'-2(\pi-\delta)|\log(1-r)|$ for the integral (29).

In conclusion, for $\delta' := \delta/(2\pi) > 0$ the real part of the right hand side of (17) is between $-C-(1-\delta')|\log(1-r)|$ and $C+(1-\delta')|\log(1-r)|$. Hence,

$$c(1-|z|)^{1-\delta'} \leq |\psi'(z)| \leq C(1-|z|)^{-1+\delta'} \tag{34}$$

This, combined with (20), yields that Ω satisfies (19). □

REFERENCES

1. D. Békollé, Projections sur des espaces de fonctions holomorphes dans des domains plans. Can.J.Math. XXXVIII, 1 (1986), 127–157.
2. K.D. Bierstedt, R. Meise, W. Summers, A projective description of weighted inductive limits. Trans.Amer.Math.Soc. 272.1(1982), 107–160.
3. J. Garnett, Bounded analytic functions. Academic Press, New York, 1981.
4. H. Jarchow, Locally convex spaces. Teubner, Stuttgart (1981).
5. G. Köthe, Topological vector spaces, Vol. 1. Second printing. Springer Verlag, Berlin–Heidelberg–New York (1983).
6. Ch. Pommerenke, Boundary behaviour of conformal maps. Grundlehren der mathematischen Wissenschaften vol. 299. Springer Verlag (1992)
7. P. Pérez Carreras, J. Bonet, Barreled locally convex spaces. North-Holland Mathematics Studies 131. North-Holland, Amsterdam (1987).
8. J. Taskinen, On the continuity of the Bergman and Szegö projections. HYMAT Reports, 1999.
9. J. Taskinen, Regulated domains and Bergman type projections. HYMAT Reports, 1999.
10. K. Zhu, Operator theory in function spaces. Marcel Dekker, New York (1995).

Recent Progress in Functional Analysis
K.D. Bierstedt, J. Bonet, M. Maestre and J. Schmets

On isomorphically equivalent extensions of quasi-Banach spaces

Jesús M. F. Castillo [a*] **and Yolanda Moreno** [b†]

[a]Departamento de Matemáticas, Universidad de Extremadura, Avenida de Elvas s/n, 06071-Badajoz, Spain. castillo@unex.es

[b]Departamento de Matemáticas, Universidad de Extremadura, Avenida de Elvas s/n, 06071-Badajoz, Spain. ymoreno@unex.es

To Professor Manuel Valdivia on the occasion of his seventieth birthday.

Abstract

We introduce the notion of isomorphically equivalent exact sequences and quasi-linear maps. We then show how this notion is closely related with the natural equivalence of some functors Ext. In particular, we make a closer inspection of the situation for certain subspaces and quotients of L_p, $0 < p < 1$, as well as for minimal extensions of quasi-Banach spaces. The applications include a complete answer to a problem of Fuchs in the domain of quasi-Banach spaces and a categorical proof of a result of Kalton and Peck.
MCS 2000 Primary 46M15; Secondary 46A16, 46M18,

1. Introduction

The theory of extensions of quasi-Banach spaces as constructed by Kalton [9] and Kalton and Peck [13] is based on the notion of equivalent quasi-linear maps, which corresponds to the classical notion of equivalent exact sequences. That approach has a lot of virtues, not the least of which is that it allows one to translate the machinery of homological algebra to the domain of quasi-Banach spaces (see [1,2]). Moreover, the notion of isomorphic quasi-Banach spaces *seems to be* clearly not adequate to handle extensions. Nevertheless, it is not too risky to guess that most authors which have worked with different aspects of exact sequences had in mind a different, perhaps intermediate, notion of equivalence of extensions. Recall, for instance, the notion of *projectively equivalent* sequences (see [13,10]) –certainly natural if one considers that what is important of a vector space is its dimension not its cardinal–, or some classical results for exact sequences involving l_1, c_0, l_∞ (see [16]) or $L_p, 0 < p < 1$ (see [14]), which establish that sometimes mere isomorphisms become a kind of weak equivalence between the sequences.

*This research has been supported in part by DGICYT project PB97-0377
†This research has been supported by a Beca Predoctoral de la Junta de Extremadura

Accordingly, we propose in this paper to study the notion of *isomorphically equivalent* exact sequences. Section 3 contains a characterization in terms of quasi-linear maps, exhibits several examples of isomorphically equivalent sequences and makes a closer inspection of minimal extensions (i.e. nontrivial extensions by a one-dimensional space). More precisely, we show that two minimal extensions of Banach spaces are isomorphic if and only if they are isomorphically equivalent: we then give an example of a minimal extension of a quasi-Banach space X isomorphic (although not isomorphically equivalent!) to the trivial extension $\mathbb{R} \oplus X$.

Section 4 contains our main results. We first transport several elements of homological algebra to the more concrete soil of quasi-Banach spaces. Thus, we obtain that given two p-Banach spaces E, H, the functors $\mathcal{Q}(E, \cdot)$ and $\mathcal{Q}(H, \cdot)$ are naturally equivalent acting on the category of p-Banach spaces if and only if $E \oplus l_p(I)$ and $H \oplus l_p(J)$ are isomorphic for some sets I, J. This yields a complete and natural answer, in the domain of quasi-Banach spaces, to the problem considered in [18]. Dually, if A, B are q-Banach subspaces of L_p, $0 < p < q \leq 1$ the functors $\mathcal{Q}(L_p/A, \cdot)$ and $\mathcal{Q}(L_p/B, \cdot)$ are naturally equivalent acting on the category of q-Banach spaces if and only if A and B are isomorphic. As applications we show: 1) If A, B are q-Banach subspaces of L_p, $0 < p < q \leq 1$, two exact sequences $0 \to A \to L_p \to L_p/A \to 0$ and $0 \to B \to L_p \to L_p/B \to 0$ are isomorphically equivalent if and only if the functors $\mathcal{Q}(L_p/A, \cdot)$ and $\mathcal{Q}(L_p/B, \cdot)$ are naturally equivalent acting on the category of p-Banach spaces, from which we deduce the result of Kalton and Peck [14, Thm. 4.4] (which is, otherwise optimal: see the final remark); 2) If Z, Z' are Banach spaces, the exact sequences $0 \to \mathbb{R} \to X \to Z \to 0$ and $0 \to \mathbb{R} \to X' \to Z' \to 0$ are isomorphically equivalent if and only if the functors $\mathcal{Q}(X, \cdot)$ and $\mathcal{Q}(X', \cdot)$ are naturally equivalent acting on the category of p-Banach spaces for some $p < 1$.

2. Preliminaries

For a sound background on homological algebra, functors and natural transformations, as well as the algebraic theory of extensions we suggest [8,17]. A comprehensive description of the theory of twisted sums of quasi-Banach spaces as developed by Kalton [9] and Kalton and Peck [13] can be found in the monograph [4]. Let us however recall briefly the basic facts the reader should have in mind for the rest of the paper.

In what follows $\mathbf{Q}$ shall denote the category of quasi-Banach spaces and operators, $\mathbf{Q_p}$ the subcategory of p-Banach spaces and $\mathbf{V}$ the category of vector spaces and linear applications. Recall that the projective spaces in $\mathbf{Q_p}$ are (depending on the dimension) the $l_p(I)$-spaces. An exact sequence $0 \to Y \to X \to Z \to 0$ in $\mathbf{Q}$ is a diagram in which the kernel of each arrow coincides with the image of the preceding; the middle space X is also called an *extension* of Z by Y. Two exact sequences $0 \to Y \to X \to Z \to 0$ and $0 \to Y \to X_1 \to Z \to 0$ in $\mathbf{Q}$ are said to be equivalent if there exists an operator $T : X \to X_1$ making commutative the diagram

$$\begin{array}{ccccccccc} 0 & \to & Y & \to & X & \to & Z & \to & 0 \\ & & \| & & \downarrow T & & \| & & \\ 0 & \to & Y & \to & X_1 & \to & Z & \to & 0. \end{array}$$

An exact sequence is said to split if it is equivalent to the trivial sequence $0 \to Y \to Y \oplus Z \to Z \to 0$. The vector space (when endowed with suitable defined operations) of all extensions of Z by Y modulo the equivalence relation is denoted $\mathrm{Ext}(Z, Y)$. Given a quasi-Banach space Z one has a covariant functor $\mathrm{Ext}(Z, \cdot) : \mathbf{Q} \to \mathbf{V}$. On the other hand, to each exact sequence $0 \to Y \to X \to Z \to 0$ in $\mathbf{Q}$ corresponds a homogeneous map $F : Z \to Y$ with the property that there exists some constant $Q(F) > 0$ such that for all $x, y \in Z$ one has

$$\|F(x+y) - Fx - Fy\| \leq Q(F)(\|x\| + \|y\|).$$

Such maps are called *quasi-linear*. The map F can be obtained taking a bounded homogeneous selection $B : Z \to X$ for the quotient map and then a linear selection $L : Z \to X$ and putting $F = B - L$. Conversely, given a quasi-linear map $F : Z \to Y$, it is possible to construct an exact sequence $0 \to Y \to Y \oplus_F Z \to Z \to 0$. The quasi-Banach space $Y \oplus_F Z$, which is the product space $Y \times Z$ endowed with the quasi-norm $\|(y,z)\|_F = \|y - Fz\| + \|z\|$, is called a *twisted sum of Y and Z*. Two quasi-linear maps $F : Z \to Y$ and $G : Z \to Y$ are said to be equivalent if $F - G = b + l$ where $b : Z \to Y$ is a homogeneous bounded map and $l : Z \to Y$ is a linear map. The vector space (with obvious operations) of quasi-linear maps modulo the equivalence relation is denoted $\mathcal{Q}(Z, Y)$. Given a quasi-Banach space Z one has a covariant functor $\mathcal{Q}(Z, \cdot) : \mathbf{Q} \to \mathbf{V}$. Let us recall that a natural transformation $\eta : \mathcal{F} \to \mathcal{G}$ between two functors $\mathcal{F}$ and $\mathcal{G}$ is a correspondence assigning to each object A a morphism $\eta_A : \mathcal{F}(A) \to \mathcal{G}(A)$ with the property that if $f : A \to B$ is a morphism then the diagram

$$\begin{array}{ccc} \mathcal{F}(A) & \xrightarrow{\eta_A} & \mathcal{G}(A) \\ {\scriptstyle \mathcal{F}(f)} \downarrow & & \downarrow {\scriptstyle \mathcal{G}(f)} \\ \mathcal{F}(B) & \xrightarrow{\eta_B} & \mathcal{G}(B) \end{array}$$

is commutative. A natural transformation is called a natural equivalence if, for each A, the arrow η_A is an isomorphism. In [2] it has been shown that the functors $\mathcal{Q}(Z, \cdot)$ and $\mathrm{Ext}(Z, \cdot)$ are naturally equivalent. In particular, every exact sequence $0 \to Y \to X \to Z \to 0$ is equivalent to an exact sequence $0 \to Y \to Y \oplus_F Z \to Z \to 0$ constructed with a quasi-linear map $F : Z \to Y$.

Finally, and when no confusion arises, given an operator ϕ we shall denote by ϕ^* either the right-composition or left-composition map with ϕ.

3. Isomorphically equivalent exact sequences, and examples

Definition. We shall say that two exact sequences $0 \to Y \to X \to Z \to 0$ and $0 \to Y_1 \to X_1 \to Z_1 \to 0$ in $\mathbf{Q}$ are *isomorphically equivalent* if there exist isomorphisms $\alpha : Y \to Y_1$, $\beta : X \to X_1$ and $\gamma : Z \to Z_1$ making commutative the diagram

$$\begin{array}{ccccccccc} 0 & \to & Y & \to & X & \to & Z & \to & 0 \\ & & \alpha \downarrow & & \downarrow \beta & & \downarrow \gamma & & \\ 0 & \to & Y_1 & \to & X_1 & \to & Z_1 & \to & 0. \end{array}$$

When α and γ have the form $\alpha = a \cdot 1_Y$ and $\gamma = c \cdot 1_Z$ then we recover the notion of *projectively equivalent* sequences. Equivalent sequences are obviously isomorphically equivalent. In terms of quasi-linear maps one has:

Proposition 3.1 *The exact sequences $0 \to Y \to Y \oplus_F Z \to Z \to 0$ and $0 \to Y_1 \to Y_1 \oplus_G Z_1 \to Z_1 \to 0$ are isomorphically equivalent if and only if there exist isomorphisms $\alpha : Y \to Y_1$ and $\gamma : Z \to Z_1$ such that αF and $G\gamma$ are equivalent quasi-linear maps.*

Proof. Let $q : Y \oplus_F Z \to Z$ and $q_1 : Y_1 \oplus_G Z_1 \to Z_1$ be the quotient maps. If $F = B - L$, where $B : Z \to Y \oplus_F Z$ and $L : Z \to Y \oplus_F Z$ are, respectively, a homogeneous bounded and a linear selection for q, then $\beta B \gamma^{-1}$ is a bounded homogeneous selection for q_1, while $\beta L \gamma^{-1}$ is a linear selection for q_1. Thus G and

$$\beta(B - L)\gamma^{-1} = \beta i (B - L)\gamma^{-1} = i_1 \alpha (B - L)\gamma^{-1} = \alpha F \gamma^{-1}$$

are equivalent quasi-linear maps. This yields that also αF and $G\gamma$ are equivalent.

Conversely, if αF and $G\gamma$ are equivalent then also G and $\alpha F\gamma^{-1}$ are equivalent. Thus, there exists an operator T making commutative the diagram

$$\begin{array}{ccccccccc} 0 & \to & Y_1 & \to & Y_1 \oplus_{\alpha F\gamma^{-1}} Z_1 & \to & Z_1 & \to & 0 \\ & & \| & & \downarrow T & & \| & & \\ 0 & \to & Y_1 & \to & Y_1 \oplus_G Z_1 & \to & Z_1 & \to & 0. \end{array}$$

The operator T has the form $T(y,z) = (y + Lz, z)$ where $L : Z_1 \to Y_1$ is linear. Let us verify that the linear map $\beta : Y \oplus_F Z \to Y_1 \oplus_G Z_1$ defined by $\beta(y,z) = (\alpha y + L\gamma z, \gamma z)$ is an operator making commutative the diagram

$$\begin{array}{ccccccccc} 0 & \to & Y & \to & Y \oplus_F Z & \to & Z & \to & 0 \\ & & \alpha\downarrow & & \downarrow\beta & & \downarrow\gamma & & \\ 0 & \to & Y_1 & \to & Y_1 \oplus_G Z_1 & \to & Z_1 & \to & 0. \end{array}$$

The linearity is obvious while the continuity follows from

$$\begin{aligned} \|\beta(y,z)\|_G &= \|(\alpha y + L\gamma z, \gamma z)\| + \|\gamma z\| \\ &= \|\alpha y + L\gamma z - G\gamma z)\| + \|\gamma z\| \\ &\leq \|T\|(\|\alpha y - \alpha F\gamma^{-1}\gamma z)\| + \|\gamma z\| \\ &\leq \|T\|\max\{\|\alpha\|, \|\gamma\|\}(\|\mathrm{y} - \mathrm{Fz}\| + \|\mathrm{z}\| \\ &= \|T\|\max\{\|\alpha\|, \|\gamma\|\}\|(\mathrm{y},\mathrm{z})\|_{\mathrm{F}}; \end{aligned}$$

the commutativity of the diagram is also clear: $\beta i(y) = \beta(y,0) = (\alpha y, 0) = i_1\alpha(y)$; and $q_1\beta(y,z) = q_1(\alpha y + L\beta z, \beta z) = \beta z$. □

Thus, we shall freely say that two quasi-linear maps are isomorphically equivalent. Some basic examples of isomorphically equivalent extensions are:

1. If A and B are separable subspaces of l_∞ then the sequences $0 \to A \to l_\infty \to l_\infty/A \to 0$ and $0 \to B \to l_\infty \to l_\infty/B \to 0$ are isomorphically equivalent if and only if A and B are isomorphic. The same is true replacing l_∞ by c_0 when the quotients are infinite dimensional (see [16, Thm.2.f.10;Thm.2.f.12]).

2. If A and B are Banach spaces not isomorphic to l_1 then the sequences $0 \to K_A \to l_1 \to A \to 0$ and $0 \to K_B \to l_1 \to B \to 0$ are isomorphically equivalent if and only if A and B are isomorphic (see [16, Thm.2.f.8]). The same is true when A and B are p-Banach spaces not isomorphic to l_p for which the kernels K_A and K_B in $0 \to K_A \to l_1 \to A \to 0$ and $0 \to K_B \to l_1 \to B \to 0$ either contain complemented (in l_p) copies of l_p or have the Hahn-Banach extension property (see [11, Thm.2.2., Cor. 2.3]).

3. Let A and B be subspaces of L_p for $p < 1$ which are either ultrasummands (i.e., complemented in some pseudo-dual) or q-Banach for some $p < q \leq 1$. The sequences $0 \to A \to L_p \to L_p/A \to 0$ and $0 \to B \to L_p \to L_p/B \to 0$ are isomorphically equivalent if and only if L_p/A and L_p/B are isomorphic (see [14, Thm. 4.4]).

4. Let A and B be Banach spaces. Two minimal extensions $\mathrm{R} \oplus_F A$ and $\mathrm{R} \oplus_G B$ are isomorphic if and only if F and G are isomorphically equivalent: when Z is a Banach space then R is the (subspace generated by the) only needle point of $\mathrm{R} \oplus_F Z$ (see [15]). Thus, an isomorphism between $\mathrm{R} \oplus_F A$ and $\mathrm{R} \oplus_G B$ induces an isomorphism between A

and B that makes F and G isomorphically equivalent. We will see in Proposition 3.2 that this is no longer true for quasi-Banach spaces.

After the characterization given in 3.1 it is clear that a nontrivial exact sequence cannot be isomorphically equivalent to the trivial sequence. However, nothing prevents a nontrivial twisted sum $Y \oplus_F Z$, even if Y and Z are totally incomparable, from being isomorphic to the direct sum $Y \oplus Z$: for instance, if $0 \to K \to l_1 \to c_0 \to 0$ is a projective presentation of c_0, since $K = K \oplus K$ and $K = l_1 \oplus K$, multiplying adequately on the left by K and on the right by c_0 one gets a nontrivial sequence $0 \to K \to K \oplus c_0 \to c_0 \to 0$.

Definition. Let us call a twisted sum $Y \oplus_F Z$ *irreducible* if it is not isomorphic to the direct sum $Y \oplus Z$. Examples of irreducible twisted sums are: *i.)* All nontrivial twisted sums of Hilbert spaces (such as those constructed in [6,13]). *ii.)* All nontrivial twisted sums $Y \oplus_F l_p(I)$ in which Y is a p-Banach space (see [5,9,13,19] for concrete examples). *iii.)* All nontrivial minimal extensions of a Banach space (such as those constructed in [9,12,19,20]). This suggests the question if every minimal extension of a quasi-Banach space must be irreducible. The answer, more or less surprisingly, is no:

Proposition 3.2 *There exists a quasi-Banach space X admitting a minimal extension $\mathbb{R} \oplus_F X$ isomorphic to the direct sum $\mathbb{R} \oplus X$.*

Proof. Let $0 \to \mathbb{R} \xrightarrow{j} E \to l_1 \to 0$ be a nontrivial sequence. Let $J : \mathbb{R} \to l_1(E)$ be the embedding $r \to (j(r), 0, 0, ...)$. Consider the sequence

$$0 \to \mathbb{R} \xrightarrow{J} l_1(E) \to l_1(E/\mathbb{R}, E, E, ...) \to 0.$$

It is not difficult to see that $l_1(l_1, E, E, ...)$ is isomorphic to $l_1(E)$. Hence, we have a nontrivial sequence $0 \to \mathbb{R} \xrightarrow{J} l_1(E) \to l_1(E) \to 0$. Since $l_1(E)$ is also isomorphic to $\mathbb{R} \oplus l_1(E)$, the result is proved. □

In [10], it is proved that the quasi-linear maps introduced by Ribe [19] and Kalton [9] are not projectively equivalent. We have been unable to find out whether they are isomorphically equivalent.

4. Main results, with applications

In [18] it is considered the following problem apparently posed by Fuchs [7]: what is the relationship between abelian groups A and B if $\mathrm{Ext}(A, C)$ is isomorphic to $\mathrm{Ext}(B, C)$ for all abelian groups C? As it stands, this problem is rather strange; after all, how could one get such isomorphisms if not with a "method"? So, in our opinion the question should be:

Problem. What is the relationship between two quasi-Banach spaces A and B if the functors $\mathrm{Ext}(A, \cdot)$ and $\mathrm{Ext}(B, \cdot)$ are naturally equivalent?

Our approach to the problem is based on the natural equivalence of some functors. The natural equivalence between the functors $\mathrm{Ext}(X, \cdot)$ and $\mathcal{Q}(X, \cdot)$ (see [2]) greatly simplifies the arguments.

It is clear that an operator $T : Y \to X$ induces a natural transformation $\eta : \mathcal{Q}(X, \cdot) \to \mathcal{Q}(Y, \cdot)$ given by $\eta_A(W) = WT$, where $W : X \to A$ is a quasi-linear map. The following lemma, which shows that all natural transformations are of that kind translates Thm. 10.1 and Prop. 10.3 of [8] to the setting of quasi-Banach spaces.

Lemma 4.1 *Let X and Y be p-Banach spaces. Consider the functors $\mathcal{Q}(X,\cdot)$ and $\mathcal{Q}(Y,\cdot)$ acting between the categories $\mathbf{Q_p}$ and $\mathbf{V}$. Every natural transformation $\eta : \mathcal{Q}(X,\cdot) \to \mathcal{Q}(Y,\cdot)$ is induced by an operator $T : Y \to X$ in the form $\eta_Z(W) = WT$*

Proof. Let $\eta : \mathcal{Q}(X,\cdot) \to \mathcal{Q}(Y,\cdot)$ be a natural transformation and let

$$0 \to K_X \stackrel{i_X}{\to} l_p(\Gamma) \stackrel{q_X}{\to} X \to 0$$

be a projective presentation of X defined by the quasi-linear map F_X. The homology sequence starting with Y (see [1]) gives an exact sequence

$$0 \to \mathcal{L}(Y, K_X) \to \mathcal{L}(Y, l_p(\Gamma)) \to \mathcal{L}(Y, X) \stackrel{\omega}{\to} \mathcal{Q}(Y, K_X) \stackrel{i_X^*}{\to} \mathcal{L}(Y, l_p(\Gamma)) \to \cdots$$

where the connecting morphism is $\omega(T) = F_X T$. The exactness of the sequence yields

$$\mathrm{Ker}\,[i_X^* : \mathcal{Q}(Y, K_X) \to \mathcal{L}(Y, l_p(\Gamma))] = \omega\,(\mathcal{L}(Y, X))\,.$$

The quasi-linear map $\eta_{K_X}(F_X)$ belongs to $\mathrm{Ker} i_X^*$ since the commutativity of the square

$$\begin{array}{ccc} \mathcal{Q}(X, K_X) & \stackrel{\eta_{K_X}}{\to} & \mathcal{Q}(Y, K_X) \\ i_X^* \downarrow & & \downarrow i_X^* \\ \mathcal{Q}(X, l_p(\Gamma)) & \stackrel{\eta_{l_p}}{\to} & \mathcal{Q}(Y, l_p(\Gamma)) \end{array}$$

yields $i_X \eta_{K_X}(F_X) = \eta_{l_p} i_X(F_X) = 0$. Let $\alpha : Y \to X$ be an operator representing $\eta_X(F_X)$; i.e., such that $\eta_{K_X}(F_X) = F_X\alpha$. Let us show that also η is represented by α in the sense given at the begining of the proof, namely: for any quasi-linear map $W : X \to Z$ one has $\eta_Z(W) = W\alpha$. The quasi-linear map $W : X \to Z$ must have the form ϕF_X for some operator $\phi : K_X \to Z$. The commutativity of the diagram

$$\begin{array}{ccc} \mathcal{Q}(X, K_X) & \stackrel{\eta_{K_X}}{\to} & \mathcal{Q}(Y, K_X) \\ \phi^* \downarrow & & \downarrow \phi^* \\ \mathcal{Q}(X, Z) & \stackrel{\eta_Z}{\to} & \mathcal{Q}(Y, Z) \end{array}$$

yields $\eta_Z(W) = \eta_Z(\phi F_X) = \eta_Z \phi^*(F_X) = \phi^* \eta_{K_X}(F_X) = \phi F_X \alpha = W\alpha$. □

With this, we obtain (compare with [8, thm. 10.4]):

Proposition 4.1 *Let X and Y be p-Banach spaces. The functors $\mathcal{Q}(X,\cdot)$ and $\mathcal{Q}(Y,\cdot)$ acting between the categories $\mathbf{Q_p}$ and $\mathbf{V}$ are naturally equivalent if and only if for some I, J the spaces $X \oplus l_p(I)$ and $Y \oplus l_p(J)$ are isomorphic.*

Proof. First, observe that a natural equivalence $\eta : \mathcal{Q}(X,\cdot) \to \mathcal{Q}(Y,\cdot)$ induces a natural equivalence $\eta^n : \mathrm{Ext}^n(X,\cdot) \to \mathrm{Ext}^n(X,\cdot)$ between the higher derived functors (see [8,1]). By the previous result, η is induced by an operator $\alpha : Y \to X$. If α is surjective then $\mathrm{K}_\alpha = \mathrm{Ker}\alpha$ is projective, as it can be deduced as follows: given any Z, the exactnes of the homology sequence

$$\mathcal{L}(Y, Z) \to \mathcal{L}(\mathrm{K}_\alpha, Z) \to \mathcal{Q}(X, Z) \stackrel{\eta_Z}{\to} \mathcal{Q}(Y, Z) \to \mathcal{Q}(\mathrm{K}_\alpha, Z) \to \mathrm{Ext}^2(X, Z) \stackrel{\eta_Z^2}{\to} \mathrm{Ext}^2(Y, Z)$$

and the fact that η_Z and η_Z^2 are isomorphisms imply that $\mathcal{Q}(\mathrm{K}_\alpha, Z) = 0$. Moreover, since η_Z is an isomorphism, $\mathcal{L}(Y, Z) \to \mathcal{L}(\mathrm{K}_\alpha, Z)$ is surjective, which implies that the sequence $0 \to \mathrm{K}_\alpha \to$

$Y \stackrel{\alpha}{\to} X \to 0$ splits, and thus $Y = \mathrm{K}_\alpha \oplus X$; that is $Y = l_p(I) \oplus X$ for some set I. If α is not surjective, take a projective presentation $0 \to K \to l_p(J) \stackrel{Q}{\to} X \to 0$ of X and consider the sequence $0 \to PB \to Y \oplus l_p(J) \stackrel{\alpha \oplus Q}{\to} X \to 0$. Applying to this sequence the previous result one gets $Y \oplus l_p(J) = X \oplus l_p(I)$. As for the converse, it is clear that if $\sigma : Y \oplus l_p(J) \to X \oplus l_p(I)$ is an isomorphism then the composition $Y \to Y \oplus l_p(J) \stackrel{\sigma}{\to} X \oplus l_p(I) \to X$ induces a natural equivalence between the functors.□

We now present a kind of dual, ad-hoc, result for $L_p = L_p(0,1), 0 < p < 1$. In what follows, given a subspace A of L_p we shall understand that F_A is a quasi-linear map inducing the sequence $0 \to A \to L_p \to L_p/A \to 0$. The injection shall be called i_A and q_A shall be the quotient map.

Lemma 4.2 *Let A and B be q-Banach subspaces of L_p, $0 < p < q \le 1$. Consider the functors $\mathcal{Q}(L_p/A, \cdot)$ and $\mathcal{Q}(L_p/B, \cdot)$ acting between the categories $\mathbf{Q_q}$ and $\mathbf{V}$. Every natural transformation $\mathcal{Q}(L_p/A, \cdot) \to \mathcal{Q}(L_p/B, \cdot)$ is induced by an operator $B \to A$.*

Proof. Let us first observe that if $0 < p < q \le 1$, given a q-Banach subspace A of L_p there is a natural equivalence $\nu^{-1,A} : \mathcal{L}(A, \cdot)$ and $\mathcal{Q}(L_p/A, \cdot)$ when those functors act between the categories $\mathbf{Q_q}$ and $\mathbf{V}$. To see this, take X a q-Banach space and apply the homology sequence in the second variable to $0 \to A \to L_p \to L_p/A \to 0$ to get:

$$0 \to \mathcal{L}(L_p/A, X) \to \mathcal{L}(L_p, X) \to \mathcal{L}(A, X) \to \mathcal{Q}(L_p/A, X) \to \mathcal{Q}(L_p, X) \to \cdots$$

Since $\mathcal{L}(L_p, X) = 0 = \mathcal{Q}(L_p, X)$ (see [14]) the vector spaces $\mathcal{L}(A, X)$ and $\mathcal{Q}(L_p/A, X)$ are isomorphic (following [1] it can be proved that hey are isomorphic even as q-Banach spaces), and the isomorphism is given by $\nu_X^{A,-1}(T) = TF_A$. It is easy to verify that $\nu^{A,-1}$ is a natural equivalence. Its inverse $\nu^A : \mathcal{Q}(L_p/A, \cdot) \to \mathcal{L}(A, \cdot)$ is slightly awkward to describe. Given a quasi-linear map $W : L_p/A \to Z$ the operator $\nu_Z^A(W) : A \to Z$ is obtained as follows: since $Wq_A = 0$ it can be written as $b - l$, where b is a homogeneous bounded and l a linear map, both defined from $L_p \to Z$; hence, the restriction $b_{|A}$ is the operator $\nu_Z^A(W)$. What we have to keep in mind about this operator is that $W = \nu_Z^A(W)F_A$.

Returning to the main proof. Let $\eta : \mathcal{Q}(L_p/A, \cdot) \to \mathcal{Q}(L_p/B, \cdot)$ be a natural transformation. Then $\eta_A(F_A) = \nu_A^B(\eta_A(F_A))\, F_B$, and $\nu_A^B(\eta_A(F_A)) : A \to Z$ is going to be the operator we are looking for. The (surprising) form in which the natural transformation η acts is as follows; since one has a commutative diagram

$$\begin{array}{ccc} \mathcal{Q}(L_p/A, A) & \stackrel{\eta_A}{\to} & \mathcal{Q}(L_p/B, A) \\ {\scriptstyle \nu_Z^A(W)^*} \downarrow & & \downarrow {\scriptstyle \nu_Z^A(W)^*} \\ \mathcal{Q}(L_p/A, Z) & \stackrel{\eta_Z}{\to} & \mathcal{Q}(L_p/B, Z), \end{array}$$

given a quasi-linear map $W : L_p/A \to Z$ one has:

$$\eta_Z(W) = \eta_Z\left(\nu_Z^A(W)F_A\right) = \nu_Z^A(W)\eta_A(F_A) = \nu_Z^A(W)\nu_A^B\left(\eta_A(F_A)\right)F_B. \ \square$$

With this we obtain:

Proposition 4.2 *Let A and B be q-Banach subspaces of L_p, $0 < p < q \le 1$. The functors $\mathcal{Q}(L_p/A, \cdot)$ and $\mathcal{Q}(L_p/B, \cdot)$ acting between the categories $\mathbf{Q_q}$ and $\mathbf{V}$ are naturally equivalent if and only if A and B are isomorphic.*

Proof. Let η be a natural equivalence with inverse η^{-1} (which has the same form as η). Thus, for all W one has $W = \eta_Z^{-1}(\eta_Z(W))$ from which :

$$\nu_Z^A(W)F_A = \eta_Z^{-1}\left(\nu_Z^A(W)\nu_A^B(\eta_A(F_A))\,F_B\right) = \nu_Z^A(W)\nu_A^B(\eta_A(F_A))\,\nu_B^A\left(\eta_B^{-1}(F_B)\right)F_A.$$

Now observe that if T is an operator and $TF_A = 0$ then $T = 0$ (simply because $\mathcal{L}(L_p,\cdot) = 0$). Hence

$$\nu_Z^A(W) = \nu_Z^A(W)\nu_A^B(\eta_A(F_A))\,\nu_B^A\left(\eta_B^{-1}(F_B)\right).$$

Since ν_Z^A is an isomorphism, choosing W so that $\nu_Z^A(W)$ is injective one gets

$$\nu_A^B(\eta_A(F_A))\,\nu_B^A\left(\eta_B^{-1}(F_B)\right) = 1;$$

and, reasoning with $\eta\eta^{-1}$, one gets $\nu_B^A\left(\eta_B^{-1}(F_B)\right)\nu_A^B(\eta_A(F_A)) = 1$, which shows that $\nu_A^B(\eta_A(F_A))$ is an isomorphism.□

We are ready to give some applications of these results:

Theorem 4.1 *Let Z and Z' be Banach spaces and let $F : Z \to \mathrm{R}$ and $G : Z' \to \mathrm{R}$ be quasi-linear maps. Then F and G are isomorphically equivalent if and only if the functors $\mathcal{Q}(\mathrm{R}\oplus_F Z,\cdot)$ and $\mathcal{Q}(\mathrm{R}\oplus_G Z',\cdot)$ are naturally equivalent acting between the categories $\mathbf{Q_p}$ and $\mathbf{V}$ for some (all) $0 < p < 1$.*

Proof. If F and G are isomorphically equivalent then the spaces $\mathrm{R}\oplus_F Z$ and $\mathrm{R}\oplus_G Z$ are isomorphic and thus the functors $\mathcal{Q}(\mathrm{R}\oplus_F Z,\cdot)$ and $\mathcal{Q}(\mathrm{R}\oplus_G Z,\cdot)$ are naturally equivalent (between no matter which categories). To obtain the reciprocal, recall from [10] that a minimal extension of a Banach space is p-Banach for all $p < 1$. Now, if $\nu : \mathcal{Q}(\mathrm{R}\oplus_F Z,\cdot) \to \mathcal{Q}(\mathrm{R}\oplus_G Z,\cdot)$ is a natural transformation then $(\mathrm{R}\oplus_F Z)\oplus l_p(I)$ and $(\mathrm{R}\oplus_G Z)\oplus l_p(J)$ must be isomorphic for all $p < 1$, which implies that $\mathrm{R}\oplus_F Z$ and $\mathrm{R}\oplus_G Z$ are isomorphic and, by 4, F and G are isomorphically equivalent.□

The example 3.2 shows that, in general, the natural equivalence of two functors $\mathcal{Q}(\mathrm{R}\oplus_F Z,\cdot)$ and $\mathcal{Q}(\mathrm{R}\oplus_G Z',\cdot)$ does not imply that F and G are isomorphically equivalent.

Theorem 4.2 *Let A and B be q-Banach subspaces of L_p for $0 < p < q \leq 1$. The sequences $0 \to A \to L_p \to L_p/A \to 0$ and $0 \to B \to L_p \to L_p/B \to 0$ are isomorphically equivalent if and only if the functors $\mathcal{Q}(L_p/A,\cdot)$ and $\mathcal{Q}(L_p/B,\cdot)$ acting between the categories $\mathbf{Q_p}$ and $\mathbf{V}$ are naturally equivalent.*

Proof. Since the functors are naturally equivalent acting on $\mathbf{Q_p}$ we get that $(L_p/A)\oplus l_p$ and $(L_p/B)\oplus l_p$ are isomorphic; a further moment of reflection shows that then also L_p/A and L_p/B are isomorphic and thus there exists an isomorphism $\psi : L_p/B \to L_p/A$ so that the natural transformation acts as $\eta_Z(W) = W\psi$. On the other hand, since the functors are naturally equivalent acting on $\mathbf{Q_q}$, for $p < q$, we get that there exists an isomorphism $\phi : B \to A$ so that the natural transformation acts as $\eta_Z(W) = \nu_Z(W)\phi F_B$. Hence

$$\eta_A(F_A) = F_A\psi = \phi F_B$$

and thus the sequences $0 \to A \to L_p \to L_p/A \to 0$ and $0 \to B \to L_p \to L_p/B \to 0$ are isomorphically equivalent.□

Observe that this result includes the result of Kalton and Peck [14, Theorem 4.4] mentioned in 3: if A, B are q-Banach subspaces of $L_p, 0 < p < q \leq 1$ and $\gamma : L_p/A \to L_p/B$ is an isomorphism then the functors $\mathcal{Q}(L_p/A, \cdot)$ and $\mathcal{Q}(L_p/B, \cdot)$ are naturally equivalent (acting wherever) and thus the sequences $0 \to A \to L_p \to L_p/A \to 0$ and $0 \to B \to L_p \to L_p/B \to 0$ are isomorphically equivalent. In the diagram

$$\begin{array}{ccccccccc} 0 & \to & A & \to & L_p & \to & L_p/A & \to & 0 \\ & & {\scriptstyle\alpha}\downarrow & & \downarrow{\scriptstyle\beta} & & \downarrow{\scriptstyle\gamma} & & \\ 0 & \to & B & \to & L_p & \to & L_p/B & \to & 0 \end{array}$$

$\beta : L_p \to L_p$ is the isomorphism such that $\beta(A) = B$.

The result is optimal. In [14] it is shown the existence of two copies of l_2 in L_0 for which the corresponding exact sequences are not isomorphically equivalent. The following example due to Kalton, and reproduced here with his kind permsission, shows that such copies also exist in L_p for $0 < p < 1$: one is the closed span R of the Rademacher functions, while the other is the closed span G of the Gaussian random variables. The point is that any operator $L_p \to L_p$ must send order-bounded sequences to order-bounded sequences due to the latttice structure of the space of operators. Since the Rademacher sequence is order-bounded and the Gaussian sequence is not, no isomorphism of L_p can extend the isomorphism $R \to G$, and thus the sequences $0 \to R \to L_p \to L_p/R \to 0$ and $0 \to G \to L_p \to L_p/G \to 0$ cannot be isomorphically equivalent.

Concluding remark. Following [2] the spaces $\mathcal{Q}(Z, Y)$ can be endowed with a natural quasi-normable and complete (but not necessarily Hausdorff) vector topology induced by the semi-quasi-norm

$$\|H\| = inf\{Q(W) : W \in [H]\}.$$

Let us call $\frac{1}{2}\mathbf{Q}$ to the category of quasi-normed complete (non-necessarily Hausdorff) spaces. Observe that all the maps appearing in the previous proof(s) are continuous with respect to this topology. Hence, we arrive to the remarkable result that

Proposition 4.3 *If the functors $\mathcal{Q}(Z, \cdot)$ and $\mathcal{Q}(Z', \cdot)$ acting from $\mathbf{Q_p} \to \mathbf{V}$ are naturally equivalent then they also are naturally equivalent acting from $\mathbf{Q_p} \to \frac{1}{2}\mathbf{Q}$.*

REFERENCES

1. F. Cabello Sánchez and J.M.F. Castillo, The long homology sequence for quasi-Banach spaces, with applications, submitted to Positivity.
2. F. Cabello Sánchez and J.M.F. Castillo, Natural equivalences for the functor Ext. I: The locally bounded case, submitted to Applied Categorical Structures.
3. F. Cabello Sánchez, J.M.F. Castillo, N. Kalton and D. Yost, Twisted sums of classical function spaces, in preparation.
4. J.M.F. Castillo and M. González, Three-space problems in banach space theory, Lecture Notes in Mathematics 1667, Springer-Verlag 1997.
5. J.M.F. Castillo and Y. Moreno, Strictly singular quasi-linear maps, Nonlinear Analysis-TMA (to appear).
6. P. Enflo, J. Lindenstrauss and G. Pisier, On the "three-space" problem for Hilbert spaces, Mathematica Scandinavica 36 (1975), 199–210.
7. L. Fuchs, Infinite abelian groups, vols. I and II, Academic Press, New York, 1970 and 1973.

8. E. Hilton and K. Stammbach, A course in homological algebra, Graduate Texts in Mathematics 4. Springer-Verlag.
9. N. Kalton, The three-space problem for locally bounded F-spaces, Compositio Mathematica 37 (1978), 243–276.
10. N. Kalton, Convexity, type and the three-space problem, Studia Mathematica 69 (1981), 247-287.
11. N. Kalton, Locally complemented subspaces and $\mathcal{L}_p$ for $p < 1$, Mathematische Nachrichten 115 (1984), 71-97.
12. N. Kalton, The basic sequence problem, Studia Mathematica 116 (1995), 167-187.
13. N. Kalton and N.T. Peck, Twisted sums of sequence spaces and thee three-space problem, Transactions of the American Mathematical Society 255 (1979), 1–30.
14. N. Kalton and N.T. Peck, Quotients of $L_p(0,1)$ for $0 \leq p < 1$, Studia Mathematica 64 (1979), 65–75.
15. N. Kalton, N.T. Peck and W. Roberts, An F-space sampler, London Mathematical Society Lecture Note Series 89.
16. J. Lindenstrauss and L. Tzafriri, Classical Banach spaces I, sequence spaces, Ergebnisse der Mathematik und ihrer Grenzgebiete 92, Springer 1977.
17. S. Mac Lane, Homology, Grundlehren der mathematischen Wissenschaften 114, Springer-Verlag 1975.
18. H. Pat Goeters, When do two groups always have isomorphic extension groups?, Rocky Mountain Journal of Mathematics 20 (1990) 129–144.
19. T. Ribe, Examples for the nonlocally convex three-space problem, Proceedings of the American Mathematical Society 237, (1979) 351-355.
20. J.W. Roberts, A nonlocally convex F-space with the Hahn-Banach extension property, in Banach spaces of Analytic functions, Springer Lecture Notes in Mathematics 604 (1977) 76–81.
21. H. P. Rosenthal. On totally incomparable Banach spaces, Journal of Functional Analysis 4, (1969) 167-175.

Recent Progress in Functional Analysis
K.D. Bierstedt, J. Bonet, M. Maestre and J. Schmets

Integrated Trigonometric Sine Functions

Pedro J. Miana*

Departamento de Matemáticas, Universidad de Zaragoza,
50.009 Zaragoza, Spain
e-mail: pjmiana@posta.unizar.es

To Professor Manuel Valdivia on the occasion of his seventieth birthday.

MCS 2000 Primary 47D09; Secondary 47D62, 26A33.

Abstract

Given $(A, D(A))$ a closed (not necessarily densely defined) linear operator in a Banach space X, a new family of bounded and linear operators, the α-times integrated trigonometric sine function (with $\alpha \geq 0$) is introduced in order to find the link between the α-times integrated cosine function (generated by $-A^2$) and the α-times integrated group (generated by iA). The particular case $\alpha = 0$ is studied in detail. As examples, we will consider Δ_p and $(-\Delta_p)^{\frac{1}{2}}$ on $L^p(\mathbb{R}^n)$ with $1 \leq p \leq +\infty$.

1. Introduction

The strongly continuous cosine function of bounded and linear operators on a Banach space X was defined by Sova in [20] using d'Alembert's functional equation

$$C(t+s) + C(t-s) = 2C(t)C(s), \qquad s, t \in \mathbb{R}.$$

The sine function is defined by $S(t)x := \int_0^t C(s)x ds$ for each $x \in X$. Some relations between cosine and sine functions can be found in [21]; one of them is the following:

$$C(t+s) - C(t-s) = 2AS(t)S(s), \qquad s, t \in \mathbb{R}, \tag{1}$$

where A is the infinitesimal generator of $(C(t))_{t\in\mathbb{R}}$ (for the definitions see, for example, in [21]). (1) shows that this pair of operator families is not an exact extension of scalar trigonometric functions. Also, it is known that $u(t) = C(t)x + S(t)y$ is the solution of the second order Cauchy problem

$$\begin{cases} u''(t) = Au(t), & \text{where } t \in \mathbb{R}, \\ u(0) = x, \quad u'(0) = y, & \text{with } x, y \in X, \end{cases}$$

with $x, y \in X$, see for example [9], [21].

---

*This research has been partially supported by the Spanish DGES,Proyecto PB97-0094.

It is well-known that if $(T(t))_{t\in\mathbb{R}}$ is a C_0-group of bounded operators in X whose infinitesimal generator is $(iA, D(iA))$, then

$$C(t) := \frac{T(t)+T(-t)}{2} \tag{2}$$

is a cosine family whose infinitesimal generator is $(-A^2, D(A^2))$; for the definitions and the proof see, for example,[9] or more recently [5]. In general, it is false that if $(C(t))_{t\in\mathbb{R}}$ is a cosine family on X then there exists a C_0-group $(T(t))_{t\in\mathbb{R}}$ of bounded operators such that the formula (2) holds; see [14], [19]. A condition ("Assumption 6.4" in [7] or "Condition (F)" in [21]) under which the representation (2) holds is the following:

if $B^2 = A$, where A is the infinitesimal generator of $(C(t))_{t\in\mathbb{R}}$, $S(t)$ maps X into $D(B)$ for $t\in\mathbb{R}$, $BS(t)$ is bounded in X for $t\in\mathbb{R}$ and $BS(t)x$ is continuous in t on $\mathbb{R}$ for each fixed $x\in X$.

The representation (2) can be set up in terms of α-times integrated groups and α-times integrated cosine functions (for the definitions see Definition 2.7 below); some results can be found in [2] and [6].

In this paper, we define a new family of bounded operators, the α-times trigonometric sine family which is the key to get the equivalence between an α-times integrated group (generated by iA) and an α-times integrated cosine function (generated by $-A^2$). Our approach has the advantage that A is not necessarily densely defined. In the particular case $\alpha = 0$, this definition is a more accurate operator-valued version of the trigonometric sine function.

In the last section, we will consider some of these families generated by Δ_p and $(-\Delta_p)^{\frac{1}{2}}$, (where $\Delta_p := \frac{\partial^2}{\partial x_1^2} + \frac{\partial^2}{\partial x_2^2} + \dots \frac{\partial^2}{\partial x_n^2}$ is the Laplacian in $\mathbb{R}^n$) and $X = L^p(\mathbb{R}^n)$ with $1 \le p \le +\infty$.

In the sequel, X is a Banach space, $\mathcal{B}(X)$ is the set of bounded and linear operators on X, A is a closed operator on X and $D(A)$ its domain; $\alpha \in [0, +\infty)$.

2. α-Times Integrated Trigonometric Sine Function

The main idea of α-times integrated families is to smoothen operator families by integrating them α times in the fractional sense of Riemann-Liouville [18]. Integrated semigroups and integrated cosine operators are two of these families.

α-times integrated semigroups with $\alpha \in \mathbb{N}$ were introduced by Arendt in [1] and they were defined by Hieber [10] in the case $\alpha \in \mathbb{R}^+$. The relationship between α-times integrated semigroups and the first order Cauchy problem can be found in [10] and [17].

The theory for n-times cosine operators and C-cosine operator functions has appeared together in several papers, see for example [15] or [22]. α-times integrated cosine functions can be introduced in the same way as α-times integrated semigroups, see [24], [25]. In these papers it is proved that α-times integrated cosine functions satisfy a certain abstract Bessel equation [24] and other singular equations [25].

In this section, we introduce a new family of integrated operators: the α-times integrated trigonometric sine family with $\alpha \ge 0$. Actually, we define an α-times trigonometric pair $(\mathcal{S}_\alpha(t), \mathcal{C}_\alpha(t))_{t\ge 0}$ where $(\mathcal{S}_\alpha(t))_{t\ge 0}$ is an α-times integrated trigonometric sine family and $(\mathcal{C}_\alpha(t))_{t\ge 0}$ is an α-times integrated cosine function. We find equivalent definitions in terms of the main condition which defines an α-times integrated trigonometric sine function and a certain equality involving its Laplace transform. We also prove the relationship between this pair and classical α-times integrated groups.

Definition 2.1. Let $(A, D(A))$ be a closed and linear operator in X, $(A, D(A))$ is called the *infinitesimal generator* of an *α-times integrated trigonometric pair* $(\mathcal{S}_\alpha(t), \mathcal{C}_\alpha(t))_{t\geq 0}$ with $\alpha \geq 0$ and $\mathcal{S}_\alpha(t), \mathcal{C}_\alpha(t) \in \mathcal{B}(X)$ for all $t \geq 0$ if
(i) the map $t \to \mathcal{S}_\alpha(t)x$ is continuous for every $x \in X$;
(ii) $\mathcal{S}_\alpha(0) = 0$ and $\mathcal{S}_\alpha(t)A(x) = A\mathcal{S}_\alpha(t)x$ for $x \in D(A)$ and $t \geq 0$;
(iii) $\mathcal{S}_{\alpha+1}(t)x := \int_0^t \mathcal{S}_\alpha(s)x ds \in D(A)$ for $x \in X$ and $\mathcal{C}_\alpha(t)x = \frac{t^\alpha}{\Gamma(\alpha+1)}x - A\int_0^t \mathcal{S}_\alpha(s)xds$ for $t \geq 0$ and $x \in X$;
(iv) $\mathcal{C}_{\alpha+1}(t)x := \int_0^t \mathcal{C}_\alpha(s)xds \in D(A)$ for $x \in X$ and $\mathcal{S}_\alpha(t)x = A\int_0^t \mathcal{C}_\alpha(s)xds$ for $t \geq 0$ and $x \in X$;
(v) $\frac{A}{\Gamma(\alpha)}(\int_t^{t+s} - \int_0^s (t+s-r)^\alpha \mathcal{S}_{\alpha+1}(r)xdr) = \mathcal{S}_{\alpha+1}(t)\mathcal{S}_\alpha(s)x + \mathcal{S}_{\alpha+1}(s)\mathcal{S}_\alpha(t)x$ for $t, s \geq 0$ and $x \in X$.

$(\mathcal{S}_\alpha(t))_{t\geq 0}$ is called the *α-times integrated trigonometric sine function* and $(\mathcal{C}_\alpha(t))_{t\geq 0}$ is called the *α-times integrated trigonometric cosine function* associated to the integrated trigonometric pair. $(\mathcal{S}_\alpha(t), \mathcal{C}_\alpha(t))_{t\geq 0}$ is said to be *non degenerate* if given $x \in X$ such that $\mathcal{S}_\alpha(t)x = 0$ for every $t \geq 0$ then $x = 0$. In all this section an α-times integrated trigonometric pair will be non degenerate. The scalar version of an α-times integrated trigonometric pair (fractional integration of scalar sine and cosine function) can be found, for example, in [18].

2.1. A particular case

If A is a densely defined operator and $\alpha = 0$, it can be proved that *(v)* is equivalent to the condition $A(\int_0^t \mathcal{S}_0(r+u)x + \mathcal{S}_0(r-u)xdu) = 2\mathcal{S}_0(t)\mathcal{S}_0(r)x$ for $t, r \in \mathbb{R}$ and $x \in X$. If we define $\mathcal{S}_0(-t) := -\mathcal{S}_0(t)$ and $\mathcal{C}_0(-t) := \mathcal{C}_0(t)$ for $t \geq 0$, then the following properties hold

(i) $\mathcal{C}_0(t-r) - \mathcal{C}_0(t+r) = 2\mathcal{S}_0(r)\mathcal{S}_0(t)$ for $t, r \in \mathbb{R}$;

(ii) $\mathcal{S}_0(t+r) + \mathcal{S}_0(t-r) = 2\mathcal{S}_0(t)\mathcal{C}(r)$ for $t, r \in \mathbb{R}$;

(iii) $\mathcal{S}_0(t+r) = \mathcal{S}_0(t)\mathcal{C}_0(r) + \mathcal{C}_0(t)\mathcal{S}_0(r)$ for $t, r \in \mathbb{R}$;

(iv) $\mathcal{S}_0(t-r) = \mathcal{S}_0(t)\mathcal{C}_0(r) - \mathcal{C}_0(t)\mathcal{S}_0(r)$ for $t, r \in \mathbb{R}$;

(v) $\mathcal{S}_0(t+r) - \mathcal{S}_0(t-r) = 2\mathcal{C}_0(t)\mathcal{S}_0(r)$ for $t, r \in \mathbb{R}$;

(vi) $(\mathcal{C}_0(t))_{t\in\mathbb{R}}$ is a cosine family, i.e., $\mathcal{C}_0(t+r) + \mathcal{C}_0(t-r) = 2\mathcal{C}_0(t)\mathcal{C}_0(r)$ for $t, r \in \mathbb{R}$;

(vii) $\mathcal{C}_0^2(t) + \mathcal{S}_0^2(t) = I$ for $t \in \mathbb{R}$.

This shows that a 0-times integrated trigonometric pair $(\mathcal{S}_0(t), \mathcal{C}_0(t))_{t\in\mathbb{R}}$ verifies the classical equalities of trigonometric functions.

A pair of operator families is usually introduced to treat the second order Cauchy problem, see for example [23]. Let $(A, D(A))$ be the generator of an 0-times integrated trigonometric pair, $(\mathcal{S}_0(t), \mathcal{C}_0(t))_{t\in\mathbb{R}}$. Then the solution of the second order Cauchy problem

$$\begin{cases} u''(t) = -A^2u(t), & \text{where } t \geq 0 \\ u(0) = x, \quad u'(0) = Ax, & \text{with } x \in D(A^2) \end{cases}$$

is given by $u(t) = \mathcal{C}_0(t)x + \mathcal{S}_0(t)x$.

2.2. Main results

It is an easy exercise to calculate the derivative of $(\mathcal{S}_\alpha(t), \mathcal{C}_\alpha(t))_{t\geq 0}$:

Proposition 2.2. *Let $(\mathcal{S}_\alpha(t), \mathcal{C}_\alpha(t))_{t\geq 0}$ be an α-times integrated trigonometric pair and $(A, D(A))$ its infinitesimal generator. Then,*

1. *the map $t \mapsto \mathcal{S}_\alpha(t)x$ is differentiable and $\frac{d}{dt}\mathcal{S}_\alpha(t)x = \mathcal{C}_\alpha(t)A(x)$ with $x \in D(A)$;*

2. *the map $t \mapsto \mathcal{C}_\alpha(t)x$ is differentiable and $\frac{d}{dt}\mathcal{C}_\alpha(t)x = \frac{t^{\alpha-1}}{\Gamma(\alpha)}x - \mathcal{S}_\alpha(t)A(x)$ if $\alpha > 0$, $\frac{d}{dt}\mathcal{C}_0(t)x = -\mathcal{S}_0(t)A(x)$ with $x \in D(A)$ and $\mathcal{C}_0''(0)x = -A^2(x)$ with $x \in D(A^2)$.*

An α-times trigonometric pair verifies the following equalities:

Proposition 2.3. *Let $(\mathcal{S}_\alpha(t), \mathcal{C}_\alpha(t))_{t\in\mathbb{R}}$ be an α-times integrated trigonometric pair and $(A, D(A))$ its infinitesimal generator. Then*

1. *$\mathcal{S}_\alpha(t)x + A^2\mathcal{S}_{\alpha+2}(t)x = \frac{t^{\alpha+1}}{\Gamma(\alpha+2)}A(x)$ for $t \geq 0$ and $x \in D(A)$;*

2. *$\mathcal{C}_\alpha(t)x + A^2\mathcal{C}_{\alpha+2}(t)x = \frac{t^\alpha}{\Gamma(\alpha+1)}x$ for $t \geq 0$ and $x \in X$.*

Proof. (1) Since $A\mathcal{S}_{\alpha+1}(t)x = \mathcal{C}_\alpha(t)x - \frac{t^\alpha}{\Gamma(\alpha+1)}x$ for $x \in X$, we get

$$A^2\mathcal{S}_{\alpha+2}(t)x = A\int_0^t \mathcal{C}_\alpha(s)xds - \frac{t^{\alpha+1}}{\Gamma(\alpha+2)}A(x) = \mathcal{S}_\alpha(s)xds - \frac{t^{\alpha+1}}{\Gamma(\alpha+1)}A(x)$$

for $x \in D(A)$. *(2)* Since $A\mathcal{C}_{\alpha+1}(t)x = \mathcal{S}_\alpha(t)x$ for $x \in X$, we get

$$A^2\mathcal{C}_{\alpha+2}(t)x = A\int_0^t \mathcal{S}_\alpha(s)xds = \frac{t^\alpha}{\Gamma(\alpha+1)}x - \mathcal{C}_\alpha(t)x.$$

■

Lemma 2.4. *Let $(A, D(A))$ the infinitesimal generator of an α-times integrated trigonometric pair $(\mathcal{S}_\alpha(t), \mathcal{C}_\alpha(t))_{t\geq 0}$.For $t \geq 0$, $\lambda \in \mathbb{C}$ and $x \in X$, we have $\int_0^t e^{-\lambda s}\mathcal{S}_\alpha(s)xds \in D(A)$ and*

$$A\int_0^t e^{-\lambda s}\mathcal{S}_\alpha(s)xds = \frac{e^{-\lambda t}t^\alpha x}{\Gamma(\alpha+1)} - e^{-\lambda t}\mathcal{C}_\alpha(t)x + \lambda t^{\alpha+1}\gamma^*(\alpha+1, \lambda t)x - \lambda\int_0^t e^{-\lambda s}\mathcal{C}_\alpha(s)xds$$

where $\gamma^*(\nu, t) = \frac{1}{\Gamma(\nu)t^\nu}\int_0^t \xi^{\nu-1}e^{-\xi}d\xi$, with $\Re\nu > 0$, is the incomplete gamma function, see [18].

Proof. Since $\int_0^t e^{-\lambda s}\mathcal{S}_\alpha(s)xds = \int_0^t \mathcal{S}_\alpha(s)xds - \lambda\int_0^t e^{-\lambda r}\int_r^t \mathcal{S}_\alpha(s)xdsdr$, then $\int_0^t e^{-\lambda s}\mathcal{S}_\alpha(s)xds \in D(A)$ for $x \in X$. Integrating by parts, we get

$$\begin{aligned}\lambda^\alpha A\int_0^t e^{-\lambda s}\mathcal{S}_\alpha(s)xds &= \lambda^\alpha e^{-\lambda t}A\int_0^t \mathcal{S}_\alpha(s)xds + \lambda^{\alpha+1}A\int_0^t e^{-\lambda s}\int_0^s \mathcal{S}_\alpha(u)xduds \\ &= \lambda^\alpha e^{-\lambda t}(\frac{t^\alpha x}{\Gamma(\alpha+1)} - \mathcal{C}_\alpha(t)x) + \lambda^{\alpha+1}\int_0^t e^{-\lambda s}(\frac{s^\alpha x}{\Gamma(\alpha+1)} - \mathcal{C}_\alpha(s)x)ds \\ &= \frac{(\lambda t)^\alpha e^{-\lambda t}}{\Gamma(\alpha+1)}x - \lambda^\alpha e^{-\lambda t}\mathcal{C}_\alpha(t)x + \int_0^{\lambda t}\frac{e^{-u}u^\alpha x}{\Gamma(\alpha+1)}du - \lambda^{\alpha+1}\int_0^t e^{-\lambda s}\mathcal{C}_\alpha(s)xds\end{aligned}$$

and we obtain the result.

■

Definition 2.5. *An α-times trigonometric pair is said to be exponentially bounded if there exist $C, \omega \in \mathbb{R}^+$ and such that $\|\mathcal{S}_\alpha(t)\|, \|\mathcal{C}_\alpha(t)\| \le Ce^{\omega t}$ for $t \ge 0$.*

The Laplace transform is introduced for exponentially bounded operator families as usual.

Theorem 2.6. *Let $(A, D(A))$ be a closed operator in X and $\mathcal{S}_\alpha : \mathbb{R}^+ \to \mathcal{B}(X)$ a strongly continuous map with $\int_0^t \mathcal{S}_\alpha(s)xds \in D(A)$ for each $x \in X$, $t \ge 0$ and $\|\mathcal{S}_\alpha(t)\| \le Ce^{\omega t}$ with $C, \omega \in \mathbb{R}^+$. Consider*

$$R_{\mathcal{S}_\alpha}(\lambda^2)x := \lambda^\alpha \int_0^{+\infty} e^{-\lambda t}\mathcal{S}_\alpha(t)xdt$$

for $\Re\lambda > \omega$ and $x \in X$. Then $R_{\mathcal{S}_\alpha}(\lambda^2) \in \mathcal{B}(X)$ and $R_{\mathcal{S}_\alpha}(\lambda^2)x \in D(A)$ for every $x \in X$. Moreover, the following statements are equivalent:
(i) $\frac{A}{\Gamma(\alpha)}(\int_t^{t+s} - \int_0^s (t+s-r)^\alpha \mathcal{S}_{\alpha+1}(r)xdr) = \mathcal{S}_{\alpha+1}(t)\mathcal{S}_\alpha(s)x + \mathcal{S}_{\alpha+1}(s)\mathcal{S}_\alpha(t)x$ for $t, s \ge 0$ and $x \in X$.
(ii) $(\mu^2 - \lambda^2)R_{\mathcal{S}_\alpha}(\lambda^2)R_{\mathcal{S}_\alpha}(\mu^2)x = A(R_{\mathcal{S}_\alpha}(\lambda^2)x - R_{\mathcal{S}_\alpha}(\mu^2)x)$ for $x \in X$ with $\Re\lambda, \Re\mu > \omega$.

Proof. As in Lemma 2.4, it holds that $R_{\mathcal{S}_\alpha}(\lambda^2)x \in D(A)$ for $\Re\lambda > \omega$ and $x \in X$, and therefore $R_{\mathcal{S}_\alpha}(\lambda^2) \in \mathcal{B}(X)$. Take $x \in X$ and consider $\mu \ne \lambda$ and $\Re\mu > \Re\lambda$. It is easy to see that

$$\frac{\lambda+\mu}{(\mu\lambda)^{\alpha+1}}R_{\mathcal{S}_\alpha}(\lambda^2)R_{\mathcal{S}_\alpha}(\mu^2)x = \int_0^{+\infty}\int_0^{+\infty} e^{-\lambda t-\mu s}\left(\mathcal{S}_{\alpha+1}(t)\mathcal{S}_\alpha(s)x + \mathcal{S}_{\alpha+1}(s)\mathcal{S}_\alpha(t)x\right)dtds.$$

Next we prove

$$\begin{aligned}&\frac{1}{(\mu-\lambda)(\lambda\mu)^{\alpha+1}}(R_{\mathcal{S}_\alpha}(\lambda^2)x - R_{\mathcal{S}_\alpha}(\mu^2)x)\\ &= \int_0^{+\infty}\int_0^{+\infty} e^{-\lambda t-\mu s}\left(\int_t^{t+s} - \int_0^s (s+t-r)^\alpha \mathcal{S}_{\alpha+1}(r)xdr\right)dtds.\end{aligned}$$

Indeed, using $R_{\mathcal{S}_\alpha}(\lambda^2)x = \lambda^{\alpha+1}\int_0^{+\infty} e^{-\lambda t}\mathcal{S}_{\alpha+1}(t)xdt$, and $\Gamma(\alpha+1) = \mu^{\alpha+1}\int_0^{+\infty} s^\alpha e^{-\mu s}ds$, we have

$$\begin{aligned}\frac{1}{(\mu-\lambda)(\lambda\mu)^{\alpha+1}}R_{\mathcal{S}_\alpha}(\lambda^2)x &= \frac{1}{(\mu-\lambda)\mu^{\alpha+1}}\int_0^{+\infty} e^{-(\lambda-\mu)t}e^{-\mu t}\mathcal{S}_{\alpha+1}(t)xdt\\ &= \frac{-1}{(\mu-\lambda)\mu^{\alpha+1}}\int_0^{+\infty} e^{-(\lambda-\mu)t}\frac{d}{dt}\int_t^{+\infty} e^{-\mu r}\mathcal{S}_{\alpha+1}(r)xdrdt\\ &= \frac{1}{(\mu-\lambda)\mu^{\alpha+1}}\int_0^{+\infty} e^{-\mu r}\mathcal{S}_{\alpha+1}(r)xdr + \frac{1}{\mu^{\alpha+1}}\int_0^{+\infty} e^{-(\lambda-\mu)t}\int_t^{+\infty} e^{-\mu r}\mathcal{S}_{\alpha+1}(r)xdrdt\\ &= \frac{R_{\mathcal{S}_\alpha}(\mu^2)x}{(\mu-\lambda)\mu^{2(\alpha+1)}} + \frac{1}{\Gamma(\alpha+1)}\int_0^{+\infty} e^{-\mu v}v^\alpha dv\int_0^{+\infty} e^{-(\lambda-\mu)t}\int_t^{+\infty} e^{-\mu r}\mathcal{S}_{\alpha+1}(r)xdrdt.\end{aligned}$$

The change of variable $v = s + t - r$ and Fubini's Theorem yield

$$\begin{aligned}&\int_0^{+\infty} e^{-(\lambda-\mu)t}\int_t^{+\infty} e^{-\mu r}\mathcal{S}_{\alpha+1}(r)x\int_0^{+\infty} e^{-\mu v}v^\alpha dvdrdt\\ &= \int_0^{+\infty}\int_t^{+\infty} e^{-\lambda t}\mathcal{S}_{\alpha+1}(r)x\int_{r-t}^{+\infty} e^{-\mu s}(s+t-r)^\alpha dsdrdt\end{aligned}$$

$$= \int_0^{+\infty}\int_0^{+\infty} e^{-\lambda t-\mu s}\int_t^{t+s} \mathcal{S}_{\alpha+1}(r)x(s+t-r)^\alpha drdsdt.$$

On other hand, we have

$$\frac{1}{(\mu-\lambda)(\lambda\mu)^{\alpha+1}}R_{\mathcal{S}_\alpha}(\mu^2)x = \frac{1}{\lambda^{\alpha+1}}\int_0^{+\infty}\int_r^{+\infty} e^{-(\mu-\lambda)s}dse^{-\lambda r}\mathcal{S}_{\alpha+1}(r)xdr$$
$$= \frac{1}{\lambda^{\alpha+1}}\int_0^{+\infty} e^{-(\mu-\lambda)s}\int_0^s e^{-\lambda r}\mathcal{S}_{\alpha+1}(r)xdrds$$
$$= \frac{1}{\Gamma(\alpha+1)}\int_0^{+\infty} e^{-(\mu-\lambda)s}\int_0^s e^{-\lambda r}\mathcal{S}_{\alpha+1}(r)x\int_0^{+\infty} t^\alpha e^{-\lambda t}dtdrds$$
$$= \frac{1}{\Gamma(\alpha+1)}\int_0^{+\infty} e^{-\mu s}\int_0^s \mathcal{S}_{\alpha+1}(r)x\int_0^{+\infty} t^\alpha e^{-\lambda(t+r-s)}dtdrds.$$

We use the change of the variable $t+r-s=u$ and apply Fubini's theorem to get

$$\frac{1}{\Gamma(\alpha+1)}\int_0^{+\infty} e^{-\mu s}\int_0^s \mathcal{S}_{\alpha+1}(r)x\int_{r-s}^{+\infty}(u+s-r)^\alpha e^{-\lambda u}dudrds$$
$$= \frac{1}{\Gamma(\alpha+1)}\int_0^{+\infty}\mathcal{S}_{\alpha+1}(r)x\int_r^{+\infty} e^{-\mu s}\int_{r-s}^{+\infty}(u+s-r)^\alpha e^{-\lambda u}dudsdr$$
$$= \frac{1}{\Gamma(\alpha+1)}\int_0^{+\infty}\mathcal{S}_{\alpha+1}(r)x\int_0^{+\infty} e^{-\lambda u}\int_r^{+\infty}(u+s-r)^\alpha e^{-\mu s}dsdudr$$
$$+\frac{1}{\Gamma(\alpha+1)}\int_0^{+\infty}\mathcal{S}_{\alpha+1}(r)x\int_{-\infty}^{0} e^{-\lambda u}\int_{r-u}^{+\infty}(u+s-r)^\alpha e^{-\mu s}dsdudr$$
$$= \frac{1}{\Gamma(\alpha+1)}\int_0^{+\infty} e^{-\lambda u}\int_0^{+\infty}\mathcal{S}_{\alpha+1}(r)x\int_r^{+\infty}(u+s-r)^\alpha e^{-\mu s}dsdrdu$$
$$+\frac{1}{\Gamma(\alpha+1)}\int_0^{+\infty}\mathcal{S}_{\alpha+1}(r)x\int_r^{+\infty} e^{-\mu s}\int_{r-s}^{0}(u+s-r)^\alpha e^{-\lambda u}dudsdr$$
$$= \frac{1}{\Gamma(\alpha+1)}\int_0^{+\infty} e^{-\lambda u}\int_0^{+\infty}\mathcal{S}_{\alpha+1}(r)x\int_r^{+\infty}(u+s-r)^\alpha e^{-\mu s}dsdrdu$$
$$+\frac{1}{\Gamma(\alpha+1)}\int_0^{+\infty}\mathcal{S}_{\alpha+1}(r)x\int_r^{+\infty} e^{-\mu s}\int_{r-s}^{0}(u+s-r)^\alpha e^{-\lambda u}dudsdr$$
$$= \frac{1}{\Gamma(\alpha+1)}\int_0^{+\infty}\int_0^{+\infty} e^{-\lambda u-\mu s}\int_0^s \mathcal{S}_{\alpha+1}(r)x(u+s-r)^\alpha drduds$$
$$+\frac{1}{\Gamma(\alpha+1)}\int_0^{+\infty}\mathcal{S}_{\alpha+1}(r)x\int_r^{+\infty} e^{-\mu s}\int_r^s (s-v)^\alpha e^{-\lambda(r-v)}dvdsdr.$$

Then

$$\frac{1}{\Gamma(\alpha+1)}\int_0^{+\infty}\mathcal{S}_{\alpha+1}(r)x\int_r^{+\infty} e^{-\mu s}\int_r^s (s-v)^\alpha e^{-\lambda(r-v)}dvdsdr$$
$$= \frac{1}{\Gamma(\alpha+1)}\int_0^{+\infty}\mathcal{S}_{\alpha+1}(r)x\int_r^{+\infty} e^{-\lambda(r-v)}\int_v^{+\infty}(s-v)^\alpha e^{-\mu s}dsdvdr$$
$$= \frac{1}{\Gamma(\alpha+1)}\int_0^{+\infty}\mathcal{S}_{\alpha+1}(r)x\int_r^{+\infty} e^{-\lambda(r-v)}\int_0^{+\infty} z^\alpha e^{-\mu(v+z)}dzdvdr$$
$$= \frac{1}{\Gamma(\alpha+1)}\int_0^{+\infty} e^{-\mu z}z^\alpha dz\int_0^{+\infty} e^{-\lambda r}\mathcal{S}_{\alpha+1}(r)x\int_r^{+\infty} e^{-(\mu-\lambda)v}dvdr$$

$$= \frac{1}{\mu^{\alpha+1}(\mu-\lambda)} \int_0^{+\infty} e^{-\mu r} \mathcal{S}_{\alpha+1}(r) x dr = \frac{1}{\mu^{2(\alpha+1)}(\mu-\lambda)} R_{\mathcal{S}_\alpha}(\mu^2) x$$

and we obtain the result. ∎

We recall now some families of integrated operators.

Definition 2.7. Let $(T_\alpha(t))_{t\geq 0}$ be a strongly continuous family in $\mathcal{B}(X)$ such that $\|T_\alpha(t)\| \leq Ce^{\omega t}$ for some $C, \omega \in \mathbb{R}^+$ and $t \geq 0$ with $\alpha > 0$. It is said to be an *α-times integrated semigroup* if

$$R_{T_\alpha}(\lambda)x := \lambda^\alpha \int_0^{+\infty} e^{-\lambda t} T_\alpha(t) x dt$$

is a pseudoresolvent for $\lambda > \omega$ (see for example [10], [17]). The operator $(B, D(B))$ such that $R_{T_\alpha}(\lambda)x = (\lambda - B)^{-1}x$, is called the *infinitesimal generator* of $(T_\alpha(t))_{t\geq 0}$. $(T_\alpha(t))_{t\in\mathbb{R}}$ is an *α-times integrated group* generated by $(B, D(B))$ if $(T_\alpha(t))_{t\geq 0}$ and $(T_\alpha(-t))_{t\geq 0}$ are α-times integrated semigroups the infinitesimal generators of which are $(B, D(B))$ and $(-B, D(B))$.

Let $(C_\alpha(t))_{t\geq 0}$ be a strongly continuous family in $\mathcal{B}(X)$ such that $\|C_\alpha(t)\| \leq Ce^{\omega t}$ for some $M, \omega \geq 0$ and $t \geq 0$ with $\alpha > 0$. It is said to be an *α-times integrated cosine function* if

$$R_{C_\alpha}(\lambda^2)x := \lambda^{\alpha-1} \int_0^{+\infty} e^{-\lambda t} C_\alpha(t) x dt$$

is a pseudoresolvent for $\Re\lambda > \omega$ (see for example [25]). The operator $(E, D(E))$ such that $R_{C_\alpha}(\lambda^2)x = (\lambda^2 - E)^{-1}x$ is called the *infinitesimal generator* of $(C_\alpha(t))_{t\geq 0}$. $S_\alpha(t) := \int_0^t C_\alpha(s)ds$ is called the *α-times integrated sine function*, see [25].

Corollary 2.8. *Let $(\mathcal{S}_\alpha(t), \mathcal{C}_\alpha(t))_{t\geq 0}$ be an α-times integrated trigonometric pair which satisfies $\|\mathcal{S}_\alpha(t)\|, \|\mathcal{C}_\alpha(t)\| \leq Ce^{\omega t}$ with $C, \omega \in \mathbb{R}^+$ and $(A, D(A))$ its infinitesimal generator. Take $\lambda \in \mathbb{C}$ with $\Re\lambda > \omega$. Then $\lambda^2 \in \rho(-A^2)$ and*

1. *$R_{\mathcal{C}_\alpha}(\lambda^2) = (\lambda^2 + A^2)^{-1}$, i.e., $(\mathcal{C}_\alpha(t))_{t\geq 0}$ is an α-times integrated cosine function generated by $(-A^2, D(A^2))$;*
2. *$R_{\mathcal{S}_\alpha}(\lambda^2) = A(\lambda^2 + A^2)^{-1}$.*

Proof. It is straighforward to show that $R_{\mathcal{S}_\alpha}(\lambda^2) = AR_{\mathcal{C}_\alpha}(\lambda^2)$, with $\Re\lambda > \omega$. Indeed,

$$\begin{aligned} R_{\mathcal{S}_\alpha}(\lambda^2)x &= \lambda^\alpha \int_0^\infty e^{-\lambda t} \mathcal{S}_\alpha(t) x dt = \lambda^\alpha \int_0^\infty e^{-\lambda t} A \int_0^t \mathcal{C}_\alpha(s) x ds dt \\ &= \lambda^\alpha A \int_0^\infty \mathcal{C}_\alpha(s) x \int_s^{+\infty} e^{-\lambda t} dt ds = \lambda^{\alpha-1} A \int_0^\infty \mathcal{C}_\alpha(s) x e^{-\lambda s} ds = AR_{\mathcal{C}_\alpha}(\lambda^2)x. \end{aligned}$$

Since $(\mu^2 - \lambda^2) R_{\mathcal{S}_\alpha}(\lambda^2) x (\lambda^2) R_{\mathcal{S}_\alpha}(\mu^2) x = A(R_{\mathcal{S}_\alpha}(\lambda^2)x - R_{\mathcal{S}_\alpha}(\mu^2)x)$ for $x \in X$ with $\Re\lambda, \Re\mu > \omega$, we get that $A^2(\mu^2 - \lambda^2) R_{\mathcal{C}_\alpha}(\lambda^2) R_{\mathcal{C}_\alpha}(\mu^2) x = A^2(R_{\mathcal{C}_\alpha}(\lambda^2)x - R_{\mathcal{C}_\alpha}(\mu^2)x)$. Since A is injective, $R_{\mathcal{C}_\alpha}$ is a pseudoresolvent and by Proposition 2.3 (ii),

$$\begin{aligned} (\lambda^2 + A^2) R_{\mathcal{C}_\alpha}(\lambda^2) x &= \lambda^2 R_{\mathcal{C}_\alpha}(\lambda^2) x + A^2 \lambda^{\alpha-1} \int_0^\infty e^{-\lambda t} \mathcal{C}_\alpha(t) x dt \\ &= \lambda^2 R_{\mathcal{C}_\alpha}(\lambda^2) x + A^2 \lambda^{\alpha+1} \int_0^\infty e^{-\lambda t} \mathcal{C}_{\alpha+2}(t) x dt = \lambda^{\alpha+1} \int_0^\infty e^{-\lambda t} \frac{t^\alpha}{\Gamma(\alpha+1)} x dt = x. \end{aligned}$$

(2) directly follows from *(1)*: $R_{\mathcal{S}_\alpha}(\lambda^2)x = A(\lambda^2 + A^2)^{-1}x$. ∎

Note that if $(\mathcal{S}_\alpha(t), \mathcal{C}_\alpha(t))_{t\geq 0}$ is an α-times integrated trigonometric pair, then the α-times integrated sine function (see definition 2.7) associated to the α-times integrated cosine function $(\mathcal{C}_\alpha(t))_{t\geq 0}$ is $(\mathcal{C}_{\alpha+1}(t))_{t\geq 0}$.

On the other hand, although $(\mathcal{C}_\alpha(t))_{t\geq 0}$ is an operator family which can be treated using integrated methods, (see [2]), or regularized resolvents ([16]), it is not possible to do the same with $(\mathcal{S}_\alpha(t))_{t\geq 0}$.

Theorem 2.9.

1. Let $(\mathcal{S}_\alpha(t), \mathcal{C}_\alpha(t))_{t\geq 0}$ be an α-times integrated trigonometric pair such that $\|\mathcal{S}_\alpha(t)\|$, $\|\mathcal{C}_\alpha(t)\| \leq Ce^{\omega t}$ with $C, \omega \in \mathbb{R}^+$, and $(A, D(A))$ its infinitesimal generator. Then $\mathcal{T}_\alpha(t) := \mathcal{C}_\alpha(t) + i\mathcal{S}_\alpha(t)$ is an α-times integrated semigroup in X generated by $(iA, D(iA))$.

2. Let $(T_\alpha(t))_{t\in\mathbb{R}}$ be an α-times integrated group generated by $(B, D(B))$. Then $(\mathcal{S}_\alpha(t), \mathcal{C}_\alpha(t))_{t\geq 0}$ is an exponentially bounded α-times integrated trigonometric pair generated by $(-iB, D(iB))$ with

$$\mathcal{S}_\alpha(t) := \frac{T_\alpha(t) - T_\alpha(-t)}{2i}, \quad \textit{and} \quad \mathcal{C}_\alpha(t) := \frac{T_\alpha(t) + T_\alpha(-t)}{2}.$$

Proof. (1) Take $x \in X$. By Corollary 2.8, we get that

$$R_{\mathcal{T}_\alpha}(\lambda)x = \lambda(\lambda^2 + A^2)^{-1}x + iA(\lambda^2 + A^2)^{-1}x = (\lambda + iA)(\lambda^2 + A^2)^{-1}x = (\lambda - iA)^{-1}x$$

for $\lambda \in \mathbb{C}$ with $\Re\lambda > \omega$. *(2)* is straighforward from Definition 2.7. ∎

3. Two Classical Examples

It is well known that if A generates an uniformly bounded holomorphic C_0-semigroup in the half plane $\{z \in \mathbb{C} : \Re z > 0\}$, the boundary value of this holomorphic semigroup is a C_0-group generated by iA ([12]). The relationship between the growth of the holomorphic semigroup and the quality of its boundary values has been investigated in several papers, for example [3], [4] and [6].

Next we consider two particular cases. Take $X = L^p(\mathbb{R}^n)$ with $1 \leq p \leq \infty$, and $\Delta_p = \frac{\partial^2}{\partial x_1^2} + \frac{\partial^2}{\partial x_2^2} + \ldots + \frac{\partial^2}{\partial x_n^2}$ is the classical Laplacian. The holomorphic C_0-semigroup generated by Δ_p is the heat semigroup $\{e^{z\Delta_p}\}_{\Re z>0}$ and satisfies the bound

$$\|e^{z\Delta_p}\|_{\mathcal{B}(L^p(\mathbb{R}^n))} \leq \left(\frac{|z|}{\Re z}\right)^{n|\frac{1}{p}-\frac{1}{2}|}$$

with $1 \leq p \leq \infty$ and $\Re z > 0$. Then ([3], [13]) $i\Delta_p$ generates an α-times integrated group with $\alpha > n|\frac{1}{p} - \frac{1}{2}|$ and also by the Theorem 2.9 (2), Δ_p generates an α-times integrated trigonometric pair.

Now, consider a second example: take $X = L^p(\mathbb{R}^n)$ with $1 \leq p \leq \infty$ and $A = i(-\Delta_p)^{\frac{1}{2}}$. We first consider the case $n = 1$. By applying Theorem 2.9 (i), we obtain the following:

A does not generate a C_0-group on $L^1(\mathbb{R})$ (see Proposition 3.4 in [6]). Therefore $(-\Delta_p)^{\frac{1}{2}}$ does not generate an 0-trigonometric pair.

A generates an α-times integrated group for all $\alpha > 0$ on $L^1(\mathbb{R})$ (see Theorem 2.1 in [4]). Therefore $(-\Delta_p)^{\frac{1}{2}}$ generates an α-times integrated trigonometric pair for all $\alpha > 0$.

A generates an α-times integrated group with $\alpha > 0$ on $L^\infty(\mathbb{R})$. Therefore $(-\Delta_p)^{\frac{1}{2}}$ generates an α-times integrated trigonometric pair for all $\alpha > 0$.

A generates a C_0-group on $L^p(\mathbb{R})$ with $1 < p < \infty$ (see [8], [6]). Therefore $(-\Delta_p)^{\frac{1}{2}}$ generates a 0-times integrated trigonometric pair.

Proposition 3.1. *Let $X = L^p(\mathbb{R}^n)$ and $A = i(-\Delta_p)^{\frac{1}{2}}$ with $1 \leq p \leq \infty$. Then $(-\Delta_p)^{\frac{1}{2}}$ generates an α-times integrated trigonometric pair for all $\alpha > (n-1)|\frac{1}{p} - \frac{1}{2}|$.*

Remark. Hieber [11] proved that Δ_p generates an α-times integrated cosine function on $L^p(\mathbb{R}^n)$ when $\alpha > (n-1)|\frac{1}{2} - \frac{1}{p}|$. He used in his proof multiplier theory. Now this result can be seen as a consequence of the fact that $(-\Delta_p)^{\frac{1}{2}}$ generates an α-times integrated trigonometric pair with $\alpha > (n-1)|\frac{1}{p} - \frac{1}{2}|$ and Corollary 2.8 (1).

REFERENCES

1. Arendt, W.: *Vector-Valued Laplace Transforms and Cauchy Problems,* Israel J. Math. **59**(3) (1987), 327–352.
2. Arendt, W., and H. Kellermann: *Integrated Solutions of Volterra Integrodifferential Equations and Applications,* Integrodifferential Equations (Proc. Conf. Trento, 1987) (G. Da Prato and M. Iannelli, eds.). Pitman Res. Notes Math. Ser., vol **190**, Longman Sci. Tech., Harlow. 1987, pp. 21–51.
3. Arendt, W., O. El Mennaoui and M. Hieber: *Boundary Values of Holomorphic Semigroups,* Proc. Amer. Math. Soc. **118**(1) (1993), 113–118.
4. Boyadzhiev, K., and R. deLaubenfels: *Boundary Values of Holomorphic Semigroups,* Proc. Amer. Math. Soc. **125**(3) (1997), 635–647.
5. Chojnacki, W.: *Group representations of bounded cosine functions,* J. Reine Angew. Math. **478** (1996), 61–84.
6. El-Mennaoui, O., and V. Kéyantuo: *Trace Theorems for Holomorphic Semigroups and the Second Order Cauchy Problem,* Proc. Amer. Math. Soc. **124**(5) (1996), 1445–1458.
7. Fattorini, H.O.: *Ordinary Differential Equations in Linear Topological Spaces, I,* J. of Diff. Equat. **5** (1968), 72–105.
8. Fattorini, H.O.: *Ordinary Differential Equations in Linear Topological Spaces, II,* J. of Diff. Equat. **6** (1969), 50–70.
9. Goldstein, J.: Semigroups of Linear Operators and Applications. Oxford Univ. Press, New York, 1985.
10. Hieber, M.: *Laplace Transforms and α-Times Integrated Semigroups,* Forum Math. **3** (1991), 595–612.
11. Hieber, M.: *Integrated Semigroups and differential operators on L^p spaces,* Math. Ann. **291** (1991), 1–16.
12. Hille, E., and R. S. Phillips: Functional Analysis and Semigroups. Amer. Math. Soc. Coll. **XXXI** 1957.
13. Kéyantuo, V.: *Integrated Semigroups and Related Partial Differential Equations,* J. Math. Anal. and Appl. **212** (1997), 135–153.

14. Kisynski, J.: *On Operator-valued Solutions of d'Alembert's Functional Equations, I,* Colloquium Math. **23** (1971), 107–114.
15. Li, Y., and S. Shaw: *On Generators of Integrated C-semigroups and C-cosine Functions,* Semigroup Forum **47** (1993), 29–35.
16. Lizama, C.: *Regularized Solutions for Abstract Volterra Equations,* J. Math. Anal. and Appl. **243** (2000), 278–292.
17. Mijatović, M., and S. Pilipović: *α-Times Integrated Semigroups ($\alpha \in \mathbb{R}^+$),* J. Math. Anal. and Appl. **210** (1997), 790–803.
18. Miller, K.S., and B. Ross: An Introduction to the Fractional Calculus and Fractional Differential Equations. Wiley, New York 1993.
19. Nagy, B.: *Cosine Operator Functions and the Abstract Cauchy Problem,* Period. Math. Hunga. **7**(3-4) (1976), 213–217.
20. Sova, M.: *Cosine Operator Functions,* Rozprawy Mat., **49** (1966), 1–47.
21. Travis, C.C., and G.F. Webb: *Cosine Families and Abstract Nonlinear Second Order Differential Equations,* Acta Math. Acad. Scien. Hunga. **32**(3-4) (1978), 75–96.
22. Wang, S., and Z. Huang.: *Strongly Continuous Integrated C-cosine Operator Functions,* Studia Math. **126**(3) (1997), 273–289.
23. Xiao, T., and Liang J.: *Differential Operators and C-wellposedness of Complete Second Order Abstract Cauchy Problems,* Pacific J. of Math. **186**(1) (1998), 167–200.
24. Yang, G.: *The Semigroup Theory and Abstract Linear Equations,* Recent Advances in Differential Equations (Proc. Conf. Kunming, 1997) (H-H. Dai and P. L. Sachdev, eds.). Pitman Res. Notes Math. Ser., vol **386**, Addison Wesley Longman Limited, 1998, pp. 101–108.
25. Yang, G.: *α-Times Integrated Cosine Function,* Recent Advances in Differential Equations (Proc. Conf. Kunming, 1997) (H-H. Dai and P. L. Sachdev, eds.). Pitman Res. Notes Math. Ser., vol **386**, Addison Wesley Longman Limited, 1998, pp. 199–212.

Recent Progress in Functional Analysis
K.D. Bierstedt, J. Bonet, M. Maestre and J. Schmets

Applications of a result of Aron, Hervés, and Valdivia to the homology of Banach algebras*

Félix Cabello Sánchez and Ricardo García

Departamento de Matemáticas, Universidad de Extremadura
Avenida de Elvas, 06071-Badajoz, España
E-mail: fcabello@unex.es, rgarcia@unex.es

Dedicated to Professor Manuel Valdivia on his 70-th birthday.

Abstract

As an application of a celebrate result of Aron, Hervés, and Valdivia about weakly continuous multilinear maps, we obtain a sequence (A_n) of finite dimensional (hence amenable) Lipschitz algebras for which the algebra $\ell_\infty(A_n)$ fails to be even weakly amenable.
MCS 2000 Primary 46H05; Secondary 46H25, 46G25

Introduction and main result

Let A be an associative Banach algebra and X a Banach bimodule over A. A derivation $D : A \to X$ is a (linear, continuous) operator satisfying Leibniz's rule:

$$D(ab) = D(a) \cdot b + a \cdot D(b).$$

The simplest derivations have the form $D(a) = a \cdot x - x \cdot a$ for some fixed $x \in X$. They are called inner. A Banach algebra is said to be amenable is every derivation $D : A \to X$ is inner for all dual bimodules X. When this holds merely for $X = A'$ we say that A is weakly amenable.

Let us recall the trivial fact that if $B \to A$ is a bounded homomorphism with dense range and B is amenable, then so is A. The same is true for weak amenability provided B (hence A) is commutative [3] (see also [10] for counterexamples in the noncommutative case). We refer the reader to [11,12,4] for background on amenability and weak amenability.

Let (A_n) be a sequence of associative Banach algebras. As usual, we write $\ell_\infty(A_n)$ for the Banach algebra of all sequences $f = (f_n)$, with $f_n \in A_n$ for all n, and $\|f\| = \sup_n \|f_n\|_{A_n}$ finite, equipped with the obvious norm and coordinatewise multiplication. If $A_n = A$ for some fixed algebra A, we simply write $\ell_\infty(A)$.

In this note, we exhibit sequences (A_n) of finite dimensional amenable Banach algebras for which the algebra $\ell_\infty(A_n)$ fails to be (weakly) amenable.

*Supported in part by DGICYT project PB97-0377.

For basic information about the Arens product in the second dual of a Banach algebra the reader can consult [8,9,6]. Here we only recall that, given a bilinear operator $B : X \times Y \to Z$ acting between Banach spaces, there is a bilinear extension $B'' : X'' \times Y'' \to Z''$ given by

$$B''(x'', y'') = \text{weak*--}\lim_x \left(\text{weak*--}\lim_y B(x, y) \right) \qquad (x'' \in X'', y'' \in Y'')$$

where the iterated limits are taken first for $y \in Y$ converging to y'' in the weak* topology of Y'' and then for $x \in X$ converging to x'' in the weak* topology of X''. The map B'' is often called the first Arens extension of B; see [1]. In particular, if A is a Banach algebra, then the bidual space A'' is always a Banach algebra under the (first) Arens product

$$a'' \cdot b'' = \text{weak*--}\lim_a \left(\text{weak*--}\lim_b (a \cdot b) \right) \qquad (a'', b'' \in A'')$$

where the iterated limits are taken for a and b in A converging respectively to a'' and b'' in the weak* topology of A''.

Our main result is the following device that allows one to obtain A'' as a quotient algebra of $\ell_\infty(A_n)$ if A_n are nicely placed linear subspaces of A, even if they cannot be embedded as subalgebras in A. We feel that the most remarkable feature of the paper is that we get homomorphisms on $\ell_\infty(A_n)$ from linear operators on A_n which are not multiplicative.

Theorem. *Let A_n and A be Banach algebras. Suppose there are linear embeddings $T_n : A_n \to A$ satisfying:*

(a) *There is a constant M such that $M^{-1}\|f\| \leq \|T_n f\| \leq M\|f\|$ for all n and every $f_n \in A_n$.*

(b) *$T_{n+1}(A_{n+1})$ contains $T_n(A_n)$ and $\cup_n T_n(A_n)$ is (strongly) dense in A.*

(c) *Given sequences (f_n) and (g_n) in $\ell_\infty(A_n)$, the sequence $T_n(f_n) \cdot T_n(g_n) - T_n(f_n \cdot g_n)$ is weakly null in A.*

Assume, moreover that

(d) *the product $A \times A \to A$ is jointly weakly continuous on bounded sets; and*

(e) *A' is a separable Banach space.*

Then there exists a surjective homomorphism from $\ell_\infty(A_n)$ onto A''.

So, if A'' fails to be amenable, then $\ell_\infty(A_n)$ cannot be amenable, even if all A_n are. Also, if A_n are commutative and A'' is not weakly amenable, then neither is $\ell_\infty(A_n)$.

Here, we are interested in the case in which A_n are finite dimensional, but note that if A satisfies (d) and (e), then the remaining conditions automatically hold for $A_n = A$ and $T_n = 1_A$ and we obtain A'' as a quotient of $\ell_\infty(A)$.

The proof of the above Theorem uses in a critical way the following result of Aron, Hervés and Valdivia [2]. See [5] for a simpler proof.

Lemma. *For a bilinear operator $B : X \times Y \to Z$ the following conditions are equivalent:*

(a) *B is jointly weakly continuous on bounded sets.*

(b) *B is jointly weakly uniformly continuous on bounded sets.*

(c) *B'' is jointly weakly* (uniformly) continuous on bounded sets.*

Proof of the Theorem. Let U be an ultrafilter on $\mathbf{N}$. Define $\Psi : \ell_\infty(A_n) \to A''$ by $\Psi(f) = \text{weak*}\!-\lim_{U(n)} T_n(f_n)$. This definition makes sense because of the weak* compactness of balls in A''. Clearly, Ψ is linear and bounded, with $\|\Psi\| \leq \sup_n \|T_n\|$.

We show that Ψ is surjective. Take $f'' \in A''$. By (b) and (e) there is a sequence (f_n), with $f_n \in A_n$ such that $T_n(f_n)$ is weakly* convergent to f'' in A'' and bounded in A. By (a) the sequence (f_n) is itself bounded, and taking $f = (f_n)$, it is clear that $\Psi(f) = f''$.

It remains to prove that Ψ is a homomorphism. Take $f, g \in \ell_\infty(A_n)$. Then,

$$\begin{aligned}
&\Psi(f) \cdot \Psi(g) - \Psi(f \cdot g)\\
&\quad = \Big(\text{weak*}\!-\lim_{U(n)} T_n f_n\Big) \cdot \Big(\text{weak*}\!-\lim_{U(n)} T_n g_n\Big) - \Big(\text{weak*}\!-\lim_{U(n)} T_n(f_n \cdot g_n)\Big)\\
&\quad = \Big(\text{weak*}\!-\lim_{U(n)} (T_n f_n \cdot T_n g_n)\Big) - \Big(\text{weak*}\!-\lim_{U(n)} T_n(f_n \cdot g_n)\Big)\\
&\quad = \text{weak*}\!-\lim_{U(n)} \left(T_n f_n \cdot T_n g_n - T_n(f_n g_n)\right) = 0.
\end{aligned}$$

This completes the proof.

Construction of the example

Example. A sequence of finite dimensional (hence amenable) Lipschitz algebras A_n such that $\ell_\infty(A_n)$ is not even weakly amenable.

Proof. Let K be a compact metric space with metric $d(\cdot,\cdot)$ and let $0 < \alpha < 1$. Then $\text{Lip}_\alpha(K)$ is the algebra of all complex-valued functions on K for which

$$\varrho_\alpha(f) = \sup_{x \neq y} \frac{|f(x) - f(y)|}{d(x,y)^\alpha}$$

is finite and $\text{lip}_\alpha(K)$ is the subalgebra of those f such that

$$\frac{|f(x) - f(y)|}{d(x,y)^\alpha} \to 0 \quad \text{as} \quad d(x,y) \to 0.$$

Both algebras are equipped with the norm $\|f\|_\alpha = \|f\|_\infty + \varrho_\alpha(f)$. Bade, Curtis and Dales proved in [3] that the algebra $\text{lip}_\alpha(K)''$ is isometrically isomorphic to $\text{Lip}_\alpha(K)$ which has point derivations for every infinite K (and, therefore, is not weakly amenable).

Take $A = \text{lip}_\alpha(I)$, where $I = [0,1]$ has the usual metric. Then the Banach space A turns out to be isomorphic (in the pure linear sense) to c_0, the space of all null sequences [7,15]. This implies that every bilinear operator from $A \times A$ into any Banach space is jointly weakly continuous on bounded sets [2] and also that A' is separable, which yields (d) and (e).

We now construct the required sequence A_n. For each n, let I_n be the (finite) subset of I consisting of all points of the form $k/2^n$, for $0 \leq k \leq 2^n$. Put $A_n = \mathrm{lip}_\alpha(I_n)$. Clearly, A_n is amenable for all n since it is isomorphic to the algebra $C(I_n)$.

There is a natural quotient homomorphism $Q_n : A \to A_n$, given by plain restriction. Obviously, $\|Q_n\| = 1$ for all n (this will be used later). Let $T_n : A_n \to A$ be defined as follows: for each $f \in A_n$, $T_n(f)$ is the polygonal interpolating f on I_n. Clearly, T_n is a linear operator, although it fails to be multiplicative. Since $Q_n \circ T_n$ is the identity on A_n it is clear that $\|T_n f\| \geq \|f\|$ for all $f \in A_n$.

Moreover, $\|T_n\| = 1$ for all n. Clearly, $\|T_n(f)\|_\infty = \|f\|_\infty$, so the point is to show that $\varrho_\alpha(T_n f)$ equals $\varrho_\alpha(f)$. It obviously suffices to see that if g is a polygonal with nodes in I_n then

$$\varrho_\alpha(g) = \sup_{x \neq y} \frac{|g(y) - g(x)|}{|y - x|^\alpha}$$

is attained at some $(x, y) \in I_n \times I_n$. This is an amusing exercise in elementary calculus. The solution appears in [13, chapter III, lemma 3.2, p. 203]. Thus, T_n is an into isometry and (a) holds.

Let us verify (b). Obviously, $T_{n+1}(A_{n+1})$ contains $T_n(A_n)$ for each n, so that $\cup_n T_n A_n$ is a (not closed) linear subspace of $\mathrm{lip}_\alpha(I)$. We show that $\cup_n T_n A_n$ is (strongly) dense in $\mathrm{lip}_\alpha(I)$. It clearly suffices to show weak density. We claim that for every $f \in \mathrm{lip}_\alpha(I)$ the sequence $T_n Q_n(f)$ converges weakly to f in $\mathrm{lip}_\alpha(I)$. We need some information about weak convergent sequences in the small space of Lipschitz functions.

Consider the operator $\Phi : \mathrm{lip}_\alpha(I) \to C_0(I^2 \backslash \Delta) \oplus_1 C(I)$ given by $\Phi(f) = (\tilde{f}, f)$, where

$$\tilde{f}(x, y) = \frac{f(y) - f(x)}{|y - x|^\alpha}$$

and Δ is the diagonal of I^2. Clearly, it is an isometric embedding, so that the weak topology in $\mathrm{lip}_\alpha(I)$ is the relative weak topology as a subspace of $C_0(I^2 \backslash \Delta) \oplus_1 C(I)$. On the other hand, weakly null sequences in $C_0(\Omega)$ spaces are bounded sequences pointwise convergent to zero. Hence $f_n \to f$ weakly in $\mathrm{lip}_\alpha(I)$ if and only if (f_n) is bounded and $f_n(x) \to f(x)$ for all $x \in I$, and this happens if and only if (f_n) is bounded and $f_n(x) \to f(x)$ for all x in some dense subset of I.

But, for $f \in \mathrm{lip}_\alpha(I)$ the sequence $(T_n Q_n(f))$ is bounded (by the norm of f) and converges pointwise to f on $\cup_n I_n$. This proves our claim. So, (b) also holds.

It only remains to verify (c). Take $(f_n), (g_n) \in \ell_\infty(A_n)$. Then $T_n(f_n) \cdot T_n(g_n) - T_n(f_n \cdot g_n)$ is weakly null in A if and only if for every ultrafilter V on $\mathbf{N}$ one has

$$\lim_{V(n)} \left(T_n(f_n) \cdot T_n(g_n) - T_n(f_n \cdot g_n)\right) = 0$$

in the weak* topology of $A'' = \mathrm{Lip}_\alpha(I)$. Take $x \in \cup_n I_n$ and let δ_x be the associated evaluation functional. Then,

$$\begin{aligned}
\left\langle \text{weak*-}\lim_{V(n)} T_n(f_n \cdot g_n), \delta_x \right\rangle &= \lim_{V(n)} \langle T_n(f_n \cdot g_n), \delta_x \rangle \\
&= \lim_{V(n)} T_n(f_n \cdot g_n)(x) \\
&= \lim_{V(n)} (f_n \cdot g_n)(x) \\
&= \lim_{V(n)} (f_n(x) g_n(x)) \\
&= \lim_{V(n)} f_n(x) \cdot \lim_{V(n)} g_n(x) \\
&= \left\langle \text{weak*-}\lim_{V(n)} T_n(f_n), \delta_x \right\rangle \left\langle \text{weak*-}\lim_{V(n)} T_n(g_n), \delta_x \right\rangle,
\end{aligned}$$

so that

$$\text{weak*-}\lim_{V(n)} T_n(f_n \cdot g_n) = \left(\text{weak*-}\lim_{V(n)} T_n(f_n) \right) \cdot \left(\text{weak*-}\lim_{V(n)} T_n(g_n) \right).$$

Since the product of $\mathrm{Lip}_\alpha(I)$ is jointly weakly* continuous on bounded sets, the right hand side of the preceding equation becomes

$$\text{weak*-}\lim_{V(n)} (T_n(f_n) \cdot T_n(g_n)),$$

which completes the proof of (c).

Thus, the Theorem yields a surjective homomorphism $\ell_\infty(A_n) \to A''$, which shows that $\ell_\infty(A_n)$ is not weakly amenable and completes the proof.

Concluding remarks

As the referee pointed out, it is implicit in [14] that there are finite dimensional (hence amenable) C^*-algebras A_n for which $\ell_\infty(A_n)$ fails to be amenable. To see this, let H be a separable Hilbert space with a fixed orthonormal basis and let H_n be the subspace spanned by the first n elements of the basis. Write i_n for the obvious inclusion of H_n into H and π_n for the obvious projection of H onto H_n. Take $A_n = L(H_n)$, the algebra of all operators on H_n and $A = K(H)$, the algebra of all compact operators on H. Then $L(H_n)$ embeds isometrically as a subalgebra in A taking $T_n(L) = i_n \circ L \circ \pi_n$. Although (d) fails, it is clear from the proof of the Theorem that Ψ is still an onto operator from $\ell_\infty(L(H_n))$ onto $K(H)'' = L(H)$. Moreover the map $\Phi : L(H) \to \ell_\infty(L(H_n))$ given by $\Phi(T) = (\pi_n \circ T \circ i_n)$ is a right inverse for Ψ and $L(H)$ is thus a complemented subspace of $\ell_\infty(L(H_n))$. This implies that $\ell_\infty(L(H_n))$ lacks the approximation property and cannot be amenable (see [14] and references therein).

Needless to say, our example is far simpler since the existence of point derivations in $\mathrm{Lip}_\alpha(I)$ is a straightforward consequence of the Banach-Alaoglu theorem.

It follows from the remarks made after the Theorem that if A is $\mathrm{lip}_\alpha(I)$, then there is a surjective homomorphism from $\ell_\infty(A)$ onto A''. Hence $\ell_\infty(A)$ fails to be amenable and the same occurs with any ultrapower A_V (with respect to a non-trivial ultrafilter V on $\mathbf{N}$) since the quotient mapping constructed in the Theorem factorizes throughout the natural homomorphism $\ell_\infty(A) \to A_V$.

It would be interesting to study Banach algebras which are "super-amenable" in the sense of having amenable ultrapowers. A reasonable conjecture appears to the that A is super-amenable if and only of A'' is amenable. Note that, in view of [14, theorem 2.5], the conjecture is true for C^*-algebras. See [9] for some (loosely) related results.

Acknowledgements

It is a pleasure to thank the anonymous referee for the correction of a serious mistake in a previous version of the paper, for helpful comments, and for calling our attention to some useful references. We also thank Jesús Jaramillo for being so amenable.

REFERENCES

1. R. Arens, The adjoint of a bilinear operation. Proceedings of the American Mathematical Society 2 (1951) 839–848.
2. R. Aron, C. Hervés, and M. Valdivia, Weakly continuous mappings on Banach spaces. Journal of Functional Analysis 52 (1983) 189–204.
3. W.G. Bade, P.C. Curtis Jr., and H.G. Dales, Amenability and weak amenability for Beurling and Lipschitz algebras. Proceedings of the London Mathematical Society 55 (3) (1987) 359–377.
4. F.F. Bonsall and J. Duncan, Complete normed algebras. Springer, 1973.
5. F. Cabello Sánchez and R. García, On the amenability of the Banach algebras $\ell_\infty(A)$. Mathematical Proceedings of the Cambridge Philosophical Society (2001), to appear.
6. F. Cabello Sánchez, R. García, and I. Villanueva, Extension of multilinear operators on Banach spaces. Extracta Mathematica, 15 (2000) 291–334.
7. Z. Ciesielski, Properties of the orthonormal Franklin system. Studia Mathematica 23 (1963) 141–157.
8. J. Duncan and S.A.R. Hosseinium, The second dual of a Banach algebra. Proceedings of the Royal Society of Edinbugh 84A (1979) 309–325.
9. G. Godefroy and B. Iochum, Arens-regularity of Banach algebras and the geometry of Banach spaces. Journal of Functional Analysis 80 (1988) 47–59.
10. N. Grønbæk, Amenability and weak amenability of tensor algebras and algebras of nuclear operators. Journal of the Australian Mathematical Society (Series A) 51 (1991) 483–488.
11. A.Ya. Helemskii, Banach and locally convex algebras. Oxford Science Publications, 1993.
12. B.E. Johnson, Cohomology of Banach algebras. Memoirs of the American Mathematical Society 127, 1972.
13. S.G. Krein, Yu.I. Petunin, and E.M. Semenov, Interpolation of linear operators. Translationn of Mathematical Monographs 45, American Mathematical Society, Providence, R.I., 1982.
14. A.T.-M. Lau, R.J. Roy, and G.A. Willis, Amenability of Banach and C^*-algebras on locally compact groups. Studia Mathematica 119 (1996) 161–178.
15. P. Wojtaszczyk, Banach spaces for analysts. Cambridge studies in advanced mathematics 25, Cambridge, 1991.

Recent Progress in Functional Analysis
K.D. Bierstedt, J. Bonet, M. Maestre and J. Schmets

On the ideal structure of some algebras with an Arens product

Mahmoud Filali

Department of Mathematical Sciences, University of Oulu,
FIN-90014 Oulu, Finland

Abstract

Let G be a non-compact locally compact group, $L^1(G)$ be its group algebra, $LUC(G)$ be the space of bounded functions on G which are uniformly continuous with respect to the right uniformity on G. Let $L^1(G)^{**}$ be the second conjugate of $L^1(G)$ with the first Arens product and $LUC(G)^*$ be the conjugate of the space $LUC(G)$ with the first Arens-type product. We see how the maximal ideals of $LUC(G)^*$ are related to those of $L^1(G)$, and give examples of weak−dense maximal ideals in $LUC(G)^*$. For a large class of locally compact groups, we compute the dimension of every right ideal in $LUC(G)^*$ and the dimension of every right ideal in $L^1(G)^{**}$ which is not generated by a right annihilator of $L^1(G)^{**}$. In particuar, we see that they are infinite dimensional; a result which was known earlier only for the radicals of these algebras. Finally, we construct new elements in the radicals of $LUC(G)^*$ and $L^1(G)^{**}$.*

MCS 2000 Primary 43A10; Secondary 22D15, 22A15

1. Introduction

Let A be a real or complex Banach algebra, A^* be its dual and A^{**} be its second dual. The *first Arens product* is defined in three stages as follows. For every μ, $\nu \in A^{**}$, $f \in A^*$ and $\phi \in A$, we define $f_\phi \in A^*$, $f_\nu \in A^*$ and $\mu\nu \in A^{**}$, respectively, by

$$f_\phi(\psi) = f(\phi\psi), \quad \psi \in A,$$

$$f_\nu(\phi) = \nu(f_\phi), \quad \phi \in A,$$

$$\mu\nu(f) = \mu(f_\nu), \quad f \in A^*.$$

When A^{**} is given the weak*-topology, we see that the mapping

$$\mu \mapsto \mu\nu : A^{**} \to A^{**}$$

is continuous for each fixed $\nu \in A^{**}$. It is not difficult to verify that the mapping

$$\nu \mapsto \mu\nu : A^{**} \to A^{**}$$

is also continuous for each fixed $\mu \in A \subseteq A^{**}$. But, in general, this mapping is not continuous. This product and the second Arens product (which is defined in a similar manner) were introduced about fifty years ago in a more general setting than that of a Banach algebra by Arens in [2] and [3]. But we are concerned in this paper with just the case where A is the group algebra $L^1(G)$ with convolution. In general the two products do not coincide. For more details, see [12].

Let G be a locally compact group with a left Haar measure λ and with a Haar modular function Δ, and let $A = L^1(G)$. Note that here

$$f_\nu(\phi) = \nu(\widehat{\phi} * f) \quad \text{for all } \nu \in L^1(G)^{**},\ f \in L^\infty(G) \text{ and } \phi \in L^1(G),$$

where $\widehat{\phi}(s) = \Delta(s^{-1})\phi(s^{-1})$ for $s \in G$.

The other algebras we are concerned with are $LUC(G)^*$ and $WAP(G)^*$ with an Arens-type product. Here $LUC(G)$ is the space of bounded functions on G which are uniformly continuous with the respect to the right uniformity on G (denoted in [19] by $C_{ru}(G)$), and $WAP(G)$ is the space of weakly almost periodic functions on G. Recall from [5] that if $C(G)$ is the space of all bounded, continuous, complex-valued functions on G and if for each $f \in C(G)$ and $s \in G$, f_s is the left translate of f by s (see below) and $f_G = \{f_s : s \in G\}$ then

$$LUC(G) = \{f \in C(G) \ : \ s \mapsto f_s \quad \text{is norm continuous}\},$$

$$WAP(G) = \{f \in C(G) : f_G \quad \text{is relatively weakly compact}\}.$$

Let F be either $LUC(G)$ or $WAP(G)$. Then the first Arens-type product in F^* is defined also in three stages: for $\mu,\ \nu \in F^*$, $f \in F$ and $s \in G$, we define $f_s \in F$, $f_\nu \in F$ and $\mu\nu \in F^*$, respectively, by

$$f_s(t) = f(st), \quad t \in G,$$
$$f_\nu(s) = \nu(f_s), \quad s \in G,$$
$$\mu\nu(f) = \mu(f_\nu), \quad f \in F.$$

This product is exactly the restriction of the first Arens product to these spaces, see for example [20]. In [19, Page 275], the Banach algebra $LUC(G)^*$ is described as "a highly legitimate object of study". It is also often less diffucult to study first $LUC(G)^*$ and then transfer the results to $L^1(G)^{**}$.

We see in the following section how the maximal ideals of F^* are related to those of $L^1(G)$. We then give two examples of weak*−dense maximal ideals in F^*.

In the third section, we extend a theorem we proved recently for G discrete ([13]) to locally compact groups of the form $I\!R^n \times H$, where H is a locally compact group containing a compact, open, normal subgroup K. We show that the dimension of a non-zero right ideal in $LUC(G)^*$ is 2^{2^κ}, where $\kappa = \max\{\omega, |H/K|\}$. Similar result is proved in $L^1(G)^{**}$ for the right ideals which are not generated by a right annihilator of $L^1(G)^{**}$.

In the last section, we provide new elements in the radical of each of the algebras $LUC(G)^*$ and $L^1(G)^{**}$. In some cases, we do not even assume that G is amenable.

A main tool used in each section is the $F-$*compactification* FG of G. This is a semigroup compactification of G and may be obtained as the maximal ideal space FG of F, i.e.

$$FG = \{x \in F^* : x \neq 0 \text{ and } x(fg) = x(f)x(g) \quad \text{for all} \quad f,\ g \in F\}.$$

The weak*−topology and the restriction of the product from F^* to FG make FG into a compact right topological semigroup which contains a homeomorphic copy of G as a dense subgroup. We denote by UG and WG the $LUC-$ and the $WAP-$compactifications of G, respectively. Note that when G is discrete, $LUC(G)$ is the space $\ell_\infty(G)$ of all bounded complex-valued functions on G, and so UG is the Stone-Čech compactification βG of G. Recall also that, in fact, WG is even a compact semitopological semigroup. See [5].

2. On the maximal ideals

The following theorem was proved by Civin in [7] for $L^1(G)^{**}$, where G is a locally compact abelian group, using some heavy harmonic analysis machinery. With the help of a theorem due to Allan [1], we proved the theorem in [9] for A^{**} for any commutative Banach algebra A. Below we see that the theorem is also true for F^* when G is abelian.

Theorem 2.1 *Let G be a locally compact abelian group, $\widehat{G}$ be the character group of G and let F be either $LUC(G)$ or $WAP(G)$. Then a maximal left, right or two-sided ideal M of F^* is either weak*−dense or there exists $\chi \in \widehat{G}$ such that $M = \{\mu \in F^* \ : \ \mu(\chi) = 0\}$.*

Proof: Let M be a maximal left, right or two-sided ideal of F^*. First it is easy to see that the weak*−closure $\overline{M}$ of M is a linear subspace of F^*. So let $\mu \in \overline{M}$ and $\nu \in F^*$ be arbitrary. Then μ and ν are the limits of some nets (μ_α) in M and (ν_β) in $L^1(G)$, respectively. It follows that if M is a right ideal then $\mu\nu = \lim_\alpha \mu_\alpha\nu \in \overline{M}$ (remember that the mapping $\mu \mapsto \mu\nu$ is weak*−continuous on F^*). Thus $\overline{M}$ is also a right ideal. If M is a left ideal, then $\nu\mu = \lim_\beta \nu_\beta\mu = \lim_\beta \mu\nu_\beta \in \overline{M}$ (this is due to the fact that $L^1(G)$ is in the centre of F^*). Thus $\overline{M}$ is a left ideal of F^*. Since M is maximal, we must have in each case either $\overline{M} = F^*$ and so M is weak*−dense in F^*, or $\overline{M} = M$ and so M is weak*−closed. Suppose that M is weak*−closed. Let $L_e = L^1(G) + \mathbb{C}e$. Then L_e is a closed subalgebra of F^*, and so by [1], $M \cap L_e$ is a maximal ideal of L_e. If $M \cap L_e = L^1(G)$ then $L^1(G) \subseteq M$, which is not possible since M is weak*−closed. It follows, by [21, page 59], that $M \cap L^1(G) = (M \cap L_e) \cap L^1(G)$ is a maximal modular ideal of $L^1(G)$, and so

$$M \cap L^1(G) = \{\mu \in L^1(G) \ : \ \mu(\chi) = \int_G \chi(s)d\mu(s) = 0\}$$

for some $\chi \in \widehat{G}$. Now as in the proof of Theorem 5.3 in [8], since $\overline{M \cap L^1(G)}$ is a maximal subspace of F^* and $\overline{M \cap L^1(G)} \subseteq \overline{M} = M$, we conclude that

$$M = \overline{M \cap L^1(G)} = \{\mu \in F^* \ : \ \mu(\chi) = 0\}.$$

Remark: Two examples of weak*−dense maximal modular left ideal in $L^1(G)^{**}$ were given by Civin in [7]. Later on, Olubummo considered also in [22] these ideals in $\ell_\infty(\mathbb{N})^*$. (These seem to be the only known examples of these type of ideals.) With the help of Lemma 2.2 below, these examples can easily be given in $LUC(G)^*$ and $WAP(G)^*$ as well: in Theorem 2.3 below, one example is obtained by letting M_c as weak*−dense in $M(G)$ (i.e. containing $L^1(G)$) and the second example is obtained by taking M_c as weak*−closed in $M(G)$ (i.e. $M_c = \{\mu \in M(G) \ : \ \mu(\chi) = 0\}$ for some $\chi \in \widehat{G}$.)

Lemma 2.2 was first proved and used by Civin and Yood in their seminal paper [8] for G discrete and $F = \ell_\infty(G)^*$. For a different proof of the lemma, see [15].

Lemma 2.2 *Let G be a locally compact group. Let $F = LUC(G)$ or $WAP(G)$, and*

$$C_0(G)^\perp = \{\nu \in F^* \ : \ \nu(f) = 0 \quad \textit{for all} \quad f \in C_0(G)\}.$$

Then $C_0(G)^\perp$ is a weak*−closed two-sided ideal of F^* and $F^* = M(G) \oplus C_0(G)^\perp$.

Proof: Let FG be the F−compactification of G. The Gelfand mapping $f \mapsto \tilde{f}$, where

$$\tilde{f}(x) = x(f) \quad \text{for} \quad x \in FG \quad \text{and} \quad f \in F,$$

gives $F = C(FG)$. Hence $F^* = C(FG)^*$ by the mapping $\mu \mapsto \tilde{\mu}$, where

$$\tilde{\mu}(\tilde{f}) = \mu(f) \quad \text{for} \quad \mu \in F^* \quad \text{and} \quad f \in F,$$

and so the Riesz representation theorem gives $F^* = M(FG)$. For $\mu \in F^*$, we use the same letter (i.e. μ) to denote the corresponding elements in $C(FG)^*$ and $M(FG)$. Since G is open in FG, we may define, for each $\mu \in F^*$, μ_c and μ_* in F^* by

$$\mu_c(B) = \mu(B \cap G) \quad \text{and} \quad \mu_*(B) = \mu(B \cap (FG \setminus G))$$

for all Borel subsets B of FG. Then clearly we have $\mu = \mu_c + \mu_*$. Furhermore, if $\mu \in M(G) \cap C_0(G)^\perp$, then $\|\mu\| = |\mu|(1) = |\mu|(FG) = |\mu|(G) + |\mu|(FG \setminus G) = 0 + 0 = 0$. So

$$M(G) \cap C_0(G)^\perp = \{0\}.$$

So F^* is the algebraic direct sum of $M(G)$ and $C_0(G)^\perp$. Since $M(G)$ and $C_0(G)^\perp$ are norm-closed in F^*, this is a topological direct sum. Next we show that $C_0(G)^\perp$ is a two-sided ideal in F^*. Let $\mu \in F^*$, $\nu \in C_0(G)^\perp$, and $f \in C_0(G)$. Then $f_\nu(s) = \nu(f_s) = 0$ for all $s \in G$ since f_s is also in $C_0(G)$. Hence $\mu\nu(f) = 0$, and so $\mu\nu \in C_0(G)^\perp$. To show that $C_0(G)^\perp$ is a right ideal, it is clearly enough to show that f_μ is in $C_0(G)$ when f has a compact support K. Since $f_s \in C_0(G)$, for each $s \in G$, we may regard μ as an element of $M(G)$ (by taking its restriction to $C_0(G)$). Let (μ_n) be a sequence of measures in $M(G)$, such that each μ_n has a compact support K_n and $\|\mu_n - \mu\| \to 0$. Then, for each n,

$$f_{\mu_n}(s) = \int_{K_n} f(st) d\mu(t) = 0 \quad \text{if} \quad s \notin KK_n^{-1},$$

and so $f_{\mu_n} \in C_0(G)$. Therefore $f_\mu \in C_0(G)$ since (f_{μ_n}) converges to f_μ uniformly on G, as required.

Theorem 2.3 *Let G be a locally compact abelian group, and $F = LUC(G)$ or $WAP(G)$. Let M_c be a maximal ideal of $M(G)$. Then $M_c \oplus C_0(G)^\perp$ is a weak*−dense maximal ideal of F^*. Furthermore, if M is a maximal ideal of F^* and $C_0(G)^\perp \subseteq M$, then $M = M_c \oplus C_0(G)^\perp$ for some maximal ideal M_c of $M(G)$.*

Proof: With the help of Lemma 2.2, $M_c \oplus C_0(G)^\perp$ is clearly an ideal in F^* whenever M_c is an ideal of $M(G)$.

We prove first the second statement. Let M be a maximal left, right or two-sided ideal of F^* which contains $C_0(G)^\perp$. Then by [1], $M_c = M \cap M(G)$ is a maximal ideal of $M(G)$.

Moreover, take $\mu \in M$, and write by Lemma 2.1, $\mu = \mu_1 + \mu_*$ with $\mu \in M(G)$ and $\mu_* \in C_0(G)^\perp$. Then since $C_0(G)^\perp \subseteq M$, we have $\mu_1 = \mu - \mu_* \in M \cap M(G) = M_c$. In other words, $M \subseteq M_c \oplus C_0(G)^\perp$. Thus $M = M_c \oplus C_0(G)^\perp$.

To show in the first statement that $M_c \oplus C_0(G)^\perp$ is weak*−dense in F^*, it is enough to notice that its closure contains $\{\mu \in F^* \ : \ \mu(\chi) = 0\}$. To show that it is maximal, let M be a left, right (or a two sided) ideal of F^* such that $M_c \oplus C_0(G)^\perp \subseteq M$. Then $C_0(G)^\perp \subseteq M$, and so by what we have just proved, $M = (M \cap M(G)) \oplus C_0(G)^\perp$. The proof is complete with the remark that $M_c = M \cap M(G)$.

3. On the dimension of right ideals

An element μ in $LUC(G)^*$, is said to be *left invariant* if

$$\mu(f_s) = \mu(f) \quad \text{for all } f \in LUC(G) \text{ and } s \in G.$$

A element μ in $L^1(G)^{**}$ is said to be *topologically left invariant* if

$$\mu(\phi * f) = \left(\int_G \phi(s)d\lambda(s)\right)\mu(f) \quad \text{for all } \phi \in L^1(G) \text{ and } f \in L^\infty(G).$$

Recall that 'topologically left invariant' implies 'left invariant', but not conversely. When $LUC(G)^*$, or equivalently $L^1(G)^{**}$, has a non-zero left invariant element, we say that G is *amenable.* Abelian groups and groups of polynomial growth are among many other examples of amenable groups. But the free group F_2 is not amenable. For all these notions see [5], [17] or [23]. Recall that $\mu \in L^1(G)^{**}$ is a right annihilator of $L^1(G)^{**}$ if it satisfies $L^1(G)^{**}\mu = \{0\}$. Using the (topologically) left invariant elements and the representations of G, we proved in [10] that finite-dimensional left ideals exist in $M(G)$ and $L^1(G)$ if and only if G is compact; they exist in $LUC(G)^*$ if and only if G is amenable; and those which are not generated by right annihilators of $L^1(G)^{**}$ exist in $L^1(G)^{**}$ if and only if G is amenable. However, in [11], the right ideals in $LUC(G)^*$ and the right ideals in $L^1(G)^{**}$, which are not generated by right annihilators of $L^1(G)^{**}$, were shown to be all of infinite dimension when G is an infinite discrete group, or a non-compact locally compact abelian group. Recently, we have managed to calculate the dimension of the non-zero right ideals when G is discrete. It is $2^{2^{|G|}}$ (see [13]). This was achieved with the help of the so-called *t-sets or thin sets* in G. These are subsets V of G which satisfy $sV \cap tV$ is finite whenever $s \neq t$ in G. In the following theorem, we calculate the dimension of the right ideals in $LUC(G)^*$ and $L^1(G)^{**}$ for a larger class of locally compact groups.

Theorem 3.1 *Let G be a non-compact, locally compact group of the form $G = \mathbb{R}^n \times H$, where H is a locally compact group containing a compact, open, normal subgroup K. Every non-zero right ideal in $LUC(G)^*$ has dimension 2^{2^κ}, where $\kappa = \max\{\omega, |H/K|\}$.*

Proof: We proceed first as in [14]. Let UG be the LUC−compactification of G. Let H/K be the discrete quotient group (the elements of H/K are the right cosets Ks), let $S = \boldsymbol{Z}^n \times H/K$ and let $\psi : \boldsymbol{Z}^n \times H \to S$ be the quotient mapping. Extend ψ to a continuous homomorphism (denoted also by the same letter) $\psi \ : \ U(\boldsymbol{Z}^n \times H) \to \beta S$. Since $LUC(\boldsymbol{Z}^n \times H) = LUC(G)_{|\boldsymbol{Z}^n \times H}$ (see [5, Theorem 5.1.11]), we may identify $U(\boldsymbol{Z}^n \times H)$

and $\overline{\mathbf{Z}^n \times H}$ and so we have $\psi : \overline{\mathbf{Z}^n \times H}$ ($\subseteq UG$) $\to \beta S$. By [6] (or see [13]), let V be a t-set in S of the same cardinality as S, i.e. $|V| = \max\{\omega, |H/K|\}$. Then by [13] each point in $\overline{V}$ is right cancellative in βS (i.e. $yx \neq zx$ whenever $y \neq z$ in βS) and $(\beta S a_1) \cap (\beta S a_2) = \emptyset$ whenever a_1 and a_2 are two distinct points in $\beta S \setminus S$. Let X be the subset of $UG \setminus G$ formed by taking, for each $a \in \overline{V} \setminus V$, one element from $K\psi^{-1}(a)$. Then X contains as many points as $\overline{V} \setminus V$, i.e. 2^{2^κ} where $\kappa = \max\{\omega, |H/K|\}$. Furthermore, by [14], each point in X is right cancellative in UG and $(UG)x_1 \cap (UG)x_2 = \emptyset$ whenever x_1 and x_2 are distinct points in X.

Now we proceed as in [11]. Let $\mu \in LUC(G)^*$, $\mu \neq 0$. With the Gelfand mapping and the Riesz representation theorem, we regard μ as a Borel measure on UG, and so we may talk about its support $supp(\mu)$. Then by [11], $supp(\mu x) = supp(\mu)x$ for each x in X since x right cancellative in UG. This implies first that $\mu x \neq 0$ for each $x \in X$, and secondly that the elements μx, where $x \in X$, have mutually disjoint supports in UG since $(UG)x_1 \cap (UG)x_2 = \emptyset$ whenever x_1 and x_2 are distinct in X. Accordingly if R is a right ideal in $LUC(G)^*$ and μ a non-zero element in R then $\{\mu x : x \in X\}$ is a set of 2^{2^κ} linearly independent elements of R, where $\kappa = \max\{\omega, |H/K|\}$, and so the proof is complete.

Theorem 3.2 *Let G be a non-compact locally compact group as in Theorem 3.1. Then every right ideal in $L^1(G)^{**}$ containing an element which is not a right annihilator of $L^1(G)^{**}$ has dimension 2^{2^κ}.*

Proof: Since $LUC(G)$ is norm-closed in $L^\infty(G)$, we may consider the natural map $\phi : L^1(G)^{**} \to LUC(G)^*$ defined by

$$\phi(\mu)(f) = \mu(f) \quad \text{for all} \quad \mu \in L^1(G)^{**} \text{ and } f \in LUC(G).$$

It follows that if R is a right ideal in $L^1(G)^{**}$ then $\phi(R)$ is a right ideal in $LUC(G)^*$. Since an element $\mu \in L^1(G)^{**}$ is a right annihilator of $L^1(G)^{**}$ if and only if $\mu(f) = 0$ for all $f \in LUC(G)$ ([16]), it follows from our assumption that R contains an element such that $\mu(f) \neq 0$ for some $f \in LUC(G)$. Accordingly, $\phi(\mu)(f) = \mu(f) \neq 0$ and so $\phi(R)$ is a non-zero right ideal of $LUC(G)^*$. The theorem follows now from Theorem 3.1.

Corollary 3.3 *Let G be as in Theorem 3.1 but not necessarily amenable. Then the dimension of the radical of $LUC(G)^*$ is either 0 or 2^{2^κ}.*

4. The radical of $LUC(G)^*$

The first elements which were found in the radicals of $L^1(G)^{**}$ and $LUC(G)^*$ were the topologically left invariant elements with $\mu(1) = 0$ when G is amenable. Then the right annihilators of $L^1(G)^{**}$ were proved to be in the radical of $L^1(G)^{**}$ when G is not discrete group and not necessarily amenable. So in each of these cases the algebras $L^1(G)^{**}$ and $LUC(G)^*$ are not semisimple. For more details and references, see [23] or [12]. In this section, we find new elements in the radicals of $LUC(G)^*$ and $L^1(G)^{**}$. This answers further questions 3 and 4 asked by Gulick in [18].

A *representation* U of G is a homomorphism of G into the group $GL(H)$ of bounded invertible operators on a Hilbert space H such that the function $s \mapsto U(s)\xi : G \to H$

is continuous for each $\xi \in H$. The *coefficient functions* are defined on G, for every pair of vectors ξ and η in H, by $u_{\xi\eta}(s) =< U(s)\xi, \eta >$. We say that U is *integrable* when $u_{\xi\eta} \in L^1(G)$ for some non-zero vectors ξ and η in H. Recall also from [5] that a function f in $C(G)$ is almost periodic if $f_G = \{f_s : \ s \in G\}$ is relatively norm-compact; let $AP(G)$ be the space of all such functions.

Proposition 4.1 *Suppose that G is not compact and amenable. Let μ be a non-zero left invariant element in $LUC(G)^*$ with $\mu(1) = 0$, and let $U : G \to GL(H)$ be a finite-dimensional representation. Let $\mu_{\xi\eta}$ be defined by*

$$\mu_{\xi\eta}(f) = \mu(u_{\xi\eta}f) \quad \textit{for all} \quad f \in LUC(G).$$

Then $\mu_{\xi\eta}$ is in the radical of $LUC(G)^*$.

Proof: We show that $(LUC(G)^*\mu_{\xi\eta})^2 = \{0\}$. let V be the anti-representation of $LUC(G)^*$ related to U^{-1}, that is

$$< V(\nu)\xi, \eta >= \nu(< U^{-1}(.)\xi, \eta >) \text{ for all } \xi,\ \eta \in H \text{ and } \nu \in LUC(G)^*.$$

Then, by [4, Lemma 2], $\nu\mu_{\xi\eta} = \mu_{\xi V(\nu)^*\eta}$ for every $\nu \in LUC(G)^*$ where $V(\nu)^*$ is the adjoint of $V(\nu)$. So it is enough to show that $\mu_{\xi\eta}\mu_{\xi\eta'} = 0$ for all $\eta' \in H$. By the same lemma, $\mu_{\xi\eta}\mu_{\xi\eta'} = \mu_{\xi V(\mu_{\xi\eta})^*\eta'}$, with

$$< V(\mu_{\xi\eta})\eta', \eta'' >= \mu_{\xi\eta}(< U^{-1}(.)\eta', \eta'' >) = \mu(u_{\xi\eta} < \eta', U(.)\eta'' >) = \mu(u_{\xi\eta}\overline{u_{\eta''\eta'}})$$

But since there is a unique (up to a multiplicative constant) left invariant element in $AP(G)^*$, and since $\mu(1) = 0$, we have $\mu(f) = 0$ for all $f \in AP(G)$. In particular,

$$\mu(u_{\xi\eta}\overline{u_{\eta''\eta'}}) = 0 \quad \text{since} \quad u_{\xi\eta}\overline{u_{\eta''\eta'}} \in AP(G)$$

(that $u_{\xi\eta} \in AP(G)$ is easy to check or see for example [5]). Therefore

$$< V(\mu_{\xi\eta})\eta', \eta'' >= 0 \quad \text{for all} \quad \eta',\ \eta'' \in H,$$

and so $V(\mu_{\xi\eta}) = 0$, and accordingly $\mu_{\xi\eta}\mu_{\xi\eta'} = 0$, as required.

Proposition 4.2 *Suppose that G is not compact and amenable. Let $\mu_{\xi\eta}$ be as defined in Proposition 4.1 but with μ as topologically left invariant. Then $\mu_{\xi\eta}$ is in the radical of $L^1(G)^{**}$.*

Proof: The proof is similar to the previous one.

Proposition 4.3 *Let G be a non-compact locally compact group (and not necessarily amenable such as the special linear group $SL(2, \mathbb{R})$). Let λ be the left Haar measure on G and let $U : G \to GL(H)$ be an integrable representation. Let $\lambda_{\xi\eta}$ be as defined in Proposition 4.1 , i.e.*

$$\lambda_{\xi\eta}(f) = \lambda(u_{\xi\eta}f) = \int_G u_{\xi\eta}(s)f(s)d\lambda(s) \quad \textit{for all} \quad f \in LUC(G).$$

Then for every $\nu \in C_0(G)^\perp$, $\mu_{\xi\eta}\nu$ and $\nu\mu_{\xi\eta}$ are contained in the radical of $LUC(G)^*$.

Proof: By [4, Proposition 2], $\frac{d}{<\xi,\eta>}\mu_{\xi\eta}$ is a minimal idempotent in $LUC(G)^*$; in fact,

$$\mu_{\xi\eta}\nu\mu_{\xi\eta} = d^{-1} < V(\nu)\xi, \eta > \mu_{\xi\eta} \quad \text{for all} \quad \nu \in LUC(G)^*,$$

where d is the formal dimension of U. Since by Lemma 2.2, $C_0(G)^\perp$ is a two-sided ideal of $LUC(G)^*$ and $L^1(G) \cap C_0(G)^\perp = \{0\}$, we must have $< V(\nu)\xi, \eta >= 0$ and so $\mu_{\xi\eta}\nu\mu_{\xi\eta} = 0$ for every $\nu \in C_0(G)^\perp$. Using again the fact that $C_0(G)^\perp$ is a two-sided ideal of $LUC(G)^*$, we deduce that

$$(LUC(G)^*\nu\mu_{\xi\eta})^2 = (LUC(G)^*\mu_{\xi\eta}\nu)^2 = \{0\}.$$

Thus $\mu_{\xi\xi}\nu$ and $\nu\mu_{\xi\xi}$ are contained in the radical.

Remark: Of course we are intereseted above in the cases when $\nu\mu_{\xi\eta}$ and $\mu_{\xi\eta}\nu$ are not trivial. For the elements $\nu \in C_0(G)^\perp$ for which $\nu\mu_{\xi\eta} \neq 0$ and $\mu_{\xi\eta}\nu \neq 0$ see [4, Propositions 3 and 4].

The author is indebted to the referee for bringing to his attention references [15] and [20].

REFERENCES

1. G. R. Allan, Ideals of vector-valued functions, Proc. London Math. Soc. (3) 18 (1968) 193–216.
2. R. Arens, Operations induced in function classes, Monatsh. Math. 155 (1951) 1–19.
3. R. Arens, The adjoint of a bilinear operation, Proc. Amer. Math. Soc. 12 (1951) 839–848.
4. J. W. Baker and M. Filali, On minimal ideals in some Banach algebras associated with a locally compact group, J. London Math. Soc. (2) 63 (2001) 83–98.
5. J. F. Berglund, H. D. Junghenn and P. Milnes, Analysis on Semigroups: Function Spaces, Compactifications, Representations, Wiley, New York, 1989.
6. C. Chou, On the size of the set of left invariant means on a semigroup, Proc. Amer. Math. Soc. 23 (1969) 199–205.
7. P. Civin, Ideals in the second conjugate algebra of a group algebra, Math. Scand. 11 (1962) 161–174.
8. P. Civin and B. Yood, The second conjugate space of a Banach algebra as an algebra, Pacific J. Math. 11 (1961) 847–870.
9. M. Filali, The ideal structure of some Banach algebras, Math. Proc. Cambridge Philos. Soc. 111 (1992) 567–576.
10. M. Filali, Finite dimensional left ideals in algebras associated to a locally compact group, Proc. Amer. Math. Soc. 127 (1999) 2325–33.
11. M. Filali, Finite dimensional right ideals in algebras associated to a locally compact group, Proc. Amer. Math. Soc. 127 (1999) 1729–34.
12. M. Filali and A. I. Singh, Recent developments on Arens regularity and ideal structure of the second dual of a group algebra and some related toplogical algebras, Conference Proceedings of International Workshop on General Topological Algebras, Tartu 1999 (to appear).
13. M. Filali, t-Sets and some algebraic properties in βS and $\ell_\infty(S)^*$, preprint 2000.

14. M. Filali, On some semigroup compactifications, Topology Proceedings 22 (Summer 1997) 111–123. Electronically published in: http:/at.yorku.ca/b/a/a/j/05.htm
15. F. Ghahramani, A. T. Lau, and V. Losert, Isometric isomophisms between Banach algebras related to locally compact groups, Trans. Amer. Math. Soc. 321 (1990), 273-283.
16. E. E. Granirer, The radical of $L^{\infty}(G)^*$, Proc. Amer. Math. Soc. 41 (1973) 321–324.
17. F. P. Greenleaf, *Invariant means on topological groups and their applications,* Van Nostrand, New York, 1969.
18. S. L. Gulick, Commutativity and ideals in the biduals of topological algebras, Pacific J. Math. 18 (1966) 121–137.
19. E. Hewitt and K. A. Ross, Abstract harmonic analysis I (2nd ed.), Springer-Verlag, Berlin, 1994.
20. A. T. Lau, Operators which commutes with convolutions on subspace of $L_{\infty}(G)^{**}$, Colloq. Math. 39 (1978), 351-359.
21. L. H. Loomis , An introduction to abstract harmonic analysis, Van Nostrand, New York, 1953.
22. A. Olubummo, Finitely additive measures on the non-negative integers, Math. Scand. 24 (1969) 186–194.
23. A. L. T. Paterson, Amenability , American Mathematical Society, 1988.

Recent Progress in Functional Analysis
K.D. Bierstedt, J. Bonet, M. Maestre and J. Schmets

Stochastic continuity algebras

Bertram M. Schreiber

Department of Mathematics, Wayne State University,
Detroit, MI 48202, USA

To Professor Manuel Valdivia on the occasion of his seventieth birthday.

Abstract

This work is concerned with the study of the aggregate of all stochastic processes which are continuous in probability, over various parameter spaces. The collection of all such random functions on compact space forms a Fréchet algebra. Our main objective is to study the closed ideals in this algebra and to relate these closed ideals to their hulls.
MSC 2000 Primary 46H10; Secondary 60G07

1. Introduction

The notion of a stochastic process which is continuous in probability (stochastically continuous in [11]) arises in numerous contexts in probability theory [2,4,5,11,15]. Indeed, the Poisson process is continuous in probability, and this notion plays a role in the study of its generalizations and, from a broader point of view, in the theory of processes with independent increments [11]. For instance, the work of X. Fernique [9] on random right-continuous functions with left-hand limits (so-called cadlag functions) involves continuity in probability in an essential way.

The study of processes continuous in probability as a generalization of the notion of a continuous function began with the approximation theorems of K. Fan [7] (cf. [5], Theorems VI.III.III and VI.III.IV) and D. Dugué ([5], Theorem VI.III.V) on the unit interval. These results were generalized to convex domains in higher dimensions in [12], where the problem of describing all compact sets in the complex plane on which every random function continuous in probability can be uniformly approximated in probability by random polynomials was raised. This problem, as well as the corresponding question for rational approximation, were taken up in [1]. Along with some stimulating examples, the authors of [1] prove, under the natural assumptions appearing below, that random polynomial approximation holds over Jordan curves and the closures of Jordan domains.

In this note we study the space of functions continuous in probability over a general topological space and develop the analogue of the space $C(K)$ for K compact. This space has the structure of a Fréchet algebra. We investigate the closed ideals of this algebra and then introduce the notion of a stochastic uniform algebra.

Just as in the deterministic, classical case, there are natural stochastic uniform algebras defined by the appropriate concept of random approximation. We shall highlight some results from [3] which show that random polynomial approximation in the plane obtains for a very large class of compact sets. For instance, if K is a compact set with the property that every continuous function on ∂K can be uniformly approximated by rational functions, then every function continuous in probability on K (with respect to a nonatomic measure) and random holomorphic on the interior of K can be uniformly approximated in probability by random polynomials.

2. Stochastic Continuity and Convergence

Consider a fixed, nonatomic, complete probability space $(\Omega, \mathcal{A}, P)$ and an index set T, which we take to be any topological space for the moment. We wish to study complex-valued stochastic processes $\varphi = \varphi(t) = \varphi(t, \omega)$, which, given the point of view of the current work, we may call random functions on T or functions on $T \times \Omega$. Identify functions φ and ψ if for every $t \in T$, $\varphi(t) = \psi(s)$ a.s. Denote by $C(T)$ the space of all continuous, complex-valued functions on T, equipped as usual with the topology of uniform convergence on compact sets.

Let $\mathcal{F}$ be a class of complex-valued functions on T. Assume that for almost all $\omega \in \Omega$, $\varphi(\cdot, \omega) \in \mathcal{F}$, and for all $t \in T$, $\varphi(t, \cdot)$ is $\mathcal{A}$-measurable. Then φ is called a *random element of* $\mathcal{F}$. Thus one speaks of random continuous functions or random polynomials.

The stochastic process φ on T is said to be *continuous in probability at* $t \in T$ if for every $\varepsilon > 0$ there is a neighborhood V of t in T such that $P[|\varphi(s) - \varphi(t)| \geq \varepsilon] < \varepsilon$ for all $s \in V$. If φ is continuous in probability at every $t \in T$, then φ is called *continuous in probability on* T (or on $T \times \Omega$). It is easy to see, using the sequential definition of continuity, that every random continuous function on a metric space is continuous in probability. Easy examples show that this is not true if the space is not metric. The converse is also false [1]. The space of all (equivalence classes of) functions on T continuous in probability with respect to P is a module over the space $L^0(P) = L^0(\Omega, \mathcal{A}, P)$ of (P-equivalence classes of) $\mathcal{A}$-measurable functions.

If T is a metric space, then the function φ is called *uniformly continuous in probability* on T if it is continuous in probability on T and for a given $\varepsilon > 0$, the neighborhoods V above can all be taken to be balls $B(t, \delta)$ for some $\delta > 0$. If T is compact and metric, then any function continuous in probability on T is uniformly continuous in probability.

Of course, continuity in measure could be introduced over any measure space. One should note, however, that no real increase in generality ensues from moving to that setting, at least if one assumes that the given measure μ is σ-finite. For in that case, there is a probability measure P such that μ and P are mutually absolutely continuous. The notions of continuity with respect to μ and P will then coincide.

We shall say that φ is *locally bounded* if for each $t \in T$ there is a neighborhood V of t such that $\varphi(V \times \Omega)$ is bounded.

Let φ_n, $n = 1, 2, \ldots$ and φ be random functions on T. We say φ_n *converges uniformly in probability* to φ if given $\varepsilon > 0$, there exists $N > 0$ such that $P[|\varphi_n(t) - \varphi(t)| \geq \varepsilon] < \varepsilon$ for all $t \in T$ and $n \geq N$. If $\mathcal{F}$ is a family of functions on T and φ is a process on T, we shall say that φ *can be approximated by random elements of* $\mathcal{F}$ if there is a sequence of random elements of $\mathcal{F}$ converging uniformly in probability to φ.

¿From the point of view of operators, the concepts introduced above can be understood as follows. For the details, see [3], Sec. 2.

Recall that a Hausdorff space S is called a *$k-$space* if every set in S which intersects every compact set of S in a closed set is itself closed. The class of k-spaces includes all locally compact spaces and all spaces that satisfy the first countability axiom, hence all metric spaces. This is the appropriate setting for the Ascoli-Arzelà Theorem ([13], Chap. 7). If X and Y are locally convex topological vector spaces, an operator $T : X \to Y$ is called *completely continuous* if it maps weakly compact sets in X into strongly compact sets in Y.

Let φ be a locally bounded random function on the space S. For $f \in L^1(\Omega, P)$ and $s \in S$, set

$$T_\varphi f(s) = E[\varphi(s, \cdot) f] = \int_\Omega \varphi(s, \omega) f(\omega)\, dP(\omega). \tag{1}$$

2.1. Theorem

Let φ be a locally bounded random function on S.

(i) *If φ is continuous in probability on S, then $T_\varphi f$ is continuous on S for every $f \in L^1(\Omega, P)$, and T_φ is a continuous operator from $L^1(\Omega, P)$ to $C(S)$.*

(ii) *If S is a k-space, then φ is continuous in probability on S if and only if T_φ is a completely continuous operator from $L^1(\Omega, P)$ to $C(S)$.*

2.2. Corollary

For S compact, the map $\varphi \mapsto T_\varphi$ defines a one-to-one linear map from the space of bounded functions continuous in probability on $S \times \Omega$ onto the space of completely continuous operators $T : L^1(\Omega, P) \to C(S)$.

2.3. Theorem

Let φ_n, $n \geq 1$, and φ be continuous in probability on S and uniformly bounded. Then φ_n converges uniformly in probability to φ if and only if for any weakly compact subset $\mathcal{W}$ in $L^1(\Omega, P)$,

$$\lim_{n\to\infty} \sup_{f\in\mathcal{W}} \sup_{t\in T} |T_{\varphi_n} f(t) - T_\varphi f(t)| = 0. \tag{2}$$

3. The Space $\mathcal{C}_P(T)$ and its Ideal Structure

In this section we introduce the analogue appropriate to the current context of the algebra $C(T)$ for T compact and study its ideal structure. We shall introduce stochastic versions of the notions of hull and kernel familiar from the classical theory.

As is well known, the topology of convergence in probability in the space $L^0(\Omega, \mathcal{A}, P)$ is metrizable. Namely, the following are two metrics for this topology:

$$d_0(f,g) = \int_\Omega \min[|f-g|, 1]\, dP \tag{3}$$

$$d_0'(f,g) = \int_\Omega \frac{|f-g|}{1+|f-g|}\, dP \tag{4}$$

When T is compact, let us employ (3) to define a metric on the space of random functions continuous in probability on T in the natural way, as follows.

Let T be compact, and denote by $\mathcal{C}_P(T)$ the space of all functions on $T \times \Omega$ that are continuous in probability on T. Equivalently, $\mathcal{C}_P(T)$ is the space of continuous maps from T into $L^0(P)$. For $\varphi, \psi \in \mathcal{C}_P(T)$, set

$$d(\varphi, \psi) = \sup_{t \in T} d_0(\varphi(t), \psi(t)) = \sup_{t \in T} \int_\Omega \min[|\varphi(t) - \psi(t)|, 1]\, dP. \tag{5}$$

Then d is a metric on $\mathcal{C}_P(T)$,

$$\sup_{t \in T} P[|\varphi(t) - \psi(t)|] \geq \varepsilon] < \varepsilon \text{ whenever } d(\varphi, \psi) < \varepsilon^2, \tag{6}$$

and

$$d(\varphi, \psi) < 2\varepsilon \text{ whenever } \sup_{t \in T} P[|\varphi(t) - \psi(t)|] \geq \varepsilon] < \varepsilon. \tag{7}$$

Since $L^0(P)$ is a Fréchet algebra under d_0, elementary arguments show that $\mathcal{C}_P(T)$, equipped with the metric d, is a Fréchet algebra. Moreover, if $\varphi, \psi \in \mathcal{C}_P(T)$ such that $|\psi(t)| \geq \alpha > 0$ a.s. for some α and all $t \in T$, then $\varphi/\psi \in \mathcal{C}_P(T)$.

Note that there are also several natural stochastic analogues of classical algebras of continuous functions defined on more general spaces T. Here are three.

On any space T we may define $\mathcal{C}_P(T)$ to be all functions continuous in probability on T, with the topology of uniform convergence in probability over all compact subsets of T. This will then be a complete topological algebra. As in the deterministic case, it is most naturally considered when T is locally compact and will be a Fréchet algebra when T is σ-compact.

For T locally compact, one can define the analogue of $C_0(T)$. Namely, a random function φ is in $\mathcal{C}_P^0(T)$ if it is continuous in probability on T and for every $\varepsilon > 0$ there exists a compact set K such that $P[|\varphi(t)| \geq \varepsilon] < \varepsilon$ for all $t \notin K$.

A random function φ on T is called *stochastically bounded* if given $\varepsilon > 0$, there exists $M > 0$ such that $P[|\varphi(t)| > M] < \varepsilon$, $t \in T$. For any space T one can consider the algebra $\mathcal{C}_P^b(T)$ of all stochastically bounded functions continuous in probability on T, in the topology of uniform convergence in probability. It is then natural to ask, in case T is completely regular, whether there is an analogue of the Stone-Čech compactification in this context.

We now proceed to study the topological ideal theory of the spaces $\mathcal{C}_P(T)$ defined at the outset; in the sequel T will always be assumed compact.

Let I be a closed ideal in $\mathcal{C}_P(T)$. For $t \in T$, let

$$\mathcal{Z}_I(t) = \{Z \in \mathcal{A} : f(t) \equiv 0 \text{ a.s. on } Z\ \forall f \in I\}. \tag{8}$$

If $Z_n \in \mathcal{Z}_I(t)$ such that $P(Z_n)$ increases to

$$\alpha_I(t) = \sup\{P(Z) : Z \in \mathcal{Z}_I(t)\}, \tag{9}$$

then clearly $Z_I(t) = Z_1 \cup Z_2 \cup \cdots \in \mathcal{Z}_I(t)$ and $P(Z_I(t)) = \alpha_I(t)$. It is easy to see that up to null sets, $Z_I(t)$ is the unique set satisfying this condition. Thus let

$$Z_I = \{(t, \omega) : \omega \in Z_I(t),\ t \in T\}. \tag{10}$$

The set Z_I will be called the *hull* of I.

3.1. Lemma

Let I be a closed ideal in $\mathcal{C}_P(T)$. For each $t \in T$ there exists $\varphi \in I$ such that $\varphi(t) \neq 0$ a.s. on $Z_I(t)^c$

Proof. First note that given $B \subset Z_I(t)^c$ such that $P(B) > 0$, there exists $\varphi_B \in I$ such that $\varphi_B(t)$ does not vanish a.s. on B. Replacing φ_B by the element

$$\frac{|\varphi_B|^2}{1+|\varphi_B|^2} = \frac{\overline{\varphi_B}}{1+|\varphi_B|^2}\varphi_B, \tag{11}$$

we see that for all $B \subset Z_I(t)^c$ with $P(B) > 0$, there exists $\varphi_B \in I$ such that $0 \leq \varphi_B \leq 1$ and $P([\varphi_B(t) > 0] \cap B) > 0$. Let

$$\mathcal{B} = \{B \subset Z_I(t) : \exists \varphi \in I,\ 0 \leq \varphi \leq 1, \varphi(t) > 0 \text{ a.s. on } B\}. \tag{12}$$

Let $\beta = \sup\{P(B) : B \in \mathcal{B}\}$, choose $\{B_n\} \subset \mathcal{B}$ with $P(B_n) \nearrow \beta$, and for each n pick $\varphi_n \in I$ such that $0 \leq \varphi_n \leq 1$ and $\varphi_n(t) > 0$ a.s. on B_n. Let $\varphi = \sum_{n=1}^{\infty} 2^{-n}\varphi_n$. Since I is closed $\varphi \in I$. We have $0 \leq \varphi \leq 1, \varphi(t) > 0$ on $B = \bigcup_1^\infty B_n$, and $P(B) = \beta$.

If $\beta < 1-\alpha_I(t)$, let $B' = (Z_I(t) \cup B)^c$ and choose $\varphi_{B'}$ as above. Setting $\varphi' = (\varphi+\varphi_{B'})/2$, we obtain a contradiction to the definition of β.

Let $A(t) \in \mathcal{A},\ t \in T$. The family $\{A(t)\}_{t\in T}$ will be called *upper semicontinuous at* t if for every $\varepsilon > 0$ there is a neighborhood U of t such that $P(A(s) \setminus A(t)) < \varepsilon$ for all $s \in U$. Call $\{A(t)\}_{t\in T}$ *upper semicontinuous* if it is upper semicontinuous at every $t \in T$. We call $\{A(t)\}_{t\in T}$ *sequentially lower semicontinuous* at t if for any sequence $t_n \to t$,

$$\liminf_{n\to\infty} A(t_n) = \bigcup_{n=1}^{\infty} \bigcap_{k=n}^{\infty} A(t_k) \subset A(t). \tag{13}$$

Let A be a subset of $T \times \Omega$ such that

$$A(t) = \{\omega : (t,\omega) \in A\} \in \mathcal{A},\ t \in T. \tag{14}$$

For convenience we shall call A upper semicontinuous [sequentially lower semicontinuous] if $\{A(t)\}_{t\in T}$ is upper semicontinuous [sequentially lower semicontinuous].

3.2. Proposition

For any closed ideal I in $\mathcal{C}_P(T)$ and $t \in T$, Z_I is upper semicontinuous at t. If T is metrizable, then Z_I is sequentially lower semicontinuous at t.

Proof. Given $t \in T$ and $\varepsilon > 0$, let $\varphi \in I$ such that $\varphi(t) \neq 0$ a.s. on $Z_I(t)^c$. Choose $\alpha > 0$ such that $P[0 < |\varphi(t)| \leq \alpha] < \varepsilon$, and let U be a neighborhood of t such that $P[|\varphi(s) - \varphi(t)| \geq \alpha/2] < \varepsilon,\ s \in U$. If $|\varphi(t)| > \alpha$ and $|\varphi(s) - \varphi(t)| < \alpha/2$, then

$$|\varphi(s| \geq |\varphi(t)| - |\varphi(s) - \varphi(t)| > \alpha/2. \tag{15}$$

Thus for $s \in U$,

$$P(A(s) \setminus A(t)) \leq P[0 < |\varphi(t)| \leq \alpha] + P[|\varphi(s) - \varphi(t)| \geq \alpha/2] < 2\varepsilon. \tag{16}$$

The second assertion follows easily from the first.

Let Z be an $\mathcal{A}$-valued function on T. Set

$$I_Z = \{\varphi \in \mathcal{C}_P(T) : \varphi(t) = 0 \text{ a.s. on } Z(t),\ t \in T\}. \tag{17}$$

Then I_Z is easily seen to be a closed ideal in $\mathcal{C}_P(T)$. This ideal will be called the *kernel* Z.

We are now in a position to state our main result on ideal theory in $\mathcal{C}_P(T)$. Namely, at least when T has finite dimension, we shall show that as in the deterministic case, there is a natural bijection between closed ideals in $\mathcal{C}_P(T)$ and their hulls.

Recall that a topological space S is said to be of *dimension* N if $N+1$ is the least integer m such that every open cover $\mathcal{U}$ of S has a refinement $\mathcal{V}$ with the property that the number of elements of $\mathcal{V}$ containing any point of S is at most m. In particular, $\mathbf{R}^n$ has dimension n, and any closed subset of $\mathbf{R}^n$ with void interior has dimension at most $n-1$. For a survey of dimension theory, see [8].

3.3. Theorem

If Z is an $\mathcal{A}$-valued function on T, then Z_{I_Z} is upper semicontinuous and $Z_{I_Z}(t) \supset Z(t)$ a.s., $t \in T$. If I is a closed ideal in $\mathcal{C}_P(T)$, then $I_{Z_I} \supset I$, and equality holds if T is finite dimensional.

Proof. Clearly $Z_{I_Z}(t) \supset Z(t)$ a.s., $t \in T$, and $I_{Z_I} \supset I$. Proposition 3.2 says Z_{I_Z} is upper semicontinuous.

Suppose now that T has dimension $N < \infty$, and let $\varphi \in I_{Z_I}$. Given $t \in T$ and $\varepsilon > 0$, use Lemma 3.1 to choose $f \in I$ such that $f(t) \neq 0$ a.s. on $Z_I(t)^c$. If

$$f_n = \frac{n|f|^2}{1+n|f|^2}, \tag{18}$$

then $f \in I$ and

$$\lim_{n\to\infty} f_n(t) = \chi_{Z_I(t)^c} \text{ a.s.} \tag{19}$$

Since $\varphi \in I_{Z_I}$,

$$\lim_{n\to\infty} f_n(t)\varphi(t) = \chi_{Z_I(t)^c}\varphi(t) = \varphi(t) \text{ a.s.} \tag{20}$$

Thus choose $g_t = f_n$ for some n so that

$$d_0\left(g_t(t)\varphi(t), \varphi(t)\right) < \frac{\varepsilon}{2}. \tag{21}$$

By continuity, there is a neighborhood U_t of t such that

$$d_0\left(g_t(s)\varphi(s), \varphi(s)\right) < \varepsilon, \qquad s \in U_t. \tag{22}$$

Let $\mathcal{V} = \{V_1, \ldots, V_n\}$ be an open subcover of $\{U_t : t \in T\}$ for which $\operatorname{card}\{j : t \in V_j\} \leq N+1$, $t \in T$, and choose a partition of unity $\alpha_1, \ldots, \alpha_n \in C(T)$ subordinate to $\mathcal{V}$ with $0 \leq \alpha_j \leq 1$. For $1 \leq j \leq n$, let $t_j \in T$ such that $V_j \subset U_{t_j}$, and set

$$\psi = \sum_{j=1}^{n} \alpha_j g_{t_j} \varphi. \tag{23}$$

If $p, q \in L^0(P)$ and $0 \le c \le 1$, then $d_0(cp, cq) \le d_0(p, q)$. Hence

$$\begin{aligned}
d_0(\psi(t), \varphi(t)) &= \int_\Omega \min\{\left|\sum_{j=1}^n \alpha_j(t)\left(g_{t_j}(t) - 1\right)\varphi(t)\right|, 1\}\, dP && (24)\\
&\le \int_\Omega \min\{\sum_{j=1}^n \alpha_j(t)\left|g_{t_j}(t) - 1\right| |\varphi(t)|, 1\}\, dP && (25)\\
&\le \int_\Omega \sum_{j=1}^n \min\{\alpha_j(t)\left|g_{t_j}(t) - 1\right| |\varphi(t)|, 1\}\, dP && (26)\\
&\le \sum_{\{j:\alpha_j(t)\neq 0\}} d_0\left(g_{t_j}(t)\varphi(t), \varphi(t)\right) && (27)\\
&< (N+1)\varepsilon. && (28)
\end{aligned}$$

Thus $d(\psi, \varphi) < (N+1)\varepsilon$. Since I is closed, the proof is complete.

We conjecture that equality holds in the first containment relation of Theorem 3.3 if Z is assumed upper semicontinuous. This would then tell us that an $\mathcal{A}$-valued function on T is the hull of a closed ideal if and only if it is upper semicontinuous. This equality can be proven under various further hypotheses on Z, but the general case is subtle.

4. Stochastic Uniform Algebras

In this section we summarize some results from [3] on closed subalgebras of $\mathcal{C}_P(T)$ for T in the plane. As indicated by these results, stochastic analogues of classical examples may differ from their deterministic forerunners.

Let $\mathcal{A}$ be a closed subalgebra of $\mathcal{C}_P(T)$. We say that $\mathcal{A}$ *separates points* of T if for all distinct $t_1, t_2 \in T$ there exists $\varphi \in \mathcal{A}$ such that $P[\varphi(t_1) \neq \varphi(t_2)] > 0$. A *stochastic uniform algebra* is a closed subalgebra of some $\mathcal{C}_P(T)$ which contains the constant functions on $T \times \Omega$ and separates points of T. If $\mathcal{A}$ contains $L^0(P)$, considered as the algebra of random constant functions on T, we may call $\mathcal{A}$ *full.*

As examples, for T a compact set in $\mathbf{C}$, consider the following analogues of classical uniform algebras. Let $\mathcal{A}_P(T)$ be the closure in $\mathcal{C}_P(T)$ of all functions which are random holomorphic functions on the interior of T. Denote the closure in $\mathcal{C}_P(T)$ of all random polynomials on T by $\mathcal{P}_P(T)$, and let $\mathcal{R}_P(T)$ be the closure in $\mathcal{C}_P(T)$ of all random rational functions on T with poles in T^c. There are obvious extensions of these algebras to algebras of functions of several complex variables.

A compact subset T of $\mathbf{C}$ is called a *stochastic Mergelyan set* if $\mathcal{P}_P(T) = \mathcal{A}_P(T)$. That is, T is a stochastic Mergelyan set if every function in $\mathcal{C}_P(T)$ which is a random holomorphic function on T^o can be approximated by random polynomials. There are stochastic Mergelyan sets with a great variety of characteristics, as the following results show.

4.1. Theorem

Let $\varphi \in \mathcal{A}_P(T)$. If the restriction of φ to the boundary ∂T is in $\mathcal{P}_P(\partial T)$, then $\varphi \in \mathcal{P}_P(T)$. Thus if ∂T is a stochastic Mergelyan set, then so is T.

In the classical setting, the Mergelyan sets ($\mathcal{P}(T) = \mathcal{A}(T)$) are those with connected complement, while the Vitushkin sets ($\mathcal{R}(T) = \mathcal{A}(T)$) are much more plentiful [16,17] (cf. [10], Chap. 5, [18]). For instance, if T^c has finitely many components, T has planar measure 0, or T is the boundary of a Vitushkin set, then T is a Vitushkin set. In the stochastic context, however, we have the following theorem.

4.2. Theorem
If ∂T is a Vitushkin set, then T is a stochastic Mergelyan set.

In conclusion, note that in a similar fashion one can study various noncommutative stochastic continuity algebras. For instance, given a Banach algebra A, the algebra $\mathcal{C}_P(T; A)$ of all functions continuous in probability on T with values in A is a Fréchet algebra. If X is a Banach space, consider the algebra $\mathcal{L}_P(X)$ of all linear maps of X to itself that are continuous in probability.

REFERENCES

1. G.F. Andrus and T. Nishiura, Stochastic approximation of random functions. Rend. di Math. 13 (1980) 593-615.
2. A. Blanc-Lapierre and R. Fortet, Theory of Random Functions, Vol. 1 (transl. by J. Gani), Gordon and Breach, New York, 1965.
3. L. Brown and B.M. Schreiber, Stochastic continuity and approximation. Studia Math. 121 (1996) 15-33.
4. J.L. Doob, Stochastic Processes, Wiley, New York, 1953.
5. D. Dugué, Traité de Statistique Théorique et Appliquée, Masson, Paris, 1958.
6. J. Dugundji, Topology, Allyn-Bacon, Boston, 1966.
7. K. Fan, Sur l'approximation et l'intégration des fonctions aléatoires. Bull. Soc. Math. France 72 (1944) 97-117.
8. V.V. Fedorchuk, The fundamentals of dimension theory. General Topology I (A.V. Arkhangel'skiĭ and L.S. Pontryagin, eds.), Encyclopaedia of Mathematical Sciences, Vol. 17, pp. 91-192, Springer-Verlag, Berlin-Heidelberg-New York, 1990.
9. X. Fernique, Les fonctions aléatoires cadlag, la compacité de leurs lois. Liet. Mat. Rink. 34 (1994) 288-306.
10. T.W. Gamelin, Uniform Algebras, Prentice-Hall, Englewood Cliffs, NJ, 1969.
11. I.I. Gikhman and A.V. Skorohod, Introduction to the Theory of Random Processes, Saunders, Philadelphia, 1969.
12. V.I. Istrătescu and O. Onicescu, Approximation theorems for random functions. Rend. Mat. e Appl. (vi) 8 (1975) 65-81.
13. J.L. Kelley, General Topology, van Nostrand, New York, 1955.
14. W. Rudin, Real and Complex Analysis, 3rd ed., McGraw-Hill, New York, 1987.
15. R. Syski, Stochastic processes. Encyclopedia of Statistical Sciences (S. Kotz and N.L. Johnson, eds.), Vol. 8, pp. 836-851, Interscience, Wiley, New York, 1988.
16. A.G. Vitushkin, Conditions on a set which are necessary and sufficient in order that any continuous function, analytic at its interior points, admit uniform approximation by rational fractions. Soviet Math. Dokl. 7 (1966) 1622-1625.

17. A.G. Vitushkin, Analytic capacity of sets in problems of approximation theory. Russian Math. Surveys 22 (1967) 139-200.
18. L. Zalcman, Analytic Capacity and Rational Approximation, Lect. Notes in Math. No. 50, Springer-Verlag, Berlin-Heidelberg-New York, 1968.

Recent Progress in Functional Analysis
K.D. Bierstedt, J. Bonet, M. Maestre and J. Schmets

Hilbert space methods in the theory of Lie triple systems

A. J. Calderón Martín [a] * and C. Martín González [b]

[a]Departamento de Matemáticas, Universidad de Cádiz,
11510 Puerto Real, Cádiz, Spain

[b]Departamento de Algebra, Geometría y Topología, Universidad de Málaga,
Apartado 59, 29080 Málaga, Spain

To Professor Manuel Valdivia on the occasion of his seventieth birthday.

Abstract

In [1], Lister introduced the concept of Lie triple system and classified the finite-dimensional simple Lie triple system over an algebraically closed field of characteristic zero. However, the classification in the infinite-dimensional case is still an open problem. In order to study infinite-dimensional Lie triple systems, we introduce in this paper the notion of two-graded L^*-algebra and L^*-triple. We obtain a structure theory of infinite dimensional two-graded L^*-algebras, and we also establish some results about L^*-triples, as a classification of L^*-triples admitting a two-graded L^*-algebra envelope, their relation with L^*-subtriples of A^-, for a ternary H^*-algebra A, and the structure of direct limits of certain systems of L^*-triples. As a tool, we develop a complete theory of direct limits of ternary H^*-structures.

MCS 2000 Primary 46K70; Secondary 16W10, 17B70, 17B05

1. On the structure of two-graded L^*-algebras

Let K be a unitary commutative ring. A *two-graded K-algebra* A is a K-algebra which splits into the direct sum $A = A_0 \oplus A_1$ of K-submodules (called the even and the odd part respectively) satisfying $A_\alpha A_\beta \subset A_{\alpha+\beta}$ for all α,β in $\mathbb{Z}_2$. If A is a two-graded algebra, its underlying algebra (forgetting the grading) will be denoted by $Un(A)$. A *homomorphism* f between two-graded algebras A and B is a homomorphism from $Un(A)$ to $Un(B)$ which preserves gradings i.e.:

$$f(A_\alpha) \subset B_\alpha$$

*Supported by the PCI of the spanish Junta de Andalucía 'Estudio analítico-algebraico de sistemas triples y de pares en diferentes clases de estructuras no asociativas', by the AECI under the proyect 'Estructuras algebraicas asociativas y no asociativas', by the PAI of the spanish Junta de Andalucía with project number FQM-0125 and by the spanish DGICYT with project number PB97-1497

for all $\alpha \in \mathbb{Z}_2$. The definitions of epimorphism, monomorphism and isomorphism of two-graded algebras are the obvious ones and the same applies to the notions of subalgebras and ideals in graded sense. If H_1 and H_2 are Hilbert spaces with scalar products $(\cdot|\cdot)_i$, $i = 1, 2$ and $f : H_1 \to H_2$ a linear map such that

$$(f(x)|f(y))_2 = k(x|y)_1$$

for any $x, y \in H_1$ and k a positive real number, then we will say that f is a *k-isogenic* map.

We recall that an *H^*-algebra* A over $\mathbb{C}$ is a nonassociative $\mathbb{C}$-algebra provided with:

1. A conjugate-linear map $* : A \to A$ such that $(x^*)^* = x$ and

$$(xy)^* = y^* x^*$$

   for any $x, y \in A$. Then $*$ is called an *involution* of the algebra A.

2. A complex Hilbert space structure whose inner product is denoted by $(\cdot|\cdot)$ and satisfies

$$(xy|z) = (x|zy^*) = (y|x^*z)$$

   for all $x, y, z \in A$.

A *two-graded H^*-algebra*, is an H^*-algebra which is a two-graded algebra whose even and odd part are selfadjoint closed orthogonal subspaces. We call the two-graded H^*-algebra A, *topologically simple* if $A^2 \neq 0$ and A has no nontrivial closed two-graded ideals. In the sequel an *L^*-algebra* will mean a Lie H^*-algebra. The classification of topologically simple L^*-algebras is given in the separable case by Schue (see [2], [3]) and later in the general case in [4] (see also [5] for an alternative approach). Following [6, Proposition 1] it is easy to prove that any two-graded H^*-algebra A with continuous involution splits into the orthogonal direct sum $A = Ann(A) \perp \overline{L(A^2)}$, where $Ann(A) := \{x \in A : xA = Ax = 0\}$, and $\overline{L(A^2)}$ is the closure of the vector span of A^2, which turns out to be a two-graded H^*-algebra with zero annihilator. Moreover, each two-graded H^*-algebra A with zero annihilator satisfies $A = \overline{\perp I_\alpha}$ where $\{I_\alpha\}_\alpha$ denotes the family of (two-graded) minimal closed ideals of A, each of them being a topologically simple two-graded H^*-algebra. This reduces the study of this two-graded algebras to the study of the topologically simple ones.

As in [1, Theorem 2.13], we have two possibilities for any topologically simple (in graded sense) two-graded H^*-algebra A:

(a) A is isomorphic to an orthogonal direct sum $B \perp B$ where B is a topologically simple H^*-algebra with $A_0 = B \perp \{0\}$, $A_1 = \{0\} \perp B$, and the product and involution are given by $(a_0, a_1)(b_0, b_1) = (a_0b_0 + a_1b_1, a_0b_1 + a_1b_0)$ and $(a_0, a_1)^* = (a_0^*, a_1^*)$, let observe that in this case, A is not topologically simple in ungraded sense.

(b) A is topologically simple as ungraded H^*-algebra.

This dichotomy reduces the study of topologically simple two-graded H^*-algebras to the study of 2-gradings of topologically simple H^*-algebras. But this is only a matter of involutive antiautomorphism since if A is a nonassociative two-graded H^*-algebra, then

$g(x_0 + x_1) = x_0 - x_1$ defines an isometry $g \in Aut(A)$, commuting with the involution $*$, satisfying $g^2 = Id$ and such that $A_0 = Sym(A, g)$ and $A_1 = Skw(A, g)$. Reciprocally every $*$-preserving involutive automorphism (necessarily isometric by [7]) induces a grading on any H^*-algebra. So, the problem on the classification of topologically simple two-graded L^*-algebras reduces to the determination of the involutive $*$-automorphisms of topologically simple L^*-algebras. These automorphisms can be found following Balachandran's techniques of [8, §5, 1235-1237]. Thus we can claim:

Theorem 1.1 *If V is an infinite-dimensional complex topologically simple two-graded L^*-algebra, then V is isometrically $*$-isomorphic to some of the following ones:*

1. *$L \perp L$ with L a complex topologically simple L^*-algebra, even part $L \perp \{0\}$, odd part $\{0\} \perp L$, involution $(a, b)^* := (a^*, b^*)$, product*

$$[(a, b), (c, d)] := ([a, c] + [b, d], [a, d] + [b, c])$$

and inner product $((a, b)|(c, d)) := (a|c) + (b|d)$ for arbitrary $a, b, c, d \in L$.

2. *A^- with A an associative two-graded H^*-algebra which is topologically simple in ungraded sense.*

3. *A^- with A an associative topologically simple H^*-algebra, even part $Skw(A, \sigma)$, odd part $Sym(A, \sigma)$, and σ being an involutive $*$-antiautomorphism of A.*

4. *$Skw(A, \tau)$ with A an associative two-graded H^*-algebra (topologically simple in ungraded sense), and τ an involutive $*$-antiautomorphism of the two-graded algebra A.*

2. Previous results on L^*-triples

Let T be a vector space over $\mathbb{C}$. We say that T is a complex *triple system* if it is endowed with a trilinear map $< \cdot, \cdot, \cdot >$ from $T \times T \times T$ onto T, called the *triple product* of T.

A triple system T is called a *Lie triple system* if its triple product, denoted by $[\cdot, \cdot, \cdot]$, satisfies

(i) $[x, x, z] = 0$

(ii) $[x, y, z] + [y, x, z] + [z, x, y] = 0$

(iii) $[x, y, [a, b, c]] - [a, b, [x, y, c]] = [[x, y, a], b, c] + [a, [x, y, b], c]$

for any $x, y, a, b, c \in T$.

We define an *H^*-triple system* as a complex triple system $(T, < \cdot, \cdot, \cdot >)$ provided with:

1. A conjugate-linear map $* : T \to T$ such that $(x^*)^*$ and

$$< x, y, z >^* = < x^*, y^*, z^* >$$

for any $x, y, z \in T$. Then we say that $*$ is an *involution* of T.

2. A complex Hilbert space structure whose inner product is denoted by $(\cdot|\cdot)$ and satisfies

$$(< x, y, z > |t) = (x| < t, z^*, y^* >) = (y| < z^*, t, x^* >) = (z| < y^*, x^*, t >)$$

for any $x, y, z, t \in T$. This identities are called *H^*-identities.*

The annihilator of an H^*-triple system T with triple product $< \cdot, \cdot, \cdot >$ is the ideal of all elements $x \in T$ such that $< x, T, T >= 0$, we shall denote it $Ann(T)$. We also say that T is topologically simple when $< T, T, T >\neq 0$ and its only closed ideals are 0 and T. The structure theorems for H^*-triples systems given in [9] reduce the interest on H^*-triple systems to the topologically simple case.

In the sequel an *L^*-triple* will mean a Lie H^*-triple system.

EXAMPLES.

1. Any closed subspace S of Hilbert-Schmidt operators on a Hilbert space H, such that $S^\sharp = S$, being $\sharp$ the adjoint operator, and $[[S, S], S] \subset S$ is an L^*-triple.

2. If L is an L^*-algebra, then it can be considered as an L^*-triple by defining the triple product $[x, y, z] = [[x, y], z]$ for any $x, y, z \in L$, and the same involution and inner product.

3. For any two-graded L^*-algebra, $L = L_0 \perp L_1$, its odd part L_1 with the triple product as above is an L^*-triple with the involution and inner product induced by the ones in L.

Last example leads us to introduce the following definition, If T is an L^*-triple isometrically $*$-isomorphic to the L^*-triple L_1 for some two-graded L^*-algebra L, we shall say that L is a two-graded L^*-algebra *envelope* of T if $L_0 := \overline{[L_1, L_1]}$.

If T is an L^*-triple with two-graded L^*-algebra envelope $L = \overline{[T,T]} \perp T$, then $\mathrm{Ann}(T) = 0$ if and only if $\mathrm{Ann}(L) = 0$. Indeed, if T has zero annihilator and $x \in \mathrm{Ann}(L)$, then $x = x_0 + x_1$ with $x_\alpha \in \mathrm{Ann}(L)$ for $\alpha = 0, 1$. Consequently $[x_1, T, T] = 0$ and $x_1 \in \mathrm{Ann}(T) = 0$. Thus $x = x_0$ is of the form $x = \sum_i [a_i, b_i]$ with $a_i, b_i \in T$ and then

$$||x||^2 = \sum_i (x|[a_i, b_i]) = \sum_i ([x, b_i^*]|a_i) = 0$$

therefore $\mathrm{Ann}(L) = 0$. Conversely if $\mathrm{Ann}(L) = 0$ and $x \in \mathrm{Ann}(T)$, then $[x, [T, T]] = 0$ by the Jacobi identity. This implies that $x \in \mathrm{Ann}(L) = 0$ hence $x = 0$. It is also easy to check that if T is an L^*-triple with two-graded L^*-algebra envelope L, then T is topologically simple if and only if L is topologically simple in graded sense.

As a consequence of the next proposition we have the uniqueness of the L^*-algebra envelope (when it exists) of an L^*-triple of zero annihilator.

Proposition 2.1 *Let T_1 and T_2 be L^*-triples with zero annihilator and consider an isometric $*$-monomorphism $f : T_1 \to T_2$. Let L_1 and L_2 be two-graded L^*-algebra envelopes of T_1 and T_2 respectively. Then there is a unique isometric $*$-monomorphism $F : L_1 \to L_2$ of two-graded algebras extending f.*

Proof. As $L_i = \overline{[T_i, T_i]} \perp T_i$ (for $i = 1, 2$), we can define first

$$F : [T_1, T_1] \perp T_1 \to L_2$$

by writing $F(x) := f(x)$ for all $x \in T_1$ and $F(\sum_j [x_j, y_j]) := \sum_j [f(x_j), f(y_j)]$ for arbitrary $x_j, y_j \in T_1$. The definition is correct since if $\sum_j [x_j, y_j] = 0$, then denoting $z := \sum_j [f(x_j), f(y_j)]$ we have

$$\|z\|^2 = (\sum_j [f(x_j), f(y_j)] | \sum_k [f(x_k), f(y_k)]) =$$

$$= \sum_j (f(x_j) | \sum_k [f(x_k), f(y_k), f(y_j^*)]) = \sum_j (f(x_j) | f(\sum_k [x_k, y_k, y_j^*])) =$$

$$= \sum_j (x_j | \sum_k [x_k, y_k, y_j^*]) = \sum_j ([x_j, y_j] | \sum_k [x_k, y_k]) = 0$$

hence $z = 0$. The fact that F is a $*$-monomorphism is easy to check and its isometric character is a consequence of the L^* conditions and of the isometric character of f. Next we can extend F to the whole L_i by continuity turning out that this extension is an isometric $*$-monomorphism of two-graded L^*-algebras as we wanted to prove.□

3. On the structure of L^*-triples

Respect to the finite dimensional case, we want to prove that any simple finite-dimensional real or complex Lie triple system is in fact an L^*-triple system. We use Lie algebra envelopes to prove this, but the reader is invited to consider also the ideas in [10].

Proposition 3.1 *Let L be a semisimple finite-dimensional complex two-graded Lie algebra. Then L admits a two-graded L^*-algebra structure. Let T be a semisimple finite-dimensional Lie triple system over $\mathbb{R}$ or $\mathbb{C}$. Then it has an L^*-triple structure.*

Proof. As L is a semisimple and finite-dimensional complex Lie algebra it has a basis $\{v_1, \ldots, v_n\}$ such that the real vector space

$$V_0 = \{\sum_{i=1}^{n} r_i v_i : r_i \in \mathbb{R}, n \in \mathbb{N}\}$$

is a real form of L (see [11, Theorem 2, p.124]). From this fact we deduce applying [12, Théorème 3, 11-11] that L has a compact real form, and then by [12, Théorème 2, 11-09], L has a Weyl basis.

Let us denote by H the Cartan subalgebra associated to the Weyl basis mentioned and let $E_\alpha \neq 0$ be the element of L_α (root space of L associated to α) that is in the Weyl basis. Let B denote the Killing form of L and σ the map described in [12, item 3 of Théorème 2, 11-09]. We have then that the inner product $< x|y > := -B(x, \sigma(y))$ and the involution $* = -\sigma$ endow L with an H^*-algebra structure. Next, we have to prove that L is a two-graded L^*-algebra. Let $G : L \to L$ denote the grading automorphism $G(x_0 + x_1) = x_0 - x_1$, by [12, 16] there is an automorphism A of L such that: (1) $A = \phi^{-1} G \phi$, (2) ϕ is an inner automorphism of L, (3) $A(H) \subset H$, and (4) $A(E_\alpha) = \pm E_{\alpha'}$ where $\alpha \mapsto \alpha'$ is a permutation in the set of roots of L.

Therefore there is an isomorphism of two-graded algebras between $L = L_0 \oplus L_1$ and the two-graded Lie algebra $L = L'_0 \oplus L'_1$ such that $L'_0 := \mathrm{Sym}(L, A)$, and $L'_1 := \mathrm{Skw}(L, A)$. It is obvious that $A \circ \sigma = \sigma \circ A$, then $(L'_\alpha)^* = L'_\alpha$ for all $\alpha = 0, 1$ and $L'_0 \perp L'_1$. Summarizing, we have proved that L is isomorphic as a two-graded algebra to a two-graded L^*-algebra hence L itself admits a two-graded L^*-algebra structure.

For further references we shall call this the standard two-graded L^*-algebra structure of L.

Suppose now that T is a semisimple finite-dimensional complex Lie triple system. Then there is a semisimple two-graded Lie algebra L whose odd part L_1, (with the triple product $[[a,b],c]$), agrees with T, so L admits a two-graded L^*-algebra structure and therefore T admits an L^*-triple structure. The real cases of both assertions in the proposition are consequences of [13, Theorem 3, p.70].□

Respect to the infinite dimensional L^*-triples, the previous classification of the topologically simple two-graded L^*-algebras given in Theorem 1.1 implies the next result:

Theorem 3.1 *Let T be an infinite dimensional topologically simple L^*-triple admitting a two-graded L^*-envelope, then T is some of the following:*

1. *The L^*-triple associated to an L^*-algebra L by defining the triple product $[a,b,c] := [[a,b],c]$ and the same involution and inner product of L.*

2. *$Skw(A, \tau)$ with A a topologically simple associative H^*-algebra, τ an involutive $*$-automorphism of A, the involution and inner products induced by the ones in A, and the triple product as in the previous case.*

3. *$Sym(A, \sigma)$ with A as in the previous case but now σ is an involutive $*$-antiautomorphism, and the triple product, involution and inner product induced by the ones in A.*

4. *$Skw(A, \tau) \cap Skw(A, \sigma)$ with A as before, τ an involutive $*$-antiautomorphism, σ an involutive $*$-automorphism such that $\sigma\tau = \tau\sigma$, and the involution and inner product induced by the ones in A.*

Last theorem classifies any infinite dimensional topologically simple L^*-triple admitting a two-graded L^*-envelope. However, the problem on the existence of L^*-algebra envelopes is still open. In the following results, we are going to describe some classes of L^*-triples admitting a two-graded L^*-envelope, and then these are classified by the last theorem.

Let A be a $\mathbb{C}$-vector space provided with a trilinear mapping

$$< \cdot, \cdot, \cdot >: A \times A \times A \to A$$

such that

$$<< x, y, z >, t, u > = < x, < y, z, t >, u > = < x, y, < z, t, u >>$$

for all $x, y, z, t, u \in A$. Then A is called a complex *ternary algebra*.

We define a *ternary H^*-algebra* as a complex ternary algebra $(A, < \cdot, \cdot, \cdot >)$ provided with:

1. A conjugate-linear map $*: A \to A$ such that $(x^*)^*$ and

$$< x, y, z >^* = < z^*, y^*, x^* >$$

for any $x, y, z \in A$. Then we say that $*$ is an *involution* of A.

2. A complex Hilbert space structure whose inner product is denoted by $(\cdot|\cdot)$ and satisfies

$$(< x, y, z > |t) = (x| < t, z^*, y^* >) = (y| < x^*, t, z^* >) = (z| < y^*, x^*, t >)$$

for any $x, y, z, t \in A$.

If $(A, < \cdot, \cdot, \cdot >)$ is a ternary H^*-algebra, then the triple systems A^- with triple product

$$[x, y, z] = < x, y, z > - < y, x, z > - < z, x, y > + < z, y, x >,$$

and A^+ with triple product $\{x, y, z\} = < x, y, z > + < z, y, x >$ and same involution and Hilbert space as A are respectively an L^*-triple and a Jordan H^*-triple system, (see [13] for definition of Jordan H^*-triple system).

Theorem 3.2 *If T is an L^*-subtriple of A^- for a topologically simple ternary H^*-algebra A, then T has a two-graded L^*-algebra envelope.*

Proof. From the classification of topologically simple ternary H^*-algebras (in the complex case) of [14, Main Theorem, p.226], one see that there is an associative topologically simple two-graded H^*-algebra $B = B_0 \perp B_1$ (see [6] for classification theorems) such that A is the ternary H^*-algebra associated to B_1 (with triple product $< xyz >= xyz$ for all $x, y, z \in B_1$). Let $L = L_0 \perp L_1$ be the two-graded L^*-subalgebra of B^- generated by T. It is easy to prove that $L_1 = T$ and $L_0 = \overline{[T, T]}$ hence the topologically simpleness of T implies that of L (recall section 2).□

We have also prove in [15] the next

Theorem 3.3 *Let $(T, [\cdot, \cdot, \cdot])$ be an infinite dimensional topologically simple L^*-triple. Write U an associative algebra such that*

$$[x, y, z] = xyz - yxz - zxy + zyx$$

for any $x, y, z \in T$. If $xyx \in T$ for every $x, y \in T$, then T has a two-graded L^*-algebra envelope.

In order to prove that the direct limit of certain systems of finite dimensional Lie triple systems is a topologically simple L^*-triple that admits a two-graded L^*-algebra envelope, we first need to study the direct limits of ternary H^*-structures.

4. Direct limits of ternary H^*-structures

Let $(I, \leq)$ be a direct set and $\{T_i\}_{i\in I}$ a family of H^*-triple systems such that for $i \leq j$ there exists an isometric $*$-monomorphism $e_{ji} : T_i \longrightarrow T_j$ satisfying $e_{ji}e_{ik} = e_{jk}$ and $e_{ii} = Id$ for all $i, j, k \in I$ with $k \leq i \leq j$. Suppose furthermore that there exists a positive real number h such that for any $i \in I$ we have:

1. $\|x^*\|_i \leq h\|x\|_i$, $x \in T_i$.
2. $\| < x, y, z > \|_i \leq h\|x\|_i \cdot \|y\|_i \cdot \|z\|_i$ for every $x, y, z \in T_i$.

Then we shall say that $S := (\{T_i\}_{i\in I}, \{e_{ji}\}_{i\leq j})$ is a *direct system* of H^*-triple systems.

Given S we define a *direct limit*, $\lim\limits_{\rightarrow} S$, as a couple $(T, \{e_i\}_{i\in I})$ where T is an H^*-triple system, $e_i : T_i \longrightarrow T$ is an isometric $*$-monomorphism that satisfies $e_i = e_j e_{ji}$ and $(T, \{e_i\}_{i\in I})$ is universal for this property in the sense that if $(B, \{t_i\}_{i\in I})$ is another such couple, then there exists a unique isometric $*$-monomorphism $\theta : T \longrightarrow B$ such that $t_i = \theta e_i$, $i \in I$. It is clear that if a direct limit exists, then it is unique up to isometric $*$-isomorphism.

Let $S = (\{T_i\}_{i\in I}, \{e_{ij}\}_{i\leq j})$ be a direct system of H^*-triple systems, let $\mathcal{U}$ be an ultrafilter on I, containing the intervals $[i, \rightarrow)$, and let $\frac{(\prod T_i)}{\mathcal{U}}$ be the ultraproduct of $\{T_i\}_{i\in I}$ respect to $\mathcal{U}$, that is, the set of equivalence classes modulo the relation $(x_i) \equiv (y_i)$ if and only if $\{i \in I : x_i = y_i\} \in \mathcal{U}$. Then, $\frac{(\prod T_i)}{\mathcal{U}}$ becomes a triple system with involution, endowed with the algebraic operations and involution induced in the quotient by the componetwise operations.

We define W as the triple subsystem with involution of $\frac{(\prod T_i)}{\mathcal{U}}$ whose elements are the equivalence classes $[(x_i)]$ such that there exists $i_0 \in I$, $x_{i_0} \in T_{i_0}$, satisfying

$$\{i \in I : x_i = e_{ii_0}(x_{i_0})\} \in \mathcal{U}.$$

We can define an inner product in W. Observe that $[(x_i)], [(y_i)] \in W$ implies the existence of x_{i_0}, y_{i_0} in some T_{i_0} so that $\{i \in I : x_i = e_{ii_0}(x_{i_0})\}$ and $\{i \in I : y_i = e_{ii_0}(y_{i_0})\}$ belong to $\mathcal{U}$, then let us define $([(x_i)]|[(y_i)]) := (x_{i_0}|y_{i_0})$. We note that all these definitions are independent of the chosen representatives in each class and of the choice of x_{i_0} and y_{i_0}.

Now we can define an H^*-triple system, denoted by T, as the completion of W, the algebraic operations, involution and inner product are extended to T by continuity.

For all $i \in I$, we define $e_i : T_i \longrightarrow T$ as $e_i(x_i) := [(y_j)]$ where $y_j = e_{ji}(x_i)$ if $j \geq i$ and $y_j = 0$ in other cases. We have that $(T, \{e_i\}_{i\in I}) = \lim\limits_{\rightarrow} S$. The conditions $e_j e_{ji} = e_i$ and the universal property of the direct limit are easy to check. As consequence, $T = \overline{\bigcup\limits_{i\in I} e_i(T_i)}$.

Theorem 4.1 *Let $S = (\{T_i\}_{i\in I}, \{e_{ji}\}_{i\leq j})$ be a direct system of simple finite-dimensional H^*-triple systems. Then $T = \lim\limits_{\rightarrow} S$ is a topologically simple H^*-triple system.*

Proof. Define $W := \bigcup\limits_{i\in I} e_i(T_i)$, then W is dense in T and we have $\overline{[W, W, W]} = W$, implying $\overline{[T, T, T]} = T$. It follows from the first structure theorem for H^*-triple systems

[9, section 1] that $Ann(T) = 0$. Let J be a minimal closed ideal and $\pi : T \longrightarrow J$, $\pi_i : e_i(T_i) \longrightarrow J$, $i \in I$ orthogonal projections. If $\pi_i \neq 0$ since T_i is simple then π_i is $*$-isogenic monomorphism [16, Corollary 5]. If $i \leq j$ the fact that $\pi_j e_j e_{ji} = \pi_i e_i$ give us that π_i and π_j have the same isogenic constant M.

If $x, y \in \bigcup_{i\in I^*} e_i(T_i) = \bigcup_{i\in I} e_i(T_i)$, $I^* = \{i \in I : \pi_i \neq 0\}$, then $(\pi(x)|\pi(y)) = M(x|y)$, and this is also true for arbitrary $x, y \in T$ by continuity, consecuently π is a $*$-isomorphism and then T is topologically simple. □

Clearly, the construction of the direct limit and Theorem 4.1 hold if we consider ternary H^*-algebras instead of H^*-triple systems.

The proof of the next theorem is immediate

Theorem 4.2 *Let $S = (\{T_i\}_{i\in I}, \{e_{ji}\}_{i\leq j})$ be a direct system of ternary H^*-algebras, then*
(a) $S^- = (\{T_i^-\}, \{e_{ji}\}_{i\leq j})$ is a direct system of L^*-triples and $\lim_{\rightarrow} S^- = (\lim_{\rightarrow} S)^-$.
(b) $S^+ = (\{T_i^+\}, \{e_{ji}\}_{i\leq j})$ is a direct system of Jordan H^*-triple systems and

$$\lim_{\rightarrow} S^+ = (\lim_{\rightarrow} S)^+.$$

Theorem 4.3 *Let $S = (\{T_i\}_{i\in I}, \{e_{ji}\}_{i\leq j})$ be a direct system of ternary H^*-algebras, with $\{\sharp_i\}_{i\in I}$ a family of isometric involutive $*$-automorphisms, $\sharp_i : T_i \longrightarrow T_i$, such that*

$$\sharp_j \circ e_{ji} = e_{ji} \circ \sharp_i$$

for $i \leq j$. Write $T = \lim_{\rightarrow} S$, then.
(a) There exists $\sharp : T \longrightarrow T$ a unique isometric involutive $*$-automorphism verifying $\sharp \circ e_i = e_i \circ \sharp_i$ for any $i \in I$.
(b) If we consider the L^*-subtriple of T_i^-, $Sym(T_i, \sharp_i)$ (resp. $Skw(T_i, \sharp_i)$), then

$$Sym(S, \sharp) := (\{Sym(T_i, \sharp_i)\}_{i\in I}, \{e_{ji}|_{Sym(T_i,\sharp_i)}\}_{i\leq j})$$

(resp. $Skw(S, \sharp)$) is a direct system of L^*-triples and

$$\lim_{\rightarrow}(Sym(S, \sharp)) = Sym(\lim_{\rightarrow} S, \sharp)$$

(resp. $\lim_{\rightarrow}(Skw(S, \sharp)) = Skw(\lim_{\rightarrow} S, \sharp)$)

Proof. (a) For $i \leq j$, we have $(e_j \circ \sharp_j) \circ e_{ji} = e_i \circ \sharp_i$. The universal property of the direct limits now shows the existence of $\sharp : T \longrightarrow T$ a unique isometric $*$-automorphism such that $\sharp \circ e_i = e_i \circ \sharp_i$ for any $i \in I$. Since $(\sharp|_{e_i(T_i)})^2 = Id$ and $T = \overline{\bigcup_{i\in I} e_i(T_i)}$, it follows that $\sharp$ is involutive.

(b) It is clear that $Sym(S, \sharp)$ is a direct system of L^*-triples. Denote by

$$(Sym(S, \sharp), \{g_i\}_{i\in I})$$

its direct limit. As $\sharp \circ e_i = e_i \circ \sharp_i$, we can consider

$$e_i|_{Sym(T_i,\sharp_i)} : Sym(T_i, \sharp_i) \longrightarrow Sym(\lim_{\rightarrow} S, \sharp),$$

$*$-isometric monomorphisms verifying $e_j|_{Sym(T_j,\sharp_j)} \circ e_{ji}|_{Sym(T_i,\sharp_i)} = e_i|_{Sym(T_i,\sharp_i)}$. By the universal property of the direct limits, we have $\Phi : \lim\limits_{\rightarrow}(Sym(S,\sharp)) \longrightarrow Sym(\lim\limits_{\rightarrow} S,\sharp)$ a unique isometric $*$-monomorphism such that $\Phi \circ g_i = e_i|_{Sym(T_i,\sharp_i)}$. Let $W = \bigcup\limits_{i\in I} e_i(T_i)$, the density of $Sym(W,\sharp|_{W,W})$ in $Sym(\lim\limits_{\rightarrow} S,\sharp)$ gives the suprayective character of Φ and completes the proof.□

Theorem 4.3 holds if we consider the Jordan H^*-subtriples $Sym(T_i,\sharp_i)$ and $Skw(T_i,\sharp_i)$ of each T_i^+.

5. Direct limits of L^*-triples

We can now formulate the following

Theorem 5.1 *Let $S = (\{T_i\}_{i\in I}, \{e_{ji}\}_{i\leq j})$ be a direct system of simple finite dimensional L^*-triples. Then, $T = \lim\limits_{\rightarrow} S$ admits a two-graded L^*-algebra envelope.*

Proof. By Section 4, T is a topologically simple L^*-triple, verifying $T = \overline{\bigcup\limits_{i\in I} T_i}$, where $\{T_i\}_{i\in I}$ is a direct family, with inclusion, of simple finite dimensional L^*-subtriples of T, and, with the notation $W := \bigcup\limits_{i\in I} T_i$, we have W is a Lie triple system with conjugate-linear involution satisfying the H^*-identities.

From [15, Section 1], every T_i has a finite-dimensional simple envelope L^*-algebra L_i, hence $0 \leq (\sum\limits_{i=1}^{n}[x_i,y_i] | \sum\limits_{i=1}^{n}[x_i,y_i]) = \sum\limits_{i,j=1}^{n}(x_i|[x_j,y_j,y_i^*])$ for $x_i, y_i \in T_i$, and the equality holds iff $\sum\limits_{i=1}^{n}[x_i,y_i] = 0$. Therefore, we can define on the even part of the two-graded Lie algebra with conjugate-linear involution $L' := [W,W] \perp W$, an inner product $(\sum\limits_{i}[x_i,y_i] | \sum\limits_{j}[x'_j,y'_j]) := \sum\limits_{i,j}(x_i|[x'_j,y'_j,y_i^*])$. We conclude that $\overline{[W,W]}\perp\overline{W}$ the completion of L' is a two-graded L^*-algebra envelope of T.□

REFERENCES

1. W.G. Lister, A structure theory of Lie triple systems. Trans. Amer. Math. Soc. 72 (1952) 217-242.
2. J. R. Schue, Hilbert Space methods in the theory of Lie algebras. Trans. Amer. Math. Soc. vol 95 (1960) 69-80.
3. J. R. Schue, Cartan decompositions for L^*-algebras. Trans. Amer. Math. Soc. vol 98 (1961) 334-349.
4. J. A. Cuenca, A. García and C. Martín, Structure theory for L*-algebras. Math-Proc. Camb. Phil. Soc. (1990) 361-365.
5. E. Neher, Generators and relations for 3-graded Lie algebras. J. Algebra 155 (1993) 1-35.
6. A. Castellón, J. A. Cuenca, and C. Martín, Applications of ternary H^*-algebras to associative H^*-superalgebras. Algebras, Groups and Geometries. 10 (1993) 181-190.
7. J. A. Cuenca and A. Rodríguez, Isomorphisms of H^*-algebras. Math. Proc. Camb. Phil. Soc. 97 (1985) 93-99.

8. V. K. Balachandran, Real L^*-algebras. Indian Journal of Pure and Applied Math. Vol. 3, No. 6 (1972) 1224-1246.
9. A. Castellón and J. A. Cuenca, Associative H^*-triple Systems. Workshop on Nonassociative Algebraic Models, Nova Science Publishers (eds. González S. and Myung H. C.), New York, (1992) 45-67.
10. E. Neher, Cartan-Involutionen von halbeinfachen rellen Jordan-Tripelsystemen. Math. Z. 169 (1979) 271-292.
11. N. Jacobson, Lie algebras. Interscience, 1962.
12. Seminaire Sophus Lie, Théorie des algèbres Lie, Ann. Ecole Norm. Sup, Paris, 1954-5.
13. A. Castellón, J. A. Cuenca, and C. Martín, Jordan H^*-triple systems. Non-Associative Algebra and Its Applications. Santos González ed. Kluwer Academic Publishers, (1994).
14. A. Castellón, J. A. Cuenca, and C. Martín, Ternary H^*-algebras. Bolletino U.M.I. (7) 6-B (1992) 217-228.
15. A. Calderón and C. Martín, On L^*-triples and Jordan H^*-pairs. To appear in Ring theory and Algebraic Geometry, Marcel Dekker, Inc.
16. A. Castellón and J. A. Cuenca, Isomorphisms of H^*-triple Systems. Annali della Scuola Normale Superiore di Pisa. Serie IV. Vol. XIX. Fasc. 4 (1992) 507-514.

Recent Progress in Functional Analysis
K.D. Bierstedt, J. Bonet, M. Maestre and J. Schmets

Truncated Hamburger moment problems with constraints

Vadym M. Adamyan [a*] and Igor M. Tkachenko [b †]

[a]Department of Theoretical Physics,
Odessa National University,
65026 Odessa, Ukraine

[b]Department of Applied Mathematics,
Polytechnic University of Valencia,
46022 Valencia, Spain

To Professor Manuel Valdivia on the occasion of his seventieth birthday.

Abstract

The interpolation problem of reconstruction of a holomorphic in the upper half-plane function with non-negative imaginary part and continuous boundary value on the real axis by the first $2n+1$ terms of its asymptotic decomposition at infinity and its values at some m points of the real axis is solved using algorithms, which are reminiscent of those of Schur and Lagrange. At the same time some algorithms are obtained for reconstruction of holomorphic in the upper half-plane contractive functions with continuous boundary values by their values at some m real points. The corresponding interpolation problems are generalized to include values of the first derivative of the sought functions at some real points.

MCS 2000 Primary 30E05, 30E10; Secondary 82C10, 82D10

1. Introduction

Fundamental quantities in system theory, quantum theory and signal processing are the frequency-dependent transfer or response functions. The imaginary parts of the latter (up to standard scalar factors) coincide with the rate of energy or particles absorption and thus are non-negative. Due to the causality principle such functions are boundary values of the functions which are holomorphic in the upper half-plane with non-negative imaginary parts. In other words they are boundary values of the Nevanlinna class functions. Explicit computation of response functions for complex systems with interacting

---

*V.M.A. is grateful to the Polytechnic University of Valencia for its hospitality during his stay in Valencia in 2000.
†The financial support of the Valencian Autonomous Government (research project N[0] INV00-01-28) is acknowledged.

particles from fundamental equations and principles is an extremely difficult problem, which is being solved so far using not so well-founded approximations. However, in many cases some frequency moments of the imaginary parts of these functions can be easily calculated by algebraic manipulations with corresponding evolution operators (Hamiltonians). Besides, the limiting values of response functions sometimes are known at some specified frequencies (energies). This information permits to limit appreciably the classes of analytic functions containing the response functions. Therefore the quest for reasonable approximations of the response functions of real systems gives rise to certain interpolation problems for holomorphic functions. Our aim here is to formulate and solve some of such problems.

In Section 2 the problem of reconstruction of a continuous in the closed upper half-plane Nevanlinna function by the given $2n+1$ first terms of its asymptotic decomposition at infinity and its values at some m points of the real axis is reduced using the Nevanlinna description formula for all solutions of the truncated Hamburger moment problem [1], [2] to the interpolation problem for the Nevanlinna functions with continuous boundary values, where all m nodes of interpolation are points on the real axis.

In the next Section this simplified problem is converted by means of the linear fractional transformation into an analogous problem for a holomorphic in the upper half-plane and continuous on its closure contractive function. The problem for contractive functions is solved then using the algorithm, which is a slight modification of the known Schur algorithm for problems like that of Nevanlinna-Pick.

An alternative method of solution of the same problem involving the Lagrange interpolation polynomials is suggested in Section 4.

Section 5 is devoted to a more sophisticated problem of reconstruction of a non-negative continuous and continuously differentiable function by its $2n+1$ moments and given m extrema. Actually this problem is solved here by reducing it as above to the interpolation problem for Nevanlinna functions in the upper half-plane with continuous and continuously differentiable boundary values, whose values together with the values of the corresponding first derivative are fixed at m nodes of interpolation on the real axis.

In such a form the last problem is a combination of

the truncated Hamburger moment problem [1–3]:

Given a set of real numbers $c_0, ..., c_{2n}$. To find a non-decreasing function $\sigma(t)$, $-\infty < t < \infty$, such that

$$\int_{-\infty}^{+\infty} t^j d\sigma(t) = c_j, \; j = 0, 1, ..., 2n. \tag{1}$$

and

the Löwner problem [4] in the class of Nevanlinna functions:

Given a finite set $\mathfrak{A}$ of points of the real axis, the upper half-plane numbers of $\varphi(t), t \in \mathfrak{A}$, and arbitrary complex numbers $\varphi'(t)$, $t \in \mathfrak{A}' \subset \mathfrak{A}$. To extrapolate φ to a matrix function $\varphi(z)$ of the Nevanlinna class with continuously differentiable boundary values on the real axis.

In this work we describe only algorithms permitting to find rational and continuous interpolations of the sought functions. Errors of such approximations depending on n, m and relative positions of the interpolation nodes as well as matrix and operator versions of the presented results will be discussed elsewhere.

2. Truncated Hamburger moment problem with point constraints

The simplest of the mentioned problems is defined as follows.

Let $\mathbb{N}$ denote the set of all holomorphic in the upper half–plane functions with non-negative imaginary parts. Functions of $\mathbb{N}$ are called Nevanlinna functions.

Problem A. *Given a finite number of points $t_1, ..., t_m$ of the real axis, a set of complex numbers $\zeta_1, ..., \zeta_m$ with positive imaginary parts, and a set of real numbers $c_0, ..., c_{2n}$. To find a set of Nevanlinna functions $\varphi(z)$ such that*

$$\varphi(z) = -\frac{c_0}{z} - \frac{c_1}{z^2} - ... - \frac{c_{2n}}{z^{2n+1}} + o\left(\frac{1}{z^{2n+1}}\right) \tag{2}$$

for $z \to \infty$ inside any angle $\varepsilon < \arg z < \pi - \varepsilon$, $0 < \varepsilon < \pi$;

$$\varphi(t_j) = \zeta_j, \; 1 \le j \le m. \tag{3}$$

By virtue of the Riesz-Herglotz theorem, the functions we are looking for, as any function $\varphi \in \mathbb{N}$, admit the integral representation

$$\varphi(z) = \alpha + \beta z + \int_{-\infty}^{+\infty} \left[\frac{1}{t-z} - \frac{t}{1+t^2}\right] d\sigma(t) \tag{4}$$

with $\operatorname{Im}\alpha \ge 0$, $\beta \ge 0$, and non-decreasing $\sigma(t)$ satisfying the condition

$$\int_{-\infty}^{+\infty} \frac{1}{1+t^2} d\sigma(t) < \infty.$$

It follows from (2) that $\alpha = \beta = 0$ in (4). Moreover, (2) is equivalent to the relations [3]

$$\int_{-\infty}^{+\infty} t^j d\sigma(t) = c_j, \; j = 0, 1, ..., 2n, \tag{5}$$

Therefore Problem **A** without the conditions (3) is nothing else but the truncated Hamburger moments problem. This problem is solvable [1,5,2] if and only if the block-Hankel matrix $(c_{k+j})_{k,j=0}^n$ is non-negative and for any set of complex numbers $\xi_0, ..., \xi_s$, $0 \le s \le n-1$, the condition

$$\sum_{j,k=0}^{s} c_{j+k}\overline{\xi}_j \xi_k = 0 \tag{6}$$

implies

$$\sum_{j,k=0}^{s} c_{j+k+2}\overline{\xi}_j \xi_k = 0. \tag{7}$$

If the conditions (6), (7) hold for a set $\xi_0, ..., \xi_s$, $0 \le s \le n-1$, such that

$$\sum_{k=0}^{s} |\xi_k| > 0,$$

then there is only one non-decreasing function $\sigma(t)$ satisfying (5), which is a step function with a finite number of discontinuity points, and hence there exists only one function

$$\varphi(z) = \int_{-\infty}^{+\infty} \frac{d\sigma(t)}{t-z}$$

satisfying (2), which is a rational Nevanlinna function. Problem **A** under these conditions is solvable only for an exceptional set of values $\zeta_1, ..., \zeta_m$ of $\varphi(z)$ at given points $t_1, ..., t_m$.

If a set of real numbers $c_0, ..., c_{2n}$ is positive definite, i.e., for any non-zero set of complex numbers $\xi_0, ..., \xi_n$ $(\max_{0\le j\le n} |\xi_j| > 0)$

$$\sum_{j,k=0}^{s} c_{j+k}\bar{\xi}_j\xi_k > 0, \tag{8}$$

then there exists an infinite set of non-negative measures σ on the real axis satisfying (5) and, thus, an infinite set of functions from $\mathbb{N}$, satisfying (2).

Let $(D_k(t))_{k=0}^{n}$ be the finite set of polynomials constructed according to the formulas

$$D_0 = \frac{1}{\Delta_0}, \quad \mathrm{D_k(t)} = \frac{1}{\Delta_\mathrm{k}} \det \begin{Vmatrix} c_0 & \cdots & c_{k-1} & 1 \\ c_1 & \cdots & c_k & t \\ \vdots & \vdots & \vdots & \vdots \\ c_k & \cdots & c_{2k-1} & t^k \end{Vmatrix},$$

$$\Delta_0 = c_0, \quad \Delta_\mathrm{k} = \det \begin{Vmatrix} c_0 & \cdots & c_k \\ \vdots & \vdots & \vdots \\ c_k & \cdots & c_{2k} \end{Vmatrix}, \quad \mathrm{k} = 1, 2, ..., \mathrm{n}. \tag{9}$$

Polynomials D_k form an orthogonal system with respect to each σ-measure satisfying (5). Let

$$E_0 \equiv 0, \; E_k(t) = \int_{-\infty}^{+\infty} \frac{D_k(t) - D_k(s)}{t-s} \, d\sigma(s), \; k = 1, ..., n,$$

be the corresponding set of conjugate polynomials. Denote by $\mathbb{N}_0$ the subset of $\mathbb{N}$ consisting of such functions $w(z)$, that $w(z)/z \to 0$ as $z \to \infty$ inside any angle $\varepsilon < \arg z < \pi - \varepsilon$, $0 < \varepsilon < \pi$. Then the formula

$$\varphi(z) = \int_{-\infty}^{+\infty} \frac{d\sigma(t)}{t-z} = -\frac{E_n(z)(w(z)+z) - E_{n-1}(z)}{D_n(z)(w(z)+z) - D_{n-1}(z)}, \quad \mathrm{Im}\, z > 0, \tag{10}$$

establishes a one-to-one correspondence between the set of all Nevanlinna functions $\varphi(z)$ satisfying (2) and the elements $w(z)$ of the subclass $\mathbb{N}_0$.

Notice that the zeros of each orthogonal polynomial $D_k(z)$ are real and by virtue of the Schwarz-Christoffel identity

$$D_{n-1}(z)E_n(z) - D_n(z)E_{n-1}(z) \equiv \Xi_n = \Delta_{n-1}\Delta_n^{-1} > 0 \quad n = 1, 2, \ldots \tag{11}$$

the zeros of $D_{n-1}(z)$ alternate with the zeros of $D_n(z)$ as well as with the zeros of $E_{n-1}(z)$. Therefore *any function $\varphi(z)$ given by the expression on the right hand side of (10) has* *a continuous boundary value on the real axis if and only if the corresponding Nevanlinna function $w \in \mathbb{N}_0$ is continuous in the closed upper half-plane and such that $w(z) + z$ has no joint zeros with $D_{n-1}(z)$.*

To meet the constraints (3) it is enough now to substitute into the right hand side of (10) any continuous in the closed upper half-plane Nevanlinna function $w(z) \in \mathbb{N}_0$ satisfying the following conditions,

$$w(t_j) = w_j \stackrel{def}{=} -t_j + \frac{D_{n-1}(t_j)\zeta_j + E_{n-1}(t_j)}{D_n(t_j)\zeta_j + E_n(t_j)}, \; j = 1, ..., m. \tag{12}$$

Note that by (5),

$$\operatorname{Im} w_j = \frac{\operatorname{Im} \zeta_j \cdot \Xi_n}{|D_n(t_j)\zeta_j + E_n(t_j)|^2} > 0, \; j = 1, ..., m.$$

Thus Problem **A** reduces to

Problem $\mathbf{A}_0$. *Given a finite number of points $t_1, ..., t_m$ of the real axis and a set of complex numbers $w_1, ..., w_m$ with positive imaginary parts. To find a set of continuous in the closed upper half-plane Nevanlinna functions $w(z) \in \mathbb{N}_0$ satisfying conditions (12).*

Each Nevanlinna function $w(z)$ in the upper half-plane admits the representation

$$w(z) = i\frac{1 + \theta(z)}{1 - \theta(z)}, \tag{13}$$

where

$$\theta(z) = \frac{w(z) - i}{w(z) + i} \tag{14}$$

is a holomorphic in the upper half-plane contractive function, i.e. $|\theta(z)| \leq 1, \; \operatorname{Im} z > 0$. The function $\theta(z)$ connected with the Nevanlinna function $w(z)$ by the linear fractional transformation (14) is continuous in the closed upper half-plane if $w(z)$ satisfies this condition. On the other hand, the Nevanlinna function $w(z)$ given as the linear fractional transformation (13) of a holomorphic in the upper half-plane and continuous in its closure contractive function $\theta(z)$ is continuous at the points of the closed upper half-plane where $\theta(z) \neq 1$. Therefore Problem $\mathbf{A}_0$ is equivalent to the following problem for contractive functions.

Let $\mathfrak{B}$ be the set of all holomorphic in the upper half-plane and continuous on its closure contractive functions.

Problem $\mathbf{A}_0'$. *Given a finite number of points $t_1, ..., t_m$ of the real axis and a set of points $\omega_1, ..., \omega_m$,*

$$\omega_j = \frac{w_j - i}{w_j + i}, \; |\omega_j| < 1, \; j = 1, ..., m. \tag{15}$$

To find a set of functions $\theta \in \mathfrak{B}$ such that

$$\theta(t_j) = \omega_j, \; j = 1, ..., m. \tag{16}$$

Problem $\mathbf{A}_0'$ is a limiting case of the Nevanlinna-Pick problem with interpolation nodes on the real axis. Two ways of its solution can be suggested.

3. Auxiliary problem. Schur algorithm

The first of them resembles the well-known Schur algorithm for interpolation problems with the nodes inside the upper half-plane or the unit circle. Note that a function $\theta \in \mathfrak{B}$ satisfies the condition

$$\theta(t_1) = \omega_1, \ |\omega_1| < 1,$$

if and only if it admits the representation

$$\theta(z) = \frac{\phi(z) + \omega_1}{\overline{\omega_1}\phi(z) + 1}, \tag{17}$$

where $\phi \in \mathfrak{B}$ and $\phi(t_1) = 0$. Taking $\gamma_1 > 0$ such that the inequalities

$$\sqrt{1 + \frac{\gamma_1^2}{(t_j - t_1)^2}} \left| \frac{\omega_j - \omega_1}{1 - \overline{\omega_1}\omega_j} \right| < 1, \ j = 2, ..., m,$$

would hold, one can choose $\phi(z)$ in (17) in the form

$$\phi(z) = \frac{z - t_1}{z - t_1 + i\gamma_1}\theta_1(z), \tag{18}$$

where θ_1 is any function from $\mathfrak{B}$ such that

$$\theta_1(t_j) = \omega_j' = \left(1 + \frac{i\gamma_1}{t_j - t_1}\right) \frac{\omega_j - \omega_1}{1 - \overline{\omega_1}\omega_j}, \ j = 2, ..., m. \tag{19}$$

Such a choice of $\theta_1(z)$ guarantees the verification of all of the conditions (16). Hence Problem $\mathbf{A}_0'$ with m nodes of interpolation on the real axis and strictly contractive values of the functions to find at these nodes, reduces to the same problem but with $m-1$ nodes of interpolation and modified values at these nodes given by (19). Repeating the above procedure $m-1$ times with a suitable choice of parameters γ_j and modifying the values of emerging contractive functions at the remaining points $t_{j+1}, ..., t_m$ according to (19), permits to obtain some solution of Problem $\mathbf{A}_0'$. Observe that contrary to the Nevanlinna-Pick problem with nodes in the open upper half-plane, our Problem $\mathbf{A}_0'$ is always solvable if the values of the function to reconstruct are strictly contractive at the nodes of interpolation.

Let $\theta_{j-1} \in \mathfrak{B}$ be a contractive function emerging after the $j-1$ step in the course of the Problem $\mathbf{A}_0'$ solution by the above method, and let $\omega_j^{(j-1)} = \theta_{j-1}(t_j)$, $\omega_1^{(0)} = \omega_1$. It follows from the above arguments that should the initial parameters $\omega_1, ..., \omega_m$ be strictly contractive, there exists a set of solutions of Problem $\mathbf{A}_0'$ described by the formula

$$\theta(z) = \frac{a(z)\epsilon(z) + b(z)}{c(z)\epsilon(z) + d(z)}, \tag{20}$$

where the elements of the matrix of the linear fractional transformation (20) are rational functions of degree $m-1$ and $\epsilon(z)$ runs the subset of all functions from $\mathfrak{B}$ satisfying the condition $\epsilon(t_m) = \omega_m^{(m-1)}$. This matrix can be calculated as

$$\begin{pmatrix} a(z) & b(z) \\ c(z) & d(z) \end{pmatrix} = \overset{\curvearrowright}{\prod_{j=1}^{m-1}} \begin{pmatrix} 1 & \omega_j^{(j-1)} \\ \overline{\omega_j^{(j-1)}} & 1 \end{pmatrix} \begin{pmatrix} \frac{z-t_j}{z-t_j+i\gamma_j} & 0 \\ 0 & 1 \end{pmatrix}, \tag{21}$$

where numbers j in matrix factors on the right hand side increase from left to right.

Observe that substituting in (20) $\epsilon(z) \equiv \omega_m^{(m-1)}$, we obtain a rational function of degree $m-1$. Hence, if initial parameters $\omega_1, ..., \omega_m$ in Problem $\mathbf{A}_0'$ are strictly contractive, then among the solutions of this problem there are rational functions of degree $m-1$.

4. Auxiliary problem. Lagrange algorithm

An alternative algorithm of solution of Problem $\mathbf{A}_0'$ employs the Lagrange interpolation polynomials. It does not require the parameters $\omega_1, ..., \omega_m$ modification, and can be used equally well in cases, where the moduli of some and even of all these parameters are equal to unity. Given the polynomials

$$P_k(t) = \prod_{j \neq k}^{m} (t - t_j), \ k = 1, ..., m,$$

consider for any $t \in \mathbb{R}$ the rational function

$$\varphi(t) = \frac{\omega_1 P_1^2(t) + ... + \omega_m P_m^2(t)}{P_1^2(t) + ... + P_m^2(t)}.$$

By construction, there are no common zeros for all polynomials $P_k(t)$. Therefore, $\varphi(t)$ is continuous on the real axis,

$$|\varphi(t)| \leq \frac{|\omega_1|\, P_1^2(t) + ... + |\omega_m|\, P_m^2(t)}{P_1^2(t) + ... + P_m^2(t)} \leq \max_j |\omega_j| \leq 1,$$

and

$$\varphi(t_j) = \frac{\omega_j P_j^2(t_j)}{P_j^2(t_j)} = \omega_j.$$

However, the rational function $\varphi(t)$ is not a boundary value of a function from $\mathfrak{B}$. Indeed, the polynomial

$$\mathcal{P}(t) = P_1^2(t) + ... + P_m^2(t) > 0, \ -\infty < t < \infty,$$

with real coefficients has only complex roots and together with each root $\alpha + i\beta$, $\beta \neq 0$, of multiplicity s it has also the root $\alpha - i\beta$ of the same multiplicity. Let $\alpha_\nu + i\beta_\nu$, $\beta_\nu > 0$, $1 \leq \nu \leq m-1$, be the set of all roots of $\mathcal{P}(t)$ in the upper half-plane and let s_ν be their multiplicities,

$$\sum_\nu s_\nu = m - 1.$$

It is evident that

$$\mathcal{P}(t) = m \prod_\nu |t - \alpha_\nu - i\beta_\nu|^2.$$

Denote by $B(t)$ the Blaeschke product

$$B(t) = \prod_\nu \left(\frac{t - \alpha_\nu - i\beta_\nu}{t - \alpha_\nu + i\beta_\nu} \right)^{s_\nu}.$$

Then the rational function of degree $2(m-1)$,

$$\theta(t) = B(t)\frac{\widetilde{\omega}_1 P_1^2(t) + ... + \widetilde{\omega}_m P_m^2(t)}{P_1^2(t) + ... + P_m^2(t)} = \frac{\widetilde{\omega}_1 P_1^2(t) + ... + \widetilde{\omega}_m P_m^2(t)}{\mathcal{Q}(t)}, \tag{22}$$

where

$$\widetilde{\omega}_j = B(t_j)^{-1}\omega_j,\ 1 \le j \le 1;\ \mathcal{Q}(t) = m\prod_{\nu}(t - \alpha_\nu + i\beta_\nu)^2, \tag{23}$$

belongs already to the class $\mathfrak{B}$, and is a rational solution of Problem $\mathbf{A}_0'$.

Notice that the rational function of degree $m-1$,

$$\theta_1(z) = \frac{\widetilde{\widetilde{\omega}}_1 P_1(z) + ... + \widetilde{\widetilde{\omega}}_m P_m(z)}{\mathcal{Q}_1(z)},$$

with

$$\mathcal{Q}_1(z) = \sqrt{m}\prod_{\nu}(z - \alpha_\nu + i\beta_\nu);\ \widetilde{\widetilde{\omega}}_j = \frac{\mathcal{Q}_1(t_j)}{P_j(t_j)}\omega_j,\ 1 \le j \le m,$$

is holomorphic in the closed upper half-plane and satisfies the conditions (16). However, in general, this function is contractive in the upper half-plane only under additional requirements imposed on the parameters ω_j. For example, it follows from the estimate

$$|\theta_1(t)| \le \frac{\left|\widetilde{\widetilde{\omega}}_1\right| |P_1(t)| + ... + \left|\widetilde{\widetilde{\omega}}_m\right| |P_m(t)|}{|\mathcal{Q}_1(t)|} =$$

$$= \frac{|\omega_1|\,|P_1(t)| + ... + |\omega_m|\,|P_m(t)|}{\sqrt{P_1^2(t) + ... + P_m^2(t)}} \le \sum_{j=1}^{m} |\omega_j|^2$$

that $\theta_1 \in \mathfrak{B}$ if

$$\sum_{j=1}^{m} |\omega_j|^2 \le 1.$$

Note that *a wide class of solutions of Problem* $\mathbf{A}_0'$ *can be obtained by substituting into (22) instead of each constant* $\widetilde{\omega}_j$ *an arbitrary function* $\widehat{\omega}_j$ *from* $\mathfrak{B}$ *satisfying the condition* $\widehat{\omega}_j(t_j) = \widetilde{\omega}_j$.

5. Inclusion of derivatives

The latter remark permits to extend the statements of the above interpolation problems with inclusion to the given data of the values of derivatives of the function under investigation at some or all interpolation nodes. One of such problems is defined as follows.

Let $\mathfrak{B}^1$ be the subset of $\mathfrak{B}$ consisting of continuously differentiable functions in the closed upper half-plane including the infinite point.

Problem $\mathbf{B}_0'$. *Given a finite number of points* $t_1, ..., t_m$ *of the real axis and two sets of complex numbers:* $\omega_1, ..., \omega_m$, $|\omega_j| < 1$, $j = 1, ..., m$ and $\omega_1', ..., \omega_m'$. *To find a set of functions* $\theta \in \mathfrak{B}^1$ *satisfying the following conditions*

$$\theta(t_j) = \omega_j,\ \theta'(t_j) = \omega_j',\ j = 1, ..., m. \tag{24}$$

A rational solution of Problem $\mathbf{B}_0'$ can be constructed by substitution into (22), instead of the parameters $\widetilde{\omega}_j$, the rational functions

$$\widehat{\omega}_j(z) = B(t_j)^{-1}\frac{\phi_j(z)+\omega_j}{1+\overline{\omega_j}\phi_j(z)}, \; \phi_j(z) = \xi_j\frac{z-t_j}{z-t_j+i\gamma_j}, \tag{25}$$

with appropriate parameters ξ_j, $|\xi_j| \le 1$ and γ_j, $\gamma_j > 0$, $j = 1, ..., m$. Note that thus obtained rational function

$$\theta(z) = \frac{\widehat{\omega}_1(z)P_1^2(z) + ... + \widehat{\omega}_m(z)P_m^2(z)}{\mathcal{Q}(z)}$$

of degree not more than $3m-2$ belongs to $\mathfrak{B}^1$ and the first equalities in (24) hold for any contractive parameters ξ_j and positive parameters γ_j. Since

$$\theta'(t_j) = 2i\omega_j \sum_\nu \frac{s_\nu\beta_\nu}{(t_j-\alpha_\nu)^2+\beta_\nu^2} - i\left(1-|\omega_j|^2\right)\frac{\xi_j}{\gamma_j}$$

and $|\omega_j| < 1$, the contractive parameters ξ_j and positive parameters γ_j can be chosen to satisfy the second set of equalities in (24).

Note that *a certain class of solutions of Problem $\mathbf{B}_0'$ can be obtained by substituting into (25) instead of contractive parameters ξ_j arbitrary functions $\widehat{\xi}_j(z)$ from $\mathfrak{B}^1$, whose values $\widehat{\xi}_j(t_j)$ at t_j satisfy the second set of equalities in (24).*

This last result permits to obtain a set of solutions of the following more sophisticated problem.

Problem C. *Given a finite number of points $t_1, ..., t_m$ of the real axis, a set of positive numbers $a_1, ..., a_m$ and a set of real numbers $c_0, ..., c_{2n}$. To find a set of non-negative continuously differentiable functions $\rho(t)$ on the real axis satisfying the conditions:*

$$\rho(t_j) = a_j, \; \rho'(t_j) = 0; \tag{26}$$

$$\int_{-\infty}^{\infty} t^k \rho(t)dt = c_k. \tag{27}$$

Let us denote by $\mathbb{N}_0^1$ the subset of $\mathbb{N}$ consisting of all Nevanlinna functions $\varphi(z)$ such that:

- the harmonic functions $\operatorname{Im}\varphi(z)$ and $\operatorname{Im}\varphi'(z)$ are continuous on any compact subset of the closed upper half-plane;
- $\varphi(z)$ and $\varphi'(z)$ are functions of the Hardy class $\mathbf{H}^2$ in the upper half-plane, i.e.

$$\sup_{\eta>0}\int_{-\infty}^{\infty} |\varphi(t+i\eta)|^2\,dt \left(=\int_{-\infty}^{\infty} |\varphi(t+i0)|^2\,dt\right) < \infty \tag{28}$$

  and the same is true for $\varphi'(z)$.

Lemma 5.1. *Let $\rho(t) \geq 0$ be a continuous function on the real axis such that*

$$\int_{-\infty}^{\infty} \frac{|\rho(t)|^2}{1+t^2} dt < \infty.$$

Then for $\rho(t)$ the following conditions are equivalent:

a) $\rho(t)$ is continuously differentiable and

$$\|\rho\|^2 \stackrel{\text{def}}{=} \int_{-\infty}^{\infty} \rho^2(t) dt < \infty, \quad \|\rho'\|^2 = \int_{-\infty}^{\infty} |\rho'(t)|^2 dt < \infty; \tag{29}$$

b)

$$\rho(t) = \lim_{\eta \downarrow 0} \frac{1}{\pi} \operatorname{Im} \varphi(t + i\eta),$$

where $\varphi(z)$ is a Nevanlinna function from $\mathbb{N}_0^1$.

Proof. Let $\rho(t)$ satisfy the conditions a). By these conditions and the Cauchy inequality

$$|\rho(t)| = \left| \rho(0) + \int_0^t \rho'(s) ds \right| \leq |\rho(0)| + \sqrt{t} \left(\int_{-\infty}^{\infty} |\rho'(s)|^2 ds \right)^{\frac{1}{2}}. \tag{30}$$

Besides the integral in

$$\varphi(z) = \int_{-\infty}^{\infty} \frac{1}{t-z} \rho(t) dt, \ \operatorname{Im} z > 0, \tag{31}$$

converges for non-real z and the function $\varphi(z)$ defined by (31) evidently belongs to the intersection $\mathbb{N} \cap \mathbf{H}^2$. Since $\varphi(z)$ is given by the Cauchy type integral and $\rho(t)$ is a continuously differentiable function, then due to the well-known properties of such integrals $\varphi(z)$ is continuous at any compact part of the closed upper half-plane. By (30) for each non-real z we have

$$\lim_{N_\pm \to \infty} \int_{-N_-}^{N_+} \left(\frac{1}{t-z} \rho'(t) - \frac{1}{(t-z)^2} \rho(t) \right) dt = \frac{1}{N_+ - z} \rho(N_+) - \frac{1}{N_- - z} \rho(N_-) = 0.$$

Hence

$$\varphi'(z) \left(= \int_{-\infty}^{\infty} \frac{1}{(t-z)^2} \rho(t) dt \right) = \int_{-\infty}^{\infty} \frac{1}{t-z} \rho'(t) dt, \ \operatorname{Im} z > 0. \tag{32}$$

According to (29) ρ' is an element of the space $\mathbf{L}^2$ on the real axis. By (32) $\varphi' \in \mathbf{H}^2$ and since ρ, ρ' are real continuous function we see that

$$\lim_{\eta \downarrow 0} \operatorname{Im} \varphi(t + i\eta) = \pi \rho(t), \lim_{\eta \downarrow 0} \operatorname{Im} \varphi'(t + i\eta) = \pi \rho'(t) \tag{33}$$

at each point of the real axis and this convergence is uniform on any finite segment of the real axis. Hence $\varphi \in \mathbb{N}_0^1$.

Let us select now an arbitrary function $\varphi \in \mathbb{N}_0^1$. Due to the continuity of φ on any compact part of the closed upper half-plane plane and the Stieltjes inversion formula we conclude that for such φ the measure $\sigma(t)$ in the Riesz-Herglotz representation (4) is absolutely continuous and

$$\rho(t) = \sigma'(t) = \lim_{\eta \downarrow 0} \frac{1}{\pi} \operatorname{Im} \varphi(t + i\eta).$$

is a continuos function from $\mathbf{L}^2$. Therefore for the given $\varphi \in \mathbb{N}_0^1$ we can write the representation (4) in the form

$$\varphi(z) = \alpha + \beta z + \int_{-\infty}^{+\infty} \frac{1}{t - z} \rho(t) dt. \tag{34}$$

Since $\rho \in \mathbf{L}^2$ the integral in (34) defines a function from $\mathbf{H}^2$ and by virtue of our assumption $\varphi \in \mathbf{H}^2$. Thus we conclude that in (34) $\alpha = \beta = 0$. By the assumption $\varphi \in \mathbb{N}_0^1$ it follows also that

$$\rho_*(t) = \lim_{\eta \downarrow 0} \frac{1}{\pi} \operatorname{Im} \varphi'(t + i\eta)$$

is a continuous function from $\mathbf{L}^2$. For each smooth compactly supported function $g(t)$ we have

$$\int_{-\infty}^{+\infty} [g'(t)\rho(t) + g(t)\rho_*(t)]\, dt =$$

$$\lim_{\eta \downarrow 0} \frac{1}{\pi} \operatorname{Im} \int -\infty^{+\infty} [g'(t)\varphi(t + i\eta) + g(t)\varphi'(t + i\eta)]\, dt = 0.$$

Then it stems from the known theorem of analysis that $\rho(t)$ is a differentiable function and $\rho'(t) = \rho_*(t)$. □

Hence Problem **C** may be treated as the following problem for the Nevanlinna functions of subclass $\mathbb{N}_0^1$.

Problem C$'$. *Given a finite number of points $t_1, ..., t_m$ of the real axis, a set of positive numbers $a_1, ..., a_m$, and a set of real numbers $c_0, ..., c_{2n}$. To find a set of Nevanlinna functions*

$$\varphi(z) = \int_{-\infty}^{\infty} \frac{1}{t - z} \rho(t) dt \tag{35}$$

from $\mathbb{N}_0^1$ by its asymptotic

$$\varphi(z) = -\frac{c_0}{z} - \frac{c_1}{z^2} - \dots - \frac{c_{2n}}{z^{2n+1}} + o\left(\frac{1}{z^{2n+1}}\right) \tag{36}$$

for $z \to \infty$ inside any angle $\varepsilon < \arg z < \pi - \varepsilon$, $0 < \varepsilon < \pi$, and some extremal values of its imaginary part on the real axis:

$$\lim_{\delta \downarrow 0} \operatorname{Im} \varphi(t_j + i\delta) = \pi\rho(t_j) = \pi a_j, \; lim_{\delta \downarrow 0} \operatorname{Im} \varphi'(t_j + i\delta) = 0, \; j = 1, ..., m. \tag{37}$$

In such a form this problem can be reduced to Problem $\mathbf{B}_0^{'}$ in the same fashion as Problem $\mathbf{A}$ was reduced to Problem $\mathbf{A}_0^{'}$.

Indeed, let us assume that the quadratic form (8) is non-degenerate. Then each Nevanlinna function $\varphi(z)$ satisfying the condition (36) can be represented in the form of the linear fractional transformation (10) over some function $w \in \mathbb{N}_0$. Those of such functions which are representable as the linear fractional transformation (10) of functions $w \in \mathbb{N}_0^1$ satisfying the unique additional condition

$$|w(t)+t| \; + |D_{n-1}(t)| \gg 0, \; -\infty < t < \infty, \tag{38}$$

belong to $\mathbb{N}_0^1$ themselves. What remains for the solution of **Problem** $\mathbf{C}^{'}$ is to single out a subset of functions from $\mathbb{N}_0^1$ which in addition to (38) would satisfy (37). But the problem of selection of functions $w(z)$ satisfying (37) with the restriction (38) by means of the linear fractional transformation

$$w(z) = i\frac{1+\theta(z)}{1-\theta(z)} \tag{39}$$

reduces to the selection of a certain subset of contractive holomorphic in the upper half-plane functions $\theta \in \mathfrak{B}^1$.

To this end we note first that (38) definitely holds if the boundary value of θ is strictly contractive ($|\theta(t)| \ll 1$) and as follows $\operatorname{Im} w(t) \gg 0$ on an interval of the real axis containing all zeros of $D_{n-1}(t)$. Then we can use relations

$$\widetilde{\varphi}(z) = \varphi(z) + \frac{E_n(z)}{D_n(z)} = \frac{\Delta_n}{D_n^2(z)} \frac{1}{D_{n-1}(z)/D_n(z) - 1 - w(z)} \tag{40}$$

and

$$\widetilde{\varphi}'(z) = \frac{2D_n'(z)}{D_n(z)}\widetilde{\varphi}(z) + \frac{1}{\Delta_n}\widetilde{\varphi}^2(z)\left(w'(z)D_n^2(z) + D_n'(z)D_{n-1}(z) - D_n(z)D_{n-1}'(z)\right)$$

connecting the values of the Nevanlinna function $\varphi(z)$ and its derivative $\varphi'(z)$with the corresponding Nevanlinna parameter $w(z)$ in (10) and its derivative $w'(z)$, respectively. They permit together with $w(t_j)$ given by (12) also to find such values of $w'(t_j)$, $j = 1, ..., m$, that satisfy (37). Afterwards it remains only to transform all these values into the data for Problem $\mathbf{B}_0^{'}$ of determination of a contractive function $\theta(z)$ connected with $w(z)$ by the linear fractional transformation (39).

Notice that for the function $\varphi(z)$ to reconstruct in Problem $\mathbf{C}^{'}$, the values of $\operatorname{Re}\varphi(t_j)$ are indefinite. Using this degree of freedom we can assume that together with $\operatorname{Im}\varphi'(t_j) = 0$ we have also $w'(t_j) = 0$ for some t_j. This assumption would result in

$$\operatorname{Re}\varphi(t_j) = \frac{D_n'(t_j)D_{n-1}(t_j)}{D_n'(t_j)D_{n-1}(t_j) - D_n(t_j)D_{n-1}'(t_j)}$$

for such t_j.

Remark The proposed way of solution of Problem $\mathbf{C}^{'}$ cannot be applied immediately in the limiting case when some or all $a_j = 0$. If in addition to the moments $c_0, ..., c_{2n}$ in this case some or all generalized moments

$$d_{j1} = \int\limits_{-\infty}^{\infty} \frac{1}{t - t_j} \rho(t)dt, \ d_{j2} = \int\limits_{-\infty}^{\infty} \frac{1}{(t - t_j)^2} \rho(t)dt$$

are given, then taking into account that the functions

$$1, t, ..., t^{2n}; \frac{1}{t - t_1}, \frac{1}{(t - t_1)^2}, \frac{1}{t - t_2}, ..., \frac{1}{(t - t_m)^2}$$

form a Chebyshev system, this limiting version of Problem $\mathbf{C}'$ can be treated and solved as the generalized moment problem [1,7].

Questions about the existence and uniqueness of the solution of namely this limiting problem were in particular elucidated recently in [6], and in the case of non-uniqueness the description of all solutions was given there. These results were obtained in [6] in general for infinite sets of interpolation points on the real axis and in the upper half-plane using the theory of self-adjoint extensions of asymmetric relations in a Hilbert space and alternatively via the theory of the reproducing kernel Hilbert space.

REFERENCES

1. M.G. Krein, A.A. Nudel'man, The Markov moment problem and extremal problems, "Nauka", Moscow, 1973 (Russian). English translation: Translation of Mathematical Monographs AMS, **50** (1977).
2. V.M. Adamyan and I.M. Tkachenko, Solution of the Truncated Hamburger Moment Problem According to M.G. Krein. Operator Theory: Advances and Applications, **118** (2000), 33-52.
3. N.I. Akhiezer, The classical moment problem and some related questions in analysis, Hafner Publishing Company, N.Y. (1965).
4. K. Löwner, Über monotone Matrixfunktionen, Math. Z. **38** (1934), 177.
5. R.E. Curto and L.A. Fialkow, Recursiveness, positivity, and truncated moment problems. Houston J. Math. **17** (1991) 603-635; see also:
   R.E. Curto and L.A. Fialkow, Solutions of the truncated moment problem. Mem. Amer. Math. Soc. **119** (1996).
6. D. Alpay, A. Dijksma and H. Langer, Classical Nevanlinna-Pick Interpolation with Real Interpolation Points. Operator Theory: Advances and Applications, **115** (2000) 1-50.
7. S. Karlin, W.S. Studden, Tschebyscheff systems: with applications in analysis and statistics, Interscience Publishers, 1966.

Recent Progress in Functional Analysis
K.D. Bierstedt, J. Bonet, M. Maestre and J. Schmets

Fourier-Bessel transformation of measures and singular differential equations

A. B. Muravnik*

Moscow Aviation Institute, Department of Differential Equations
RUSSIA 125871, Moscow, Volokolamskoe shosse 4

To Professor Manuel Valdivia on the occasion of his seventieth birthday.

Abstract

This paper is devoted to the investigation of Fourier-Bessel transformation (see [2]) for non-negative f: $\tilde{f}(\xi,\eta) = \frac{1}{\eta^\nu}\int\limits_0^\infty \int\limits_{\mathbf{R}^n} y^{\nu+1} J_\nu(\eta y) f(x,y) e^{-i\xi\cdot x} dx dy;\ \nu > -\frac{1}{2}$. *We apply the method of [5] which provides the estimate for weighted* L_∞*-norm of the spherical mean of* $|\hat{f}|^2$ *via its weighted* L_1*-norm (generally it is wrong without the requirement of the non-negativity of f). We prove that (unlike in the classical case of Fourier transformation) a similar estimate is valid for the one-dimensional case too: a weighted* L_∞*-norm of* $\tilde{f}$ *is estimated by its weighted* L_2*-norm. The obtained result and the estimates for the multi-dimensional case (see [6] and [7]) are applied to the investigation of singular differential equations containing Bessel operator (where the parameter at the singularity equals to* $2\nu + 1$*); equations of such kind arise in models of mathematical physics with degenerative heterogeneities and in axially symmetric problems. We obtain estimates for weighted* L_∞*-norms of solutions (for ordinary differential equations) and of weighted hemispherical means of squared solutions (for partial differential equations).*
MCS 2000 Primary 42B10; Secondary 44A45

1. Introduction

It is proved in [5] that if $f \geq 0$, then for any $\alpha \in (0, (n-1)/2]$

$$\|r^\alpha \sigma(f)\|_\infty \leq C\|r^{\alpha-1}\sigma(f)\|_1, \tag{1}$$

where $\sigma(f)(r)$ is the mean value of $|\hat{f}|^2$ over the sphere of radius r with the centre at the origin, and C depends only on the dimension of the space.

We note that, generally, (1) does not hold because we can construct a sequence $\{f_m\}_{m=1}^\infty$ such that $\|r^{\alpha-1}\sigma(f_m)\|_1$ does not depend on m but $\sigma(f_m)(1)$ tends to infinity as $m \to \infty$.

*E-mail: muravnik@postmaster.co.uk

Thus, the requirement that f be non-negative prohibits the above type of behaviour. Actually, it represents a certain restriction on the shape of the graph of $\hat{f}$.

One can expect that in the one-dimensional case (1) gives the similar estimate with replacing the mean by the function itself. But it turns out that in this case the integral in the right-hand side of (1) diverges for any non-negative f: $\hat{f}(0)$ is equal to the integral of f over the whole real line so there is a non-integrable singularity at the origin.

In this work we investigate Fourier-Bessel transformation, which is applied in the theory of differential equations containing singular Bessel operator with respect to a selected variable (called the *special* variable). These equations arise in models of mathematical physics with degenerative space heterogeneities. We prove that, unlike the classical regular case, in the above-mentioned singular case the estimate of the claimed type is valid for one-dimensional integral transformation (so-called *pure* Fourier-Bessel transformation): a weighted L_∞-norm of the transform is estimated from above by its weighted L_2-norm.

Then we apply the obtained estimate to singular ordinary differential equations containing Bessel operator. Using the fact that in Fourier-Bessel images Bessel operator acts as a multiplier, we find the following estimates for norms of their solutions:

$$\|u\|_\infty \leq u(0)\,,$$

$$\|r^{\frac{\alpha}{2}}u\|_\infty \leq C\|r^{\frac{\alpha-1}{2}}u\|_2\,.$$

We also use the estimates for *mixed* Fourier-Bessel transforms (i.e. for the multi-dimensional case) of non-negative functions, obtained in [6], and find the following estimates for solutions of partial differential equations containing Bessel operator:

$$\|r^\gamma S_{p,q}u\|_\infty \leq \|r^{\gamma-1}S_{\alpha,\beta}u\|_1\,,$$

if the number of *non-special* variables is more than 1;

$$\|r^\gamma S_{p,q}u\|_\infty \leq \|r^{\gamma-1}u^2(0,y)\|_1\,,$$

if the non-special variable x is single.

Here $S_{p,q}$ denotes the weighted hemispherical mean value of $|\cdot|^2$ with the weight $|x|^p y^q$, constant C and the allowed values of parameters p, q, α, β, γ are determined by the dimension of the space and the index of the Bessel function from the kernel of the transformation (or the parameter at the singularity of the Bessel operator contained in the equation).

2. Preliminaries

In this section we introduce the necessary notations and definitions; we also recall the properties of the Fourier-Bessel transformation.

Let $k \stackrel{\text{def}}{=} 2\nu + 1$ be a positive parameter. In what follows, all the absolute constants generally depend on ν and n. We write

$$\mathbf{R}^{n+1}_+ \stackrel{\text{def}}{=} \left\{(x,y)\,\middle|\, x \in \mathbf{R}^n,\ y > 0\right\}.$$

$S_+(r)$ denotes the upper hemisphere in $\mathbf{R}_+^{n+1}$ with radius r, centred at the origin; dS_z denotes the surface measure with respect to the (vector) variable z. Also, let for $\mu > 0$

$$L_{p,\mu}(\mathbf{R}_+^{n+1}) \stackrel{\text{def}}{=} \Big\{ f \,\Big|\, \|f\| = \Big(\int\limits_{\mathbf{R}_+^{n+1}} y^\mu |f(x,y)|^p \, dxdy \Big)^{\frac{1}{p}} < \infty \Big\}, \quad p \text{ finite};$$

$$L_{\infty,\mu}(\mathbf{R}_+^{n+1}) \stackrel{\text{def}}{=} \Big\{ f \,\Big|\, \|f\| = \operatorname{vrai\,sup} y^\mu |f(x,y)| < \infty \Big\}.$$

Further, let

$$S_{p,q} f(r) \stackrel{\text{def}}{=} \int\limits_{S_+(1)} |x|^p y^q |f(rx, ry)|^2 \, dS_{x,y}.$$

The set of infinitely smooth functions with compact support defined on $\mathbf{R}^{n+1}$ is denoted by $C_0^\infty(\mathbf{R}^{n+1})$. We consider the subset of $C_0^\infty(\mathbf{R}^{n+1})$ consisting of even functions with respect to y, and denote by $C_{0,\text{even}}^\infty(\mathbf{R}_+^{n+1})$ the set of restrictions of elements of that subset to $\mathbf{R}_+^{n+1}$. The space $C_{0,\text{even}}^\infty(\mathbf{R}_+^{n+1})$ is known as *the space of test functions*.
Distributions on $C_{0,\text{even}}^\infty(\mathbf{R}_+^{n+1})$ are introduced (following, for example, [1]) with respect to the degenerative measure $y^k dxdy$ by

$$\langle f, \varphi \rangle \stackrel{\text{def}}{=} \int\limits_{\mathbf{R}_+^{n+1}} y^k f(x,y) \varphi(x,y) \, dxdy \quad \text{for any } \varphi \in C_{0,\text{even}}^\infty(\mathbf{R}_+^{n+1}). \tag{2}$$

Thus, all linear continuous functionals on $C_{0,\text{even}}^\infty(\mathbf{R}_+^{n+1})$ which are given by (2) (with $f \in L_{1,k,loc}(\mathbf{R}_+^{n+1})$) are called *regular* (and the corresponding function f is called *ordinary*).
The Fourier-Bessel transformation is given by [2]:

$$\tilde{f}(\xi_1, \ldots, \xi_n, \eta) \stackrel{\text{def}}{=} \mathcal{F}_\text{b}(f)(\xi, \eta) \stackrel{\text{def}}{=} \int\limits_{\mathbf{R}_+^{n+1}} y^k j_\nu(\eta y) e^{-i\xi \cdot x} f(x,y) \, dxdy,$$

where $j_\nu(z) = J_\nu(z)/z^\nu$ is the normalised (in the uniform sense) Bessel function.
We note that

$$f(x,y) = C \int\limits_{\mathbf{R}_+^{n+1}} \eta^k j_\nu(\eta y) e^{i\xi \cdot x} \tilde{f}(\xi, \eta) \, d\xi d\eta.$$

In the one-dimensional case, naturally

$$\tilde{f}(\eta) \stackrel{\text{def}}{=} \mathcal{F}_\text{b}(f)(\eta) \stackrel{\text{def}}{=} \int\limits_{\mathbf{R}_+} y^k j_\nu(\eta y) f(y) \, dy$$

(pure Fourier-Bessel transformation).
Correspondingly,

$$f(y) = C \int\limits_{\mathbf{R}_+} \eta^k j_\nu(\eta y) \tilde{f}(\eta) \, d\eta.$$

The generalized shift operator corresponding to the considered degenerative measure is found (for the one-dimensional case) in [4]:

$$T_y^h f(y) \overset{\text{def}}{=} C \int_0^\pi f\left(\sqrt{y^2+h^2-2yh\cos\theta}\right)\sin^{k-1}\theta d\theta \tag{3}$$

such that

$$\int_0^\infty \eta^k g(\eta) T_y^\eta f(y) d\eta = \int_0^\infty \eta^k f(\eta) T_y^\eta g(y) d\eta,$$

and therefore one can introduce the generalized convolution:

$$(f * g)(y) \overset{\text{def}}{=} \int_0^\infty \eta^k f(\eta) T_y^\eta g(y) d\eta$$

such that $\widetilde{f*g} = \tilde{f}\tilde{g}$ (see also [3]).
In the general case of mixed Fourier-Bessel transformation the generalized shift operator is constructed as superposition of (3) with respect to the special variable y and classical shift operators with respect to the remaining variables.

$B \overset{\text{def}}{=} B_{k,y}$ denotes Bessel operator:

$$Bu = \frac{1}{y^k}\frac{d}{dy}\left(y^k \frac{d}{dy}\right).$$

Also, let

$$\Delta_B \overset{\text{def}}{=} \sum_{j=1}^n \frac{\partial^2}{\partial x_j^2} + \frac{1}{y^k}\frac{\partial}{\partial y}\left(y^k \frac{\partial}{\partial y}\right).$$

3. Estimates for Fourier-Bessel transforms of non-negative functions

We start the investigation from the case of pure Fourier-Bessel transformation (i.e. from the one-dimensional case). So let a non-negative f belongs to $f \in L_{1,k}(\mathbf{R}_+) \cap L_{2,k}(\mathbf{R}_+)$.

Our claim is to prove the inequality

$$\|r^{\frac{\alpha}{2}}\tilde{f}\|_\infty \le C\|r^{\frac{\alpha-1}{2}}\tilde{f}\|_2 \tag{4}$$

for any $\alpha \in (0, \frac{k}{2}]$.
First of all we will prove (4) for the largest claimed value of α; then the result will be extended for the whole interval $(0, \frac{k}{2})$.
Under our assumptions $\tilde{f}$ is an ordinary function belonging to $L_{2,k}(\mathbf{R}_+)$ (see [2]) and then the integral $\int_0^\infty y^k f(y) j_\nu(ry) dy$ converges absolutely for almost every positive r. Therefore

$$\tilde{f}^2(r) = \int_0^\infty y^k f(y) j_\nu(ry) dy \int_0^\infty \eta^k f(\eta) j_\nu(r\eta) d\eta =$$

$$= \int_0^\infty \int_0^\infty y^k \eta^k f(y) f(\eta) j_\nu(ry) j_\nu(r\eta) dy d\eta.$$

Taking into account that $j_\nu(ry)j_\nu(r\eta) = T_y^\eta j_\nu(ry)$ (see [4]) we obtain:

$$\tilde{f}^2(r) = \int_0^\infty \int_0^\infty y^k \eta^k f(y) f(\eta) T_y^\eta j_\nu(ry)\,dy\,d\eta =$$

$$= \int_0^\infty \eta^k f(\eta) \int_0^\infty y^k f(y) T_y^\eta j_\nu(ry)\,dy\,d\eta = \int_0^\infty \eta^k f(\eta) \int_0^\infty y^k j_\nu(ry) T_y^\eta f(y)\,dy\,d\eta.$$

We note that $f \geq 0$ and the generalized shift operator preserves the sign; on the other hand $|j_\nu(ry)| = |\frac{J_\nu(ry)}{(ry)^\nu}| \leq \frac{C}{(ry)^{\nu+\frac{1}{2}}} = \frac{C}{(ry)^{\frac{k}{2}}}$. Hence

$$\tilde{f}^2(r) \leq C \int_0^\infty \eta^k f(\eta) \int_0^\infty y^k \frac{1}{(ry)^{\frac{k}{2}}} T_y^\eta f(y)\,dy\,d\eta \tag{5}$$

under the assumption that the integral at the right-hand side of (5) converges.

Now we will prove that it converges indeed.

$$\int_0^\infty \eta^k f(\eta) \int_0^\infty y^k \frac{1}{(ry)^{\frac{k}{2}}} T_y^\eta f(y)\,dy\,d\eta = \frac{1}{r^{\frac{k}{2}}} \int_0^\infty \eta^k f(\eta) \int_0^\infty y^k f(y) T_y^\eta y^{-\frac{k}{2}}\,dy\,d\eta =$$

$$= \frac{1}{r^{\frac{k}{2}}} \int_0^\infty \eta^k f(\eta) \big(f * y^{-\frac{k}{2}}\big)(\eta)\,d\eta = r^{-\frac{k}{2}} \langle f, f * y^{-\frac{k}{2}} \rangle =$$

$$= r^{-\frac{k}{2}} \langle \tilde{f}, \mathcal{F}_b(f * y^{-\frac{k}{2}}) \rangle = r^{-\frac{k}{2}} \langle \tilde{f}\widetilde{y^{-\frac{k}{2}}}, \tilde{f} \rangle = r^{-\frac{k}{2}} \langle \widetilde{y^{-\frac{k}{2}}}, \tilde{f}^2 \rangle.$$

The formulas for the Fourier transform of the Riesz kernel and for the Fourier transform of a radial function (see for instance [10], p. 155) trivially imply that $\widetilde{y^{-s}} = C_s y^{s-k-1}$ for $s \in (0, k+1)$. Therefore

$$r^{-\frac{k}{2}} \langle \widetilde{y^{-\frac{k}{2}}}, \tilde{f}^2 \rangle = r^{-\frac{k}{2}} \langle y^{\frac{k}{2}-k-1}, \tilde{f}^2 \rangle = \frac{1}{r^{\frac{k}{2}}} \int_0^\infty y^{\frac{k}{2}-1} \tilde{f}^2(y)\,dy. \tag{6}$$

The last integral converges by the virtue of the following reasons:

$\tilde{f}(0) = \int_0^\infty y^k f(y)\,dy < \infty$ (because $f \in L_{1,k}(\mathbf{R}_+)$) and $\frac{k}{2} - 1 > -1$ hence the singularity at the origin is integrable;

$f \in L_{2,k}(\mathbf{R}_+)$ therefore $\tilde{f} \in L_{2,k}(\mathbf{R}_+)$ (see [2]) and $\frac{k}{2} - 1 < k$ so there is a sufficient rate of decay at infinity.

The proved convergence means that all the operations leading from (5) to (6) are really legible i.e.

$$\tilde{f}^2(r) \leq \frac{C}{r^{\frac{k}{2}}} \int_0^\infty y^{\frac{k}{2}-1} \tilde{f}^2(y)\,dy \quad \text{on } (0, +\infty).$$

The following statement is therefore true:

Lemma 3.1 *There exists C such that for any non-negative f from $L_{1,k}(\mathbf{R}_+) \cap L_{2,k}(\mathbf{R}_+)$*

$$\sup_{\mathbf{R}_+} r^{\frac{k}{2}} \tilde{f}^2(r) \leq C \int_0^\infty y^{\frac{k}{2}-1} \tilde{f}^2(y) dy. \tag{7}$$

The inequality (7) is actually the claimed inequality (4) with the particular value of the parameter α: $\alpha = \frac{k}{2}$. Now we will extend (4) to the whole claimed interval.

We define f_γ as follows: $f_\gamma \stackrel{\text{def}}{=} f * y^{\gamma-k-1}$, where $\gamma \stackrel{\text{def}}{=} \frac{1}{2}\left(\frac{k}{2} - \alpha\right) > 0$.
One cannot apply (7) to f_γ immediately because it is not proved that f_γ satisfies the conditions of Lemma 3.1.

Let us prove the validity of (7) for f_γ.

A formal application of the formula for the Fourier-Bessel transformation to the generalized convolution gives: $\widetilde{f_\gamma} = C_\gamma \tilde{f} y^{-\gamma}$. The right-hand side of the last equality belongs to $L_{2,k}(\mathbf{R}_+)$ because $\int_0^\infty y^{k-2\gamma} \tilde{f}^2(y) dy$ converges.

Really
$k - 2\gamma > \frac{k}{2} > 0$ and $\tilde{f}(0) = \int_0^\infty y^k f(y) dy < \infty \Rightarrow$ there is no singularity at the origin at all;
$k - 2\gamma < k$ and $\tilde{f} \in L_{2,k}(\mathbf{R}_+)$ $\big($since $f \in L_{2,k}(\mathbf{R}_+)$, see [2]$\big)$ $\Rightarrow$ there is a sufficient rate of decay at infinity.
Hence $\tilde{f} y^{-\gamma}$ is an ordinary function and therefore (see [3] and [4]) the above-mentioned formal application is valid i.e. $\widetilde{f_\gamma}$ is really equal to $C_\gamma \tilde{f} y^{-\gamma}$ and $\widetilde{f_\gamma}(r) = \int_0^\infty y^k f_\gamma(y) j_\nu(ry) dy$ for almost every positive r.

Then similarly to $\tilde{f}^2(r)$

$$\widetilde{f_\gamma}^2(r) \leq C r^{-\frac{k}{2}} \int_0^\infty \eta^k f_\gamma(\eta) \int_0^\infty y^k f_\gamma(y) T_y^\eta y^{-\frac{k}{2}} dy d\eta \tag{8}$$

under the assumption that the integral at the right-hand side of (8) converges.
Let us prove that it converges indeed.

After the formal changing of the order of the integration and the formal transport of T_y^η to another factor we obtain that the mentioned integral is equal to

$$\int_0^\infty y^{\frac{k}{2}-1} \widetilde{f_\gamma}^2(y) dy = C_\gamma^2 \int_0^\infty y^{\frac{k}{2}-1} \tilde{f}^2(y) y^{-2\gamma} dy = C_\gamma^2 \int_0^\infty y^{\alpha-1} \tilde{f}^2(y) dy.$$

It converges by the virtue of the following reasons:
since α is less than $k+1$ then it converges at infinity because $\tilde{f} \in L_{2,k}(\mathbf{R}_+)$;
since α is positive then it converges at the origin because $\tilde{f}(0)$ equals to the convergent integral $\int_0^\infty y^k f(y) dy$.

So

$$\widetilde{f_\gamma}^2(r) \le \frac{C}{r^{\frac{k}{2}}} \int_0^\infty y^{\frac{k}{2}-1} \widetilde{f_\gamma}^2(y)dy \quad \text{on } (0,+\infty)$$

that is Lemma 3.1 is valid for f_γ indeed.

Now we can substitute f_γ instead of f to (7).
It yields:

$$C_\gamma^2 \sup_{\mathbf{R}_+} r^{\frac{k}{2}-2\gamma} \tilde{f}^2(r) \le C C_\gamma^2 \int_0^\infty y^{\frac{k}{2}-1-2\gamma} \tilde{f}^2(y)dy.$$

Thus the following statement is true:

Theorem 3.2 *There exists C such that for any $\alpha \in (0, \frac{k}{2}]$ and for any non-negative f from $L_{1,k}(\mathbf{R}_+) \cap L_{2,k}(\mathbf{R}_+)$*

$$\sup_{\mathbf{R}_+} r^\alpha \tilde{f}^2(r) \le C \int_0^\infty y^{\alpha-1} \tilde{f}^2(y)dy.$$

Remark 3.3 *Note that in the last inequality C does not depend on α.*

The estimates for the case of mixed Fourier-Bessel transformation (i.e. for the multi-dimensional case) are found in [6]; we quote them here for completeness:
If $n > 1$, then for any $p > -n$ and any $q > -1$ there exists C such that, for any $\alpha \in (0,(n-1)/2)$, any $\beta \in (0,k/2)$ and for $(\alpha,\beta) = ((n-1)/2, k/2)$

$$\|r^{\alpha+\beta} S_{p+\alpha,q+\beta} \tilde{f}\|_\infty \le C \|r^{\alpha+\beta-1} S_{\alpha-n,\beta-1} \tilde{f}\|_1. \tag{9}$$

If $n = 1$, then for any $p > -1$ and any $q > k/2 - 1$ there exists C such that

$$\|r^{\frac{k}{2}} S_{p,q} \tilde{f}\|_\infty \le C \|\eta^{\frac{k}{2}-1} |\tilde{f}_{|\xi=0}|^2\|_1. \tag{10}$$

Remark 3.4 *Belongness of f to $L_{1,k} \cap L_{2,k}$ is assumed to provide the convergence of the right-hand side of the inequality. However, even without this asumption the inequality still holds formally. So we may keep only the assumption of non-negativity, that is (4), (9) and (10) are valid for measures.*

4. Estimates for solutions of singular equations

In this section we apply the above results to estimate norms of solutions of differential equations with singular Bessel operator.
We will start from the case of ordinary equations.
Let u from $L_{2,k}(\mathbf{R}_+)$ satisfies (at least in the sense of distributions) the following equation:

$$P(-B)u = f, \tag{11}$$

where $P(t)$ is a polynomial with real coefficients.

Then $\tilde{u}$ also belongs to $L_{2,k}(\mathbf{R}_+)$ (see [2]) and

$$P(\eta^2)\tilde{u}(\eta) = \tilde{f}(\eta). \tag{12}$$

$P(\eta^2) \in L_{2,k,loc}(\mathbf{R}_+)$, $\tilde{u}(\eta) \in L_{2,k,loc}(\mathbf{R}_+) \Rightarrow \tilde{f}(\eta) \in L_{1,k,loc}(\mathbf{R}_+)$ that is $\tilde{f}(\eta)$ is an ordinary function.
Thus (12) is an equality of ordinary functions and hence the following division is legible:

$$\tilde{u}(\eta) = \frac{\tilde{f}(\eta)}{P(\eta^2)} \in L_{2,k}(\mathbf{R}_+).$$

Now we denote $\frac{\tilde{f}(\eta)}{P(\eta^2)}$ by $g(\eta)$ and assume that g is non-negative ang belongs to $L_{1,k}(\mathbf{R}_+)$.
Then g satisfies the conditions of Theorem 3.1 and $u = \tilde{g}$.
Therefore there exists C such that for any α from $(0, \frac{k}{2}]$

$$\sup_{\mathbf{R}_+} r^\alpha u^2(r) \leq C \int_0^\infty y^{\alpha-1} u^2(y)dy.$$

Thus the following statement is proved:

Theorem 4.1 *There exists C such that if $\frac{\tilde{f}(\eta)}{P(\eta^2)}$ is non-negative and belongs to $L_{1,k}(\mathbf{R}_+)$ then for any solution (at least in the sense of distributions) of (11) belonging to $L_{2,k}(\mathbf{R}_+)$, for any $\alpha \in (0, \frac{k}{2}]$*

$$\|u\|_{\infty,\frac{\alpha}{2}} \leq C\|u\|_{2,\alpha-1}. \tag{13}$$

Under the assumptions of Theorem 4.1 the L_∞-norm of the solution also could be estimated:

$$|u(y)| = |\tilde{g}(y)| = \Big|\int_0^\infty \eta^k j_\nu(y\eta)g(\eta)d\eta\Big| \leq \int_0^\infty \eta^k |g(\eta)|d\eta = \int_0^\infty \eta^k g(\eta)d\eta$$

because g is assumed to be non-negative. The last integral converges (since we assume that g belongs to $\in L_{1,k}(\mathbf{R}_+)$) and equals to

$$\int_0^\infty \eta^k j_\nu(0 \cdot \eta)g(\eta)d\eta = \tilde{g}(0) = u(0)$$

hence $u(0) \geq 0$ and for any non-negative y

$$|u(y)| \leq u(0).$$

Thus the following statement is valid:

Theorem 4.2 *If* $\dfrac{\tilde{f}(\eta)}{P(\eta^2)}$ *is non-negative and belongs to* $L_{1,k}(\mathbf{R}_+)$ *then for any solution (at least in the sense of distributions) of (11) belonging to* $L_{2,k}(\mathbf{R}_+)$

$$u(0) \in [0, +\infty)$$

and

$$\|u\|_\infty \leq u(0). \tag{14}$$

Now we can go to the case of partial differential equations.
We will deal with the following equation:

$$P(-\Delta_B) = f, \tag{15}$$

where $P(t)$ is a polynomial with real coefficients.
Similarly to the proof of Theorem 4.1 inequality (9) yields the following statement:

Theorem 4.3 *Let* $n > 1$, $p > -n$, $q > -1$; *let* $\dfrac{\tilde{f}(\xi, \eta)}{P(|\xi|^2 + \eta^2)}$ *is non-negative and belongs to* $L_{1,k}(\mathbf{R}_+^{n+1})$. *Then there exists* C *such that, if* u *from* $L_{2,k}(\mathbf{R}_+^{n+1})$ *satisfies (15) (at least in the sense of distributions) then for any* α *from* $(p, p+(n-1)/2)$, *any* β *from* $(q, q+k/2)$ *and for* $(\alpha, \beta) = (p + (n-1)/2, q + k/2)$

$$\|r^{\alpha+\beta-p-q} S_{\alpha,\beta} u\|_\infty \leq C \|r^{\alpha+\beta-p-q-1} S_{\alpha-n-p,\beta-q-1} u\|_1 \tag{16}$$

On the same way (10) leads to

Theorem 4.4 *Let* $n = 1$, $p > -1$, $q > k/2 - 1$; *let* $\dfrac{\tilde{f}(\xi, \eta)}{P(\xi^2 + \eta^2)}$ *is non-negative and belongs to* $L_{1,k}(\mathbf{R}_+^2)$. *Then there exists* C *such that, if* u *from* $L_{2,k}(\mathbf{R}_+^2)$ *satisfies (15) (at least in the sense of distributions) then*

$$\|r^{\frac{k}{2}} S_{p,q} u\|_\infty \leq C \|y^{\frac{k}{2}-1} u^2(0, y)\|_1. \tag{17}$$

Remark 4.5 *Note that under the assumptions of Theorem 4.1 (Theorem 4.3, Theorem 4.4 correspondingly) the right-hand side of (13) ((16), (17) correspondingly) always converges (similarly to inequality (4) under the assumptions of Theorem 3.1).*

Remark 4.6 *In the regular case of* $k = 0$ *we have (instead of (4)) a well-known property of the cosine-Fourier transform: the cosine-Fourier transform of a non-negative function achieves its supremum at the origin.*
And the corresponding property of the regular equation $P\Big(-\dfrac{d^2}{dt^2}\Big)u = f$ *follows: if* $\dfrac{\hat{f}(\tau)}{P(\tau^2)}$ *is non-negative and summable (here* ˆ *denotes the cosine-Fourier transform) then any square-summable soulution of the last equation achieves its supremum at the origin (cf. (14)).*

Remark 4.7 *Note, however, that Theorem 4.1 has no analog in the regular case: since $u(0)$ is positive by the virtue of Theorem 4.2 (unless the solution is trivial) then the integral in the right-hand side of (13) would diverge.*

It should be mentioned that although estimate (1) is sharp for weight α in $(0, (n-1)/2]$, additional (besides the non-negativity) assumptions about f can improve that result. P.Sjölin in [9] proved that if f is also radial and compactly supported then weights in the right-hand side and left-hand side of (1) are connected by *an inequality* and belong to a wider interval than the one in [5]; the sharpness of the strengthened results is also proved in [9]. Sjölin's approach (which is different from the basic idea of [5]) is applicable in the non-classical case too: in [8] we obtain the mentioned strengthened estimates for pure Fourier-Bessel transformation and prove their sharpness.
Note, however, that Sjölin's assumption about radiality in fact restricts the consideration to one-variable functions. In the general multi-dimensional case the question about strengthened estimates and their sharpness is still opened even the classical case of Fourier transformation; up to now that problem is solved only for the case of two dimensions (see [11]).

Acknowledgement. The author is grateful to P. Mattila for his attentive concern and useful considerations.

REFERENCES

1. V.V. Katrakhov, On the theory of partial differential equations with singular coefficients, *Sov. Math. Dokl.* **15** (1974), 1230–1234.
2. I.A. Kipriyanov, Fourier-Bessel transforms and imbedding theorems for weight classes, *Proc. Steklov Inst. Math.* **89** (1967), 149–246.
3. I.A. Kipriyanov and A.A. Kulikov, The Paley-Wiener-Schwartz theorem for the Fourier-Bessel transform, *Sov. Math. Dokl.* **37** (1988), 13–17.
4. B.M. Levitan, Expansions in Fourier series and integrals with Bessel functions, *Uspekhi Mat. Nauk* **6** (1951), no. 2, 102–143.
5. P. Mattila, Spherical averages of Fourier transforms of measures with finite energy; dimensions of intersections and distance sets, *Mathematika* **34** (1987), 207–228.
6. A.B. Muravnik, On weighted norm estimates for mixed Fourier-Bessel transforms of non-negative functions, *Integral methods in science and engineering 1996: analytic methods*, Pitman Res. Notes Math. Ser. **374**, Addison Wesley Longman, Harlow, 1997, 119–123.
7. A.B. Muravnik, Fourier-Bessel transformation of measures with several special variables and properties of singular differential equations, *J. of Korean Math. Soc.* **37** (2000), no. 6, 1043–1057.
8. A.B. Muravnik, Fourier-Bessel transformation of compactly supported non-negative functions and estimates of solutions of singular differential equations, to appear in *Functional Differential Equations.*
9. P. Sjölin, Estimates of averages of Fourier transforms of measures with finite energy, *Ann. Acad. Sci. Fenn. Math.* **22** (1997), no. 1, 227–236.

10. E.M. Stein and G. Weiss, *Introduction to Fourier analysis on Euclidean spaces*, Princeton Univ. Press, Princeton, 1971.
11. T. Wolff, Decay of circular means of Fourier transforms of measures, *Internat. Math. Res. Notices* (1999), no. 10, 547–567.

Recent Progress in Functional Analysis
K.D. Bierstedt, J. Bonet, M. Maestre and J. Schmets

A trace theorem for normal boundary conditions

Markus Poppenberg

Fachbereich Mathematik, Universität Dortmund, D-44221 Dortmund, Germany

To Professor Manuel Valdivia on the occasion of his seventieth birthday.

Abstract

Using splitting theory of Vogt we show that any system of normal boundary operators admits a tame linear right inverse in the space of smooth functions on a bounded domain.
MSC 2000 Primary 46E35; Secondary 34B05, 46A04, 46A45

Let $\Omega \subset \mathcal{R}^n$ be a bounded open set with C^∞–boundary. Consider differential operators

$$B_j = B_j(x,\partial) = \sum_{|\beta|\le m_j} b_{j,\beta}(x)\partial^\beta, \qquad j = 1,\dots,p \tag{1}$$

where $b_{j,\beta} \in C^\infty(\overline{\Omega})$. The set $\{B_j\}_{j=1}^p$ is called normal (cf. [3], [8]) if $m_j \neq m_i$ for $j \neq i$ and if $B_j^P(x,\nu) \neq 0$ for $j = 1,\dots,p$ and any $x \in \partial\Omega$ where $\nu = \nu(x)$ denotes the inward normal vector to $\partial\Omega$ at x and B_j^P denotes the principal part of B_j. A normal set $\{B_j\}_{j=1}^p$ is called a Dirichlet system if $m_j = j-1, j = 1,\dots,p$. We can e.g. consider the Dirichlet boundary conditions $u \mapsto (\frac{\partial^{j-1}u}{\partial\nu^{j-1}}{}_{|\partial\Omega})_{j=1}^p$ which give for each $k \geq p$ a trace

$$T_k^p : H^k(\Omega) \to \prod_{i=1}^p H^{k-i+1/2}(\partial\Omega), \quad T_k^p u = \Big\{\Big(\frac{\partial^{j-1}u}{\partial\nu^{j-1}}\Big)_{|\partial\Omega}\Big\}_{j=1}^p. \tag{2}$$

The operators T_k^p are surjective admitting a continuous linear right inverse Z_k^p depending on k ([3], [8]). We shall construct a tame linear right inverse for the induced trace $T^p : H^\infty(\Omega) \to H^\infty(\partial\Omega)^p$ using splitting theory of Vogt (cf. [5], [6], [7]). Here $H^\infty(\Omega), H^\infty(\partial\Omega)$ denote the intersection of all Sobolev spaces $H^k(\Omega), H^k(\partial\Omega)$, respectively. A linear map $A : E \to F$ between Fréchet spaces equipped with fixed systems $|\ |_k$ of seminorms is called tame if there is b such that for any k there is C such that $|Ax|_k \leq C|x|_{k+b}$ for all x.

Let $(F_k)_k, (G_k)_k$ be Hilbert spaces with continuous imbeddings $F_{k+1} \hookrightarrow F_k, G_{k+1} \hookrightarrow G_k$. Let $T_k : F_k \to G_k$ be surjective continuous linear maps such that $(T_k)_{|F_{k+1}} = T_{k+1}$. Let $E_k = N(T_k)$ denote the kernel of T_k, thus $E_{k+1} \hookrightarrow E_k$. We have exact sequences

$$0 \longrightarrow E_k \hookrightarrow F_k \xrightarrow{T_k} G_k \longrightarrow 0. \tag{3}$$

We equip the Fréchet spaces $E = \bigcap_k E_k, F = \bigcap_k F_k, G = \bigcap_k G_k$ with the induced norms. We then have a mapping $T : F \to G$ defined by $Tx = T_k x, x \in F$ where $N(T) = E$.

Lemma. Let E_k, F_k, G_k, T_k and E, F, G, T be as above where (3) is an exact sequence of Hilbert spaces for each k. Assume that there are tame isomorphisms $E \cong \Lambda_1, G \cong \Lambda_2$ for certain power series spaces of infinite type Λ_1, Λ_2. Then the sequence of Fréchet spaces

$$0 \longrightarrow E \hookrightarrow F \xrightarrow{T} G \longrightarrow 0 \tag{4}$$

is exact and splits tamely, i.e., there is a tame linear map $Z : G \to F$ such that $T \circ Z = \mathrm{Id}_G$.

Proof. This follows from the tame splitting theorem [5, Theorem 6.1] (cf. [7]).

Theorem. Let $\{B_j\}_{j=1}^p$ be a normal system on $\partial\Omega$. Then there exists a tame linear map $R : H^\infty(\partial\Omega)^p \to H^\infty(\Omega)$ such that $B_j R g = g_j, j = 1, \ldots, p$, for each $g = \{g_j\}_{j=1}^p$.

Proof. The trace operator T_k^p induces for $k \geq p$ an exact sequences of Hilbert spaces

$$0 \longrightarrow N(T_k^p) \hookrightarrow H^k(\Omega) \xrightarrow{T_k^p} \prod_{i=1}^{p} H^{k-i+1/2}(\partial\Omega) \longrightarrow 0 \tag{5}$$

and a corresponding sequence of Fréchet spaces

$$0 \longrightarrow N(T^p) \hookrightarrow H^\infty(\Omega) \xrightarrow{T^p} H^\infty(\partial\Omega)^p \longrightarrow 0. \tag{6}$$

By usual methods (cf. [8, Theorem 14.1]) we may assume that $\{B_j\}_{j=1}^p = T^p$. By [4, 4.10,4.14] the spaces $H^\infty(\Omega), H^\infty(\partial\Omega)^p$ are each tamely isomorphic to power series spaces of infinite type. We consider Δ^p (Δ the Laplacian) as an operator in $L^2(\Omega)$ with domain $D_p = N(T_{2p}^p) = \{u \in H^{2p}(\Omega) : T_{2p}^p u = 0\}$. The spectrum of Δ^p is discrete (cf. [2, Theorem 17]). We choose λ such that $\Delta^p - \lambda$ is an isomorphism $D_p \to L^2(\Omega)$. Then $\Delta^p - \lambda : N(T^p) \to H^\infty(\Omega)$ is an isomorphism (cf. [8]) which is a tame isomorphism by classical elliptic estimates (cf. [1, Theorem 15.2]). Hence $N(T^p) \cong H^\infty(\Omega)$ tamely isomorphic. By the Lemma the sequence (6) splits tamely. This gives the result.

REFERENCES

1. S. Agmon and A. Douglis and L. Nirenberg, Estimates near the boundary for solutions of elliptic partial differential equations satisfying general boundary conditions, Comm. Pure Appl. Math. 12 (1959) 623-727.
2. F.E. Browder, On the spectral theory of elliptic differential operators I, Math. Ann. 142 (1961) 22-130.
3. J.L. Lions and E. Magenes, Non-homogeneous boundary value problems and applications Vol. I–II, Springer, Berlin, 1972.
4. M. Poppenberg, Tame sequence space representations of spaces of C^∞–functions, Results Math. 29 (1996) 317-334.
5. M. Poppenberg and D. Vogt, A tame splitting theorem for exact sequences of Fréchet spaces, Math. Z. 219 (1995) 141-161.
6. M. Tidten, Fortsetzungen von C^∞–Funktionen, welche auf einer abgeschlossenen Menge in $\mathcal{R}^n$ definiert sind, Manuscripta Math. 27 (1979) 291-312.
7. D. Vogt, Tame spaces and power series spaces, Math. Z. 196 (1987) 523-536.
8. J. Wloka, Partial differential equations, Cambridge University Press, London, 1987.

Recent Progress in Functional Analysis
K.D. Bierstedt, J. Bonet, M. Maestre and J. Schmets

Operators into Hardy spaces and analytic Pettis integrable functions *

Francisco J. Freniche, Juan Carlos García-Vázquez and Luis Rodríguez-Piazza

Departamento de Análisis Matemático, Universidad de Sevilla,
P.O. Box 41080, Sevilla, Spain.

To Professor Manuel Valdivia on the occasion of his seventieth birthday.

Abstract

We present some recent results on operators with values in Hardy spaces and vector valued functions. Some compact operators are constructed which are not representable by functions and the non coincidence of the vector valued Hardy spaces with either the projective or the injective tensor product is proved. We improve some previous results on the failure of Fatou's theorem on radial almost everywhere convergence for analytic Pettis integrable functions.

MCS 2000 Primary 46B28; Secondary 46G10, 46E40, 31A20, 28B05.

1. Introduction

In this work we collect some recent contributions of the authors to the representability of operators and to the Pettis integrability. We shall focus our attention on Hardy spaces. Most of the results we present are contained in [8] and [9], but we have included new constructions improving some of them.

A natural question that arises in the context of operators with values in function spaces is that of the representability by a vector valued function. Namely, if $\mathcal{F}(S)$ is a Banach space of functions on a set S, X is a Banach space and $u : X \to \mathcal{F}(S)$ is an operator, the question is to find a function $F : S \to X^*$ satisfying $ux(\cdot) = \langle F(\cdot), x\rangle$ for all $x \in X$. If the evaluation δ_s at each point of S is a continuous functional on $\mathcal{F}(S)$ then the function F does exist: it is enough to define $\langle F(s), x\rangle = \delta_s(ux)$. This is the case when the function space is a space $\mathcal{C}(S)$ of continuous functions on a compact Hausdorff space. On the other hand, the identity operator on $L^1[0,1]$ is an example of an operator which is not representable.

We are interested in the case when the space of functions is the Hardy space H^p. This space can be viewed in two different ways as a function space: as a space of analytic

*This research has been supported by DGICYT grant PB96-1327.

functions on the unit disk $\mathbf{D}$, denoted by $H^p(\mathbf{D})$, and as a subspace of $L^p(\mathbf{T})$, where $\mathbf{T}$ is the unit circle, denoted this time by $H^p(\mathbf{T})$. It is well known that if $f \in H^p(\mathbf{D})$, then its radial limit $\lim_{r \to 1^-} f(rz)$ exists for almost every $z \in \mathbf{T}$ and defines a function in $H^p(\mathbf{T})$. Conversely, every $f \in H^p(\mathbf{T})$ is the boundary value of its Poisson integral, namely, of the analytic function on the disk given by $f(re^{it}) = P_r * f(e^{it})$, where P_r is the Poisson kernel. This provides the natural identification between these two spaces.

If we think of H^p as $H^p(\mathbf{D})$ then every operator $u : X \to H^p(\mathbf{D})$ can be represented by a function $F : \mathbf{D} \to X^*$ defined by $\langle F(z), x \rangle = ux(z)$, which turns out to be analytic. If we think now the operator u takes values into $H^p(\mathbf{T})$, then we can associate also an analytic X^* valued function F on $\mathbf{D}$, defined by convolution with the Poisson kernel $\langle F(re^{it}), x \rangle = P_r * ux(t)$ [1]. If F had radial limits almost everywhere, which is not always the case [4], then the function on $\mathbf{T}$, still denoted by F, would represent u because it would satisfy $\langle F(\cdot), x \rangle = ux(\cdot)$ for every $x \in X$, on the basis of the classical Fatou theorem. We prove in Section 2 that for any infinite dimensional Banach space X, there exists a compact operator $u : X \to H^p(\mathbf{T})$ which is not representable by any function on $\mathbf{T}$, therefore the induced analytic function does not have radial limits. This result will be derived from a general result stated in Theorem 3, characterizing when every approximable operator $u : X \to \mathcal{F}(S)$ is representable. In Theorem 3, $\mathcal{F}(S)$ is a Banach space of measurable functions on a complete finite measure space, such that the convergence in the norm of $\mathcal{F}(S)$ implies the convergence in measure.

In the third Section we shall see that even in the case that the operator is representable on $\mathbf{T}$ by a Pettis integrable function $F : \mathbf{T} \to X$, the corresponding function on $\mathbf{D}$, the Poisson integral of F, may not have radial limits. This result shows the failure of Fatou's theorem on radial almost everywhere convergence in the setting of the Pettis integral.

The question of extending classical theorems in Harmonic Analysis to the setting of vector valued functions has been considered by a number of authors [1], [2], [3], [13], [12], etc. Section 3 is devoted to the problem of whether the Poisson integral of a Pettis integrable function on the torus $\mathbf{T}$ has radial limits almost everywhere and if there exists a conjugate function. Regarding the Bochner integral, it is well known that the question on Fatou's theorem has a positive answer, by using the extension of Lebesgue's differentiation theorem. On the other hand, some examples of Bochner integrable functions without conjugate functions are known (see for instance [13]). Nevertheless, for Bochner integrable functions taking values into a Banach space with the UMD property, the conjugate function does exist [21]. In [6], some examples are given of Pettis integrable functions not satisfying Lebesgue's differentiation theorem.

Our first contribution [8, Section 1] is the proof that in every infinite dimensional Banach space X, for every $p \in [1, +\infty)$, a strongly measurable p-Pettis integrable function can be constructed which fails Fatou's theorem and, at the same time, does not have a conjugate function, that is, the conjugate operator is not representable by a function. Of course, the Poisson integral of this function can not be analytic, thus the question of constructing analytic Pettis integrable functions without radial limits arises. In [8] and [9] we gave some partial answer to this problem, and we present in Section 3 of this paper the following new improvement of these previous results: for every infinite dimensional Banach space there exists a p-Pettis integrable analytic function $F : \mathbf{T} \to X$ such that $\lim_{r \to 1^-} \|P_r * F(z)\| = +\infty$ for every $z \in \mathbf{T}$. The operator $u : x^* \mapsto x^* \circ F$ takes values in

$H^p(\mathbf{T})$ since F is p-Pettis integrable and analytic. Moreover, the function F represents this operator and its Poisson integral $G(re^{it}) = P_r * F(e^{it})$ is the analytic function on $\mathbf{D}$ representing the operator u. The analytic function G does not have radial limits even in the weak topology.

At the end of Section 2 we give an application to the vector valued Hardy space $H^p(\mathbf{T}, X)$ which is defined as the subspace of the Bochner space $L^p(\mathbf{T}, X)$ consisting of those F whose Fourier coefficients

$$\widehat{F}(k) = \frac{1}{2\pi}\int_0^{2\pi} F(t)\mathrm{e}^{-ikt}\,dt = 0$$

for any frequency $k < 0$. It is shown in Corollary 6 that the injective tensor product $H^p(\mathbf{T})\widetilde{\otimes}_\epsilon X$ does not coincide with $H^p(\mathbf{T}, X)$, for infinite dimensional X. Considering the same question for the projective tensor product $H^p(\mathbf{T})\widetilde{\otimes}_\pi X$, we present in the last Section of this paper our improvements of some results in [2]. We prove in Corollary 14 that the projective norm is strictly finer than the L^p-norm whenever the space X is infinite dimensional. The techniques we use apply to some other classes of closed subspaces of L^p instead of H^p, and so it is in this general setting that the results are presented.

In this article we shall use standard notations as can be found in [5], [18], [19].

2. Representability of operators into function spaces

In this Section, (S, Σ, σ) will be a finite complete measure space and $\mathcal{F}(S)$ will be a Banach space which is a linear subspace of $L^0(S)$, the linear space of (classes of) measurable scalar functions on S. We shall assume that the canonical inclusion from $\mathcal{F}(S)$ into $L^0(S)$ is continuous when $L^0(S)$ is endowed with the topology of the convergence in measure.

First of all we define the concept of representable operator:

Definition 1 *An operator $u : X \longrightarrow \mathcal{F}(S)$ is representable by an X^*-valued function if there exists a function $F : S \longrightarrow X^*$ such that for every $x \in X$, $\langle F(\cdot), x\rangle = ux(\cdot)$ holds almost everywhere.*

Let us observe that the function F which represents the operator u is weak*-scalarly measurable. We shall also say that u is the associated operator to the function F.

Now we state a result which shows the connection between representability and order boundedness. That the condition in Proposition 2 below is sufficient is a consequence of the existence of a lifting [14], as it was pointed out in [10]. That the condition is necessary follows from an easy modification of a result by Stefansson [22, Lemma 2.6].

Proposition 2 *Let X be a Banach space. Let $u : X \longrightarrow \mathcal{F}(S)$ be a bounded operator. Then u is representable by an X^*-valued function if and only if there exists a measurable function $h : S \longrightarrow \mathbf{R}$ such that, for every $x \in X$, $|ux(\cdot)| \le \|x\|h(\cdot)$ almost everywhere.*

In the next result, $L^0(S, Z)$ is the space of strongly measurable functions with values in the Banach space Z, endowed with the convergence in measure. We recall that $L^0(S, Z)$

is metrizable and complete. For instance, a translation invariant metric generating this topology is given by

$$\rho(F,G) = \int_S \inf\{1, \|F(s) - G(s)\|\} \, d\sigma(s),$$

for $F, G \in L^0(S, Z)$.

Let us recall that, given two Banach spaces X and Y, every tensor $u \in Y \otimes X^*$ can be viewed as a finite rank operator $u : X \to Y$ (if $u = y \otimes x^*$ then $ux = \langle x^*, x\rangle y$, for every $x \in X$). The injective norm coincides with the operator norm under this identification. When endowed with this norm the tensor product will be denoted by $Y \otimes_\epsilon X^*$, and the completion $Y \widetilde{\otimes}_\epsilon X^*$ is a subspace of the space of compact operators from X into Y. If X^* or Y has the approximation property, then $Y \widetilde{\otimes}_\epsilon X^*$ is the whole space of compact operators.

Let us consider the natural inclusion I taking $f \otimes x^*$ into $f(\cdot)x^*$ from $\mathcal{F}(S) \otimes_\epsilon X^*$ into $L^0(S, X^*)$. If it is continuous, then as $L^0(S, X^*)$ is complete, it can be extended to the whole $\mathcal{F}(S) \widetilde{\otimes}_\epsilon X^*$. In this case every approximable operator $u : X \to \mathcal{F}(S)$ is representable by a function in $L^0(S, X^*)$. It turns out that the continuity of this injection characterizes when every approximable operator with values in $\mathcal{F}(S)$ is representable, and this is the content of the following theorem. We give here just a sketch of its proof; the details can be found in [9].

Theorem 3 *Let X be a Banach space and let Z be a closed subspace of X^*. Then the following conditions are equivalent:*

1. *The natural inclusion from $\mathcal{F}(S) \otimes Z$ with the injective norm into $L^0(S, Z)$ is a continuous operator.*

2. *Every operator $u \in \mathcal{F}(S) \widetilde{\otimes}_\epsilon Z$ is representable by a strongly measurable Z-valued function.*

3. *Every operator $u \in \mathcal{F}(S) \widetilde{\otimes}_\epsilon Z$ is representable by a X^*-valued function.*

Proof. The argument to prove that (1) implies (2) has already been given. That (3) follows from (2) is obvious. Thus we need only to prove that (3) implies (1). We argue by contradiction, assuming that the inclusion I from $\mathcal{F}(S) \otimes_\epsilon Z$ into $L^0(S, Z)$ is not continuous. Let ρ be the given metric of $L^0(S, Z)$. We have:

Claim. There exists $C > 0$ such that for every closed subspace $Y \subset Z$ with $\dim(Z/Y) < \infty$ and for every $\delta > 0$, there exists an operator $u \in \mathcal{F}(S) \otimes Y$ satisfying $\|u\| < \delta$ and $\rho(Iu, 0) > C$.

The proof of this claim is based upon the facts that Y is complemented in Z and that, if Y_0 is a complement of Y, the inclusion $\mathcal{F}(S) \otimes_\epsilon Y_0 \to L^0(S, Z)$ is continuous, since Y_0 is finite dimensional and the inclusion from $\mathcal{F}(S)$ into $L^0(S)$ is continuous.

Using the claim we construct, by induction, a sequence (E_k) of finite dimensional subspaces of Z, a sequence (u_k) of operators in $\mathcal{F}(S) \otimes E_k$, and an increasing sequence (D_k) of finite subsets of the unit ball B_X such that:

(a) $\rho(Iu_k, 0) > C$,

(b) $\|u_k\| < 1/k^3$,

(c) $\|x^*\| \leq 2\sup_{x\in D_k} |\langle x^*, x\rangle|$, for every $x^* \in E_1 + \ldots + E_k$, and,

(d) $E_k \subset D_n^\perp = \{x^* \in Z : \langle x^*, x\rangle = 0 \text{ for every } x \in D_n\}$, for every $k > n$.

Let $u = \sum_{k=1}^\infty ku_k$ in $\mathcal{F}(S)\widetilde{\otimes}_\epsilon Z$. If u were representable by an X^*-valued function, by Proposition 2, there would exist a measurable real valued function h on S such that $|ux(\cdot)| \leq h(\cdot)$ almost everywhere, for every $x \in B_X$.

We have that, by (d), $u_k x = 0$ for every $x \in D_n$ and every $k > n$. Then

$$ux = \sum_{k=1}^n ku_k x$$

for every $x \in D_n$. As the function $\sum_{k=1}^n kIu_k$ takes values in $E_1 + \ldots + E_n$ we have that

$$\|\sum_{k=1}^n kIu_k(s)\| \leq 2 \sup_{x\in D_n} |ux(s)| \leq 2h(s)$$

almost everywhere for every $n \in \mathbf{N}$.

It follows that $\|nIu_n(s)\| \leq 4h(s)$ almost everywhere for every $n \in \mathbf{N}$, and so (Iu_n) tends to zero in measure, a contradiction to (a). □

Let us observe that there exist spaces $\mathcal{F}(S)$ such that for every Banach space Z the inclusion from $\mathcal{F}(S)\otimes_\epsilon Z$ into $L^0(S, Z)$ is continuous. For instance, the space $\mathcal{C}(S)$ for S a compact space and σ a positive Radon measure, satisfies $\mathcal{C}(S)\widetilde{\otimes}_\epsilon Z = \mathcal{C}(S, Z) \subset L^0(S, Z)$. Also the space $L^\infty(S)$ has the same property since $L^\infty(S)\otimes_\epsilon Z$ is isometrically a subspace of $L^\infty(S, X)$.

This is also the case of the inclusion from $H^p(\mathbf{D})\otimes_\epsilon Z$ into $L^0(\mathbf{D}, Z)$. For the torus $\mathbf{T}$ the situation is different, as we see in the following proposition, which will be needed in order to apply Theorem 3 to the Hardy space $H^p(\mathbf{T})$.

Proposition 4 *Let $1 \leq p < +\infty$. For every infinite dimensional Banach space X the natural inclusion from $H^p(\mathbf{T})\otimes_\epsilon X$ into $L^0(\mathbf{T}, X)$ is not continuous.*

Proof. Let $m_k \geq 1$ be a Hadamard lacunary sequence of non negative integers. The sequence of exponential functions $\mathrm{e}^{im_k t}$ expands a copy of the Hilbert space ℓ^2 inside $H^p(\mathbf{T})$ [23, I.8.20].

On the basis of Dvoretzky's theorem, we can choose in X, for each n, a normalized basic sequence $(x_k)_{k=1}^n$ 2-equivalent to the canonical basis of ℓ_2^n. Consider the function $F_n(t) = \sum_{k=1}^n \mathrm{e}^{im_k t} x_k$ and the operator represented by F_n, $u_n = \sum_{k=1}^n \mathrm{e}^{im_k t}\otimes x_k$.

For every $t \in \mathbf{T}$ we have $\|F_n(t)\| = \|\sum_{k=1}^n \mathrm{e}^{im_k t} x_k\|$ which behaves as $\sqrt{n}$. Thus the sequence (F_n) is not bounded in measure.

On the other hand, for every $x^* \in X^*$ with $\|x^*\| \leq 1$, we have

$$\begin{aligned}\|u_n x^*\|_p &= \|\langle F_n(\cdot), x^*\rangle\|_p = \|\sum_{k=1}^n \langle x_k, x^*\rangle \mathrm{e}^{im_k t}\|_p \\ &\leq C\Big(\sum_{k=1}^n |\langle x_k, x^*\rangle|^2\Big)^{1/2} \leq C \sup_{\|(\alpha_k)\|_2\leq 1} \Big|\sum_{k=1}^n \alpha_k\langle x_k, x^*\rangle\Big| \leq C_1\end{aligned}$$

for some finite constants C and C_1. Thus the sequence (u_n) is bounded in the operator norm. □

Corollary 5 *Let $1 \leq p < +\infty$. Let X be an infinite dimensional Banach space. Then:*

1. *There exists a compact operator $u : X \to H^p(\mathbf{T})$ which is not representable.*
2. *There exists a weak*-to-weak continuous compact operator $u : X^* \to H^p(\mathbf{T})$ which is not representable.*

Proof. It suffices to apply Theorem 3 and Proposition 4. □

From this corollary we shall derive the non coincidence of the vector valued Hardy space $H^p(\mathbf{T}, X)$ with the injective tensor product $H^p(\mathbf{T})\tilde{\otimes}_\epsilon X$. Indeed, the space $H^p(\mathbf{T}, X)$ can be identified with a linear subspace of $H^p(\mathbf{T})\tilde{\otimes}_\epsilon X$ under the map taking $F \in H^p(\mathbf{T}, X)$ into the operator $u : x^* \mapsto x^* \circ F$ which is in $H^p(\mathbf{T})\tilde{\otimes}_\epsilon X$. It is plain that the operator u is representable precisely by the function F. Thus, it suffices to take $u \in H^p(\mathbf{T})\tilde{\otimes}_\epsilon X$ which is not representable to obtain the next result:

Corollary 6 *If X is an infinite dimensional Banach space and $1 \leq p < +\infty$ then $H^p(\mathbf{T}, X) \neq H^p(\mathbf{T})\tilde{\otimes}_\epsilon X$.*

In fact, this result can be also derived directly from Proposition 4, without using Theorem 3. Indeed, if $H^p(\mathbf{T}, X) = H^p(\mathbf{T})\tilde{\otimes}_\epsilon X$, then the L^p-norm and the injective norm will be equivalent, as these spaces are complete. As the convergence in L^p implies the convergence in measure, we would obtain that the inclusion from $H^p(\mathbf{T}) \otimes_\epsilon X$ into $L^0(\mathbf{T}, X)$ is continuous, contradicting Proposition 4.

To finish this Section we shall mention that Theorem 3 can also be applied to the case of $\mathcal{F}(S)$ is an order continuous Köthe function space defined on (S, Σ, σ) in the sense of [18, 1.b.17]. As we always assumed σ to be finite, we recall that $\mathcal{F}(S)$ is order continuous if and only if

$$\lim_{\sigma(A)\to 0} \|f\chi_A\|_{\mathcal{F}(S)} = 0, \text{ for every } f \in \mathcal{F}(S).$$

Actually, the weaker condition

$$\lim_{\sigma(A)\to 0} \|\chi_A\|_{\mathcal{F}(S)} = 0$$

is sufficient to obtain the following theorem whose proof is given in [9]. So the Theorem applies not only to order continuous Köthe function spaces but also to some Orlicz spaces satisfying this condition which are not order continuous.

Theorem 7 *Let $\mathcal{F}(S)$ be a Köthe function space satisfying $\lim_{\sigma(A)\to 0} \|\chi_A\|_{\mathcal{F}(S)} = 0$. If (S, Σ, σ) is not purely atomic and X is an infinite dimensional Banach space, then the natural inclusion from $\mathcal{F}(S) \otimes_\epsilon X$ into $L^0(S, X)$ is not continuous. Thus there exists $u \in \mathcal{F}(S)\tilde{\otimes}_\epsilon X^*$ which is not representable.*

This result improves the one by Robert in [20] where it is shown that if L is an order continuous Banach lattice and X is an infinite dimensional Banach space, then there exists a bounded operator $u : X \to L$ which is not order bounded, that is, there is no $h \in L$ satisfying $|ux| \leq \|x\|h$ for every $x \in X$. Indeed, for non purely atomic order continuous Köthe function spaces, it follows from our results and the fact that every order bounded operator is representable, that there exists a operator $u : X \to L$ which is not order bounded even when considered with values in $L^0(S)$.

3. Analytic p-Pettis integrable functions failing Fatou's theorem

Given a finite measure space (S, Σ, σ) and a Banach space X, it is said that a function $F : S \to X$ is Pettis integrable when:

1. The function $x^* \circ F$ is in $L^1(S)$, for every $x^* \in X^*$, and,
2. for every $A \in \Sigma$, there exists $\int_A F\, d\sigma \in X$, called the Pettis integral of F on A, satisfying
$$\langle x^*, \int_A F(\cdot)\, d\sigma\rangle = \int_A \langle x^*, F(\cdot)\rangle\, d\sigma,$$
for every $x^* \in X^*$.

If F is Pettis integrable, then, given $f \in L^\infty(S)$, the function fF is also Pettis integrable. Therefore, for a Pettis integrable function F on the torus $\mathbf{T}$, the Fourier coefficients $\widehat{F}(k) \in X$ makes sense as a Pettis integral. Also the Poisson integral $P_r * F$ of F have sense as an X-valued function.

Given $p \in [1, +\infty)$, a Pettis integrable function F is said to be p-Pettis integrable when $x^* \circ F \in L^p(S)$ for every $x^* \in X^*$, and the operator $x^* \in X^* \mapsto x^* \circ F \in L^p(S)$ is compact.

In this Section we shall prove that for every infinite dimensional Banach space X, there exists an X-valued, analytic, strongly measurable, Pettis integrable function $F : \mathbf{T} \to X$ failing both Fatou's and Lebesgue's theorems on almost everywhere convergence of Poisson and Fejér means, respectively. We notice that we say that F is analytic whenever $\widehat{F}(k) = 0$ for every $k < 0$, or, equivalently, when the extension of F to the unit disk $\mathbf{D}$, obtained by convolution with the Poisson kernel, namely $F(rz) = P_r * F(z)$ for every $z \in \mathbf{T}$, is an analytic function. We shall show that the funcion F also satisfies to be p-Pettis integrable for every $p \in [1, +\infty)$. Observe that in the case of F being analytic and p-Pettis integrable then $x^* \mapsto x^* \circ F$ takes X^* into $H^p(\mathbf{T})$.

Let $\mathcal{H}$ be the Hilbert transform taking $f \in L^1(\mathbf{T})$ into $\mathcal{H}(f) = \widetilde{f}$, the conjugate function of f. Given an operator $u : X^* \to L^p(\mathbf{T})$, we shall say that the operator $\widetilde{u} : X^* \to L^p(\mathbf{T})$ is the conjugate operator of u if it satisfies $\widetilde{u}(x^*) = \mathcal{H}(ux^*)$ for every $x^* \in X^*$. Observe that if $p \neq 1$ then every operator has a conjugate operator because of the Riesz Theorem on the L^p-boundedness of the Hilbert transform.

Let F be p-Pettis integrable, and let $u_F : X^* \to L^p(\mathbf{T})$ be the operator induced by F, that is, satisfying $u_F x^* = x^* \circ F$ for every $X^* \in X^*$; by definition, $u_F \in L^p(\mathbf{T})\widetilde{\otimes}_\epsilon X$. If u_F has a conjugate operator and it is representable by a function, in the sense of Definition 1, we will say that F admits a conjugate function. The following theorem was obtained in [8]:

Theorem 8 *Let $1 \leq p < +\infty$ and let X be an infinite dimensional Banach space. There exists a strongly measurable p-Pettis integrable function $F : \mathbf{T} \to X$ such that*

$$\lim_{r \to 1^-} \|P_r * F(z)\| = +\infty, \quad \text{uniformly on } z \in \mathbf{T},$$

and F does not admit conjugate function.

As it was remarked there, the operator u_F represented by the function F in Theorem 8, has a conjugate operator, even for $p = 1$, but the conjugate operator is not representable. Actually, there exists an X-valued vector measure μ satisfying that $x^* \circ \mu$ is the measure with density $\mathcal{H}(x^* \circ F)$, for every $x^* \in X^*$.

As an application of Theorem 8 we obtain the non coincidence of two spaces of vector valued harmonic functions on the disk $\mathbf{D}$, the spaces $h_p^w(\mathbf{D}, X)$ and $h_p(\mathbf{D}, X)$. These spaces where considered in [1]. Let us recall that $h_p(\mathbf{D}, X)$ is the Banach space of harmonic X-valued functions F on the unit disk such that

$$\|F\|_{h^p} = \sup_{0 \leq r < 1} \left(\frac{1}{2\pi} \int_0^{2\pi} \|F(re^{it})\|^p \, dt \right)^{1/p} < +\infty.$$

The space $h_w^p(\mathbf{D}, X)$ is the Banach space of harmonic X-valued functions F on the unit disk such that

$$\|F\|_{h_w^p} = \sup_{0 \leq r < 1} \sup_{\|x^*\| \leq 1} \left(\frac{1}{2\pi} \int_0^{2\pi} |x^* \circ F(re^{it})|^p \, dt \right)^{1/p} < +\infty.$$

Corollary 9 *If X is infinite dimensional and $1 \leq p < \infty$ then $h_w^p(\mathbf{D}, X) \neq h^p(\mathbf{D}, X)$.*

The Poisson integral $P_r * F(e^{it})$ of the function F in Theorem 8 is a harmonic non analytic function in $h_w^p(\mathbf{D}, X)$ but not in $h^p(\mathbf{D}, X)$, which does not have radial limits at any point. Thus the question of constructing an analytic Pettis integrable function with the last property arises. Partial results were given in both [8] and [9] and we give here a stronger result with a complete proof.

In the following lemma we shall denote by λ the normalized Lebesgue measure on $\mathbf{T}$, and by Σ the σ-algebra of Borel sets of $\mathbf{T}$. We shall make use of the fact that, when $\mathbf{T}$ is identified with the unit circle, the map $z \mapsto z^M$ preserves the measure λ for every integer $M \neq 0$. We shall also denote by K_l the Fejér kernel and by $\sigma_l(F, z)$ the convolution $K_l * F(z)$, for $z \in \mathbf{T}$ and for F Pettis integrable on $\mathbf{T}$.

Let us notice that we shall regard a bounded analytic function G on $\mathbf{T}$ as the extension to the torus of the analytic function on the unit disk $\mathbf{D}$, denoted also by G, defined by $G(rz) = P_r * G(z)$.

Lemma 10 *Given $a \in (0, 1)$, $\beta > 1$ and natural numbers N, M with $N < M$, there exists a bounded analytic function $G : \mathbf{T} \to \ell_2^N$ satisfying the following conditions:*

1. $\lambda\{z \in \mathbf{T} : \|G(z)\| > \sqrt{N}\} = a$.
2. $\|G(z)\| \leq \beta\sqrt{N}$ *for every* $z \in \mathbf{T}$.

3. $\| \int_A G d\lambda \| \leq \sqrt{1 + a\beta^2}$ *for every* $A \in \Sigma$.

4. For every $z \in \mathbf{T}$ *and every* $r \in [(1/2)^{1/N}, (1/2)^{1/M}]$ *we have*

$$\|G(rz)\| \geq \frac{\sqrt{N}}{2} \exp\left(\frac{a \log \beta}{3}\right).$$

5. For every $z \in \mathbf{T}$ *and every integer* l *such that* $N \leq l < M$ *we have*

$$\|\sigma_l(G, z)\| \geq \frac{\sqrt{N}}{4} \exp\left(a \log \beta\right).$$

Proof. Let us identify the interval $I = (-\pi a, \pi a)$ with an arc in $\mathbf{T}$, so that $\lambda(I) = a$. There exists $\varphi \in H^\infty(\mathbf{D})$ such that its boundary values satisfy almost everywhere $|\varphi(\mathrm{e}^{it})| = \beta$ for $t \in I$, and $|\varphi(\mathrm{e}^{it})| = 1$ otherwise, as well as

$$\log|\varphi(r\mathrm{e}^{it})| = P_r * \log|\varphi|(\mathrm{e}^{it})$$

for all $t \in \mathbf{R}$ and all $r \in [0, 1)$ [7].

Consider the ℓ_2^N-valued function $G(z) = \big(\varphi(z^M), z\varphi(z^M), \cdots, z^{N-1}\varphi(z^M)\big)$ and let us put $f_j(z) = z^{j-1}\varphi(z^M)$ for $j = 1, \cdots, N$. It is clear that

$$\|f_j\|_2 = \left(\int_{\mathbf{T}} |\varphi(z^M)|^2 \, d\lambda(z)\right)^{1/2} = \left(\int_{\mathbf{T}} |\varphi(z)|^2 \, d\lambda(z)\right)^{1/2} \leq \sqrt{1 + \beta^2 a}.$$

Let us notice that $(f_j)_{j=1}^N$ is orthogonal in $H^2(\mathbf{T})$ since their Fourier series have disjoint supports. Indeed, $\widehat{f_j}(k) \neq 0$ implies that k is congruent with $j - 1$ modulus M. We have

$$\Big\| \int_A G d\lambda \Big\|^2 = \sum_{j=1}^N \Big| \int_A f_j d\lambda \Big|^2 = \sum_{j=1}^N \|f_j\|_2^2 \, \Big|\Big\langle \chi_A, \frac{f_j}{\|f_j\|_2} \Big\rangle\Big|^2 \leq \|\chi_A\|_2^2 (1 + \beta^2 a),$$

for all measurable set A and we have shown (3).

To prove (2) and (1), let us observe that $\|G(z)\| = \sqrt{N}|\varphi(z^M)|$ for every $z \in \mathbf{T}$, therefore $\|G(z)\| \leq \beta\sqrt{N}$ and

$$\lambda\left\{z \in \mathbf{T} : \|G(z)\| > \sqrt{N}\right\} = \lambda(I) = a.$$

It is easy to see that $P_r(t) \geq 1/3$ for every $r \in [0, 1/2]$ and $t \in \mathbf{T}$. It follows that if $|z| \leq 1/2$ then $\log|\varphi(z)| \geq (a \log \beta)/3$, because of the choice of the function φ.

Hence, if $r^M \leq 1/2$ and $z \in \mathbf{T}$ then

$$|\varphi((rz)^M))| \geq \exp\left(\frac{a \log \beta}{3}\right),$$

and so we obtain (4), since if we also have $r^N \geq 1/2$, then

$$\|G(rz)\| \geq \left(\sum_{j=0}^{N-1} r^{2j}\right)^{1/2} \exp\left(\frac{a \log \beta}{3}\right) \geq \frac{\sqrt{N}}{2} \exp\left(\frac{a \log \beta}{3}\right).$$

In order to prove (5), we notice that $\widehat{G}(j) = \widehat{\varphi}(0)e_{j+1}$ if $0 \leq j < N$ and $\widehat{G}(j) = 0$ if $N \leq j < M$, where $(e_j)_{j=1}^n$ is the usual basis in ℓ_2^N. Then, if $z \in \mathbf{T}$ and l is an integer with $N \leq l < M$, then

$$\begin{aligned} \|\sigma_l(G,z)\| &= \|\sum_{j=0}^{N-1} \widehat{\varphi}(0)\widehat{K_l}(j)z^j e_{j+1}\| = |\widehat{\varphi}(0)|\Big(\sum_{j=0}^{N-1} \widehat{K_l}(j)^2\Big)^{1/2} \\ &= |\widehat{\varphi}(0)|\Big(\sum_{j=0}^{N-1}(1-\frac{j}{l+1})^2\Big)^{1/2} \geq \frac{|\widehat{\varphi}(0)|\sqrt{N}}{4}. \end{aligned}$$

To finish the proof, let us observe that the Fourier coefficient $\widehat{\varphi}(0)$ is the constant term of the analytic function φ, hence $|\widehat{\varphi}(0)| = \exp(a\log\beta)$. □

Theorem 11 *Let X be an infinite dimensional Banach space. There exists a strongly measurable function $F : \mathbf{T} \to X$ which is analytic and p-Pettis integrable, for every $p \in [1,+\infty)$, satisfying that*

$$\lim_{r\to 1^-} \|P_r * F(z)\| = +\infty, \quad \textit{uniformly in } z \in \mathbf{T}.$$

In particular, for every $z \in \mathbf{T}$, $P_r * F(z)$ does not converge to $F(z)$ as $r \to 1^-$, even in the weak topology.

Moreover, the function F can be constructed satisfying also

$$\lim_{l\to\infty} \|\sigma_l(F,z)\| = +\infty, \quad \textit{uniformly on } z \in \mathbf{T}.$$

Proof. We set $a_k = 1/k^2$, $\beta_k = \exp(k^3)$, $N_k = 2^{k^4}$ and $M_k = N_{k+1}$. Then we apply the former lemma to this choice of the parameters, getting functions $G_k \in H^\infty(\mathbf{T}, \ell_2^{N_k})$ satisfying the conditions listed in the statement of that lemma.

On the other hand, by the Mazur theorem, X contains a closed subspace Y with a Schauder basis with basis constant 2. Applying Dvoretzky's theorem we can split Y as a direct sum $Y = \oplus_{k=1}^\infty Y_k$ where each Y_k is a 4-complemented subspace of Y which contains a subspace 2-isomorphic to $\ell_2^{N_k}$. That is, there exists $i_k : \ell_2^{N_k} \to Y_k$ satisfying $\|h\| \leq \|i_k h\| \leq 2\|h\|$ for every $h \in \ell_2^{N_k}$.

Let $F_k = (a_k/\sqrt{N_k})(i_k \circ G_k) : \mathbf{T} \to Y$.

Since $\|F_k(z)\| > 2a_k$ implies that $\|G_k(z)\| > \sqrt{N_k}$, by condition (1) in the lemma, we have $\lambda\{z \in \mathbf{T} : \|F_k(z)\| > 2/k^2\} \leq 1/k^2$. It follows that $\sum_{k=1}^\infty \|F_k(z)\| < +\infty$ for almost every $z \in \mathbf{T}$. Therefore, we can define a strongly measurable function as the sum $F(z) = \sum_{k=1}^\infty F_k(z)$.

For every Borel set $A \subset \mathbf{T}$, we have

$$\|\int_A F_k\, d\lambda\| = \frac{a_k}{\sqrt{N_k}}\|i_k(\int_A G_k\, d\lambda)\| \leq \frac{2a_k}{\sqrt{N_k}}\sqrt{1+a_k\beta_k^2} \leq C_1\frac{\beta_k}{\sqrt{N_k}}$$

for some finite constant C_1, because of condition (3) in the lemma. It follows that the series $\sum_{k=1}^\infty \|\int_A F_k\, d\lambda\|$ converges.

Given $x^* \in X^*$ with $\|x^*\| \leq 1$, we have that

$$\|x^* \circ F_k\|_1 \leq 4 \sup_{A\in\Sigma} |\int_A x^* \circ F_k \, d\lambda| \leq 4 \sup_{A\in\Sigma} \|\int_A F_k \, d\lambda\| \leq 4C_1 \frac{\beta_k}{\sqrt{N_k}}.$$

We obtain that F is Pettis integrable, and $\int_{\mathbf{T}} fF \, d\lambda = \sum_{k=1}^{\infty} \int_{\mathbf{T}} fG_k \, d\lambda$ for every $f \in L^\infty(\mathbf{T})$. If follows that F is analytic because every F_k is analytic, and

$$F(rz) = P_r * F(z) = \sum_{k=1}^{\infty} P_r * F_k(z) = \sum_{k=1}^{\infty} F_k(rz)$$

for every $r \in [0, 1)$ and every $z \in \mathbf{T}$.

From condition (2) in the lemma and from we have seen above, there exists a finite constant C such that, for every x^* with $\|x^*\| \leq 1$,

$$\|x^* \circ F_k\|_\infty \leq C\beta_k \text{ and } \|x^* \circ F_k\|_1 \leq C\frac{\beta_k}{\sqrt{N_k}}.$$

Interpolating these two inequalities, if $p \in [1, +\infty)$, then for some $\theta = \theta(p) > 0$, we obtain

$$\|x^* \circ F_k\|_p \leq (C\frac{\beta_k}{\sqrt{N_k}})^\theta (C\beta_k)^{1-\theta} = C\frac{\beta_k}{N_k^{\theta/2}}.$$

This shows that the series of associated operators to F_k converges in $H^p(\mathbf{T})\widetilde{\otimes}_\varepsilon X$, yielding that F is p-Pettis integrable.

Condition (4) in the lemma implies that, for r such that $r^{N_k} \geq 1/2$ and $r^{N_{k+1}} \leq 1/2$, and for every $z \in \mathbf{T}$,

$$4\|F(rz)\| \geq \|F_k(rz)\| \geq \frac{a_k}{\sqrt{N_k}}\|G_k(rz)\| \geq \frac{a_k}{2} \exp\left(\frac{a_k \log \beta_k}{3}\right) = \frac{e^{k/3}}{2k^2}$$

hence we obtain that $\lim_{r\to 1^-} \|P_r * F(z)\| = +\infty$ uniformly in $z \in \mathbf{T}$.

That the Fejér means of F also diverge can be derived in the same way, this time from condition (5) in the former lemma. □

Now we give an application of this theorem to spaces of vector valued analytic functions. Recall that $H^p(\mathbf{D}, X)$ is the subspace of $h^p(\mathbf{D}, X)$ which consists of those F which are analytic; and the corresponding subspace of $h^p_w(\mathbf{D}, X)$ will be denoted by $H^p_w(\mathbf{D}, X)$.

Corollary 12 *If X is infinite dimensional and $1 \leq p < \infty$ then $H^p(\mathbf{D}, X) \neq H^p_w(\mathbf{D}, X)$.*

Proof. Let F the function constructed in Theorem 11. Let us consider the analytic function, still denoted by F, defined by $F(rz) = P_r * F(z)$, that is, the Poisson integral of F. As $\|P_r * F(z)\| \to +\infty$ if $r \to 1^-$ uniformly in $z \in \mathbf{T}$, it follows that F is not in $H^p(\mathbf{D}, X)$. On the other hand, F is in $H^p_w(\mathbf{D}, X)$ since $\|x^* \circ (P_r * F)\|_p = \|P_r * (x^* \circ F)\|_p \leq \|x^* \circ F\|_p \leq C_p\|x^*\|$ for some finite constant C_p. □

4. Projective tensor products

Let us recall that, given two Banach spaces X and Y, the projective norm in $X \otimes Y$ generates the finest topology which makes the bilinear map $(x, y) \in X \times Y \mapsto x \otimes y \in X \otimes Y$ continuous [5].

Let (S, Σ, σ) be a finite measure space. Let us recall the classical result by Grothendieck [11] that the Bochner space $L^1(S, X)$ can be identified with $L^1(S)\widetilde{\otimes}_\pi X$ for every Banach space X. For $p > 1$ the situation is very different; in [16] for a class of Banach spaces it is shown to fail that $L^p(S, X)$ and $L^p(S)\widetilde{\otimes}_\pi X$ coincide. Moreover, in [17] a "norm" was introduced so that the completion of $L^p(S) \otimes X$ is isometric to $L^p(S, X)$.

In the space $H^p(\mathbf{T}) \otimes X$ we shall consider the following three norms: the injective norm, the norm which is induced by $H^p(\mathbf{T}, X)$, actually the L^p-norm, and the projective norm. The space $H^p(\mathbf{T}) \otimes X$ is dense in $H^p(\mathbf{T}, X)$ as, for instance, every $F \in H^p(\mathbf{T}, X)$ can be approximated by its Fejér means $\sigma_l(F, \cdot)$ which are in $H^p(\mathbf{T}) \otimes X$, the Fejér kernel K_l being a trigonometric polynomial. Of course, $H^p(\mathbf{T}) \otimes X$ is also dense in $H^p(\mathbf{T})\widetilde{\otimes}_\epsilon X$ and $H^p(\mathbf{T})\widetilde{\otimes}_\pi X$. Therefore, the coincidence of two of these topologies is equivalent to the coincidence of the corresponding completion of $H^p(\mathbf{T}) \otimes X$. Thus, what we have shown in Corollary 6 above is that the injective norm does not coincide with the L^p-norm.

It was proved in [2] that the coincidence between $H^p(\mathbf{D})\widetilde{\otimes}_\pi X$ and $H^p(\mathbf{D}, X)$ implies that X has the analytic Radon Nikodym property. It was also shown in [2] that, if $1 < p \leq 2$, then the spaces $H^p(\mathbf{D})\widetilde{\otimes}_\pi \ell_p$ and $H^p(\mathbf{D}, \ell_p)$ are different. Let us recall that the space $H^p(\mathbf{T}, X)$ can be regarded as a subspace of $H^p(\mathbf{D}, X)$ via convolution with the Poisson kernel and X is said to have the analytic Radon Nikodym property whenever $H^p(\mathbf{T}, X) = H^p(\mathbf{D}, X)$ with this identification. Thus, $H^p(\mathbf{D})\widetilde{\otimes}_\pi X = H^p(\mathbf{D}, X)$ implies that $H^p(\mathbf{T})\widetilde{\otimes}_\pi X = H^p(\mathbf{T}, X)$.

With more generality, let us consider the problem of the equivalence of the L^p-norm and the projective norm, for any closed linear subspace $\mathcal{F}(S)$ of $L^p(S)$, instead of $H^p(\mathbf{T})$. The following characterization result holds:

Theorem 13 *Let $\mathcal{F}(S)$ be a closed subspace of $L^p(S)$. The following conditions are equivalent:*

1. *The L^p-norm and the L^1-norm are not equivalent on $\mathcal{F}(S)$.*
2. *For every infinite dimensional Banach space X, the L^p norm and the projective norm are not equivalent on $\mathcal{F}(S) \otimes X$.*
3. *The L^p norm and the projective norm are not equivalent on $\mathcal{F}(S) \otimes \ell_1$.*

The proof can be found in [9]. It makes use of a result included in [15] characterizing the closed subspaces of $L^p(S)$ for which the L^1-norm is equivalent to the L^p-norm.

As a direct consequence of Theorem 13 and the fact that on $H^p(\mathbf{T})$ the L^p-norm and the L^1-norm are not equivalent, we obtain the following result:

Corollary 14 *If $1 < p < +\infty$ and X is infinite dimensional then $H^p(\mathbf{T}, X)$ does not coincide with $H^p(\mathbf{T})\widetilde{\otimes}_\pi X$.*

In the general case of $\mathcal{F}(S)$ a closed subspace of $L^p(S)$, we can define the analogue of $H^p(\mathbf{T}, X)$ in the following way:

$$\mathcal{F}(S, X) = \{F \in L^p(S, X) : x^* \circ F \in \mathcal{F}(S) \text{ for every } x^* \in X^*\},$$

endowed with the L^p-norm. As the Fourier coefficients $\widehat{F}(k)$ with $k < 0$, of every $F \in L^p(\mathbf{T}, X)$, are null if and only if each $x^* \circ F$ is in $H^p(\mathbf{T})$, it follows that $\mathcal{F}(S, X)$ is $H^p(\mathbf{T}, X)$ if we assume $\mathcal{F}(S)$ to be $H^p(\mathbf{T})$.

It is clear that $\mathcal{F}(S) \otimes X$ can be regarded as a subspace of $\mathcal{F}(S, X)$ by identifying the tensor $f \otimes x$ with the function $f(\cdot)x$. Nevertheless, although in many of the natural situations, $\mathcal{F}(S) \otimes X$ is dense in $\mathcal{F}(S, X)$, as it happens in the case $\mathcal{F}(S) = H^p(\mathbf{T})$, in general we will not have that density property. Actually, it could be a bit surprising that this property, for every $\mathcal{F}(S)$, characterizes the approximation property of X (see [9] for the proof):

Theorem 15 *Let $p \in [1, +\infty)$. A Banach space X has the approximation property if and only if $\mathcal{F}(S) \otimes X$ is dense in $\mathcal{F}(S, X)$ for every finite measure space (S, Σ, σ) and every closed linear subspace $\mathcal{F}(S)$ of $L^p(S)$.*

Regarding functions defined on the disk $\mathbf{D}$, it should be remarked that the Hardy space $H^p(\mathbf{D}, X)$ does not appear as an space $\mathcal{F}(\mathbf{D}, X)$. Nevertheless, as we said before, $H^p(\mathbf{D}, X) = H^p(\mathbf{D})\widetilde{\otimes}_\pi X$ implies $H^p(\mathbf{T}, X) = H^p(\mathbf{T})\widetilde{\otimes}_\pi X$. Thus, from Corollary 14 we obtain $H^p(\mathbf{D}, X) \neq H^p(\mathbf{D})\widetilde{\otimes}_\pi X$, for $1 < p < +\infty$ and X infinite dimensional, improving the results in [2].

Let us mention finally that the same result is proved in [9] for some other spaces of vector valued analytic functions, such as the vector valued Bergman space, which is obtained as an $\mathcal{F}(S, X)$, just by taking $S = \mathbf{D}$ and $\mathcal{F}(S)$ the classical Bergman space of complex valued analytic functions.

REFERENCES

1. O. Blasco, *Boundary values of vector valued harmonic functions considered as operators*, Studia Math. **LXXXVI** (1987), 19-33.
2. O. Blasco, *Boundary values of functions in vector-valued Hardy spaces and geometry of Banach spaces*, J. Funct. Anal. **78** (1988), 346-364.
3. A. V. Bukhvalov, *Hardy spaces of vector valued functions*, J. Soviet Math. **16** (1981), 1051-1059.
4. A. V. Bukhvalov and A. A. Danilevich, *Boundary properties of analytic and harmonic functions with values in Banach spaces*, Math. Notes **32** (1982), 104-110.
5. J. Diestel and J.J. Uhl, *Vector measures*, Math. Surveys **15**, Amer. Math. Soc., Providence, R.I., 1977.
6. S. J. Dilworth and M. Girardi, *Nowhere weak differentiability of the Pettis integral*, Quaestiones. Math. **18** (1995), 365-380.
7. P. L. Duren, *Theory of H^p spaces*, Pure Appl. Math. **38**, Academic Press, New York, 1970.

8. F.J. Freniche, J.C. García-Vázquez and L. Rodríguez-Piazza, *The failure of Fatou's theorem on Poisson integrals of Pettis integrable functions*, J. Funct. Anal. **160** (1998), 28-41.
9. F.J. Freniche, J.C. García-Vázquez and L. Rodríguez-Piazza, *Tensor products and operators in spaces of analytic functions,* J. London Math. Soc., to appear.
10. N. E. Gretsky and J. J. Uhl, *Carleman and Korotkov operators on Banach spaces,* Acta Sci. Math. **43** (1981), 207-218.
11. A. Grothendieck, *Produits tensoriels topologiques et espaces nucléaires*, Mem. Amer. Math. Soc. **16**, 1955.
12. W. Hensgen, *On complementation of vector valued Hardy spaces*, Proc. Amer. Math. Soc. **104** (1988), 1153-1162.
13. W. Hensgen, *On the dual space of $H^p(X)$, $1 < p < \infty$*, J. Funct. Anal. **92** (1990), 348-371.
14. A. Ionescu Tulcea and C. Ionescu Tulcea, *Topics in the theory of lifting*, Ergebnisse der Mathematik und ihrer Grenzgebiete **48**, Springer-Verlag, Berlin, 1969.
15. M. I. Kadec and A. Pelczyński, *Absolutely summing operators in $\mathcal{L}^p$-spaces and their applications*, Studia Math. **29** (1968), 275-326.
16. S. Kwapień, *On operators factorizable through L_p space*, Bull. Soc. Math. France (Mém.) **31-32** (1972), 215-225.
17. V. L. Levin, *Tensor products and functors in categories of Banach spaces defined by KB-lineals*, Trans. Moscow Math. Soc. **20** (1969), 41-77.
18. J. Lindenstrauss and L. Tzafriri, *Classical Banach Spaces vol. I,* Ergebnisse der Mathematik und ihrer Grenzgebiete **92**, Springer Verlag, Berlin, 1977.
19. J. Lindenstrauss and L. Tzafriri, *Classical Banach Spaces vol. II,* Ergebnisse der Mathematik und ihrer Grenzgebiete **97**, Springer Verlag, Berlin, 1979.
20. D. Robert, *Sur les operateurs linéaires qui transforment la boule unité d'un espace de Banach en une partie latticiellement bornée d'un espace de Banach reticulé*, Israel J. Math. **22** (1975), 354-360.
21. J. L. Rubio de Francia, *Probability and Banach Spaces*, Lecture Notes in Math. **1221**, Springer Verlag, Berlin, 1973.
22. G.F. Stefansson, *Pettis integrability*, Trans. Amer. Math. Soc. **330** (1992), 401-418.
23. A. Zygmund, *Trigonometric series, second edition*, Cambridge Univ. Press, 1968.

Recent Progress in Functional Analysis
K.D. Bierstedt, J. Bonet, M. Maestre and J. Schmets

The norm problem for elementary operators

Martin Mathieu

Department of Pure Mathematics, Queen's University Belfast,
Belfast BT7 1NN, Northern Ireland; e-mail: m.m@qub.ac.uk

To Professor Manuel Valdivia on the occasion of his seventieth birthday.

Abstract

Among the outstanding problems in the theory of elementary operators on Banach algebras is the task to find a formula which describes the norm of an elementary operator in terms of the norms of its coefficients. Here we report on the state-of-the-art of the knowledge on this problem along the lines of our talk at the Functional Analysis Valencia 2000 Conference in July 2000.

MCS 2000 Primary 47B47; Secondary 46L07, 47A30

1. Setting the scene

Throughout we denote by A a complex unital Banach algebra and by $\mathcal{E}\ell(A)$ the algebra of its elementary operators. By definition, $S \in \mathcal{E}\ell(A)$ if S is a linear mapping on A of the form $Sx = \sum_{j=1}^{n} a_j x b_j$, where $a = (a_1, \dots, a_n)$ and $b = (b_1, \dots, b_n)$ are n-tuples of elements of A. Clearly, every elementary operator S is bounded, and it is easy to give upper bounds for the norm of S, e.g. the projective tensor norm of $\sum_{j=1}^{n} a_j \otimes b_j$. Many important classes of bounded linear operators on Banach algebras, for instance inner derivations, are included in $\mathcal{E}\ell(A)$, and elementary operators also form the building blocks of more general classes of operators. Hence, they have been studied under a variety of aspects but until now no satisfactory *lower bounds* for the norm of an arbitrary elementary operator, or even a formula describing its norm precisely, have been found.

Take $a, b \in A$ and let $L_a: x \mapsto ax$ and $R_b: x \mapsto xb$ denote left and right multiplication, respectively. It is trivial to show that

$$\|L_a\| = \|a\|, \ \|R_b\| = \|b\|, \quad \text{and} \quad \|L_a + R_a\| = 2\|a\|.$$

In the estimate $\|a^2\| \le \|L_a R_a\| \le \|a\|^2$ both inequalities can be strict, in general. It is *very non-trivial* to describe $\|L_a - R_a\|$; in fact, no formula for the norm of an inner derivation on an arbitrary Banach algebra is known.

What is the problem? Why does the complexity of the question increase so dramatically immediately? The reason seems to be that, although the n-tuples $a = (a_1, \dots, a_n)$, $b = (b_1, \dots, b_n)$ obviously determine the elementary operator S uniquely, S does not determine its coefficients uniquely. Let us pursue this observation in the next section by considering the two-sided multiplication $L_a R_b$.

2. Special cases on special algebras

Suppose $A = C(X)$, the algebra of complex-valued continuous functions on a compact Hausdorff space X with more than one point. Then, there are $a, b \in A$, both non-zero, such that $L_a R_b = 0$. Clearly, this rules out the possibility of a lower bound for the norm $\|L_a R_b\|$ in terms of $\|a\|$ and $\|b\|$. The other extreme is the case $A = B(E)$, the bounded operators on a Banach space E, or slightly more general a closed subalgebra A of $B(E)$ which contains all finite rank operators. In this case, the norm of $L_a R_b$ is maximal, that is, $\|L_a R_b\| = \|a\| \, \|b\|$. Banach algebras A for which there exists a constant $\kappa > 0$, possibly depending on A, such that, for all $a, b \in A$, $\|L_a R_b\| \geq \kappa \, \|a\| \, \|b\|$ have been termed *ultraprime*; elementary operators defined on them turned out to be very well behaved, see [13], [14]. The understanding of ultraprime Banach algebras has been developed rather far (for a recent account on various generalisations see [18]), but it also emerged that some open questions on them revert back to the norm problem itself.

Consequently, from this point of our discussions onward, we will focus our attention on C^*-algebras. Here, the fundamental observation is that, for a C^*-algebra A, the existence of some constant $\kappa > 0$ with the above property entails that

$$\|L_a R_b\| = \|a\| \, \|b\| \qquad (a, b \in A),$$

that is, the maximal possible constant $\kappa = 1$; this in turn is equivalent to the purely algebraic property of A being *prime* (i.e., A has no non-zero orthogonal ideals), see [12, Part I]. From this observation it is but an exercise to show that, for every C^*-algebra A and any $a, b \in A$,

$$\|L_a R_b\| = \sup_{\pi} \|\pi(a)\| \, \|\pi(b)\|$$

where the supremum runs over all irreducible representations π of A.

Interestingly enough, the norm of a generalised inner derivation $L_a - R_b$ can also be determined in the case of a prime C^*-algebra. Based on the theorem by Stampfli from 1970 [22], it is possible to deduce that

$$\|L_a - R_b\| = \inf_{\lambda \in \mathbf{C}} \{\|a - \lambda\| + \|b - \lambda\|\}$$

for all a, b in a prime C^*-algebra A. (For the argument see Fialkow's survey [7], pp. 68–69, which is also relevant to our discussion in other respects.) However, the transition to general C^*-algebras is not so smooth as in the case of a two-sided multiplication. Indeed, the subsequent result has been published so far only in the case $a = b$, that is, of inner derivations [16].

Theorem 2.1. *Let A be a boundedly centrally closed C^*-algebra. For all $a, b \in A$,*

$$\|L_a - R_b\| = \inf_{z \in Z(A)} \sup_{\pi} \{\|\pi(a - z)\| + \|\pi(b - z)\|\},$$

where the supremum is taken over all irreducible representations π of A.

Starting from Stampfli's theorem, it is the same sort of exercise as in the case of two-sided multiplications to obtain an identity as above *with* '$\sup_\pi \inf_{z \in Z(\pi(A))}$' *instead*; indeed, no assumption on the C^*-algebra A is then needed. The improved formula above, however, uses the additional property of A being *boundedly centrally closed*. One way to define this property is to require that the primitive spectrum of A is extremally disconnected (though not necessarily Hausdorff); an equivalent one is the existence of sufficiently many central projections in A in the sense that the annihilator of each ideal in A is of the form eA for some central projection e in A. From these descriptions it follows immediately that all prime C^*-algebras and all von Neumann algebras fall into this class. It is quite non-trivial, however, that every hereditary C^*-subalgebra of a boundedly centrally closed C^*-algebra is boundedly centrally closed itself. All these results, including a proof of Theorem 2.1, are presented in full detail in [3], in particular Chapters 3 and 4. The extension from the case of boundedly centrally closed C^*-algebras to arbitrary C^*-algebras is achieved through the concept of the *bounded central closure* of a C^*-algebra. Combining this extension of Theorem 2.1 with a result of Somerset [20], yields the following definite answer to the norm problem for inner derivations. (For a full account on the historical development, see [3], Section 4.6.)

Corollary 2.2. *Let A be a C^*-algebra. For every element $a \in A$, there exists a local multiplier a' of A such that $a - a'$ is central and the norm of the inner derivation $L_a - R_a$ on A is given by $\|L_a - R_a\| = 2\|a'\|$.*

The approach via local multipliers has substantially contributed to the clarification of a number of problems on elementary operators. For example, the above-mentioned non-uniqueness of the coefficients of an elementary operator S is now completely understood (see [3], Section 5.1). As illustrated in Theorem 2.1, it also helped to advance in the norm problem for elementary operators, if only in simple cases (but compare Section 4 below). Beyond the basic examples of a two-sided multiplication L_aR_b and a generalised inner derivation $L_a - R_b$, very little however is known. Probably the next step would be to determine the norm of the elementary operator $L_aR_b + L_bR_a$, $a, b \in A$. The precise conditions under which the upper bound $2\|a\|\,\|b\|$ is achieved, if A is a prime C^*-algebra, are not known. Indeed, it was shown in [15] that, for a, b elements in a prime C^*-algebra A, a lower bound is provided by $\frac{2}{3}\|a\|\,\|b\|$. Only under additional hypotheses, the better bound $\|a\|\,\|b\|$ could be established so far [6], [21].

3. General case on very special algebras

The understanding of the behaviour of the norm of elementary operators led to other insights into their structural properties. The Fong-Sourour conjecture [8] stated that there are no non-zero compact elementary operators on $C(H)$ for a separable Hilbert space H, where, for every Banach space E, $K(E)$ denotes the ideal of compact operators and $C(E) = B(E)/K(E)$ stands for the Calkin algebra on E. This was confirmed by different methods in [1], [9], and [12, Part II]. The solution provided in [1] uses a strong rigidity property of the norm of an elementary operator on $B(H)$. This was recently extended by Saksman and Tylli [19] in the following theorem.

Theorem 3.1. *Let $A = C(\ell^p)$ denote the Calkin algebra on ℓ^p for $1 < p < \infty$. Let S be an elementary operator on A. Then*

$$\|S\| = \|S\|_e = \|S\|_w,$$

where $\|S\|_e$ and $\|S\|_w$ denote the essential and the weak essential norm of S, respectively, that is, the distance from S to the compact respectively the weakly compact operators on A. Furthermore, the weak essential norm of every elementary operator on $B(\ell^p)$ and the elementary operator it induces on A coincide.

The Saksman-Tylli theorem, whose proof relies on techniques from Banach space geometry, generalises the Apostol-Fialkow theorem in two directions: from the case $p = 2$ to the full range of reflexive ℓ^p's and at the same time removing all additional commutativity assumptions on the coefficients. It also contributes to the generalised Fong-Sourour conjecture which asks for a description of those Banach spaces E with the property that there are no non-zero weakly compact elementary operators on $C(E)$.

Corollary 3.2. *There are no non-zero weakly compact elementary operators on $C(\ell^p)$ for $1 < p < \infty$.*

The exceptional behaviour of elementary operators on Calkin algebras has been observed by many authors in a number of instances over the past decades. An account of this can be found in [17]. In fact, there is a full answer to the norm problem in the case of Hilbert space, but the way it has been achieved is another surprise concerning the properties of the Calkin algebra. This will be discussed in the last section.

4. General case on general C^*-algebras

Within the extended Grothendieck programme, elementary operators arise as follows. There is a canonical mapping from the algebraic tensor product $A \otimes A$ into $B(A)$ defined as follows

$$\Theta: A \otimes A \to B(A), \quad \sum_{j=1}^{n} a_j \otimes b_j \mapsto \sum_{j=1}^{n} L_{a_j} R_{b_j},$$

which extends to a contraction from the projective tensor product $A \hat{\otimes} A$ onto the closure of $\mathcal{E}\ell(A)$ in $B(A)$. In general, Θ is not injective. Let us again confine ourselves with C^*-algebras. Then, Θ is injective if and only if A is prime (see e.g. [3], Section 5.1). However, even in this situation, Θ is no isometry. The reason for this is that one uses the wrong tensor norm on $A \otimes A$. To understand the proper norm, we need to look at the operator space structure of a C^*-algebra.

Let us denote by $S \otimes \mathrm{id}$ the extension of $S \in \mathcal{E}\ell(A)$ to an elementary operator on $A \otimes K(\ell^2)$. The *completely bounded norm* $\|S\|_{cb}$ is defined to be the norm of $S \otimes \mathrm{id}$. The *Haagerup tensor norm* on $A \otimes A$ is defined by

$$\|u\|_h = \inf\Big\{\Big\|\sum_{j=1}^{n} a_j a_j^*\Big\|^{1/2} \Big\|\sum_{j=1}^{n} b_j^* b_j\Big\|^{1/2}\Big\},$$

where the infimum is taken over all possible representations of an element $u \in A \otimes A$ as $u = \sum_{j=1}^{n} a_j \otimes b_j$. It is easily seen that Θ is a contraction from $A \otimes_h A$ into $CB(A)$, the Banach algebra of all completely bounded operators on A. Haagerup proved that Θ is an isometry from $A \otimes_h A$ into $CB(A)$ in the case $A = B(H)$, and this was extended to arbitrary prime C^*-algebras by the author [11]. Chatterjee, Sinclair, and Smith extended these results further and the definite answer was obtained in [2] as follows.

Theorem 4.1. *Let A be a boundedly centrally closed C^*-algebra, and let $A \otimes_{Z,h} A$ denote the central Haagerup tensor product of A with itself. Then Θ induces an isometry from $A \otimes_{Z,h} A$ onto $(\mathcal{E}\ell(A), \|\cdot\|_{cb})$.*

Using the bounded central closure ${}^cA$ of a C^*-algebra A and the fact that the *cb*-norm of $S \in \mathcal{E}\ell(A)$ coincides with the *cb*-norm of its extension to ${}^cA$, we thus obtain a formula for the completely bounded norm $\|S\|_{cb}$ of every elementary operator S on an arbitrary C^*-algebra. This formula is quite satisfactory in that it takes care of the ambiguity in the choice of the coefficients of an elementary operator (the non-injectivity of Θ) as well as the non-commutative structure of a general C^*-algebra, both through the central Haagerup tensor product. The answer to our problem, however, is achieved in a different category.

5. General case on good C^*-algebras

In view of the results in the previous section the question when the norm and the *cb*-norm of elementary operators coincide is close at hand. That is, we intend to find a class of 'good' C^*-algebras A distinguished by the property that $\|S\| = \|S\|_{cb}$ for *every* $S \in \mathcal{E}\ell(A)$. From the general theory we know that commutative C^*-algebras are in this class. Magajna showed that very non-commutative C^*-algebras can share this property. In [10] he proved that the Calkin algebra on a separable Hilbert space has this property, from which it can be deduced that, whenever A is an antiliminal C^*-algebra, then the norm and the *cb*-norm of every elementary operator on A agree. Recall that a C^*-algebra A is said to be *antiliminal* if, for every non-zero positive element $a \in A$, the hereditary C^*-subalgebra $\overline{aAa}$ generated by a is non-abelian. As a consequence of the Glimm-Sakai theorem, for a dense set of irreducible representations π of an antiliminal C^*-algebra A there are no non-zero compact operators contained in $\pi(A)$. This observation, combined with Magajna's theorem, leads to the following characterisation of the 'good' C^*-algebras in our sense.

Theorem 5.1. *For every C^*-algebra A, the following conditions are equivalent.*

(a) *For all $S \in \mathcal{E}\ell(A)$, $\|S\| = \|S\|_{cb}$;*

(b) *There is an exact sequence of C^*-algebras*

$$0 \longrightarrow J \longrightarrow A \longrightarrow B \longrightarrow 0$$

such that J is abelian and B is antiliminal.

This theorem is one of the main results in [5]; the C^*-algebras with the above property are called *antiliminal-by-abelian*. Incidentally, they had already appeared in connection with the study of factorial states in the work by Archbold and Batty [4]. Theorem 5.1 tells us precisely how far the approach via the *cb*-norm leads towards a solution to our original problem.

6. The challenge

The above discussion on the norm problem for elementary operators sheds considerable light on the present situation. In addition, it emerges that the solution to the following problem would yield a complete answer, at least in the case of general C^*-algebras.

Determine the norm for every elementary operator on $B(H)$, H a Hilbert space!

REFERENCES

1. C. Apostol and L. A. Fialkow, Structural properties of elementary operators, Can. J. Math. **38** (1986), 1485–1524.
2. P. Ara and M. Mathieu, On the central Haagerup tensor product, Proc. Edinburgh Math. Soc. **37** (1994), 161–174.
3. P. Ara and M. Mathieu, "Local multipliers of C^*-algebras", Springer-Verlag, London, 2001, to appear.
4. R. J. Archbold and C. J. K. Batty, On factorial states of operator algebras, II, J. Operator Th. **13** (1985), 131–142.
5. R. J. Archbold, M. Mathieu and D. W. B. Somerset, Elementary operators on antiliminal C^*-algebras, Math. Ann. **313** (1999), 609–616.
6. M. Boumazgour, Normes d'opérateurs elementaires, PhD Thesis, Univ. Cadi Ayyad, Marrakech 2000.
7. L. A. Fialkow, Structural properties of elementary operators, in: "Elementary operators and applications", M. Mathieu (ed.), (Proc. Int. Workshop, Blaubeuren 1991), World Scientific, Singapore, 1992; pp. 55–113.
8. C. K. Fong and A. R. Sourour, On the operator identity $\sum A_k X B_k \equiv 0$, Can. J. Math. **31** (1979), 845–857.
9. B. Magajna, On a system of operator equations, Can. Math. Bull. **30** (1987), 200–209.
10. B. Magajna, The Haagerup norm on the tensor product of operator modules, J. Funct. Anal. **129** (1995), 325–348.
11. M. Mathieu, Generalising elementary operators, Semesterbericht Funktionalanalysis **14**, Univ. Tübingen 1988, 133–153.
12. M. Mathieu, Elementary operators on prime C^*-algebras, I, Math. Ann. **284** (1989), 223–244; II, Glasgow Math. J. **30** (1988), 275–284.
13. M. Mathieu, Rings of quotients of ultraprime Banach algebras. With applications to elementary operators, Proc. Centre Math. Anal. Austral. Nat. Univ. **21** (1989), 297–317.
14. M. Mathieu, How to use primeness to describe properties of elementary operators, Proc. Symposia Pure Math., Part II **51** (1990), 195–199.
15. M. Mathieu, More properties of the product of two derivations of a C^*-algebra, Bull. Austr. Math. Soc. **40** (1990), 115–120.
16. M. Mathieu, The *cb*-norm of a derivation, in: "Algebraic methods in operator theory", (R. E. Curto and P. E. T. Jørgensen, eds.), Birkhäuser, Basel, 1994; pp. 144–152.
17. M. Mathieu, Elementary operators on Calkin algebras, in preparation.
18. A. A. Mohammed, Algebras multiplicativamente primas: visión algebráica y analítica, PhD Thesis, Univ. Granada, Granada 2000.
19. E. Saksman and H.-O. Tylli, The Apostol-Fialkow formula for elementary operators on Banach spaces, J. Funct. Anal. **161** (1999), 1–26.
20. D. W. B. Somerset, The proximinality of the centre of a C^*-algebra, J. Approx. Theory **89** (1997), 114–117.
21. L. L. Stachó and B. Zalar, Uniform primeness of the Jordan algebra of symmetric operators, Proc. Amer. Math. Soc. **126** (1998), 2241–2247.
22. J. G. Stampfli, The norm of a derivation, Pacific J. Math. **33** (1970), 737–747.

Recent Progress in Functional Analysis
K.D. Bierstedt, J. Bonet, M. Maestre and J. Schmets

Problems on Boolean algebras of projections in locally convex spaces

W.J. Ricker *

School of Mathematics, University of New South Wales,
Sydney, NSW, 2052, Australia

To Professor Manuel Valdivia on the occasion of his seventieth birthday.

Abstract

The theory of Boolean algebras of projections $\mathcal{B}$ in a Banach space X was developed by W. Bade, N. Dunford and others, and is by now well understood. For X a non-normable locally convex space the situation is fundamentally different. Genuinely new phenomena occur which cannot be overcome by simply replacing a norm with a family of seminorms and mimicking the Banach space arguments. Although many of the Banach space results have been successfully extended to the locally convex setting over the past 10-15 years, there remain several major "results" which remain resistant. We discuss some of these open problems and highlight the intimate connections between topological and geometric properties of X and order and completeness properties of $\mathcal{B}$. The solution to some of these problems will invariably rely on methods and techniques coming from the theory of locally convex spaces.

MCS 2000 Primary 47L10; Secondary 46A03, 47L05

Introduction

Let X be a locally convex Hausdorff space (briefly, lcHs) and $L(X)$ be the space of all continuous linear operators of X into itself. To stress when $L(X)$ is equipped with the strong (resp. weak) operator topology τ_s (resp. τ_w) we write $L_s(X)$, (resp. $L_w(X)$). The zero and identity operator on X are denoted by 0 and I, respectively. The space of all continuous linear functionals on any lcHs X is denoted by X'. A family $\mathcal{B} \subseteq L(X)$ of commuting projections which contains 0 and I is a *Boolean algebra* (briefly, B.a.) if it contains $I - Q_1$ and $Q_1 \wedge Q_2 := Q_1Q_2$ and $Q_1 \vee Q_2 := Q_1 + Q_2 - Q_1Q_2$ whenever $Q_1, Q_2 \in \mathcal{B}$. The partial order $\leq$ in $\mathcal{B}$ is then given by $Q_1 \leq Q_2$ (i.e. $Q_1Q_2 = Q_1$) iff $Q_1X \subseteq Q_2X$. The Stone representation theorem guarantees a compact Hausdorff space $\Omega_{\mathcal{B}}$ and a B.a. isomorphism P from $Co(\Omega_{\mathcal{B}})$, the algebra of all closed-open sets in $\Omega_{\mathcal{B}}$, onto $\mathcal{B}$. That is, P is multiplicative (i.e. $P(E \cap F) = P(E)P(F)$ for $E, F \in Co(\Omega_{\mathcal{B}})$),

*Research supported by the Australian Research Council.

satisfies $P(\Omega_{\mathcal{B}}) = I$, and is finitely additive (i.e. $P(\cup_{j=1}^{n} E_j) = \sum_{j=1}^{n} P(E_j)$ for every finite, pairwise disjoint family $\{E_j\}_{j=1}^{n} \subseteq Co(\Omega_{\mathcal{B}})$). Such a P is also called a (finitely additive) spectral measure. If $Co(\Omega_{\mathcal{B}})$ is replaced by an arbitrary algebra of subsets Σ of some non-empty set Ω, then a *(finitely additive) spectral measure* is any multiplicative, finitely additive map $P : \Sigma \longrightarrow L(X)$ satisfying $P(\Omega) = I$. Its range $P(\Sigma) := \{P(E) : E \in \Sigma\}$ is a B.a. of projections in X.Throughout this note by a B.a. $\mathcal{B} \subseteq L(X)$ we always mean a B.a. of projections in the above sense.

In the above generality little can be expected. The theory takes on significance only when $\mathcal{B}$ has additional *completeness* properties (in the B.a. sense) or, if Σ is a *σ-algebra* of sets and P is *τ_s-countably additive*; in this case P is simply called a *spectral measure*. If X is a Hilbert space, then the resolution of the identity of any normal operator T fits into this scheme, with Σ the σ-algebra of all Borel subsets of the spectrum $\Omega := \sigma(T)$ of T. For a Banach space X the above framework incorporates the theory of scalar-type spectral operators as developed by N. Dunford and others. The theory of (complete) B.a.'s of projections and (σ-additive) spectral measures in Banach spaces is by now well understood; see the monographs [8, 9] and the references therein.

The situation in a *non-Banach* lcHs is fundamentally different. Such a basic property as equicontinuity of $P(\Sigma) \subseteq L(X)$, which is automatic if X is a Banach space and P is a τ_s-countably additive spectral measure defined on a σ-algebra Σ, fails in a general lcHs X. The relative τ_w-compactness of $P(\Sigma) \subseteq L_w(X)$, which is always satisfied if X is a Banach space and P is a τ_s-countably additive spectral measure defined on a σ-algebra Σ, fails in a general lcHs X. And, so on. However, various results from the Banach space setting do carry over to the lcHs setting provided certain restrictions are placed on $\mathcal{B}$, P, X and/or $L_s(X)$; see the works of C. Ionescu-Tulcea, F. Maeda and H.H. Schaefer listed in the bibliography of [9] and [4, 5, 6, 7, 20, 21, 22, 31, 32, 34] for more recent results. Whether such restrictions are genuinely needed is often difficult to determine. The ability to provide relevant examples is limited by the depths of ones knowledge concerning the finer points of the theory of lc-spaces.

It seems appropriate at a conference which includes the theory of lc-spaces (and at which many authorities on the topic are present) to take the opportunity to present a series of open problems in the theory of B.a.'s of projections in lcH-spaces. This is especially so since the solution to many of the problems will invariably rely on the techniques of lc-spaces. Equally important is the fact that operator and order theoretic aspects of such problems, on occasions, also stimulate questions in the theory of lc-spaces. This is illustrated by some recent work of J. Bonet, [3]. Not all the problems are of equal difficulty or equal importance. The order in which they are listed is based on ease of presentation and economy of writing.

Some open problems

A B.a. $\mathcal{B} \subseteq L(X)$ is *Bade complete* (resp. *Bade σ-complete*) if it is complete (resp. σ-complete) as an abstract B.a. and if, for every family (resp. countable family) $\{B_\alpha\} \subseteq \mathcal{B}$ we have $(\wedge_\alpha B_\alpha)X = \cap_\alpha B_\alpha X$ and $(\vee_\alpha B_\alpha)X = \overline{\text{span}(\cup_\alpha B_\alpha X)}$, where the bar denotes closure in X. These notions were introduced by W. Bade (without the terminology "Bade"), for Banach spaces, [1,2]. A basic result states any abstractly σ-complete B.a. of projections in a Banach space is equicontinuous (i.e. uniformly bounded), [1]. The same holds in Fréchet lc-spaces, [34; Proposition 1.2], and in the strong dual of any reflexive Fréchet space, [32; Proposition 3.1]. Since the range of a spectral measure is always a Bade σ-complete B.a. [21; Proposition 4.1(iii)], it follows, by considering Banach spaces in their weak topology, that this result fails in arbitrary lcH-spaces, [18; Proposition 4]. All such known counterexamples occur in non-barrelled spaces.

Question 1 *Let X be a barrelled lcHs. Is every abstractly σ-complete (or perhaps, every Bade σ-complete) B.a. of projections $\mathcal{B} \subseteq L(X)$ necessarily equicontinuous ?*

There do exist some *non-barrelled* lcH-spaces X for which *every* abstractly σ-complete B.a. of projections in $L(X)$ is equicontinuous. This is the case for metrizable lcH-spaces, [34; Proposition 1.2]. Not all such spaces are barrelled. There also exist non-barrelled, *non-metrizable*, quasicomplete spaces X with the same property. Indeed, let Y be a reflexive, hereditarily indecomposable Banach space, [11], and X be Y with its weak topology. If $\mathcal{B} \subseteq L(X)$ is any abstractly σ-complete B.a., then $\mathcal{B}$ is also abstractly σ-complete when interpreted as a subset of $L(Y)$. Since a hereditarily indecomposable Banach space cannot contain a copy of c_0 it follows that the closure $\overline{\mathcal{B}}$, of $\mathcal{B}$, in $L_s(Y)$, a Bade complete B.a. of projections, [30; Corollary 1.1]. So, $\overline{\mathcal{B}}$ is a finite subset of $L(Y)$, [27; Proposition 1]. Then $\mathcal{B}$ is also a finite subset of $L(X) = L(Y)$. In particular, $\mathcal{B}$ is equicontinuous in $L(X)$.

In view of the comments prior to Qu.1 it is imperative to understand the intimate connection between Bade complete and Bade σ-complete B.a. 's of projections and ranges of spectral measures. A B.a. $\mathcal{B} \subseteq L(X)$ has the *monotone property* (resp. *σ-monotone property*) if $\lim_\alpha B_\alpha$ exists in $L_w(X)$ and belongs to $\mathcal{B}$ whenever $\{B_\alpha\} \subseteq \mathcal{B}$ is a monotone net (resp. monotone sequence) with respect to the partial order of $\mathcal{B}$. If we require the (apriori) stronger condition that $\lim_\alpha B_\alpha$ exists in $L_s(X)$ and belongs to $\mathcal{B}$, then $\mathcal{B}$ is said to have the *ordered convergence property* (resp. *σ-ordered convergence property*). These notions are discussed in [20, 21]. They are actually equivalent and imply Bade completeness (resp. Bade σ-completeness), [20; Theorem 2]. If, *in addition*, $\mathcal{B} \subseteq L(X)$ is *equicontinuous*, then the above two properties are equivalent to Bade completeness (resp. Bade σ-completeness), [20; p.211].

Question 2 *Does there exist a lcHs X and a non-equicontinuous, Bade complete (respect. Bade σ-complete) B.a. $\mathcal{B} \subseteq L(X)$ which fails to have the monotone (resp. σ-monotone) property?*

There is a "missing condition" making all three notions equivalent. Let $\mathcal{B} \subseteq L(X)$ be a B.a.. A monotone net $\{B_\alpha\} \subseteq \mathcal{B}$ is *small on small sets* if, for every neighbourhood U of 0 in X there is a neighbourhood V of 0 in X such that, for every $x \in V$ there is an index $\alpha(x)$ with $B_\alpha x \in U$ whenever $\alpha \geq \alpha(x)$. If $\{B_\alpha\}$ is convergent in $L_s(X)$ then it is small on small sets. This concept was introduced by B. Nagy in [17]. A B.a. $\mathcal{B} \subseteq L(X)$ is *small* (resp. *σ-small*) if every monotone net (resp. monotone sequence) from $\mathcal{B}$ is small on small sets. If $\mathcal{B}$ happens to be equicontinuous or have the property that every monotone net in $\mathcal{B}$ is τ_s-convergent (e.g. $\mathcal{B}$ has the ordered convergence property), then $\mathcal{B}$ is small. For an example which *fails* to be σ-small, fix $p \in (1,2)$ and let $X := L^p(\mathbb{R})$. Any operator from $L(X)$ which commutes with all translations $\{T_t\}_{t\in\mathbb{R}}$ given by $T_t f : s \mapsto f(s+t)$ for a.e. $s \in \mathbb{R}$ and $f \in X$, is a p-multiplier operator. The collection $\mathcal{B}$ of all projections from $L(X)$ which are p-multipliers is a B.a.. There exists an increasing sequence $\{P_n\}_{n=1}^{\infty} \subseteq \mathcal{B}$ with sup $\{\|P_n\| : n \in \mathbb{N}\} = \infty$, [16]. By taking U to be the unit ball of X it follows from the Uniform Boundedness Principle that $\{P_n\}_{n=1}^{\infty}$ is not small on small sets. So, $\mathcal{B}$ is not σ-small.

Question 3 *Let $\mathcal{B} \subseteq L(X)$ be a B.a. with the property that every monotone net (respect. monotone sequence) from $\mathcal{B}$ is Cauchy in $L_s(X)$. Is $\mathcal{B}$ necessarily small (resp. σ-small)?*

Examples exist which do *not* have the ordered convergence property, but still satisfy the hypothesis of Qu.3. For instance, let $X := \ell^2([0,1])$ and Σ be the Borel subsets of $[0,1]$. For $E \in \Sigma$, let $P(E) \in L(X)$ be the operator of multiplication (pointwise on $[0,1]$) by χ_E. Then $P : \Sigma \longrightarrow L_s(X)$ is a spectral measure and so $\mathcal{B} := P(\Sigma)$ is a B.a. with the σ-monotone property. Since P is a B.a. isomorphism of Σ onto $\mathcal{B}$, it follows if $\{P(E_\alpha)\} \subseteq \mathcal{B}$ is a monotone net, then $P(E_\alpha) \longrightarrow Q_E$ in $L_s(X)$, where $E := \{t \in [0,1] : \lim_\alpha \chi_{E_\alpha}(t) = 1\}$ and $Q_E \in L(X)$ is the operator of multiplication by χ_E. Of course, Q_E belongs to $\mathcal{B}$ iff $E \in \Sigma$, which is *not* always so. Nevertheless, $\{P(E_\alpha)\}$ is always Cauchy in $L_s(X)$. Since $\mathcal{B}$ is equicontinuous, it is small. On the other hand, if $X := \ell^\infty, \Sigma := 2^{\mathbb{N}}$ and $P(E)x = \chi_E x$ for $x \in X$ and $E \in \Sigma$, then $\mathcal{B} := \{P(E) : E \in \Sigma\}$ is an abstractly complete B.a. which fails the hypothesis of Qu.3. Being equicontinuous, $\mathcal{B}$ is small.

The following conditions, for a B.a. $\mathcal{B} \subseteq L(X)$, are equivalent, [20; Theorem 2].

(i) *$\mathcal{B}$ has the σ-monotone (resp. monotone) property.*

(ii) *$\mathcal{B}$ has the σ-ordered (resp. ordered) convergence property.*

(iii) *$\mathcal{B}$ is Bade σ-complete (resp. Bade complete) and σ-small (resp. small).*

Note that the first example discussed after Qu.3 is both Bade σ-complete and small, but is not Bade complete (as it is not even abstractly complete).

To make the precise connection with spectral measures $P : \Sigma \longrightarrow L_s(X)$ we require a further concept. A set $E \in \Sigma$ is *P-null* if $P(E) = 0$. By multiplicativity of P this coincides with $P(F) = 0$ for every $F \in \Sigma$ with $F \subseteq E$. Two sets $E, F \in \Sigma$ are P-equivalent if $E\triangle F := (E\backslash F) \cup (F\backslash E)$ is P-null. The equivalence class of $E \in \Sigma$ is denoted by $[E]$. Let $\Sigma(P) := \{[E] : E \in \Sigma\}$. Since the B.a. operations of Σ transfer to well defined operations in $\Sigma(P)$, it turns out $\Sigma(P)$ is also a B.a.. Moreover, the induced map $\widetilde{P} : \Sigma(P) \longrightarrow P(\Sigma)$ is a B.a. isomorphism. For each continuous seminorm ρ on $L_s(X)$ define a pseudometric d_ρ by

$$d_\rho([E],[F]) = \sup\{\rho(P(A)) : A \in \Sigma,\ A \subseteq E\triangle F\}, \qquad [E],[F] \in \Sigma(P).$$

The topology and uniform structure on $\Sigma(P)$ defined by this family of pseudometrics is denoted by $\tau_s(P)$. Then P is called a *closed spectral measure* if $(\Sigma(P), \tau_s(P))$ is a complete uniform space. This agrees with I. Kluvánek's notion of closedness for arbitrary lc-space valued vector measures, [15; Ch.IV]. Examples of spectral measures which *fail* to be closed can be found in [21, 22], for instance; see also the first example after Qu.3. For a B.a. of projections $\mathcal{B} \subseteq L(X)$ the above equivalences (i)-(iii) are in turn equivalent (see [20]) to;

(iv) *$\mathcal{B}$ is the range of some spectral (resp. closed spectral) measure.*

In view of the equivalences (i)-(iv) it follows that an example of a Bade complete or Bade σ-complete B.a. of projections of the type required by Qu.2 (if it exists) cannot be the range of any spectral measure. In particular, it must fail to be σ-small.

The range of a spectral measure P is a bounded subset of $L_s(X)$. By the Nikodym boundedness theorem, under the additional assumption that X is *quasi-barrelled*, it follows the range of P is equicontinuous in $L(X)$. This is even true for *finitely additive* spectral measures with *bounded range* and domain a *σ-algebra*; see the proof of [19; Proposition 2.5] which is based on Lemma 1.3 of [19].

Question 4 *Let X be a lcHs. If every finitely additive, $L_s(X)$-valued spectral measure with bounded range and defined on a σ-algebra has equicontinuous range is X quasi-barrelled?*

A B.a. $\mathcal{B} \subseteq L(X)$ is *countably decomposable* if every pairwise disjoint family of elements from $\mathcal{B}$ is countable. This implies if $\mathcal{B}$ is Bade σ-complete, then it is Bade complete, [22; Proposition 2.9]. Every Bade σ-complete B.a. $\mathcal{B}$ in a separable, metrizable lc-space is countably decomposable. Indeed, by the comments immediately after Qu.1 we know $\mathcal{B}$ is equicontinuous. So, the equivalences (i)-(iv) above imply (as equicontinuity implies $\mathcal{B}$ is small) that $\mathcal{B} = P(\Sigma)$ for some spectral measure $P : \Sigma \longrightarrow L_s(X)$. By separability of X, the B.a. $\mathcal{B}$ is countably decomposable; see [22; Lemma 2.10] and its proof. An equicontinuous spectral measure (in any lcHs X) whose range is countably decomposable is a closed spectral measure, [22; Corollary 2.9.1]. Equivalences (i)-(iv) then suggest the following problem.

Question 5 *Does every countably decomposable B.a. of projections with the σ-monotone property have the monotone property? Equivalently, is every spectral measure whose range is countably decomposable necessarily closed?*

Let $\mathbb{P}(X)$ be the family of all projections on X belonging to $L(X)$. If $\mathcal{B} \subseteq L(X)$ is an *equicontinuous* B.a. and $\overline{\mathcal{B}}^s$ (resp. $\overline{\mathcal{B}}^w$) denotes the closure of $\mathcal{B}$ in $L_s(X)$ (resp. $L_w(X)$), then $\overline{\mathcal{B}}^s \subseteq \mathbb{P}(X)$ and $\overline{\mathcal{B}}^s$ is again a B.a.. If, in addition, X is quasicomplete and $\mathcal{B}$ is Bade σ-complete, then $\overline{\mathcal{B}}^s$ is actually Bade complete; this follows from [21; Proposition 4.2] after noting $\mathcal{B}$ is the range of some spectral measure (as equicontinuity of $\mathcal{B}$ implies $\mathcal{B}$ is σ-small). For X only sequentially complete, even with $\mathcal{B}$ equicontinuous and Bade σ-complete, it can happen that the B.a. $\overline{\mathcal{B}}^s$ *fails* to be Bade complete, [21; Example 3.7]. For non-equicontinuous Bade σ-complete $\mathcal{B}$ other problems arise; neither $\overline{\mathcal{B}}^s$ nor $\overline{\mathcal{B}}^w$ need even be B.a. 's! Indeed, let H be a Hilbert space and $\mathcal{B} \subseteq L(H)$ be a Bade complete B.a. of selfadjoint projections which contains no atoms ($B \in \mathcal{B}$ is an *atom* if $C \in \mathcal{B}$ with $C \leq B$ implies that $C = 0$ or $C = B$). Let X be the quasicomplete lcHs H equipped with its weak topology $\sigma(H, H')$. Then $\mathcal{B} \subseteq L(X)$ is still Bade complete. However, $\mathcal{B}$ is not a closed subset of $L_s(X)$. Indeed, the closure $\overline{\mathcal{B}}^s$ (i.e. the weak operator topology closure of $\mathcal{B}$ in $L(H)$) is not even contained in $\mathbb{P}(X)$,[10; Lemma 2.3]. So, there cannot exist any B.a. of projections in $L(X) = L(H_w)$ which is τ_s-closed and contains $\mathcal{B}$.

If X is a lcHs, $\mathcal{B} \subseteq L(X)$ is a B.a. with the σ-monotone property and $L_s(X)$ is quasi-complete, then $\overline{\mathcal{B}}^s \cap \mathbb{P}(X)$ has the monotone property; see the proof of [21; Proposition 4.11].

Question 6 (a) *Suppose that $\mathcal{B} \subseteq L(X)$ is a Bade σ-complete B.a. of projections such that $\overline{\mathcal{B}}^s \subseteq \mathbb{P}(X)$. Is $\overline{\mathcal{B}}^s$ Bade complete or, at least, abstractly complete as a B.a.?*

(b) *Does there exist a quasicomplete lcHs X and a Bade σ-complete B.a. of projections $\mathcal{B} \subseteq L(X)$ such that $\mathcal{B}$ is a closed subset of the lcHs $L_s(X)$, but $\mathcal{B}$ is not Bade complete?*

(c) *Does there exist a lcHs X and a Bade σ-complete B.a. of projections $\mathcal{B} \subseteq L(X)$ which is not a bounded subset of $L_s(X)$?*

For X a Fréchet lc-space no example as required in Qu.6(b) can exist; see Proposition 3.5 and Corollary 3.5.2 of [21]. The same is true for Qu.6(c), [34; Proposition 1.2]. If an example exists as required in Qu.6(c), then $\mathcal{B}$ cannot be σ-small. The existence of such an example would also answer Qu.2 as it would fail the σ-monotone property (otherwise $\mathcal{B}$ would be the range of a spectral measure and hence, would be bounded in $L_s(X)$).

Any vector measure with values in a quasicomplete lcHs has relatively weakly compact range, [33]. So, if $L_s(X)$ is quasicomplete, then every spectral measure $P : \Sigma \longrightarrow L_s(X)$ has relatively τ_w-compact range. Without this restriction on $L_s(X)$ the question of which $L_s(X)$-valued spectral measures have relatively τ_w-compact range is delicate, [21; Section 2]. A result of A. Grothendieck, [14; pp.97-98], implies an *equicontinuous* subset $\mathcal{M} \subseteq L(X)$ is relatively τ_w-compact iff $\mathcal{M}(x) := \{Tx : T \in \mathcal{M}\}$ is relatively weakly compact in X, for each $x \in X$. Of course, the "if direction" is valid without equicontinuity of $\mathcal{M}$ as the map $T \mapsto Tx$ is continuous from $L_w(X)$ into $(X, \sigma(X, X'))$, for $x \in X$. The above characterization fails for a general subset $\mathcal{M} \subseteq L(X)$ without the equicontinuity condition on $\mathcal{M}$, [26]. However, it seems not to be known for sets $\mathcal{M}$ of the form $P(\Sigma)$.

Question 7 *Do there exist a lcHs X and a non-equicontinuous spectral measure $P : \Sigma \longrightarrow L_s(X)$ such that $P(\Sigma)(x) \subseteq X$ is relatively weakly compact, for each $x \in X$, but $P(\Sigma)$ is not relatively τ_w-compact in $L(X)$?*

There are sufficient conditions on a lcHs X and a B.a. $\mathcal{B} \subseteq L(X)$ which imply that $\mathcal{B}$ is τ_s-closed, [21, 22]. Indeed, if $\mathcal{B}$ is Bade σ-complete, then in every separable Fréchet lc-space this is the case. An example is given in [13] of an equicontinuous, τ_s-closed B.a. of projections in the separable Banach space c which is not abstractly σ-complete (and hence, cannot be Bade σ-complete). There exist abstractly complete B.a. 's of projections in Hilbert spaces (necessarily non-separable, [30; Corollary 3.1]) which are *not* τ_s-closed, [30; Remark 2]. There also exist Bade complete B.a. 's of projections (necessarily in non-Fréchet lc-spaces) which fail to be τ_s-closed. This was already observed for the example mentioned (in a Hilbert space with its weak topology) just before Qu.6. Further examples can be found in [29; p.364] and [22; Example 2.4]. But, for each one of these "counter-examples" which are also *equicontinuous*, $\mathcal{B}$ is at least *sequentially τ_s-closed* in $L(X)$. That is, if $\{B_n\}_{n=1}^{\infty} \subseteq \mathcal{B}$ is any τ_s-convergent sequence in $L(X)$, say to $B \in L(X)$, then actually $B \in \mathcal{B}$. Is it always the case that every equicontinuous Bade σ-complete B.a. $\mathcal{B} \subseteq L(X)$ is sequentially τ_s-closed? Recall $\mathcal{B}$ is *atomic* if there is a family of atoms $\{B_\alpha\}_{\alpha \in A}$ in $\mathcal{B}$ such that, for every $B \in \mathcal{B}$ there is a subset $C \subseteq A$ with $\sum_{\alpha \in C} B_\alpha = B$. The summability of the series $\sum_{\alpha \in C} B_\alpha$ is meant as the τ_s-limit in $L(X)$ of the net of partial sums over all finite subsets of C (directed by inclusion). It is shown in [29; p.367] if $L_s(X)$ is quasicomplete and $\mathcal{B} \subseteq L(X)$ is any *atomic* B.a. with the σ-monotone property (not necessarily equicontinuous), then $\mathcal{B}$ is sequentially τ_s-closed. An example in [29; pp.369-370] shows the conclusion fails for general non-atomic B.a. 's; for this example the B.a. is *not* equicontinuous.

Question 8 *Is every equicontinuous B.a.* $\mathcal{B} \subseteq L(X)$ *with the* σ*-monotone property sequentially* τ_s*-closed? Even more specific, is there a Bade* σ*-complete B.a.* $\mathcal{B}$ *acting in a non-separable Banach (or Hilbert) space which is not sequentially* τ_s*-closed?*

The counter-example referred to prior to Qu.8 is not equicontinuous; its essential feature is the existence of a *sequence* of projections from $\mathcal{B}$ whose limit is not in $\mathbb{P}(X)$.

Question 9 *Is a B.a. of projections* $\mathcal{B} \subseteq L(X)$ *with the* σ*-monotone property necessarily a sequentially closed subset in* $\mathbb{P}(X)$ *for the relative topology from* $L_s(X)$*?*

It is shown in [29; pp.370-371] that the counter-example mentioned prior to Qu.8 does *not* answer Qu.9. All of the known examples of spectral measures $P : \Sigma \longrightarrow L_s(X)$ which fail to be closed measures have the property that the B.a. of projections $P(\Sigma)$ contains a net of atoms which converges to some element of $\mathbb{P}(X)\backslash P(\Sigma)$.

Question 10 *Does there exist a lcHs* X *and a non-atomic spectral measure* $P : \Sigma \longrightarrow L_s(X)$ *(i.e. the B.a.* $P(\Sigma)$ *is atom free) which fails to be a closed measure?*

If an example of the type required for Qu.10 exists, then there cannot exist any localizable measure $\mu : \Sigma \longrightarrow [0, \infty]$ satisfying $P \ll \mu$, [22; Proposition 2.23].

Let $P : \Sigma \longrightarrow L_s(X)$ be a spectral measure, with Σ a σ-algebra of subsets of some set Ω. A Σ-measurable function $f : \Omega \longrightarrow \mathbb{C}$ is *P-integrable* if $\int_\Omega |f|\, d|\langle Px, x'\rangle| < \infty$ for $x \in X$ and $x' \in X'$, and if there exists $\int_\Omega f dP \in L(X)$ such that $\langle(\int_\Omega f\, dP)x, x'\rangle = \int_\Omega f\, d\langle Px, x'\rangle$ for $x \in X$, $x' \in X'$. Here $\langle Px, x'\rangle$ is the complex measure $E \mapsto \langle P(E)x, x'\rangle$, for $E \in \Sigma$, and $|\langle Px, x'\rangle|$ is its variation measure. This definition (for spectral measures) agrees with that for general lc-space valued vector measures, [15]; see [19; Lemma 1.2]. The space of P-integrable functions is denoted by $\mathcal{L}^1(P)$. Two P-integrable functions f and g are P-equivalent if $\{w \in \Omega : f(w) \neq g(w)\}$ is a P-null set. The quotient space of $\mathcal{L}^1(P)$ modulo P-equivalence is denoted by $L^1(P)$.

Let X be a Banach space. The *only* P-integrable functions are the *P-essentially bounded* ones, [31; Section 4, Remark (1)], that is, those measurable functions f with $|f|_P := \inf\{\|f\chi_E\|_\infty : E \in \Sigma,\, P(E) = I\} < \infty$. The space of all ($P$-equivalence classes) of such functions is denoted by $L^\infty(P)$. The previous fact is false in Fréchet lc-spaces. Indeed, let ω be the Fréchet space of all complex sequences with the topology of pointwise convergence on $\mathbb{N}$, and $P : 2^{\mathbb{N}} \longrightarrow L(\omega)$ be the spectral measure given by $P(E)x = (x_1\chi_E(1), x_2\chi_E(2), \ldots)$ for $E \in 2^{\mathbb{N}}$ and $x = (x_1, x_2, \ldots) \in \omega$. An example of $f \in L^1(P)\backslash L^\infty(P)$ is given by $f(n) = n$, for $n \in \mathbb{N}$, where $\int_{\mathbb{N}} f dP \in L(\omega)$ is specified by $x \mapsto (x_1, 2x_2, 3x_3, \ldots)$ for $x \in \omega$. There also exist examples of (non-normable) Fréchet lc-spaces and non-trivial spectral measures P in such spaces, whose *only* P-integrable functions are those in $L^\infty(P)$; two examples occur in [28; Section 2]. However, for each of the Fréchet spaces in these two examples there exist *other* spectral measures Q for which the inclusion $L^\infty(Q) \subseteq L^1(Q)$ is proper.

Question 11 *Let X be a Fréchet lc-space with the property that* **every** *$L_s(X)$-valued spectral measure P satisfies $L^1(P) = L^\infty(P)$. Is X isomorphic to a Banach space?*

Any Fréchet space containing a copy of ω contains a complemented copy of ω. Taking into account the properties of $P : 2^{\mathbb{N}} \longrightarrow L(\omega)$ described prior to Qu.11, it is clear no Fréchet lc-space containing a copy of ω can fulfill the hypotheses of Qu.11. Beyond the class of Fréchet lc-spaces no "result" along the lines suggested by Qu.11 is possible. Indeed, there exist non-metrizable lcH-spaces X (even quasicomplete) with the property that $L^1(P) = L^\infty(P)$ for every spectral measure P acting on X, [25].

Given a B.a. $\mathcal{B} \subseteq L(X)$, let $\overline{\langle\mathcal{B}\rangle}^s$ denote the closed subalgebra of $L(X)$ generated by $\mathcal{B}$ in $L_s(X)$. If X is quasicomplete, $L_s(X)$ is sequentially complete and $\mathcal{B}$ is equicontinuous, then $\overline{\langle\mathcal{B}\rangle}^s$ has the structure of a Dedekind complete, complex *f-algebra* with separately continuous multiplication. With respect to this order structure there is a family of seminorms on $L_s(X)$ which generate the restriction of τ_s to $\overline{\langle\mathcal{B}\rangle}^s$ and such that $\overline{\langle\mathcal{B}\rangle}^s$ is locally solid, complete and Lebesgue with respect to this topology, [6; Section 2]. As a consequence, the restriction to $\overline{\langle\mathcal{B}\rangle}^s$ of every $\xi \in (L_s(X))'$ has the form

$$T \mapsto \langle Tx, x'\rangle, \qquad T \in \overline{\langle\mathcal{B}\rangle}^s, \tag{$*$}$$

for some $x \in X$ and $x' \in X'$, [6; Proposition 3.2]. For X a Banach space this is due to T.A. Gillespie, [12]. Let $(\overline{\langle\mathcal{B}\rangle}^s)^\sim$ be the *order dual* of the complex Riesz space $\overline{\langle\mathcal{B}\rangle}^s$. A linear functional $\varphi \in (\overline{\langle\mathcal{B}\rangle}^s)^\sim$ is *order continuous* if $\inf|\varphi(T_\alpha)| = 0$ for every decreasing net $T_\alpha \downarrow 0$ in $\overline{\langle\mathcal{B}\rangle}^s$. Denote the set of all order continuous functionals by $(\overline{\langle\mathcal{B}\rangle}^s)_n^\sim$. For X a Banach space and $\mathcal{B} \subseteq L(X)$ a Bade σ-complete B.a. it is known $(\overline{\langle\mathcal{B}\rangle}^s)' = (\overline{\langle\mathcal{B}\rangle}^s)_n^\sim$, that is, every $\varphi \in (\overline{\langle\mathcal{B}\rangle}^s)_n^\sim$ is of the form $(*)$ for some $x \in X$ and $x' \in X'$, [6; Proposition 3.10]. In particular, every order continuous functional on $\overline{\langle\mathcal{B}\rangle}^s$ is automatically τ_s-continuous. This is a classical result of R. Pallu de la Barriére, [24], when $\overline{\langle\mathcal{B}\rangle}^s$ is an abelian W^*-algebra.

Question 12 *Let X be a quasicomplete lcHs with $L_s(X)$ sequentially complete and $\mathcal{B} \subseteq L(X)$ be any equicontinuous, Bade σ-complete B.a.. Is it still the case that $(\overline{\langle\mathcal{B}\rangle}^s)' = (\overline{\langle\mathcal{B}\rangle}^s)_n^\sim$?*

Given a lcHs X let X'_β be the strong dual space of X. Then $L_b(X'_\beta)$ denotes $L(X'_\beta)$ equipped with the topology of uniform convergence on the bounded subsets of X'_β. Let $L_{w^*}(X'_\beta)$ denote $L(X'_\beta)$ equipped with its *weak-star operator topology*, that is, the topology generated by the seminorms $q_{x,x'}(S) = \langle x, Sx'\rangle$, for $S \in L(X'_\beta)$, as x varies through X and x' varies through X'. Given $T \in L(X)$, its dual operator T' is an element of $L(X'_\beta)$.

A result of M. Orhon, [23; Theorem 2], states if X is a Banach space and $\mathcal{B} \subseteq L(X)$ is a Bade complete B.a., then the subalgebra $\{T' : T \in \overline{\langle\mathcal{B}\rangle}^s\} \subseteq L(X')$ is closed in $L_{w^*}(X'_{\|\cdot\|})$. Since $\overline{\langle\mathcal{B}\rangle}^s$ coincides with the closed subalgebra $\overline{\langle\mathcal{B}\rangle}^b$ of $L_b(X)$ generated by $\mathcal{B}$ with respect to the *operator norm*, [9; Ch.XVII], and $\|S\| = \|S'\|$ for all $S \in L(X)$, it follows the subalgebra $\overline{\langle\mathcal{B}'\rangle}^b$ of $L_b(X'_{\|\cdot\|})$ generated by $\mathcal{B}' := \{B' : B \in \mathcal{B}\}$ with respect to the operator norm in $L_b(X'_{\|\cdot\|})$ coincides with $\overline{\langle\mathcal{B}'\rangle}^{w^*}$, i.e. with that generated by $\mathcal{B}'$ in $L_{w^*}(X'_{\|\cdot\|})$. This is surprising as the B.a. of projections $\mathcal{B}'$ need not be Bade σ-complete in $L(X'_{\|\cdot\|})$.

Question 13 *Let X be a Fréchet lc-space and $\mathcal{B} \subseteq L(X)$ be a Bade complete B.a. of projections. Is $\{T' : T \in \overline{\langle \mathcal{B} \rangle^s}\} \subseteq L(X'_\beta)$ a closed subalgebra of $L_{w^*}(X'_\beta)$? Moreover, is it the case that $\overline{\langle \mathcal{B}' \rangle^b} \subseteq L_b(X'_\beta)$ coincides with $\overline{\langle \mathcal{B}' \rangle^{w^*}} \subseteq L_{w^*}(X'_\beta)$?*

REFERENCES

1. Bade, W.G., On Boolean algebras of projections and algebras of operators, *Trans. Amer. Math. Soc.* **80** (1955), 345-359.
2. Bade, W.G., A multiplicity theory for Boolean algebras of projections in Banach spaces, *Trans. Amer. Math. Soc.* **92** (1959), 508-530.
3. Bonet, J., Closed linear maps from a barrelled normed space into itself need not be continuous, *Bull. Austral. Math. Soc.* **57** (1998), 177-179.
4. Dodds, P.G. and de Pagter, B., Orthomorphisms and Boolean algebras of projections, *Math. Z.* **187** (1984), 361-381.
5. Dodds, P.G. and de Pagter, B., Algebras of unbounded scalar-type spectral operators, *Pacific J. Math.* **130** (1987), 41-74.
6. Dodds, P.G., de Pagter, B. and Ricker, W.J., Reflexivity and order properties of scalar-type spectral operators in locally convex spaces, *Trans. Amer. Math. Soc.* **293** (1986), 355-380.
7. Dodds, P.G. and Ricker, W.J., Spectral measures and the Bade reflexivity theorem, *J. Funct. Anal.* **61** (1985), 136-163.
8. Dowson, H.R., *Spectral theory of linear operators*, Academic Press, London, 1978.
9. Dunford, N. and Schwartz, J.T., *Linear operators III: Spectral operators,* Wiley-Interscience, New York, 1971.
10. Dye, H., The unitary structure in finite rings of operators, *Duke Math. J.* **20** (1953), 55-70.
11. Ferenzci, V., A uniformly convex hereditarily indecomposable Banach space, *Israel J. Math.* **102** (1997), 199-225.
12. Gillespie, T.A., Boolean algebras of projections and reflexive algebras of operators, *Proc. London Math. Soc.* **37** (1978), 56-74.
13. Gillespie, T.A., Strongly closed bounded Boolean algebras of projections, *Glasgow Math. J.* **22** (1981), 73-75.
14. Grothendieck, A., Produits tensoriels et éspaces nucléaires, *Mem. Amer. Math. Soc.* No.16, 1955.
15. Kluvánek, I. and Knowles, G., *Vector measures and control systems*, North Holland, Amsterdam, 1976.
16. Mockenhaupt, G. and Ricker, W.J., Idempotent multipliers for $L^p(\mathbb{R})$, *Arch. Math.* **74** (2000), 61-65.
17. Nagy, B., On Boolean algebras of projections and prespectral operators, in: *Operator Theory Adv. Appl.* 6, Birkhäuser, Basel, 1982, pp.145-162.
18. Okada, S. and Ricker, W.J., Spectral measures which fail to be equicontinuous, *Period.*

Math. Hungar. **28** (1994), 55-61.

19. Okada, S. and Ricker, W.J., Continuous extensions of spectral measures, *Colloq. Math.* **71** (1996), 115-132.
20. Okada, S. and Ricker, W.J., Representation of complete Boolean algebras of projections as ranges of spectral measures (also errata in same journal; **63** (1997), 689-693), *Acta Sci. Math. (Szeged)*, **63** (1997), 209-227.
21. Okada, S. and Ricker, W.J., Boolean algebras of projections and ranges of spectral measures, *Dissertationes Math.*, **365**, 33pp., 1997.
22. Okada, S. and Ricker, W.J., Criteria for closedness of spectral measures and completeness of Boolean algebras of projections, *J. Math. Anal. Appl.* **232** (1999), 197-221.
23. Orhon, M., Boolean algebras of commuting projections, *Math. Z.* **183** (1983), 531-537.
24. Pallu de la Barriére, R., Sur les algébres d'opérateurs dans les espaces hilbertiens, *Bull. Soc. Math. France*, **82** (1954), 1-51.
25. Ricker, W.J., Spectral measures and integration: counterexamples, *Semesterbericht Funktionalanalysis Tübingen, Sommersemester*, **16** (1989), 123-129.
26. Ricker, W.J., Weak compactness in spaces of linear operators, Miniconf. on Probability and Analysis, University of New South Wales, 1991, *Proc. Centre Math. Anal. (Canberra),* **29** (1992), 212-221.
27. Ricker, W.J., Well-bounded operators of type (B) in H.I. spaces, *Acta Sci. Math. (Szeged)*, **59** (1994), 475-488.
28. Ricker, W.J., Weak compactness of the integration map associated with a spectral measure, *Indag. Math. (New Series)*, **5** (1994), 353-364.
29. Ricker, W.J., The sequential closedness of σ-complete Boolean algebras of projections, *J. Math. Anal. Appl.* **208** (1997), 364-371.
30. Ricker, W.J., The strong closure of σ-complete Boolean algebras of projections, *Arch. Math. (Basel),* **72** (1999), 282-288.
31. Ricker, W.J., and Schaefer, H.H., The uniformly closed algebra generated by a complete Boolean algebra of projections, *Math. Z.* **201** (1989), 429-439.
32. Ricker, W.J., Resolutions of the identity in Fréchet spaces, *Integral Equations Operator Theory*, to appear .
33. Tweddle, I., Weak compactness in locally convex spaces, *Glasgow Math. J.* **9** (1968), 123-127.
34. Walsh, B., Structure of spectral measures on locally convex spaces, *Trans. Amer. Math. Soc.* **120** (1965), 295-326.

Recent Progress in Functional Analysis
K.D. Bierstedt, J. Bonet, M. Maestre and J. Schmets

Non associative C^*-algebras revisited

Kaidi El Amin and Antonio Morales Campoy[a], and Angel Rodríguez Palacios*[b]

[a]Departamento de Algebra y Análisis Matemático,
Universidad de Almería, Facultad de Ciencias Experimentales
04120-Almería, Spain
elamin@ual.es and amorales@ual.es

[b]Departamento de Análisis Matemático, Facultad de Ciencias,
Universidad de Granada, 18071 Granada, Spain.
apalacio@goliat.ugr.es

To Professor Manuel Valdivia on the occasion of his seventieth birthday.

Abstract
We give a detailed survey of some recent developments of non-associative C^*-algebras. Moreover, we prove new results concerning multipliers and isometries of non-associative C^*-algebras.
MCS 2000 46K70, 46L70

1. Introduction

In this paper we are dealing with non-associative generalizations of C^*-algebras. In relation to this matter, a first question arises, namely how associativity can be removed in C^*-algebras. Since C^*-algebras are originally defined as certain algebras of operators on complex Hilbert spaces, it seems that they are "essentially" associative. However, fortunately, the abstract characterizations of (associative) C^*-algebras given by either Gelfand-Naimark or Vidav-Palmer theorems allows us to consider the working of such abstract systems of axioms in a general non-associative setting.

To be more precise, for a norm-unital complete normed (possibly non associative) complex algebra A, we consider the following conditions:

(VP) (**VIDAV-PALMER AXIOM**). $A = H(A, \mathbf{1}) + iH(A, \mathbf{1})$.

(GN) (**GELFAND-NAIMARK AXIOM**). *There is a conjugate-linear vector space involution $*$ on A satisfying $\mathbf{1}^* = \mathbf{1}$ and $\| a^*a \| = \| a \|^2$ for every a in A.*

*Partially supported by Junta de Andalucía grant FQM 0199

In both conditions, $\mathbf{1}$ denotes the unit of A, whereas, in *(VP)*, $H(A, \mathbf{1})$ stands for the closed real subspace of A consisting of those element h in A such that $f(h)$ belongs to $\mathbb{R}$ whenever f is a bounded linear functional on A satisfying $\| f \| = f(\mathbf{1}) = 1$.

If the norm-unital complete normed complex algebra A above is associative, then *(GN)* and *(VP)* are equivalent conditions, both providing nice characterizations of unital C^*-algebras (see for instance [10, Section 38]). In the general non-associative case we are considering, things begin to be funnier. Indeed, it is easily seen that *(GN)* implies *(VP)* (argue as in the proof of [10, Proposition 12.20]), but the converse implication is not true (take A equal to the Banach space of all 2×2-matrices over $\mathbb{C}$, regarded as operators on the two-dimensional complex Hilbert space, and endow A with the product $a \circ b := \frac{1}{2}(ab+ba)$).

The funny aspect of the non-associative consideration of Vidav-Palmer and Gelfand-Naimark axioms greatly increases thanks to the fact, which is explained in what follows, that Condition *(VP)* (respectively, *(GN)*) on a norm-unital complete normed complex algebra A implies that A is "nearly" (respectively, "very nearly") associative. To specify our last assertion, let us recall some elemental concepts of non-associative algebra. **Alternative algebras** are defined as those algebras A satisfying $a^2b = a(ab)$ and $ba^2 = (ba)a$ for all a, b in A. By Artin's theorem [65, p. 29], an algebra A is alternative (if and) only if, for all a, b in A, the subalgebra of A generated by $\{a, b\}$ is associative. Following [65, p. 141], we define **non-commutative Jordan algebras** as those algebras A satisfying the **Jordan identity** $(ab)a^2 = a(ba^2)$ and the **flexibility** condition $(ab)a = a(ba)$. Non-commutative Jordan algebras are **power-associative** [65, p. 141] (i.e., all single-generated subalgebras are associative) and, as a consequence of Artin's theorem, alternative algebras are non-commutative Jordan algebras. For an element a in a non-commutative Jordan algebra A, we denote by U_a the mapping $b \to a(ab+ba) - a^2b$ from A to A. In Definitions 1.1 and 1.2 immediately below we provide the algebraic notions just introduced with analytic robes.

Definition 1.1. By a **non-commutative JB^*-algebra** we mean a complete normed non-commutative Jordan complex algebra (say A) with a conjugate-linear algebra-involution $*$ satisfying

$$\| U_a(a^*) \| = \| a \|^3$$

for every a in A.

Definition 1.2. By an **alternative C^*-algebra** we mean a complete normed alternative complex algebra (say A) with a conjugate-linear algebra-involution $*$ satisfying

$$\| a^*a \| = \| a \|^2$$

for all a in A.

Since, for elements a, b in an alternative algebra, the equality $U_a(b) = aba$ holds, it is not difficult to realize that alternative C^*-algebras become particular examples of non-commutative JB^*-algebras. In fact alternative C^*-algebras are nothing but those non-commutative JB^*-algebras which are alternative [48, Proposition 1.3]. Now the behaviour of Vidav-Palmer and Gelfand-Naimark axioms in the non-associative setting are clarified by means of Theorems 1.3 and 1.4, respectively, which follow.

Theorem 1.3 ([54, Theorem 12]). *Norm-unital complete normed complex algebras fulfilling Vidav-Palmer axiom are nothing but unital non-commutative JB^*-algebras.*

Theorem 1.4 ([53, Theorem 14]). *Norm-unital complete normed complex algebras fulfilling Gelfand-Naimark axiom are nothing but unital alternative C^*-algebras.*

After Theorems 1.3 and 1.4 above, there is no doubt that both alternative C^*-algebras and non-commutative JB^*-algebras become reasonable non-associative generalizations (the second containing the former) of (possibly non unital) classical C^*-algebras.

The basic structure theory for non-commutative JB^*-algebras is concluded about 1984 (see [3], [11], [48], and [49]). In these papers a precise classification of certain prime non-commutative JB^*-algebras (the so-called "non-commutative JBW^*-factors") is obtained, and the fact that every non-commutative JB^*-algebra has a faithful family of the so-called "Type I" factor representations is proven. When these results specialize for classical C^*-algebras, Type I non-commutative JBW^*-factors are nothing but the (associative) W^*-factors consisting of all bounded linear operators on some complex Hilbert space, and, consequently, Type I factor representations are precisely irreducible representations on Hilbert spaces. Alternative C^*-algebras are specifically considered in [12] and [48], where it is shown that alternative W^*-factors are either associative or equal to the (essentially unique) alternative C^*-algebra $\mathbb{O}_{\mathbb{C}}$ of complex octonions. In fact, as noticed in [59, p. 103], it follows easily from [76, Theorem 9, p. 194] that every prime alternative C^*-algebra is either associative or equal to $\mathbb{O}_{\mathbb{C}}$.

In recent years we have revisited the theory of non-commutative JB^*-algebras and alternative C^*-algebras with the aim of refining some previously known facts, as well as of developing some previously unexplored aspects. Most results got in this goal appear in [39], [40], and [41]. In the present paper we review the main results obtained in the papers just quoted, and prove some new facts.

Section 2 deals with the theorem in [39] that the product p_A of every non-zero alternative C^*-algebra A is a vertex of the closed unit ball of the Banach space of all continuous bilinear mappings from $A \times A$ into A. We note that this result seems to be new even in the particular case that the alternative C^*-algebra A above is in fact associative. If A is only assumed to be a non-commutative JB^*-algebra, then it is easily seen that the above vertex property for p_A can fail. The question whether the vertex property for p_A characterizes alternative C^*-algebras A among non-commutative JB^*-algebras remains an open problem. In any case, if the vertex property for p_A is relaxed to the extreme point property, then the answer to the above question is negative.

In Section 3 we collect a classification of prime non-commutative JB^*-algebras, which generalizes that of non-commutative JBW^*-factors. According to the main result of [40], if A is a prime non-commutative JB^*-algebra, and if A is neither quadratic nor commutative, then there exists a prime C^*-algebra B, and a real number λ with $\frac{1}{2} < \lambda \leq 1$ such that $A = B$ as involutive Banach spaces, and the product of A is related to that of B (denoted by $\square$, say) by means of the equality $ab = \lambda a \square b + (1 - \lambda) b \square a$. We note that prime JB^*-algebras which are either quadratic or commutative are well-understood (see [49, Section 3] and the Zel'manov-type prime theorem for JB^*-algebras [26, Theorem 2.3], respectively).

Following [67, Definition 20.18], we say that a bounded domain Ω in a complex Banach space is **symmetric** if for each x in Ω there exists an involutive holomorphic mapping $\varphi : \Omega \to \Omega$ having x as an isolated fixed point. It is well-known that the open unit balls of C^*-algebras are bounded symmetric domains. It is also folklore that C^*-algebras have approximate units bounded by one. In Section 4 we review the result obtained in [41] asserting that the above two properties characterize C^*-algebras among complete normed associative complex algebras. The key tools in the proof are W. Kaup's materialization (up to biholomorphic equivalence) of bounded symmetric domains as open unit balls of JB^*-triples [43], the Braun-Kaup-Upmeier holomorphic characterization of the Banach spaces underlying unital JB^*-algebras [13], and the Vidav-Palmer theorem (both in its original form [8, Theorem 6.9] and in Moore's reformulation [9, Theorem 31.10]). Actually, applying the non-associative versions of the Vidav-Palmer and Moore's theorems (see Theorem 1.3 and [44], respectively), it is shown in [41] that a complete normed complex algebra is a non-commutative JB^*-algebra if and only if it has an approximate unit bounded by one, and its open unit ball is a bounded symmetric domain.

Sections 5 and 6 are devoted to prove new results. In Section 5 we introduce multipliers on non-commutative JB^*-algebras, and prove that the set $M(A)$ of all multipliers on a given non-commutative JB^*-algebra A becomes a new non-commutative JB^*-algebra. Actually, in a precise categorical sense, $M(A)$ is the largest non-commutative JB^*-algebra which contains A as a closed essential ideal (Theorem 5.6). We note that, if A is in fact an alternative C^*-algebra, then so is $M(A)$.

Section 6 deals with the non-associative discussion of the Kadison-Paterson-Sinclair theorem [47] asserting that surjective linear isometries between C^*-algebras are precisely the compositions of Jordan-$*$-isomorphisms (between the given algebras) with left multiplications by unitary elements in the multiplier C^*-algebra of the range algebra. In this direction we prove (see Propositions 6.3 and 6.8, and Theorem 6.7) that, for a non-commutative JB^*-algebra A, the following assertions are equivalent:

1. Left multiplications on A by unitary elements of $M(A)$ are isometries.
2. A is an alternative C^*-algebra.
3. For every non-commutative JB^*-algebra B, and every surjective linear isometry $F : B \to A$, there exists a Jordan-$*$-isomorphism $G : B \to A$, and a unitary element u in $M(A)$ satisfying $F(b) = uG(b)$ for all b in B.

Section 6 also contains a discussion of the question whether linearly isometric non-commutative JB^*-algebras are Jordan-$*$-isomorphic (see Theorem 6.10 and Corollary 6.12). A similar discussion in the particular unital case can be found in [13, Section 5]. Moreover, we prove that hermitian operators on a non-commutative JB^*-algebra A are nothing but those operators on A which can be expressed as the sum of a left multiplication by a self-adjoint element of $M(A)$, and a Jordan-derivation of A anticommuting with the JB^*-involution of A (Theorem 6.13).

The concluding section of the paper (Section 7) is devoted to notes and remarks.

We devote the last part of the present section to briefly review the relation between non-commutative JB^*-algebras and other close mathematical models. First we note that, by the power-associativity of non-commutative Jordan algebras, every self-adjoint element of a non-commutative JB^*-algebra A is contained in a commutative C^*-algebra. Analogously, by Artin's theorem, every element of an alternative C^*-algebra is contained in a C^*-algebra. Let us also note that, in questions and results concerning a given non-commutative JB^*-algebra A, we often can assume that A is commutative (called then simply a **JB^*-algebra**). The sentence just formulated merits some explanation. For every algebra A, let us denote by A^+ the algebra whose vector space is the one of A and whose product $\circ$ is defined by $a \circ b := \frac{1}{2}(ab + ba)$. With this convention of symbols, the fact is that, if A is a non-commutative JB^*-algebra, then A^+ becomes a JB^*-algebra under the norm and the involution of A. JB^*-algebras were introduced by I. Kaplansky, and studied first by J. D. M. Wright [69] (in the unital case) and M. A. Youngson [74] (in the general case). By the main results in those papers, JB^*-algebras are in a bijective categorical correspondence with the so-called JB-algebras. The correspondence is obtained by passing from each JB^*-algebra A to its self-adjoint part A_{sa}. **JB-algebras** are defined as those complete normed Jordan real algebras B satisfying $\| x \|^2 \leq \| x^2 + y^2 \|$ for all x, y in B. They were introduced by E. M. Alfsen, F. W. Shultz, and E. Stormer [2], and their basic theory is today nicely collected in [29]. Finally, let us shortly comment on the relation between non-commutative JB^*-algebras and JB^*-triples (see Section 4 for a definition). Every non-commutative JB^*-algebra is a JB^*-triple under a triple product naturally derived from its binary product and its JB^*-involution (see [13], [67], and [74]). As a partial converse, every JB^*-triple can be seen as a JB^*-subtriple of a suitable JB^*-algebra [27]. Moreover, alternative C^*-algebras have shown useful in the structure theory of JB^*-triples [31]. JB^*-triples were introduced by W. Kaup [42] in the search of an algebraic setting for the study of bounded symmetric domains in complex Banach spaces.

2. Geometric properties of the products of alternative C^*-algebras

As in the case of C^*-algebras, the algebraic structure of non-commutative JB^*-algebras is closely related to the geometry of the Banach spaces underlying them. Let us therefore begin our work by fixing notation and recalling some basic concepts in the setting of normed spaces.

Let X be a normed space. We denote by S_X, B_X, and X' the unit sphere, the closed unit ball, and the dual space, respectively, of X. $BL(X)$ will denote the normed algebra of all bounded linear operators on X, and I_X will stand for the identity operator on X. Each continuous bilinear mapping from $X \times X$ into X will be called a **product** on X. Each product f on X has a natural norm $\| f \|$ given by $\| f \| := \sup\{\| f(x,y) \| \ : \ x, y \in B_X\}$. We denote by $\Pi(X)$ the normed space of all products on X.

Now, let u be a norm-one element in the normed space X. The set of **states** of X relative to u, $D(X,u)$, is defined as the non empty, convex, and weak*-compact subset of X' given by

$$D(X,u) := \{\phi \in B_{X'} \ : \ \phi(u) = 1\} \ .$$

For x in X, the **numerical range** of x relative to u, $V(X,u,x)$, is given by

$$V(X,u,x) := \{\phi(x) \ : \ \phi \in D(X,u)\} \ .$$

We say that u is a **vertex** of B_X if the conditions $x \in X$ and $\phi(x) = 0$ for all ϕ in $D(X,u)$ imply $x = 0$. It is well-known and easy to see that the vertex property for u implies that

u is an extreme point of B_X. For x in X, we define the **numerical radius** of x relative to u, $v(X,u,x)$, by

$$v(X,u,x) := Max\{\mid \rho \mid \ : \ \rho \in V(X,u,x)\} \ .$$

The **numerical index** of X relative to u, $n(X,u)$, is the number given by

$$n(X,u) := Max\{r \geq 0 \ : \ r \parallel x \parallel \leq v(X,u,x) \text{ for all } x \text{ in } X\} \ .$$

We note that $0 \leq n(X,u) \leq 1$ and that the condition $n(X,u) > 0$ implies that u is a vertex of B_X. Note also that, if Y is a subspace of X containing u, then $n(Y,u) \geq n(X,u)$.

The study of the geometry of norm-unital complex Banach algebras at their units ([8], [9]) takes its first impetus from the celebrated Bohnenblust-Karlin theorem [7] asserting that the unit $\mathbf{1}$ of such an algebra A is a vertex of the closed unit ball of A. As observed in [8, pp. 33-34], the Bohnenblust-Karlin paper actually contains the stronger result that, for such an algebra A, the inequality $n(A,\mathbf{1}) \geq \frac{1}{e}$ holds.

Now let A be a (possibly non unital and/or non associative) complete normed complex algebra. Then the product p_A of A becomes a natural distinguished element of the Banach space $\Pi(A)$ of all products on the Banach space underlying A. Moreover, in most natural examples (for instance, if A has a norm-one unit or is a non-zero non-commutative JB^*-algebra), we have $\parallel p_A \parallel = 1$. In these cases one can naturally wonder if p_A is a vertex of the closed unit ball of $\Pi(A)$. Even if A is a non-commutative JB^*-algebra, the answer to the above question can be negative. Indeed, if B is a C^*-algebra which fails to be commutative, if λ is a real number with $0 < \lambda < 1$, and if we replace the product xy of B with the one $(x,y) \to \lambda xy + (1-\lambda)yx$, then we obtain a non-commutative JB^*-algebra (say A) whose product is not an extreme point (much less a vertex) of $B_{\Pi(A)}$. With $\lambda = \frac{1}{2}$ in the above construction, we even obtain a (commutative) JB^*-algebra with such a pathology. However, in the case that A is in fact an alternative C^*-algebra, the answer to the question we are considering is more than affirmative. Precisely, we have the following theorem.

Theorem 2.1 ([39, Theorem 2.5]). *Let A be a non zero alternative C^*-algebra. Then $n(\Pi(A), p_A)$ is equal to 1 or $\frac{1}{2}$ depending on whether or not A is commutative.*

In order to provide the reader with a sketch of proof of Theorem 2.1, we comment on the background needed in such a proof, putting special emphasis in those results which will be applied later in the present paper. Among them, the more important one is the following.

Theorem 2.2 ([48, Theorem 1.7]). *Let A be a non-commutative JB^*-algebra. Then the bidual A'' of A becomes naturally a unital non-commutative JB^*-algebra containing A as a $*$-invariant subalgebra. Moreover A'' satisfies all multilinear identities satisfied by A.*

In fact, in the proof of Theorem 2.1 we only need the next straightforward consequence of Theorem 2.2.

Corollary 2.3 ([48, Corollary 1.9]). *If A is an alternative C^*-algebra, then A'' becomes naturally a unital alternative C^*-algebra containing A as a $*$-invariant subalgebra.*

It follows from Theorem 2.2 (respectively, Corollary 2.3) that thc unital hull A_1 of a non-commutative JB^*- (respectively, alternative C^*-) algebra A can be seen as a non-commutative JB^*- (respectively, alternative C^*-) algebra for suitable norm and involution extending those of A [48, Corollary 1.10]. In fact we have had to refine this result by proving the following lemma (compare [10, Lemma 12.19 and its proof]).

Lemma 2.4 ([39, Lemma 2.3]). *Let A be a non-commutative JB^*-algebra. For x in A_1, let T_x denote the operator on A defined by $T_x(a) := xa$. Then A_1, endowed with the unique conjugate-linear algebra involution extending that of A and the norm $\| \cdot \|$ given by $\| x \| := \| T_x \|$ for all x in A_1, is a non-commutative JB^*-algebra containing A isometrically.*

Theorem 2.2 (respectively, Corollary 2.3) gives rise naturally to the so-called **non-commutative JBW^*- (respectively, alternative W^*-) algebras**, namely non-commutative JB^*- (respectively, alternative C^*-) algebras which are dual Banach spaces. The fact that *the product of every non-commutative JBW^*-algebra is separately w^*-continuous* [48, Theorem 3.5] will be often applied along this paper. For instance, such a fact, together with Theorem 2.2, yields easily Lemma 2.4 as well as the result that, *if A is a non-commutative JB^*-algebra, and if a is an element of A, then a belongs to the norm-closure of aB_A* [39, Lemma 2.4].

Another background result applied in the proof of Theorem 2.1 is a non-associative generalization of [19, Theorem 1] asserting that, *if A is a non zero non-commutative JB^*-algebra with a unit $\mathbf{1}$, then $n(A, \mathbf{1})$ is equal to 1 or $\frac{1}{2}$ depending on whether or not A is associative and commutative* [53, Theorem 26] (see also [33, Theorem 4]). Since commutative alternative complex algebras are associative [76, Corollary 7.1.2], it follows from the above that, *if A is a non zero alternative C^*-algebra with a unit $\mathbf{1}$, then $n(A, \mathbf{1})$ is equal to 1 or $\frac{1}{2}$ depending on whether or not A is commutative.*

Before to formally attack a sketch of proof of Theorem 2.1, let us note that, given a unital alternative C^*-algebra A, unitary elements of A are defined verbatim as in the associative particular case, that left multiplications on A by unitary elements of A are surjective linear isometries (a consequence of [65, p. 38]), and that, easily (see for instance [12, Theorem 2.10]), the Russo-Dye-Palmer equalities

$$B_A = \overline{co}\{u : u \text{ unitary in } A\} = \overline{co}\{e^{ih} : h \in A_{sa}\}$$

hold for A. Here $\overline{co}$ means closed convex hull and, to be brief, we have written e^{ih} instead of $\exp(ih)$.

Sketch of proof of Theorem 2.1.- Given a set E and a normed algebra B, let us denote by $B(E, B)$ the normed algebra of all bounded functions from E into B (with point-wise operations and the supremum norm). Now, for the non-zero alternative C^*-algebra A, let us consider the chain of linear mappings

$$A_1 \xrightarrow{F_1} BL(A) \xrightarrow{F_2} \Pi(A) \xrightarrow{F_3} \Pi(A'') \xrightarrow{F_4} B((A'')_{sa} \times (A'')_{sa}, A''),$$

where $F_1(z) := T_z$ for every z in A_1, $F_2(T)(a, b) := T(ab)$ for every T in $BL(A)$ and all a, b in A, $F_3(f) := f'''$ (the third Arens transpose of f [4]) for every f in $\Pi(A)$, and $F_4(g)(h,k) := e^{-ih}(g(e^{ih}, e^{ik})e^{-ik})$ for every g in $\Pi(A'')$ and all h, k in $(A'')_{sa}$. It follows easily from the information collected above that, for $i = 1, ..., 4$, F_i is a linear isometry. Moreover, we have $F_1(\mathbf{1}) = I_A$, $F_2(I_A) = p_A$, $F_3(p_A) = p_{A''}$, and $F_4(p_{A''}) = \mathbb{I}$, where $\mathbb{I}$ denotes the constant mapping equal to the unit of A'' on $(A'')_{sa} \times (A'')_{sa}$. Let δ denote either 1 or $\frac{1}{2}$ depending on whether or not A is commutative. Since A_1 and $B((A'')_{sa} \times (A'')_{sa}, A'')$ are alternative C^*-algebras with units $\mathbf{1}$ and $\mathbb{I}$, respectively, and they are commutative if and only if A is, it follows

$$\delta = n(A_1, \mathbf{1}) \geq n(BL(A), I_A) \geq n(\Pi(A), p_A)$$

$$\geq n(\Pi(A''), P_{A''}) \geq n(B((A'')_{sa} \times (A'')_{sa}, A''), \mathbb{I}) = \delta \ . \ \blacksquare$$

The **normed space numerical index**, $N(X)$, of a non-zero normed space X is defined by the equality $N(X) := n(BL(X), I_X)$. The above argument clarifies the proof of Huruya's theorem [32] that, *if A is a non zero C^*-algebra, then $N(A)$ is equal to* 1 *or* $\frac{1}{2}$ *depending on whether or not A is commutative*, and generalizes Huruya's result to the setting of alternative C^*-algebras. In fact, with methods rather similar to those in the proof of Theorem 2.1, we have been able to prove the stronger result that, *if A is a non zero non-commutative JB^*-algebra, then $N(A)$ is equal to* 1 *or* $\frac{1}{2}$ *depending on whether or not A is associative and commutative* [39, Proposition 2.6]. This result was already formulated in [33, Theorem 5] as a direct consequence of Theorem 2.2, the particular case of such a result for unital non-commutative JB^*-algebras [53, Corollary 33], and the claim in [21] that, for every normed space X, the equality $N(X') = N(X)$ holds. However, as a matter of fact, the proof of the claim in [21] never appeared, and the question if for an arbitrary normed space X the equality $N(X') = N(X)$ holds remains an open problem among people interested in the field. In view of this open problem, we investigated about the normed space numerical indexes of preduals of non-commutative JBW^*-algebras, and proved that, *if A is a non zero non-commutative JBW^*-algebra (with predual denoted by A_*), then $N(A_*)$ is equal to* 1 *or* $\frac{1}{2}$ *depending on whether or not A is associative and commutative* [39, Proposition 2.8].

In relation to Theorem 2.1, we conjecture that, if A is a non-commutative JB^*-algebra such that p_A is a vertex of $B_{\Pi(A)}$, then A is an alternative C^*-algebra. We know that, if in the above conjecture we relax the condition that p_A is a vertex of $B_{\Pi(A)}$ to the one that p_A is an extreme point of $B_{\Pi(A)}$, then the answer is negative [39, Example 3.2].

3. Prime non-commutative JB^*-algebras

By a **non-commutative JBW^*-factor** we mean a prime non-commutative JBW^*-algebra. A non-commutative JBW^*-factor is said to be of **Type I** if the closed unit ball of its predual has some extreme point (compare [49, Theorem 1.11]). As we commented in Section 1, one of the main results in the structure theory of non-commutative JB^*-algebras is the following.

Theorem 3.1 ([49, Theorem 2.7]). *Type I non-commutative JBW^*-factors are either commutative, quadratic, or of the form $BL(H)^{(\lambda)}$ for some complex Hilbert space H and some $\frac{1}{2} < \lambda \leq 1$.*

We recall that, according to [65, pp. 49-50], an algebra A over a field $\mathbb{F}$ is called **quadratic** over $\mathbb{F}$ if it has a unit $\mathbf{1}, A \neq \mathbb{F}\mathbf{1}$, and, for each a in A, there are elements $t(a)$ and $n(a)$ of $\mathbb{F}$ such that $a^2 - t(a)a + n(a)\mathbf{1} = 0$. We also recall that, if A is a non-commutative JB^*-algebra, and if λ is a real number with $0 \leq \lambda \leq 1$, then the involutive Banach space of A, endowed with the product $(a,b) \to \lambda ab + (1-\lambda)ba$, becomes a non-commutative JB^*-algebra which is usually denoted by $A^{(\lambda)}$.

In [40] we obtain a reasonable generalization of Theorem 3.1, which reads as follows.

Theorem 3.2 ([40, Theorem 4]). *Prime non-commutative JB^*-algebras are either commutative, quadratic, or of the form $C^{(\lambda)}$ for some prime C^*-algebra C and some $\frac{1}{2} < \lambda \leq 1$.*

In fact we derived Theorem 3.2 from Theorem 3.1 and the fact that *every JB^*-algebra has a faithful family of Type I factor representations* [49, Corollary 1.13]. In what follows we provide the reader with an outline of the argument. First we recall that a **factor representation** of a given non-commutative JB^*-algebra A is a w^*-dense range *-homomorphism from A into some non-commutative JBW^*-factor. For convenience, let us say that a factor representation $\varphi : A \to B$ is commutative, quadratic, or quasi-associative whenever the non-commutative JBW^*-factor B is commutative, quadratic, or of the form $B^{(\lambda)}$ for some W^*-factor B and some $\frac{1}{2} < \lambda \leq 1$, respectively. Now, if the non-commutative JB^*-algebra A is prime, then it follows easily from the information collected above that at least one of the following families of factor representations of A is faithful:

1. The family of all commutative Type I factor representations of A.

2. The family of all quadratic Type I factor representations of A.

3. The family of all quasi-associative Type I factor representations of A.

Since clearly A is commutative whenever the family in (1) is faithful, the unique remaining problem is to show that, if the family in (2) (respectively, (3)) is faithful, then A is quadratic (respectively, of the form $C^{(\lambda)}$ for some prime C^*-algebra C and some $\frac{1}{2} < \lambda \leq 1$). To overcome this obstacle we replaced algebraic ultraproducts with Banach ultraproducts [30] in an argument of E. Zel'manov [75] in his determination of prime nondegenerate Jordan triples of Clifford type, to obtain the proposition which follows. We note that, if $\{A_i\}_{i \in I}$ is a family of non-commutative JB^*-algebras, and if $\mathcal{U}$ is an ultrafilter on I, then the Banach ultraproduct $(A_i)_{\mathcal{U}}$ is a non-commutative JB^*-algebra in a natural way.

Proposition 3.3 ([40, Proposition 2]). *Let A be a prime non-commutative JB^*-algebra, I a non-empty set, and, for each i in I, let φ_i be a $*$-homomorphism from A into a non-commutative JB^*-algebra A_i. Assume that $\cap_{i \in I} Ker(\varphi_i) = 0$. Then there exists an ultrafilter $\mathcal{U}$ on I such that the $*$-homomorphism $\varphi : x \to (\varphi_i(x))$ from A to $(A_i)_{\mathcal{U}}$ is injective.*

When in the above proposition the family $\{\varphi_i\}_{i \in I}$ actually consists of quadratic (respectively, quasi-associative) factor representations of the prime non-commutative JB^*-algebra A, it is not difficult to see that the non-commutative JB^*-algebra $(A_i)_{\mathcal{U}}$ is quadratic (respectively, of the form $C^{(\lambda)}$ for some prime C^*-algebra C and some $\frac{1}{2} < \lambda \leq 1$), so that, with some additional effort, it follows from the proposition that A is quadratic (respectively, of the form $C^{(\lambda)}$ for some prime C^*-algebra C and some $\frac{1}{2} < \lambda \leq 1$), thus concluding the proof of Theorem 3.2.

In relation to Theorem 3.2, we note that prime JB^*-algebras which are either quadratic or commutative are well-understood. Quadratic prime non-commutative JB^*-algebras have been precisely described in [49, Section 3]. According to that description, they are in fact Type I non-commutative JBW^*-factors. Commutative prime JB^*-algebras are classified in the Zel'manov-type theorem for such algebras [26, Theorem 2.3].

We recall that a **W^*-algebra** is a C^*-algebra which is a dual Banach space, and that a **W^*-factor** is a prime W^*-algebra. The next result follows directly from Theorem 3.2.

Corollary 3.4 ([3] [11]). *Non-commutative JBW^*-factors are either commutative, quadratic, or of the form $B^{(\lambda)}$ for some W^*-factor B and some real number λ with $\frac{1}{2} < \lambda \leq 1$.*

For (commutative) JBW^*-factors, the reader is referred to [26, Proposition 1.1]. According to Theorem 3.1, for non-commutative JBW^*-factors of Type I, the W^*-factor B arising in the above Corollary is equal to the algebra $BL(H)$ of all bounded linear operators on some complex Hilbert space H. This result follows from Corollary 3.4 and the fact that the algebras of the form $BL(H)$, with H a complex Hilbert space, are the unique W^*-factors of Type I [29, Proposition 7.5.2]. A classification of (commutative) JBW^*-factors of Type I can be obtained from the categorical correspondence between JBW-algebras and JBW^*-algebras [22] and the structure theorem for JBW-factors of Type I [29, Corollary 5.3.7, and Theorems 5.3.8, 6.1.8, and 7.5.11]. The precise formulation of such a classification can be found in [40, Proposition 6].

A normed algebra A is called *topologically simple* if $A^2 \neq 0$ and the unique closed ideals of A are $\{0\}$ and A. Since topologically simple normed algebras are prime, the following corollary follows with minor effort from Theorem 3.2.

Corollary 3.5 ([40, Corollary 7]). *Topologically simple non-commutative JB^*-algebras are either commutative, quadratic, or of the form $B^{(\lambda)}$ for some topologically simple C^*-algebra B and some real number λ with $\frac{1}{2} < \lambda \leq 1$.*

We note that every quadratic prime JB^*-algebra is algebraically (hence topologically) simple. For topologically simple (commutative) JB^*-algebras, the reader is referred to [26, Corollary 3.1].

4. Holomorphic characterization of non-commutative JB^*-algebras

An **approximate unit** of a normed algebra A is a net $\{b_\lambda\}_{\lambda\in\Lambda}$ in A satisfying

$$\lim\{ab_\lambda\}_{\lambda\in\Lambda} = a \text{ and } \lim\{b_\lambda a\}_{\lambda\in\Lambda} = a$$

for every a in A. If A is a non-commutative JB^*-algebra, the self-adjoint part A_{sa} of A (regarded as a closed real subalgebra of A^+) is a JB-algebra [29, Proposition 3.8.2]. In this way, the self-adjoint part of any non-commutative JB^*-algebra A is endowed with the order induced by the positive cone $\{a^2 : a \in A_{sa}\}$ [29, Section 3.3]. The following proposition is proved in [68]. We include here the proof because reference [68] is not easily available. We recall that *every JB-algebra has an increasing approximate unit consisting of positive elements with norm* ≤ 1 [29, Proposition 3.5.4].

Proposition 4.1. *Every non-commutative JB^*-algebra has an increasing approximate unit consisting of positive elements with norm ≤ 1.*

Proof. Let A be a non-commutative JB^*-algebra, and let $\{b_\lambda\}_{\lambda\in\Lambda}$ be an increasing approximate unit of the JB-algebra A_{sa} consisting of positive elements with norm ≤ 1. We are proving that $\{b_\lambda\}_{\lambda\in\Lambda}$ is in fact an approximate unit of A. Since $\{b_\lambda\}_{\lambda\in\Lambda}$ is clearly an approximate unit of A^+, it is enough to show that $\lim\{[a,b_\lambda]\}_{\lambda\in\Lambda} = 0$ for every a in A. Here $[.,.]$ denotes the usual commutator on A. But, keeping in mind that the commutator is a derivation of A^+ in each of its variables [65, p. 146], for a in A we obtain

$$\lim\{[a^2,b_\lambda]\}_{\lambda\in\Lambda} = 2\lim\{a\circ[a,b_\lambda]\}_{\lambda\in\Lambda} = 2\lim\{[a,a\circ b_\lambda]\}_{\lambda\in\Lambda} = 0 \ .$$

Now the proof is concluded by applying the well-known fact that every non-commutative JB^*-algebra is the linear hull of the set of squares of its elements. ∎

In [41] we rediscover the above result as a consequence of the following remarkable inequality for non-commutative JB^*-algebras.

Theorem 4.2 ([41, Theorem 1.3]). *Let A be a non-commutative JB^*-algebra, and a be in A_{sa}. Then, for all b in A we have*

$$\| [a,b] \|^2 \leq 16 \| b \| \| a^2\circ b - a\circ(a\circ b) \| \ .$$

The proof of Theorem 4.2 above involves the whole theory of Type I factor representations of non-commutative JB^*-algebras outlined in Section 3. If one is only interested in the specialization of Theorem 4.2 in the case that A is an alternative C^*-algebra, then the proof is much easier. Indeed, in such a case the argument given in [41, Lemma 1.1] for classical C^*-algebras works verbatim.

Together with Proposition 4.1, the following result becomes of special interest for the matter we are developing in the present section.

Proposition 4.3. *The open unit ball of every non-commutative JB^*-algebra is a bounded symmetric domain.*

The proof of the above proposition consists of the facts that non-commutative JB^*-algebras are JB^*-triples in a natural way ([13], [74]) and that *open unit balls of JB^*-triples are bounded symmetric domains* [42]. We recall that a JB^*-**triple** is a complex Banach space J with a continuous triple product $\{...\} : J\times J\times J \to J$ which is linear and symmetric in the outer variables, and conjugate-linear in the middle variable, and satisfies:

1. For all x in J, the mapping $y\to\{xxy\}$ from J to J is a hermitian operator on J and has nonnegative spectrum.

2. The **main identity**

$$\{ab\{xyz\}\} = \{\{abx\}yz\} - \{x\{bay\}z\} + \{xy\{abz\}\}$$

   holds for all a, b, x, y, z in J.

3. $\| \{xxx\} \| = \| x \|^3$ for every x in J.

Concerning Condition (1) above, we also recall that a bounded linear operator T on a complex Banach space X is said to be **hermitian** if it belongs to $H(BL(X), I_X)$ (equivalently, if $\| \exp(irT) \| = 1$ for every r in $\mathbb{R}$ [10, Corollary 10.13]).

For a vector space E, let $L(E)$ denote the associative algebra of all linear mappings from E to E, and for a non-commutative Jordan algebra A, let $(a,b) \to U_{a,b}$ be the unique symmetric bilinear mapping from $A \times A$ to $L(A)$ satisfying $U_{a,a} = U_a$ for every a in A. Now we can specify the result in [13] and [74] pointed out above. Indeed, *every non-commutative JB^*-algebra A is a JB^*-triple under the triple product $\{...\}$ defined by $\{abc\} := U_{a,c}(b^*)$ for all a, b, c in A.*

In [41] we prove that the properties given by Propositions 4.1 and 4.3 characterize non-commutative JB^*-algebras among complete normed complex algebras. This is emphasized in the theorem which follows.

Theorem 4.4 ([41, Theorem 3.3]). *Let A be a complete normed complex algebra. Then A is a non-commutative JB^*-algebra (for some involution $*$) if (and only if) A has an approximate unit bounded by one and the open unit ball of A is a bounded symmetric domain.*

Many old and new auxiliary results have been needed to prove Theorem 4.4 above. Concerning new ones, we make to stand out for the moment Lemma 4.5 which follows. We begin by recalling some concepts taken from [13, p. 285]. Let X be a normed space, u an element in X, and Q a subset of X. We define the **tangent cone to Q at u**, $T_u(Q)$, as the set of all $x \in X$ such that

$$x = \lim \frac{x_n - u}{t_n}$$

for some sequence $\{x_n\}$ in Q with $\lim\{x_n\} = u$ and some sequence $\{t_n\}$ of positive real numbers. When X is complex, the **holomorphic tangent cone to Q at u**, $\hat{T}_u(Q)$, is defined as

$$\hat{T}_u(Q) := \cap_{\lambda \in \mathbb{C} \setminus \{0\}} \lambda T_u(Q) .$$

From now on, for a normed space X, Δ_X will denote the open unit ball of X.

Lemma 4.5 ([41, Lemma 3.1]). *Let X be a complex normed space, and u an element in X. Then u is a vertex of B_X if and only if $\hat{T}_u(\Delta_X) = 0$.*

Now, speaking about old results applied in the proof of Theorem 4.4, let us enumerate the following:

1. Kaup's algebraic characterization of bounded symmetric domains ([42] and [43]), namely *a complex Banach space X is a JB^*-triple (for some triple product) if and only if Δ_X is a bounded symmetric domain.*

2. The Chu-Iochum-Loupias result [17] that, *if X is a JB^*-triple, and if $T : X \to X'$ is a bounded linear mapping, then T is weakly compact.* In fact we are applying the reformulation of this result (via [55]) that *every product on a JB^*-triple is Arens regular.*

3. The non-associative version of the Bohnenblust-Karlin theorem (see for instance [60, Theorem 1.5]), namely, *if A is a norm-unital complete normed complex algebra, then the unit of A is a vertex of B_A.*

4. The celebrated Dineen's result [20] that *the bidual of every JB^*-triple is a JB^*-triple.*

5. The Braun-Kaup-Upmeier holomorphic characterization of Banach spaces underlying unital JB^*-algebras [13] (reformulated via Lemma 4.5): *a complex Banach space X underlies a JB^*-algebra with unit u if and only if u is a vertex of B_X and Δ_X is a bounded symmetric domain.*

6. The non-associative version of Vidav-Palmer theorem (Theorem 1.3).

Turning back to new results applied in the proof of Theorem 4.4, the main one is Theorem 4.6 which follows.

Theorem 4.6 ([41, Theorem 2.4]). *Let A be a complete normed complex algebra such that A'', endowed with the Arens product and a suitable involution $*$, is a non-commutative JB^*-algebra. Then A is a $*$-invariant subset of A'', and hence a non-commutative JB^*-algebra.*

In the case that A has a unit, Theorem 4.6 follows easily from a dual version of Theorem 1.3, proved in [44], asserting that *a norm-unital complete normed complex algebra A is a non-commutative JB^*-algebra if and only if $S \cap iS = 0$, where S denotes the real linear hull of $D(A, \mathbf{1})$.* The non unital case of Theorem 4.6 is reduced to the unital one after a lot of work (see [41, Section 2] for details).

Now, let us provide the reader with the following

Sketch of proof of the "if" part of Theorem 4.4.- By the second assumption and Result 1 above, there exists a JB^*-triple X and a surjective linear isometry $\Phi : A \to X$. By the first assumption and Result 2, A'' (endowed with the Arens product) has a unit $\mathbf{1}$ with $\| \mathbf{1} \| = 1$. By Result 3, $\mathbf{1}$ is a vertex of $B_{A''}$, and hence $\Phi''(\mathbf{1})$ is a vertex of $B_{X''}$. By Results 4, 1, and 5, X'' underlies a JB^*-algebra with unit $\Phi''(\mathbf{1})$, and therefore (by an easy observation of M. A. Youngson in [72]) we have

$$X'' = H(X'', \Phi''(\mathbf{1})) + iH(X'', \Phi''(\mathbf{1})),$$

and hence $A'' = H(A'', \mathbf{1}) + iH(A'', \mathbf{1})$. By Result 6, A'' is a non-commutative JB^*-algebra. Finally, by Theorem 4.6, A is a non-commutative JB^*-algebra. ∎

The following corollaries follow straightforwardly from Theorem 4.4.

Corollary 4.7 ([41, Corollary 3.4]). *An associative complete normed complex algebra is a C^*-algebra if and only if it has an approximate unit bounded by one and its open unit ball is a bounded symmetric domain.*

Corollary 4.8. *An alternative complete normed complex algebra is an alternative C^*-algebra if and only if it has an approximate unit bounded by one and its open unit ball is a bounded symmetric domain.*

Corollary 4.9 ([41, Corollary 3.5]). *A normed complex algebra is a non-commutative JB^*-algebra if and only if it is linearly isometric to a non-commutative JB^*-algebra and has an approximate unit bounded by one.*

Corollary 4.10. *An alternative normed complex algebra is an alternative C^*-algebra if and only if it is linearly isometric to a non-commutative JB^*-algebra and has an approximate unit bounded by one.*

Corollary 4.11. *An alternative normed complex algebra is an alternative C^*-algebra if and only if it is linearly isometric to an alternative C^*-algebra and has an approximate unit bounded by one.*

Corollary 4.12. *An associative normed complex algebra is a C^*-algebra if and only if it is linearly isometric to a non-commutative JB^*-algebra and has an approximate unit bounded by one.*

Corollary 4.13 ([62, Corollary 1.3]). *An associative normed complex algebra is a C^*-algebra if and only if it is linearly isometric to a C^*-algebra and has an approximate unit bounded by one.*

For surjective linear isometries between non-commutative JB^*-algebras the reader is referred to Section 6 of the present paper.

5. Multipliers of non-commutative JB^*-algebras

Every semiprime associative algebra A has a natural enlargement, namely the so-called multiplier algebra $M(A)$ of A, which can be characterized as the largest semiprime associative algebra containing A as an essential ideal. In the case that A is an (associative) C^*-algebra, $M(A)$ becomes naturally a C^*-algebra which contains A as a closed (essential) ideal. More precisely, in this case $M(A)$ can be rediscovered as the closed $*$-invariant subalgebra of A'' given by $\{x \in A'' : xA + Ax \subseteq A\}$ (see for instance [50, Propositions 3.12.3 and 3.7.8]). Now let A be a non-commutative JB^*-algebra. The fact just commented suggests to define the set of multipliers, $M(A)$, of A by the equality $M(A) := \{x \in A'' : xA + Ax \subseteq A\}$. It is clear that $M(A)$ is a closed $*$-invariant subspace of A'' containing A and the unit of A''. In this way, the equality $M(A) = A$ holds if and only if A has a unit. It is also clear that, if $M(A)$ were a subalgebra of A'', then A would be an ideal of $M(A)$. We are showing in the present section that $M(A)$ is in fact a subalgebra of A'' (and hence a non-commutative JB^*-algebra) which contains A as an essential ideal. We also will show that, in a categorical sense, $M(A)$ is the largest non-commutative JB^*-algebra containing A as a closed essential ideal.

Our argument begins by invoking the next result, which is taken from C. M. Edwards' paper [23].

Lemma 5.1. *Let B be a JB-algebra. Then $M(B) := \{x \in B'' : x \circ B \subseteq B\}$ is a subalgebra of B''.*

It will be also useful the following lemma, whose verification can be made following the lines of the proof of [34, Lemma 4.2]. Let X be a complex Banach space. By a **conjugation** on X we mean an involutive conjugate-linear isometry on X. Conjugations τ on a complex Banach space X give rise by natural transposition to conjugations τ' on X'. Given a conjugation τ on X, X^τ will stand for the closed real subspace of X given by $X^\tau := \{x \in X : \tau(x) = x\}$.

Lemma 5.2. *Let X be a complex Banach space, and τ a conjugation on X. Then, up to a natural identification, we have $(X^\tau)'' = (X'')^{\tau''}$.*

Taking in the above lemma X equal to a non-commutative JB^*-algebra A, and τ equal to the JB^*-involution of A, we obtain the following corollary.

Corollary 5.3. *Let A be a non-commutative JB^*-algebra. Then the Banach space identification $(A_{sa})'' = (A'')_{sa}$ is also a JB-algebra identification.*

Proof. The Banach space identification $(A_{sa})'' = (A'')_{sa}$ is the identity on A_{sa}, A_{sa} is w^*-dense in both $(A_{sa})''$ and $(A'')_{sa}$, and the products of $(A_{sa})''$ and $(A'')_{sa}$ are separately w^*-continuous. ∎

Now, putting together Lemma 5.1 and Corollary 5.3, the next result follows.

Corollary 5.4. *Let A be a (commutative) JB^*-algebra. Then $M(A)$ is a subalgebra of A''.*

Now, to obtain the non-commutative generalization of the above corollary we only need a single new fact, which is proved in the next lemma. We recall that every derivation of a JB^*-algebra is automatically continuous [73].

Lemma 5.5. *Let A be a JB^*-algebra, and D a derivation of A. Then $M(A)$ is D''-invariant.*

Proof. Let a, x be in A and $M(A)$, respectively. We have

$$D''(x) \circ a = D(x \circ a) - x \circ D(a) \in A .$$

Since a is arbitrary in A, it follows that $D''(x)$ lies in $M(A)$. ∎

Theorem 5.6. *Let A be a non-commutative JB^*-algebra. Then $M(A)$ is a closed $*$-invariant subalgebra of A'' containing A as an essential ideal. Moreover, if B is another non-commutative JB^*-algebra containing A as a closed essential ideal, then B can be seen as a closed $*$-subalgebra of $M(A)$ containing A. In addition we have $M(A) = M(A^+)$.*

Proof. Keeping in mind the equality $(A'')^+ = (A^+)''$, the inclusion $M(A) \subseteq M(A^+)$ is clear. Let b be in A. Then the mapping $D : a \to [b,a]$ from A to A is a derivation of A^+, and we have $D''(x) = [b,x]$ for every x in A''. It follows from Lemma 5.5 that $D''(M(A^+)) \subseteq M(A^+)$, or equivalently $[b,x] \in M(A^+)$ for every x in $M(A^+)$. Therefore, for x in $M(A^+)$ we have

$$[b^2, x] = 2b \circ [b,x] \in A \circ M(A^+) \subseteq A \,.$$

Since b is arbitrary in A, and A is the linear hull of the set of squares of its elements, we deduce $[A,x] \subseteq A$ for every x in $M(A^+)$. It follows

$$M(A^+)A + AM(A^+) \subseteq A \,,$$

and hence $M(A^+) \subseteq M(A)$. Then the equality $M(A^+) = M(A)$ is proved.

Now, for x, y in $M(A)$ and A in A we have

$$[x,y] \circ a = [x, y \circ a] - [x,a] \circ y \in [M(A), A] + A \circ M(A) \subseteq A \,,$$

and hence $[M(A), M(A)] \subseteq M(A^+)$. On the other hand, Corollary 5.4 applies to A^+ giving $M(A) \circ M(A) \subseteq M(A^+) \circ M(A^+) \subseteq M(A^+)$. It follows from the first paragraph of the proof that $M(A)$ is a subalgebra of A''.

Assume that P is an ideal of $M(A)$ with $P \cap A = 0$. Then, since A is an ideal of $M(A)$, we actually have $AP = 0$, so $A''P = 0$, and so $P = 0$. Therefore A is an essential ideal of $M(A)$.

Let B be a non-commutative JB^*-algebra, and $\varphi : A \to B$ a one-to-one (automatically isometric) *-homomorphism such that $\varphi(A)$ is an essential ideal of B. Then φ'' is a one-to-one $*$-homomorphism from A'' to B'' whose range is a w^*-closed ideal of B'' (apply that $*$-homomorphisms between non-commutative JB^*-algebras have norm-closed range [48, Corollary 1.11 and Proposition 2.1], and that w^*-continuous linear operators with norm-closed range have in fact w^*-closed range [49, Lemma 1.3]). By [48, Theorem 3.9] we have $\varphi''(A'') = B''e$ for a suitable central projection e in B''. Now $(B''(1-e)) \cap B$ is an ideal of B whose intersection with $\varphi(A)$ is zero, and hence $(B''(1-e)) \cap B = 0$ because $\varphi(A)$ is an essential ideal of B. Therefore the mapping $\psi : b \to be$ from B to $\varphi''(A'')$ is a one-to-one $*$-homomorphism. Then $\eta := (\varphi'')^{-1}\psi$ is a one-to-one $*$-homomorphism from B to A'' satisfying $\eta(\varphi(a)) = a$ for every a in A. In this way we can see B as a closed $*$-invariant subalgebra of A'' containing A as an ideal. In this regarding we have clearly $B \subseteq M(A)$. ∎

Let A be a non-commutative JB^*-algebra. The above theorem allows us to say that $M(A)$ is the **multiplier non-commutative JB^*-algebra** of A. The equality $M(A) = M(A^+)$ in the theorem can be understood in the sense that, if we consider the JB-algebra A_{sa}, then the multiplier JB-algebra of A_{sa} (in the sense of [23]) is nothing but the self-adjoint part of the multiplier JB^*-algebra of A. It is worth mentioning that the concluding paragraph of the proof of Theorem 5.6 is quite standard (see [26, Proposition 1.3], [14, Lemma 2.3], and [18, Proposition 1.2] for forerunners).

Let X be a JB^*-triple. We recall that the bidual X'' of X is a JB^*-triple under a triple product extending that of X [20], and that the set

$$\mathcal{M}(X) := \{x \in X'' : \{xXX\} \subseteq X\}$$

is a JB^*-subtriple of X'' containing X as a triple ideal [14]. The JB^*-triple $\mathcal{M}(X)$ just defined is called the **multiplier JB^*-triple** of X. Therefore, for a given non-commutative JB^*-algebra A, we can consider the multiplier non-commutative JB^*-algebra $M(A)$ of A, and the multiplier JB^*-triple $\mathcal{M}(A)$ of the JB^*-triple underlying A. Actually, the following result holds.

Proposition 5.7. *Let A be a non-commutative JB^*-algebra. Then we have $M(A) = \mathcal{M}(A)$.*

Proof. The inclusion $M(A) \subseteq \mathcal{M}(A)$ is clear. To prove the converse inclusion, we start by noticing that, clearly, the JB^*-triples underlying A and A^+ coincide, and that, by Theorem 5.6, the equality $M(A) = M(A^+)$ holds, so that we may assume that A is commutative. We note also that, since the equality $\{xyz\}^* = \{x^*y^*z^*\}$ is true for all x, y, z in A'', and A is a $*$-invariant subset of A'', $\mathcal{M}(A)$ is $*$-invariant too, and therefore it is enough to show that $a \circ x$ lies in A whenever a and x are self-adjoint elements of A and $\mathcal{M}(A)$, respectively. But, for such a and x, we can find a self-adjoint element b in A satisfying $b^3 = a$ (see for instance [45, Proposition 1.2]), and apply Shirshov's theorem [76, p.71] to obtain that

$$x \circ a = x \circ b^3 = b \circ (2U_{x,b}(b) - U_{b,b}(x)) = b \circ (2\{xbb\} - \{bxb\})$$

belongs to A. ∎

6. Isometries of non-commutative JB^*-algebras

This section is devoted to the non-associative discussion of the following Paterson-Sinclair refinement of Kadison's classical theorem [37] on isometries of C^*-algebras.

Theorem 6.1 ([47, Theorem 1]). *Let A and B be C^*-algebras, and F a mapping from B to A. Then F is a surjective linear isometry (if and) only if there exists a Jordan-$*$-isomorphism $G : B \to A$, and a unitary element u in the multiplier C^*-algebra of A satisfying $F(b) = uG(b)$ for every b in B.*

We recall that, given algebras A and B, **Jordan homomorphisms** from B to A are defined as homomorphisms from B^+ to A^+. By an **isomorphism** between algebras we mean a one-to-one surjective homomorphism. If follows from Theorem 6.1 that unit-preserving surjective linear isometries between unital C^*-algebras are in fact Jordan-$*$-isomorphisms. This particular case of Theorem 6.1 remains true in the setting of non-commutative JB^*-algebras thanks to the following Wright-Youngson theorem.

Theorem 6.2 ([71, Theorem 6]). *Let A and B be unital non-commutative JB^*-algebras, and $F : B \to A$ a unit-preserving surjective linear isometry. Then F is a Jordan-$*$-isomorphism.*

The above theorem can be also derived from the fact that non-commutative JB^*-algebras are JB^*-triples in a natural way, and Kaup's theorem [42] that surjective linear isometries between JB^*-triples preserve triple products. In any case, the easiest known proof of Theorem 6.2 seems to be the one provided by the implication $(i) \Rightarrow (ii)$ in [38, Lemma 6].

Concerning concepts involved in the statement, the general formulation of Theorem 6.1 could have a sense in the more general setting of non-commutative JB^*-algebras. Indeed, in the previous section we introduced (automatically unital) multiplier non-commutative JB^*-algebras of arbitrary non-commutative JB^*-algebras. On the other hand, a reasonable notion of unitary element in a unital non-commutative JB^*-algebra A can be given, by invoking McCrimmon's definition of invertible elements in unital non-commutative Jordan algebras [46], and saying that an element a in A is **unitary** whenever it is invertible and satisfies $a^* = a^{-1}$. Let A be a unital non-commutative Jordan algebra, and a an element of A. We recall that a is said to be **invertible** in A if there exists b in A such that the equalities $ab = ba = \mathbf{1}$ and $a^2b = ba^2 = a$ hold. If a is invertible in A, then the element b above is unique, is called the inverse of a , and is denoted by a^{-1}. Moreover a is invertible in A if and only if it is invertible in the Jordan algebra A^+. This reduces most questions and results on inverses in non-commutative Jordan algebras to the commutative case. For this particular case, the reader is referred to [36, Section I.11].

Despite the above comments, even the "if" part of Theorem 6.1 does not remain true in the setting of non-commutative JB^*-algebras. Indeed, Jordan-$*$-isomorphisms between non-commutative JB^*-algebras are isometries [69] but, unfortunately, left multiplications by unitary elements of a unital non-commutative JB^*-algebra need not be isometries. This handicap becomes more than an anecdote in view of the following result. Given an element x in the multiplier non-commutative JB^*-algebra of a non-commutative JB^*-algebra A, we denote by T_x the operator on A defined by $T_x(a) := xa$ for every a in A. By a **Jordan-derivation** of an algebra A we mean a derivation of A^+.

Proposition 6.3. *Let A be a non-commutative JB^*-algebra. Then A is an alternative C^*-algebra if and only if, for every unitary element u of $M(A)$, T_u is an isometry.*

Proof. The "only if" part is very easy. Assume that A is an alternative C^*-algebra. Then it follows from Theorem 5.6 and Corollary 2.3 that $M(A)$ is a unital alternative C^*-algebra. Therefore, as we have seen in Section 2, left multiplications on $M(A)$ by unitary elements of $M(A)$ are surjective linear isometries. Since A is invariant under such isometries, it follows that $T_u : A \to A$ is an isometry whenever u is a unitary element in $M(A)$.

Now assume that A is a non-commutative JB^*-algebra such that T_u is an isometry whenever u is a unitary element in $M(A)$. For x in A'', denote by $L_x^{A''}$ the operator of left multiplication by x on A''. We remark that $L_x^{A''} = (T_x)''$ whenever x belongs to $M(A)$ (indeed, both sides of the equality are w^*-continuous operators on A'' coinciding on A). Let h be in $(M(A))_{sa}$ and r be a real number. Then $\exp(irh)$ is a unitary element of $M(A)$, and therefore, by the assumption on A and the equality $L_{\exp(irh)}^{A''} = (T_{\exp(irh)})''$ just established, $L_{\exp(irh)}^{A''}$ is an isometry on A''. Now put $G_r := L_{\exp(irh)}^{A''} L_{\exp(-irh)}^{A''}$. Then G_r is an isometry on A'' preserving the unit of A''. Since $G_0 = I_{A''}$ and the mapping $r \to G_r$ is continuous, there exists a positive number k such that G_r is surjective whenever $|r| < k$.

It follows from Theorem 6.2 that, for $|r| < k$, G_r is a Jordan-$*$-automorphism of A''. If $\sum(1/n!)r^n F_n$ is the power series development of G_r, then we easily obtain $F_0 = I_{A''}$, $F_1 = 0$, and $F_2 = 2((L_h^{A''})^2 - L_{h^2}^{A''})$. By [53, Lemma 13], $(L_h^{A''})^2 - L_{h^2}^{A''}$ is a Jordan-derivation of A'' commuting with the JB^*-involution of A''. Now, arguing as in the conclusion of the proof of [53, Theorem 14], we realize that actually the equality $(L_h^{A''})^2 - L_{h^2}^{A''} = 0$ holds. In particular, for x in $M(A)$ we have $h(hx) = h^2x$. Since h is an arbitrary element of $(M(A))_{sa}$, an easy linearization argument gives $y(yx) = y^2x$ for all x, y in $M(A)$. By applying the JB^*-involution of $M(A)$ to both sides of the above equality, it follows that $M(A)$ (and hence A) is alternative. ∎

In relation to the above proposition, we note that, if A is an alternative C^*-algebra, then, for every unitary element u of $M(A), T_u$ is in fact a SURJECTIVE linear isometry on A (with inverse mapping equal to T_{u^*}). Now that we know that alternative C^*-algebras are the unique non-commutative JB^*-algebras which can play the role of A in a reasonable non-associative generalization of the "if" part of Theorem 6.1, we proceed to prove that they are also "good" for the non-associative generalization of the "only if" part of that theorem.

Lemma 6.4. *Let A be a unital alternative C^*-algebra. Then vertices of B_A and unitary elements of A coincide.*

Proof. Let u be a unitary element of A. Then u is a vertex of B_A because $\mathbf{1}$ is a vertex of B_A [60, Theorem 1.5] and the mapping $a \to ua$ from A to A is a surjective linear isometry sending $\mathbf{1}$ into u.

Now, let u be a vertex of B_A. Then the closed subalgebra B of A generated by $\{\mathbf{1}, u, u^*\}$ is a unital (associative) C^*-algebra. Since the vertex property is hereditary, it follows from [7, Example 4.1] that u is a unitary element of B, and hence also of A. ∎

Remark 6.5. Actually the assertion in the above lemma remains true if A is only assumed to be a unital non-commutative JB^*-algebra. This follows straightforwardly from Lemma 4.5 and the equivalence $(i) \Leftrightarrow (iii)$ in [13, Proposition 4.3]. The proof we have given of this fact in the particular case of alternative C^*-algebras has however its methodological own interest.

Let X and Y be JB^*-triples, and $F : X \to Y$ a surjective linear isometry. It follows easily from the already quoted Kaup's Kadison type theorem that $F''(\mathcal{M}(X)) = \mathcal{M}(Y)$. In the particular case of non-commutative JB^*-algebras, we can apply Proposition 5.7 to arrive in the following non-associative generalization of [47, Theorem 2].

Lemma 6.6. *Let A and B be non-commutative JB^*-algebras, and $F : B \to A$ a surjective linear isometry. Then we have $F''(M(B)) = M(A)$. In particular, Jordan-$*$-isomorphisms from B to A extend uniquely to Jordan-$*$-isomorphisms from $M(B)$ to $M(A)$.*

Proof. In view of the previous comments, we only must prove the uniqueness of Jordan-$*$-isomorphisms from $M(B)$ to $M(A)$ extending a given Jordan-$*$-isomorphism (say G)

from B to A. But, if R and S are Jordan-$*$-isomorphisms from $M(B)$ to $M(A)$ extending G, then for b in B and x in $M(B)$ we have

$$(R(x) - S(x)) \circ G(b) = R(x) \circ R(b) - S(x) \circ S(b)$$

$$= R(x \circ b) - S(x \circ b) = G(x \circ b) - G(x \circ b) = 0.$$

■

Theorem 6.7. *Let A be an alternative C^*-algebra, B a non-commutative JB^*-algebra, and F a mapping from B to A. Then F is a surjective linear isometry (if and) only if there exists a Jordan-$*$-isomorphism $G : B \to A$, and a unitary element u in $M(A)$ satisfying $F = T_u G$.*

Proof. Assume that F is a surjective linear isometry. Put $u := F''(\mathbf{1})$. By Lemma 6.4, u is a unitary element of A'', and, by Lemma 6.6, u lies in $M(A)$. Write $G := T_{u^*}T$. Then G is a Jordan-$*$-isomorphism from B to A because it is a surjective linear isometry satisfying $G''(\mathbf{1}) = \mathbf{1}$, and therefore Theorem 6.2 successfully applies. Finally, the equality $F = T_u G$ is clear. ■

To conclude the discussion about verbatim non-associative versions of Theorem 6.1, we show that alternative C^*-algebras are also the unique non-commutative JB^*-algebras which can play the role of A in the "only if" part of such versions. Let A be a non-commutative JB^*-algebra, and u a unitary element of $M(A)$. It is easily deducible from [36, Section I.12] and [70, Corollary 2.5] that the Banach space of A with product $\circ_u$ and involution $*_u$ defined by $a \circ_u b := U_{a,b}(u^*)$ and $a^{*_u} := U_u(a^*)$, respectively, becomes a (commutative) JB^*-algebra. Such a JB^*-algebra will be denoted by $A(u)$.

Proposition 6.8. *Let A be a non-commutative JB^*-algebra which is not an alternative C^*-algebra. Then there exists a non-commutative JB^*-algebra B, and a surjective linear isometry $F : B \to A$ which cannot be written in the form $T_u G$ with u a unitary element in $M(A)$ and G a Jordan-$*$-isomorphism from B to A.*

Proof. By Proposition 6.3, there is a unitary element v in $M(A)$ such that T_v is not an isometry on A. Take B equal to $A(v)$, and $F : B \to A$ equal to the identity mapping. Assume that $F = T_u G$ for some unitary u in $M(A)$ and some Jordan-$*$-isomorphism G from B to A. Noticing that the JB^*-algebra B'' is nothing but $A''(v)$ (by the w^*-continuity of the JB^*-involutions and the separate w^*-continuity of the products on JBW^*-algebras), we have $G''(v) = \mathbf{1}$ (because v is the unit of $A''(v)$ and G'' is a Jordan-$*$-isomorphism), and hence $F''(v) = u$. Therefore $v = u$ (because F is the identity mapping). Finally, the equality $F = T_v G$ implies that T_v is an isometry, contrarily to the choice of v. ■

Now that the verbatim non-associative variant of Theorem 6.1 has been altogether discussed, we pass to consider the consequence of that theorem that *linearly isometric C^*-algebras are Jordan-$*$-isomorphic.* In the unital case, the non-associative variant of such an assertion has been fully discussed in [13, Section 5]. In fact, as the next example shows, *linearly isometric non-commutative JB^*-algebras need not be Jordan-$*$-isomorphic.*

Example 6.9 ([13, Example 5.7]). **JC^*-algebras** are defined as those JB^*-algebras which can be seen as closed $*$-invariant subalgebras of A^+ for some C^*-algebra A. Let C be the unital simple JC^*-algebra of all symmetric 2×2-matrices over $\mathbb{C}$, put $S := \{z \in \mathbb{C} : |z| = 1\}$, let A stand for the unital JC^*-algebra of all continuous complex-valued functions from S to C, consider the unitary element u of A defined by $u(s) := diag\{s, 1\}$ for every s in S, and put $B := A(u)$. Then A and B are linearly isometric JB^*-algebras, but they are not Jordan-$*$-isomorphic.

For a non-commutative JB^*-algebra A, consider the property *(P)* which follows.

(P) Non-commutative JB^*-algebras which are linearly isometric to A are in fact Jordan-$*$-isomorphic to A.

Despite the above example, the class of those non-commutative JB^*-algebras A satisfying Property *(P)* is reasonably wide, and in fact much larger than that of alternative C^*-algebras. The verification of this fact relies on the next theorem. We remark that, if u is a unitary element in the multiplier non-commutative JB^*-algebra of a non-commutative JB^*-algebra A, then the operator U_u (acting on A) is a surjective linear isometry on A.

Theorem 6.10. *Let A be a non-commutative JB^*-algebra. The following assertions are equivalent:*

1. *For every non-commutative JB^*-algebra B, and every surjective linear isometry $F : B \to A$, there exists a Jordan-$*$-isomorphism $G : B \to A$, and a unitary element u in $M(A)$ satisfying $F = U_u G$.*

2. *For each unitary element v of $M(A)$ there is a unitary element u in $M(A)$ such that $u^2 = v$.*

Proof. $1 \Rightarrow 2$. Let v be a unitary element of $M(A)$. Take B equal to $A(v)$, and $F : B \to A$ equal to the identity mapping. By the assumption 1, we have $F = U_u G$ for some unitary u in $M(A)$ and some Jordan-$*$-isomorphism G from B to A. Arguing as in the proof of Proposition 6.8, we find $G''(v) = \mathbf{1}$, and hence $F''(v) = u^2$. Therefore $v = u^2$ (because F is the identity mapping).

$2 \Rightarrow 1$. Let B be a non-commutative JB^*-algebra, and $F : B \to A$ a surjective linear isometry. Put $v := F''(\mathbf{1})$. By Remark 6.5, v is a unitary element of A'', and, by Lemma 6.6, v belongs to $M(A)$. By the assumption 2, there is a unitary element u in $M(A)$ with $u^2 = v$. Write $G := U_{u^*} F$. Then G is a Jordan-$*$-isomorphism from B to A because it is a surjective linear isometry satisfying $G''(\mathbf{1}) = \mathbf{1}$, and Theorem 6.2 applies. On the other hand, the equality $F = U_u G$ is clear. ∎

Remark 6.11. An argument similar to the one in the above proof allows us to obtain the following variant of Theorem 6.10. Indeed, *given a non-commutative JB^*-algebra A, the following assertions on A are equivalent:*

1. *For every non-commutative JB^*-algebra B, and every surjective linear isometry $F : B \to A$, there exists a Jordan-$*$-isomorphism $G : B \to A$, together with unitary elements $u_1, ..., u_n$ in $M(A)$, satisfying $F = U_{u_1} U_{u_2} ... U_{u_n} G$.*

2. *For each unitary element v of $M(A)$ there are unitary elements $u_1, ..., u_n$ in $M(A)$ such that $U_{u_1}U_{u_2}...U_{u_n}(\mathbf{1}) = v$.*

The next corollary extends [13, Lemma 5.2] in several directions.

Corollary 6.12. *A non-commutative JB^*-algebra A satisfies Property (P) whenever one of the following conditions is fulfilled:*

1. *A is of the form $B^{(\lambda)}$ for some alternative C^*-algebra B and some $0 \leq \lambda \leq 1$.*
2. *For each unitary element v of $M(A)$ there is a unitary element u in $M(A)$ such that $u^2 = v$.*
3. *A is a non-commutative JBW^*-algebra.*

Proof. Both Conditions 1 and 2 are sufficient for Property *(P)* in view of Theorems 6.7 and 6.10, respectively. To conclude the proof, we realize that Condition 3 implies Condition 2. Indeed, if A is a non-commutative JBW^*-algebra, and if v is a unitary element in A, then the w^*-closure of the subalgebra of A generated by $\{v, v^*\}$ is an (associative) W^*-algebra, and it is well-known that W^*-algebras fulfill Condition ii). ■

We conclude this section by determining the hermitian operators on a non-commutative JB^*-algebra. Our determination generalizes and unifies both that of Paterson-Sinclair [47] for the associative case and that of M. A. Youngson [73] for the unital non-associative case.

Theorem 6.13. *Let A be a non-commutative JB^*-algebra, and R a bounded linear operator on A. Then R is hermitian if and only if it can be expressed in the form $T_x + D$ for some self-adjoint element x of $M(A)$ and some Jordan-derivation D of A anticommuting with the JB^*-involution of A.*

Proof. Let x be in $(M(A))_{sa}$. Since the mapping $y \to T_y$ from $M(A)$ to $BL(A)$ is a linear isometry sending $\mathbf{1}$ to I_A, and the equality $(M(A))_{sa} = H(M(A), \mathbf{1})$ holds, we obtain that T_x belongs to $H(BL(A), I_A)$, i.e., T_x is an hermitian operator on A. Now let D be a Jordan-derivation of A anticommuting with the JB^*-involution of A. Then, for every λ in $\mathbb{R}$, $\exp(i\lambda D)$ is a Jordan-$*$-automorphism of A, and hence we have $\| \exp(i\lambda D) \| = 1$, i.e., D is a hermitian operator on A.

Conversely, let R be an hermitian operator on A. Then, for λ in $\mathbb{R}$, $\exp(i\lambda R)$ is a surjective linear isometry on A, so, by Lemma 6.6, we have

$$\exp(i\lambda R'')(M(A)) = (\exp(i\lambda R))''(M(A)) = M(A) ,$$

and so

$$x := R''(\mathbf{1}) = \lim_{\lambda \to 0} \frac{\exp(i\lambda R'')(\mathbf{1}) - \mathbf{1}}{i\lambda}$$

lies in $M(A)$. On the other hand, since R'' is a hermitian operator on A'', and the mapping $S \to S(\mathbf{1})$ from $BL(A'')$ to A'' is a linear contraction sending $I_{A''}$ to $\mathbf{1}$, we deduce that x belongs to $H(A'', \mathbf{1})$. It follows that x lies in $(M(A))_{sa}$. Put $D := R - T_x$. By the first paragraph of the proof, D is a hermitian operator on A. Now D'' is a hermitian operator on A'' with $D''(\mathbf{1}) = 0$, so that, by [73, Theorem 11], D'' is a Jordan-derivation of A'' anticommuting with the JB^*-involution of A''. Therefore D is a Jordan-derivation of A anticommuting with the JB^*-involution of A. Since the equality $R = T_x + D$ is obvious, the proof is concluded. ∎

7. Notes and remarks

7.1.- The following refinement of Theorem 1.4 is proved in [16]. *If A is a unital complete normed complex algebra, and if there exists a conjugate-linear vector space involution $\square$ on A satisfying $\mathbf{1}^{\square} = \mathbf{1}$ and*

$$\| a^{\square} a \| = \| a^{\square} \| \| a \|$$

for every a in A, then A is an alternative C^*-algebra for some C^*-involution $*$. If in addition the dimension of A is different from 2, then we have $\square = *$.

7.2.- As noticed in [58, Corollary 1.2], the proof of Theorem 2.2 given in [48] allows us to realize that, *if a normed complex algebra B is isometrically Jordan-isomorphic to a non-commutative JB^*-algebra, then B is a non-commutative JB^*-algebra.* On the other hand, it is known that *the norm of every non-commutative JB^*-algebra A is minimal*, i.e., if $|||.|||$ is an algebra-norm on A satisfying $|||.||| \leq \| \cdot \|$, then we have in fact $|||.||| = \| \cdot \|$ [51, Proposition 11]. Now, keeping in mind the above results, we can show that, *if a normed complex algebra B is the range of a contractive Jordan-homomorphism from a non-commutative JB^*-algebra, then B is a non-commutative JB^*-algebra.* The proof goes as follows. Let A be a non-commutative JB^*-algebra and φ a contractive Jordan-homomorphism from A onto the normed complex algebra B. Since closed ideals of A^+ are ideals of A (a consequence of [48, Theorem 4.3]), and quotients of non-commutative JB^*-algebras by closed ideals are non-commutative JB^*-algebras [48, Corollary 1.11], we may assume that φ is injective. Then we can define an algebra norm $|||.|||$ on A^+ by $|||a||| := \| \varphi(a) \|$. Since φ is contractive, and the norm of A^+ is minimal we obtain that $|||.||| = \| \cdot \|$ on A. Now B is isometrically Jordan-isomorphic to A, and hence B is a non-commutative JB^*-algebra.

The result just proved implies that, *if a normed complex alternative algebra B is the range of a contractive Jordan-homomorphism from a non-commutative JB^*-algebra, then B is a an alternative C^*-algebra* (compare [51, Corollary 12]).

7.3.- *Every non-commutative JB^*-algebra A has minimum norm topology*, i.e., the topology of an arbitrary algebra norm on A is always stronger than that of the natural norm (see [5], [51], and [24]). As pointed out in [51, Remark 14], this fact can be applied, together with [49, Theorem 3.5], to derive the theorem, originally due to J. E. Galé [28], asserting that, *if a normed complex algebra B is the range of a weakly compact homomorphism from a non-commutative JB^*-algebra, then B is bicontinuously isomorphic to a finite direct sum of simple non-commutative JB^*-algebras which are either quadratic or finite-dimensional.*

Now, *let B be a normed complex alternative algebra, and assume that B is the range of a weakly compact Jordan-homomorphism from a non-commutative JB^*-algebra.* By the above, B^+ is a finite direct sum of simple ideals which are either quadratic or finite-dimensional. Moreover, such ideals of B^+ are in fact ideals of B (use that, for b in B, the mapping $x \to [b,x]$ is a derivation of B^+, and the folklore fact that direct summands of semiprime algebras are invariant under derivations [58, Lemma 7.5]). Therefore those simple direct summands of B^+ which are quadratic actually are simple quadratic alternative algebras, and hence finite-dimensional [76, Theorems 2.3.4 and 2.2.1]. *Then B is finite-dimensional.*

We note also that *the range of any weakly compact Jordan-homomorphism from an alternative C^*-algebra into a complex normed algebra is finite-dimensional* (compare [51, Corollary 13]). The proof of this assertion involves no new idea, and hence is left to the reader.

7.4.- Most criteria of associativity and commutativity for non-commutative JB^*-algebras reviewed in Section 2 rely in the fact that *a non-commutative JB^*-algebra is associative and commutative if (and only if) it has no non-zero nilpotent element* [33]. A recent related result is the one in [6] that *a non-commutative JB^*-algebra A is commutative if (and only if) there exits a positive constant k satisfying* $\| ab \| \leq k \| ba \|$ for all a, b in A.

7.5.- The structure theorem for prime non-commutative JB^*-algebras (Theorem 3.2) becomes a natural analytical variant of the classification theorem for prime nondegenerate non-commutative Jordan algebras, proved by W. G. Skosyrskii [66]. We recall that a non-commutative Jordan algebra A is said to be **nondegenerate** if the conditions $a \in A$ and $U_a = 0$ imply $a = 0$. As it always happens whenever people work with quite general assumptions, the conclusion in Skosyrskii's theorem becomes lightly rough, and involves some complicated notions, like that of a "central order in an algebra", or that of a "quasi-associative algebra over its extended centroid". However, in a "tour de force", Theorem 3.2 actually can be derived from Skosyrskii's classification and some early known results on non-commutative JB^*-algebras. This is explained in what follows.

We begin by establishing a purely algebraic corollary to Skosyrskii's theorem, whose formulation avoids the "complications" quoted above. According to [25] (see also [15]) a prime algebra A over a field $\mathbb{F}$ is called **centrally closed** over $\mathbb{F}$ if, for every non zero ideal M of A and for every linear mapping $f : M \to A$ satisfying $f(ax) = af(x)$ and $f(xa) = f(x)a$ for all x in M and a in A, there exists λ in $\mathbb{F}$ such that $f(x) = \lambda x$ for every x in M. Now it follows from the main result in [66] that, *if A is a centrally closed prime nondegenerate non-commutative Jordan algebra over an algebraically closed field $\mathbb{F}$, then at least one of the following assertions hold:*

1. *A is commutative.*
2. *A is quadratic (over $\mathbb{F}$).*
3. *A^+ is associative.*
4. *A is* **split quasi-associative** *over $\mathbb{F}$, i.e., there exists an associative algebra B over $\mathbb{F}$, and some λ in $\mathbb{F} \setminus \{1/2\}$ such that A and B coincide as vector spaces, but the product ab of A is related to the one $a\Box b$ of B by means of the equality $ab = \lambda a\Box b + (1-\lambda) b\Box a$.*

We do not know whether the result just formulated is or not explicitly stated in Skosyrskii's paper (since it is written in Russian, and we only know about its main result thanks to the appropriate note in Mathematical Reviews). In any case, the steps to derive the corollary above from the main result of [66] are not difficult, and therefore are left to the reader.

Now let A be a prime non-commutative JB^*-algebra which is neither commutative nor quadratic. Since, clearly, non-commutative JB^*-algebras are nondegenerate, and prime non-commutative JB^*-algebras are centrally closed [63], the above paragraph applies, so that either A^+ is associative or A is split quasi-associative over $\mathbb{C}$. In the last case, it follows easily from [57, Theorem 2] and [56, Lemma] that A is of the form $C^{(\lambda)}$ for some prime C^*-algebra C and some $\frac{1}{2} < \lambda \leq 1$. Assume that A^+ is associative. Then, since A^+ is a JB^*-algebra, A^+ actually is a commutative C^*-algebra. Since commutative C^*-algebras have no non zero derivations [64, Lemma 4.1.2], and for a in A the mapping $b \to ab - ba$ from A to A is a derivation of A^+ [65, p. 146], we have $A = A^+$, and therefore A is commutative. Now Theorem 3.2 has been re-proved.

7.6.- Prime non-commutative JB^*-algebras with non zero socle were precisely classified in [52, Theorem 1.4]. As a consequence of such a classification, *prime non-commutative JB^*-algebras with non zero socle are either commutative, quadratic, or of the form $C^{(\lambda)}$ for some prime C^*-algebra C with non zero socle, and some $\frac{1}{2} < \lambda \leq 1$.* We note that all quadratic non-commutative JB^*-algebras have non zero socle.

7.7.- It is shown in [62, Theorem A] that, *given a C^*-algebra A, a (possibly non-associative) normed complex algebra B having an approximate unit bounded by one, and a surjective linear isometry $F : B \to A$, there exists an isometric Jordan-isomorphism $G : B \to A$, and a unitary element u in $M(A)$ satisfying $F = T_u G$.* It is worth mentioning that the above result follows straightforwardly from Corollary 4.9 and Theorem 6.7. Even Corollary 4.9 and Theorem 6.7 give rise to the result in [62] just quoted with "alternative C^*-algebra" instead of "C^*-algebra". This "alternative" generalization of [62, Theorem A] motivated most results collected in Sections 4, 5, and 6 of the present paper. In a first attempt we tried to obtain such a generalization by replacing associativity with alternativity in the original arguments of [62]. All things worked without relevant problems until the application made in [62] of a result of C. A. Akemann and G. K. Pedersen [1] asserting that, *if A is a C^*-algebra, and if u is a unitary element in A'' such that au^*a lies in A for every a in A, then u belongs to $M(A)$.* In that time we were unable to avoid associativity in the proof of the result of [1] just mentioned, and were obliged to get round the handicap and design a new strategy, which apparently avoided the "alternative" version of the result of [1], and whose main ingredient was Theorem 4.6. Now we claim that Theorem 4.6 germinally contains the "alternative" generalization of the Akemann-Pedersen result, and even its general non-associative extension. The proof of the claim goes as follows.

Let A be a non-commutative JB^*-algebra, and u a unitary element in A'' such that $U_a(u^*)$ lies in A for every a in A. Regarding A (and hence A'') as a JB^*-triple in its canonical way, we have that $\{aub\} = U_{a,b}(u^*)$ belongs to A whenever a and b are in A. Therefore we can consider the complete normed complex algebra B consisting of the Banach space of A and the product $a \circ b := \{aub\}$ (apply either [70, Corollary 2.5] or [27, Corollary 3] for the submultiplicativity of the norm). Let us also consider the JB^*-algebra $A''(u)$ in the meaning explained immediately before Proposition 6.8. Then the product of $A''(u)$ is a product on the Banach space of B'', which extends the product of B and is w^*-continuous in its first variable. Therefore B'' (with the Arens product relative to that of B) coincides with $A''(u)$, and hence is a JB^*-algebra. Now Theorem 4.6 applies, so that B is invariant under the JB^*-involution of $B'' = A''(u)$. Therefore $\{uau\} = U_u(a*)$ lies in A whenever a is in A. By the main identity of JB^*-triples, for a, b in A we have

$$\{uab\} = 2\{uab\} - \{u\{uba\}u\} = 2\{ua\{buu\}\} - \{u\{uba\}u\} = \{bu\{uau\}\} ,$$

and hence $\{uab\}$ lies in A. This proves that u lies in $\mathcal{M}(A)$. *Then,* by Proposition 5.7, *u belongs to $M(A)$.* As a consequence of the fact just proved, *if A is an alternative C^*-algebra, and if u is a unitary element in A'' such that au^*a lies in A for every a in A, then u belongs to $M(A)$.*

7.8.- In the proof of Theorem 4.4 we applied a characterization of Banach spaces underlying unital JB^*-algebras proved in [13]. An independent characterization of such Banach spaces is the one given in [61] that *a non zero complex Banach space X underlies a JB^*-algebra with unit u if and only if $\|u\| = 1$, $X = H(X, u) + iH(X, u)$, and*

$$\inf\{\|f\| : f \in \Pi(X) \text{ and } f(x, u) = f(u, x) = x \ \forall x \in X\} = 1.$$

Here, as in the case that X is a norm-unital complete normed complex algebra and u is the unit of X, $H(X, u)$ stands for the set of those elements h in X such that $V(X, u, h) \subset \mathbb{R}$. A refinement of the result of [61] just quoted can be found in [60, Theorem 4.4].

7.9.- *If A and B are non-commutative JB^*-algebras, and if $F : A \to B$ is an isomorphism, then there exists a $*$-isomorphism $G : A \to B$, and a derivation D of A anticommuting with the JB^*-involution of A, such that $F = G\exp(D)$* [48, Theorem 2.9]. It follows from this result and Example 6.9 that *linearly isometric non-commutative JB^*-algebras need not be Jordan-isomorphic.* A similar pathology does not occur for JB-algebras [35].

Acknowledgments.- Part of this work was done while the third author was visiting the University of Almería. He is grateful to the Department of Algebra and Mathematical Analysis of that university for its hospitality and support.

REFERENCES

[1] C. A. AKEMANN and G. K. PEDERSEN, Complications of semicontinuity in C^*-algebra theory. *Duke Math. J.* **40** (1973), 785-795.

[2] E. M. ALFSEN, F. W. SHULTZ, and E. STORMER, A Gelfand-Neumark theorem for Jordan algebras. *Adv. Math.* **28** (1978), 11-56.

[3] K. ALVERMANN and G. JANSSEN, Real and complex non-commutative Jordan Banach algebras. *Math. Z.* **185** (1984), 105-113.

[4] R. ARENS, The adjoint of a bilinear operation. *Proc. Amer. Math. Soc.* **2** (1951), 839-848.

[5] A. BENSEBAH, Weakness of the topology of a JB^*-algebra. *Canad. Math. Bull.* **35** (1992), 449-454.

[6] M. BENSLINANE, L. MESMOUDI, and A. RODRIGUEZ, A characterization of commutativity for non-associative normed algebras. *Extracta Math.* **15** (2000), 121-143.

[7] H. F. BOHNENBLUST and S. KARLIN, Geometrical properties of the unit sphere of a Banach algebra. *Ann. Math.* **62** (1955), 217-229.

[8] F. F. BONSALL and J. DUNCAN, *Numerical ranges of operators on normed spaces and of elements of normed algebras.* London Math. Soc. Lecture Note Series **2**, Cambridge, 1971.

[9] F. F. BONSALL and J. DUNCAN, *Numerical ranges II.* London Math. Soc. Lecture Note Series **10**, Cambridge, 1973.

[10] F. F. BONSALL and J. DUNCAN, *Complete normed algebras.* Springer-Verlag, Berlin, 1973.

[11] R. B. BRAUN, Structure and representations of non-commutative C^*-Jordan algebras. *Manuscripta Math.* **41** (1983), 139-171.

[12] R. B. BRAUN, A Gelfand-Neumark theorem for C^*-alternative algebras. *Math. Z.* **185** (1984), 225-242.

[13] R. B. BRAUN, W. KAUP, and H. UPMEIER, A holomorphic characterization of Jordan C^*-algebras. *Math. Z.* **161** (1978), 277-290.

[14] L. J. BUNCE and C. H. CHU, Compact operations, multipliers and Radon-Nikodym property in JB^*-triples. *Pacific J. Math.* **153** (1992), 249-265.

[15] M. CABRERA and A. RODRIGUEZ, Extended centroid and central closure of semiprime normed algebras: a first approach. *Commun. Algebra* **18** (1990), 2293-2326.

[16] M. CABRERA and A. RODRIGUEZ, New associative and nonassociative Gelfand-Naimark theorems. *Manuscripta Math.* **79** (1993), 197-208.

[17] C. H. CHU, B. IOCUM, and G. LOUPIAS, Grothendieck's theorem and factorization of operators in Jordan triples. *Math. Ann.* **284** (1989), 41-53.

[18] C. H. CHU, A. MORENO, and A. RODRIGUEZ, On prime JB^*-triples. In *Function Spaces; Proceedings of the Third Conference on Function Spaces, May 19-23, 1998, Southern Illinois University at Edwardsville* (Ed. K. Jarosz), *Contemporary Math.* **232** (1999), 105-109.

[19] M. J. CRABB, J. DUNCAN, and C. M. McGREGOR, Characterizations of commutativity for C^*-algebras. *Glasgow Math. J.* **15** (1974), 172-175.

[20] S. DINEEN, The second dual of a JB^*-triple system. In *Complex Analysis, Functional Analysis and Approximation Theory* (ed. by J. Múgica), 67-69, North-Holland Math. Stud. **125**, North-Holland, Amsterdam-New York, 1986.

[21] J. DUNCAN, Review of [44]. *Math. Rev.* **87e**:46067.

[22] C. M. EDWARDS, On Jordan W^*-algebras. *Bull. Sci. Math.* **104** (1980), 393-403.

[23] C. M. EDWARDS, Multipliers of JB-algebras. *Math. Ann.* **249** (1980), 265-272.

[24] S. EJAZIAN and A. NIKNAM, A Kaplansky theorem for JB^*-algebras. *Rocky Mt. J. Math.* **28** (1988), 977-962.

[25] T. S. ERICKSON, W. S. MARTINDALE *3rd*, and J. M. OSBORN, Prime nonassociative algebras. *Pacific J. Math.* **60** (1975), 49-63.

[26] A. FERNANDEZ, E. GARCIA, and A. RODRIGUEZ, A Zel'manov prime theorem for JB^*-algebras. *J. London Math. Soc.* **46** (1992), 319-335.

[27] Y. FRIEDMAN and B. RUSSO, The Gelfand-Naimark theorem for JB^*-triples. *Duke Math. J.* **53** (1986), 139-148.

[28] J. E. GALE, Weakly compact homomorphisms in non-associative algebras. In *Nonassociative algebraic models* (Ed. S. González and H. Ch. Myung), 167-171, Nova Science Publishers, New York, 1992.

[29] H. HANCHE-OLSEN and E. STORMER, *Jordan operator algebras.* Monograph Stud. Math. **21**, Pitman, 1984.

[30] S. HEINRICH, Ultraproducts in Banach space theory. *J. Reine Angew. Math.* **313** (1980), 72-104.

[31] G. HORN, Coordinatization theorem for JBW^*-triples. *Quart. J. Math. Oxford* **38** (1987), 321-335.

[32] T. HURUYA, The normed space numerical index of C^*-algebras. *Proc. Amer. Math. Soc.* **63** (1977), 289-290.

[33] B. IOCHUM, G. LOUPIAS, and A. RODRIGUEZ, Commutativity of C^*-algebras and associativity of JB^*-algebras. *Math. Proc. Cambridge Phil. Soc.* **106** (1989), 281-291.

[34] J. M. ISIDRO, W. KAUP, and A. RODRIGUEZ, On real forms of JB^*-triples. *Manuscripta Math.* **86** (1995), 311-335.

[35] J. M. ISIDRO and A. RODRIGUEZ, Isometries of JB-algebras. *Manuscripta Math.* **86** (1995), 337-348.

[36] N. JACOBSON, *Structure and representations of Jordan algebras.* Amer. Math. Soc. Coll. Publ. **39**, Providence, Rhode Island 1968.

[37] R. V. KADISON, Isometries of operator algebras. *Ann. Math.* **54** (1951), 325-338.

[38] A. M. KAIDI, J. MARTINEZ, and A. RODRIGUEZ, On a non-associative Vidav-Palmer theorem. *Quart. J. Math. Oxford* **32** (1981), 435-442.

[39] A. KAIDI, A MORALES, and A. RODRIGUEZ, Geometric properties of the product of a C^*-algebra. *Rocky Mt. J. Math.* (to appear).

[40] A. KAIDI, A MORALES, and A. RODRIGUEZ, Prime non-commutative JB^*-algebras. *Bull. London Math. Soc.* **32** (2000), 703-708.

[41] A. KAIDI, A MORALES, and A. RODRIGUEZ, A holomorphic characterization of C^*- and JB^*-algebras. *Manuscripta Math.* (to appear).

[42] W. KAUP, Algebraic characterization of symmetric complex Banach manifolds. *Math. Ann.* **228** (1977), 39-64.

[43] W. KAUP, A Riemann mapping theorem for bounded symmetric domains in complex Banach spaces. *Math. Z.* **183** (1983), 503-529.

[44] J. MARTINEZ, J. F. MENA, R. PAYA, and A. RODRIGUEZ, An approach to numerical ranges without Banach algebra theory. *Illinois J. Math.* **29** (1985), 609-625.

[45] A. MORENO and A. RODRIGUEZ, On the Zel'manovian classification of prime JB^*-triples. *J. Algebra* **226** (2000), 577-613.

[46] K. McCRIMMON, Noncommutative Jordan rings. *Trans. Amer. Math. Soc.* **158** (1971), 1-33.

[47] A. L. T. PATERSON and A. M. SINCLAIR, Characterisation of isometries between C^*-algebras. *J. London Math. Soc.* **5** (1972), 755-761.

[48] R. PAYA, J. PEREZ, and A. RODRIGUEZ, Non-commutative Jordan C^*-algebras. *Manuscripta Math.* **37** (1982), 87-120.

[49] R. PAYA, J. PEREZ, and A. RODRIGUEZ, Type I factor representations of non-

commutative JB^*-algebras. *Proc. London Math. Soc.* **48** (1984), 428-444.
[50] G. K. PEDERSEN, *C^*-algebras and their automorphism groups.* Academic Press, London 1979.
[51] J. PEREZ, L. RICO, and A. RODRIGUEZ, Full subalgebras of Jordan-Banach algebras and algebra norms on JB^*-algebras. *Proc. Amer. Math. Soc.* **121** (1994), 1133-1143.
[52] J. PEREZ, L. RICO, A. RODRIGUEZ, and A. R. VILLENA, Prime Jordan-Banach algebras with non-zero socle. *Commun. Algebra* **20** (1992), 17-53.
[53] A. RODRIGUEZ, A Vidav-Palmer theorem for Jordan C^*-algebras and related topics. *J. London Math. Soc.* **22** (1980), 318-332.
[54] A. RODRIGUEZ, Non-associative normed algebras spanned by hermitian elements. *Proc. London Math. Soc.* **47** (1983), 258-274.
[55] A. RODRIGUEZ, A note on Arens regularity. *Quart. J. Math. (Oxford)* **38** (1987), 91-93.
[56] A. RODRIGUEZ, Mutations of C^*-algebras and quasiassociative JB^*-algebras. *Collectanea Math.* **38** (1987), 131-135.
[57] A. RODRIGUEZ, Jordan axioms for C^*-algebras. *Manuscripta Math.* **61** (1988), 279-314.
[58] A. RODRIGUEZ, An approach to Jordan-Banach algebras from the theory of nonassociative complete normed algebras. In *Algèbres de Jordan and algèbres normées non associatives, Ann. Sci. Univ. "Blaise Pascal", Clermont II, Sér. Math.* **27** (1991), 1-57.
[59] A. RODRIGUEZ, Jordan structures in Analysis. In *Jordan algebras, Proceedings of the Conference held in Oberwolfach, Germany, August 9-15, 1992* (Ed. W. Kaup, K. McCrimmon, and H. P. Petersson), 97-186, Walter de Gruyter, Berlin-New York 1994.
[60] A. RODRIGUEZ, Multiplicative characterization of Hilbert spaces and other interesting classes of Banach spaces. In *Encuentro de Análisis Matemático,* Homenaje al profesor B. Rodríguez-Salinas (Ed. F. Bombal, F. L. Hernández, P. Jiménez y J. L. de María), *Rev. Mat. Univ. Complutense Madrid* **9** (1996), 149-189.
[61] A. RODRIGUEZ, A Banach space characterization of JB^*-algebras. *Math. Japon.* **45** (1997), 217-221.
[62] A. RODRIGUEZ, Isometries and Jordan-isomorphisms onto C^*-algebras. *J. Operator Theory* **40** (1998), 71-85.
[63] A. RODRIGUEZ and A. R. VILLENA, Centroid and extended centroid of JB^*-algebras. In *Nonassociative algebraic models* (Ed. S. González and H. Ch. Myung), 223-232, Nova Science Publishers, New York, 1992.
[64] S. SAKAI, *C^*- and W^*-algebras.* Springer-Verlag, Berlin, 1971.
[65] R. D. SCHAFER, *An introduction to nonassociative algebras.* Academic Press, New York, 1966.
[66] V. G. SKOSYRSKII, Strongly prime noncommutative Jordan algebras (in Russian). *Trudy Inst. Mat. (Novosibirk)* **16** (1989), 131-164, 198-199, *Math. Rev.* **91b**:17001.
[67] H. UPMEIER, *Symmetric Banach Manifolds and Jordan C^*-algebras.* North-Holland, Amsterdam, 1985.
[68] A. R. VILLENA, *Algunas cuestiones sobre las JB^*-álgebras: centroide, centroide extendido y zócalo.* Tesis Doctoral, Universidad de Granada 1990.

[69] J. D. M. WRIGHT, Jordan C^*-algebras. *Michigan Math. J.* **24** (1977), 291-302.

[70] J. D. M. WRIGHT and M. A. YOUNGSON, A Russo Dye theorem for Jordan C^*-algebras. In *Functional analysis: surveys and recent results* (Ed. by K. D. Bierstedt and F. Fuchssteiner), 279-282, North Holland Math. Studies **27**, Amsterdam, 1977.

[71] J. D. M. WRIGHT and M. A. YOUNGSON, On isometries of Jordan algebras. *J. London Math. Soc.* **17** (1978), 339-344.

[72] M. A. YOUNGSON, A Vidav theorem for Banach Jordan algebras. *Math. Proc. Cambridge Philos. Soc.* **84** (1978), 263-272.

[73] M. A. YOUNGSON, Hermitian operators on Banach Jordan algebras, *Proc. Edinburgh Math. Soc.* **22** (1979), 93-104.

[74] M. A. YOUNGSON, Non unital Banach Jordan algebras and C^*-triple systems. *Proc. Edinburgh Math. Soc.* **24** (1981), 19-31.

[75] E. I. ZEL'MANOV, On prime Jordan triple systems III. *Siberian Math. J.* **26** (1985), 71-82.

[76] K. A. ZHEVLAKOV, A. M. SLIN'KO, I. P. SHESTAKOV, and A. I. SHIRSHOV, *Rings that are nearly associative.* Academic Press, New York, 1982.

Recent Progress in Functional Analysis
K.D. Bierstedt, J. Bonet, M. Maestre and J. Schmets

Grothendieck's inequalities revisited

Antonio M. Peralta[*] and Angel Rodríguez Palacios[†]

Departamento de Análisis Matemático, Facultad de Ciencias,
Universidad de Granada, 18071 Granada, Spain.
aperalta@goliat.ugr.es and apalacio@goliat.ugr.es

To Professor Manuel Valdivia on the occasion of his seventieth birthday.

Abstract

We review the main results obtained in 2 other papers concerning Grothendieck's inequalities for real and complex JB-triples. We improve the constants involved in this inequalities. We show that for every complex (respectively, real) JB*-triples* $\mathcal{E}$, $\mathcal{F}$, $M = 3 + 2\sqrt{3}$ *(respectively,* $M = 2(3 + 2\sqrt{3})$*), and every bounded bilinear form* U *on* $\mathcal{E} \times \mathcal{F}$, *there exist states* $\Phi \in D(BL(\mathcal{E}), I_{\mathcal{E}})$ *and* $\Psi \in D(BL(\mathcal{F}), I_{\mathcal{F}})$ *such that*

$$|U(x,y)| \leq M \|U\| |||x|||_{\Phi} |||y|||_{\Psi}$$

for all $(x, y) \in \mathcal{E} \times \mathcal{F}$, *where* $D(BL(\mathcal{E}), I_{\mathcal{E}})$ *is the set of states of* $BL(\mathcal{E})$ *relative to the identity map on* $\mathcal{E}$ *and* $|||x|||_{\Phi}^2 := \Phi(L(x,x))$.
MCS 2000 17C65, 46K70, 46L05, 46L10, and 46L70

Introduction

A celebrated result of A. Grothendieck [11] asserts that there is a universal constant K such that, if Ω is a compact Hausdorff space and T is a bounded linear operator from $C(\Omega)$ to a complex Hilbert space H, then there exists a probability measure μ on Ω such that

$$\|T(f)\|^2 \leq K^2 \|T\|^2 \left(\int_{\Omega} |f|^2 d\mu \right)$$

for all $f \in C(\Omega)$. This result is called "Commutative Little Grothendieck's inequality". Actually the result just quoted is a consequence of the so called "Commutative Big Grothendieck's Inequality" assuring the existence of a universal constant $M > 0$ such that for every pair of compact Hausdorff spaces (Ω_1, Ω_2) and every bounded bilinear form U

[*]Supported by Ministry of Education and Science D.G.I.C.Y.T. project no. PB 98-1371, and Junta de Andalucía grant FQM 0199
[†]Partially supported by Junta de Andalucía grant FQM 0199

on $C(\Omega_1) \times C(\Omega_2)$ there are probability measures μ_1 and μ_2 on Ω_1 and Ω_2, respectively, satisfying

$$|U(f,g)|^2 \leq M^2 \|U\|^2 \left(\int_{\Omega_1} |f|^2 d\mu_1 \right) \left(\int_{\Omega_2} |g|^2 d\mu_2 \right)$$

for all $(f,g) \in C(\Omega_1) \times C(\Omega_2)$.

Reasonable non-commutative generalizations of the original little and big Grothendieck's inequalities have been obtained by G. Pisier ([23], [24]) and U. Haagerup ([12],[13]). In these generalizations non-commutative C*-algebras replace $C(\Omega)$-spaces, norm-one positive linear functionals replace probability measures, and the module $|a|$ of an element a in a C*-algebra is defined as $(\frac{aa^*+a^*a}{2})^{\frac{1}{2}}$.

At the end of 80's, the important works of T. Barton and Y. Friedman [2] and C-H. Chu, B. Iochum, and G. Loupias [8] on Grothendieck's inequalities for the so-called complex JB*-triples appeared. Complex JB*-triples are natural generalizations of C*-algebras, although they need not have a natural order structure. One of the most important ideas contained in the Barton-Friedman paper is the construction of "natural" prehilbertian seminorms $\|.\|_\varphi$, associated to norm-one continuous linear functionals φ on complex JB*-triples, in order to play, in Grothendieck's inequalities, the same role as that of the prehilbertian seminorms derived from norm-one positive linear functionals in the case of C*-algebras.

Real JB*-triples have been recently introduced in [16], and their theory has been quickly developed. The class of real JB*-triples includes all JB-algebras [14], all real C*-algebras [10], all J*B-algebras [1], and all complex JB*-triples (regarded as real Banach spaces). We have studied in deep the papers [2] and [8], cited in the previous paragraph, with the aim of extending their results to the context of real JB*-triples, as well as obtaining weak* versions of Grothendieck's inequalities for the so-called real or complex JBW*-triples. The last goal follows the line of [13, Proposition 2.3] in the case of von Neumann algebras. The results obtained by us in these directions appear in [21] and [22]. In fact we have found some gaps in the proofs of the results of [2] and [8], and given partial solutions to them (see [21, Introduction] and [22, Section 1]). In words of L. J. Bunce [6], "the articles [21] and [22] provide antidotes to some subtle difficulties in [2] and subsequence works, including certain results on the important strong* topology of a JBW*-triple".

In the present paper we review the main results in [21] and [22], and prove some new related facts. Most novelties consist in getting better values of the constants involved in Grothendieck's inequalities. In some case (see for instance Theorem 2.6) such an improvement need a completely new proof. As shown in [21, Introduction] and [22, Section 1], the actual formulations of Grothendieck's inequalities for complex JB*-triples in [2] and [8] remain up to date mere conjectures. However, Grothendieck's inequalities remain valid for JB*-algebras as shown in [8]. We show in Theorems 1.2, 1.8 and 2.2 that those conjectures are valid (even for real JB*-triples) whenever we allow a small enlargement of the family of prehilbert seminorms $\{\|.\|_\varphi\}$.

Notation

Let X be a normed space. We denote by S_X, B_X, X^*, and I_X the unit sphere, the closed unit ball, the dual space, and the identity operator, respectively, of X. When necessary we will use the symbol J_X for the natural embedding of X in its bidual X^{**}. If

Y is another normed space, then $BL(X, Y)$ will stand for the normed space of all bounded linear operators form X to Y. Of course we write $BL(X)$ instead of $BL(X, X)$. Now assume that the normed space X is complex. A **conjugation** on X will be a conjugate-linear isometry on X of period 2. If τ is a conjugation on X, then X^τ will stand for the real normed space of all τ-fixed elements of X. Real normed spaces which can be written as X^τ, for some conjugation τ on X, are called **real forms** of X. By $X_{\mathbb{R}}$ we denote the real Banach space underlying X.

1. Little Grothendieck's inequality

A complex JB*-triple is a complex Banach space $\mathcal{E}$ with a continuous triple product $\{.,.,.\} : \mathcal{E} \times \mathcal{E} \times \mathcal{E} \to \mathcal{E}$ which is bilinear and symmetric in the outer variables and conjugate linear in the middle variable, and satisfies:

1. (Jordan Identity) $L(a,b)\{x,y,z\} = \{L(a,b)x,y,z\} - \{x,L(b,a)y,z\} + \{x,y,L(a,b)z\}$ for all a,b,c,x,y,z in $\mathcal{E}$, where $L(a,b)x := \{a,b,x\}$;
2. The map $L(a,a)$ from $\mathcal{E}$ to $\mathcal{E}$ is an hermitian operator with nonnegative spectrum for all a in $\mathcal{E}$;
3. $\|\{a,a,a\}\| = \|a\|^3$ for all a in $\mathcal{E}$.

Concerning condition 2 above, we recall that a bounded linear operator T on a complex Banach space X is said to be **hermitian** is $\|\exp(i\lambda T)\| = 1$, for every $\lambda \in \mathbb{R}$.

Complex JB*-triples were introduced by W. Kaup in order to provide an algebraic setting for the study of bounded symmetric domains in complex Banach spaces (see [17], [18] and [29]).

By a complex JBW*-triple we mean a complex JB*-triple which is a dual Banach space. We recall that the triple product of every complex JBW*-triple is separately weak*-continuous [4], and that the bidual $\mathcal{E}^{**}$ of a complex JB*-triple $\mathcal{E}$ is a JBW*-triple whose triple product extends the one of $\mathcal{E}$ [9].

Given a complex JBW*-triple $\mathcal{W}$ and a norm-one element φ in the predual $\mathcal{W}_*$ of $\mathcal{W}$, we can construct a prehilbert seminorn $\|.\|_\varphi$ as follows (see [2, Proposition 1.2]). By the Hahn-Banach theorem there exists $z \in \mathcal{W}$ such that $\varphi(z) = \|z\| = 1$. Then $(x,y) \mapsto \varphi\{x,y,z\}$ becomes a positive sesquilinear form on $\mathcal{W}$ which does not depend on the point of support z for φ. The prehilbert seminorm $\|.\|_\varphi$ is then defined by $\|x\|_\varphi^2 := \varphi\{x,x,z\}$ for all $x \in \mathcal{W}$. If $\mathcal{E}$ is a complex JB*-triple and φ is a norm-one element in $\mathcal{E}^*$, then $\|.\|_\varphi$ acts on $\mathcal{E}^{**}$, hence in particular it acts on $\mathcal{E}$.

In [2, Theorem 1.3], Barton and Friedman established a "Little Grothendieck's inequality" for complex JB*-triples, assuring that if T is a bounded linear operator from a complex JB*-triple $\mathcal{E}$ to a complex Hilbert space $\mathcal{H}$ whose second transpose T^{**} attains its norm at a so-called "complete tripotent", then there exists a norm-one functional $\varphi \in \mathcal{E}^*$ such that

$$\|T(x)\| \leq \sqrt{2}\|T\|\|x\|_\varphi$$

for all $x \in \mathcal{E}$. However, although assumed in the proof, the hypothesis that T^{**} attains its norm at a complete tripotent does not arise in the statement of [2, Theorem 1.3] (compare

[21]). Since by [21, proof of Theorem 4.3] we know that T^{**} attains its norm at a complete tripotent whenever it attains its norm, and the set of all operators $T \in BL(\mathcal{E}, \mathcal{H})$ such that T^{**} attains its norm is norm dense in $BL(\mathcal{E}, \mathcal{H})$ [19, Theorem 1], we have the following theorem.

Theorem 1.1. [21, Theorem 1.1] *Let $\mathcal{E}$ be a complex JB*-triple and $\mathcal{H}$ a complex Hilbert space. Then the set of those bounded linear operators T from $\mathcal{E}$ to $\mathcal{H}$ such that there exists a norm-one functional $\varphi \in \mathcal{E}^*$ satisfying*

$$\|T(x)\| \leq \sqrt{2}\|T\|\|x\|_\varphi$$

for all $x \in \mathcal{E}$, is norm dense in $BL(\mathcal{E}, \mathcal{H})$.

Let $\mathcal{E}$ and $\mathcal{H}$ be as in Theorem 1.1. The question if for every T in $BL(\mathcal{E}, \mathcal{H})$ there exists $\varphi \in S_{\mathcal{E}^*}$ satisfying

$$\|T(x)\| \leq \sqrt{2}\|T\|\|x\|_\varphi$$

for all $x \in \mathcal{E}$, remains an open problem. In any case, if we allow a slightly enlargement of the family of prehilbert seminorms $\{\|.\|_\varphi : \varphi \in S_{\mathcal{E}^*}\}$, then, as we are showing in what follows, the answer to the above question becomes affirmative. We note that the new prehilber seminorms we are building are as naturally derived from the structure of $\mathcal{E}$ as those in the family $\{\|.\|_\varphi : \varphi \in S_{\mathcal{E}^*}\}$.

Let X be a Banach space, and u a norm-one element in X. The set of **states** of X relative to u, $D(X, u)$, is defined as the non empty, convex, and weak*-compact subset of X^* given by

$$D(X, u) := \{\Phi \in B_{X^*} : \Phi(u) = 1\}.$$

For $x \in X$, the **numerical range** of x relative to u, $V(X, u, x)$, is given by $V(X, u, x) := \{\Phi(x) : \Phi \in D(X, u)\}$. It is well known that a bounded linear operator T on a complex Banach space X is hermitian if and only if $V(BL(X), I_X, T) \subseteq \mathbb{R}$ (compare [5, Corollary 10.13]).

Let $\mathcal{E}$ be a complex JB*-triple and $\Phi \in D(BL(\mathcal{E}), I_\mathcal{E})$. Since for every $x \in \mathcal{E}$, the operator $L(x, x)$ is hermitian and has non-negative spectrum, it follows from [5, Lemma 38.3] that the mapping $(x, y) \to \Phi(L(x, y))$ from $\mathcal{E} \times \mathcal{E}$ to $\mathbb{C}$ becomes a positive sesquilinear form on $\mathcal{E}$. Then we define the prehilbert seminorm $|||.|||_\Phi$ on $\mathcal{E}$ by $|||x|||_\Phi^2 := \Phi(L(x, x))$.

Let $\varphi \in S_{\mathcal{E}^*}$ and let $e \in S_{\mathcal{E}^{**}}$ such that $\varphi(e) = 1$. We consider the element $\Phi_{\varphi,e}$ of $D(BL(\mathcal{E}), I_\mathcal{E})$ given by $\Phi_{\varphi,e}(T) := \varphi T^{**}(e)$ for all $T \in BL(\mathcal{E})$, so that we have $|||.|||_{\Phi_{\varphi,e}} = \|.\|_\varphi$ on $\mathcal{E}$.

Theorem 1.2. *Let $\mathcal{E}$ be a complex JB*-triple, $\mathcal{H}$ a complex Hilbert space and $T : \mathcal{E} \to \mathcal{H}$ a bounded linear operator. Then there exists $\Phi \in D(BL(\mathcal{E}), \mathcal{I}_\mathcal{E})$ such that*

$$\|T(x)\| \leq \sqrt{2}\ \|T\|\ |||x|||_\Phi$$

for all $x \in \mathcal{E}$.

Proof. By Theorem 1.1, for every $n \in \mathbb{N}$ there is a bounded linear operator $T_n : \mathcal{E} \to \mathcal{H}$ and a norm-one functional $\varphi_n \in \mathcal{E}^*$ satisfying

$$\|T_n - T\| \leq \frac{1}{n},$$

and

$$\|T_n(x)\| \leq \sqrt{2}\|T_n\|\|x\|_{\varphi_n} = \sqrt{2}\|T_n\| \; |||x|||_{\Phi_{\varphi_n,e_n}}$$

for all $x \in \mathcal{E}$, where $e_n \in S_{\mathcal{E}^{**}}$ with $\varphi_n(e_n) = 1$ $(n \in \mathbb{N})$.

Since $D(BL(\mathcal{E}), I_{\mathcal{E}})$ is weak*-compact, we can take a weak* cluster point $\Phi \in D(BL(\mathcal{E}), I_{\mathcal{E}})$ of the sequence Φ_{φ_n,e_n} to obtain

$$\|T(x)\| \leq \sqrt{2}\; \|T\| \; |||x|||_{\Phi}$$

for all $x \in \mathcal{E}$. □

From the previous Theorem we can now derive a remarkable result of U. Haagerup.

Corollary 1.3. [12, Theorem 3.2] *Let $\mathcal{A}$ be a C^*-algebra, $\mathcal{H}$ a complex Hilbert space, and $T : \mathcal{A} \to \mathcal{H}$ a bounded linear operator. Then there exist two norm-one positive linear functionals φ and ψ on $\mathcal{A}$, such that*

$$\|T(x)\|^2 \leq \|T\|^2(\varphi(x^*x) + \psi(xx^*)),$$

for all $x \in \mathcal{A}$.

Proof. By Theorem 1.2 there exists $\Phi \in D(BL(\mathcal{A}), I_{\mathcal{A}})$ such that

$$\|T(x)\|^2 \leq 2\|T\|^2\Phi(L(x,x))$$

for all $x \in \mathcal{A}$. Since for every $x \in \mathcal{A}$ the equality $L(x,x) = \dfrac{1}{2}(L_{xx^*} + L_{x^*x})$ holds (where, for $a \in \mathcal{A}$, L_a and R_a stands for the left and right multiplication by a, respectively), we have

$$\|T(x)\|^2 \leq \|T\|^2\Phi(L_{xx^*} + R_{x^*x})$$

for all $x \in \mathcal{A}$.

Now, denoting by $\widehat{\varphi}$ and $\widehat{\psi}$ the positive linear functionals on $\mathcal{A}$ given by $\widehat{\varphi}(x) := \Phi(L_x)$, and $\widehat{\psi}(x) := \Phi(R_x)$, respectively, and choosing norm-one positive linear functionals φ, ψ on $\mathcal{A}$ satisfying $\widehat{\varphi} \leq \varphi$ and $\widehat{\psi} \leq \psi$ (which is posible because $\widehat{\varphi}$ and $\widehat{\psi}$ belong to $B_{\mathcal{A}^*}$), we get

$$\|T(x)\|^2 \leq \|T\|^2(\varphi(x^*x) + \psi(xx^*)),$$

for all $x \in \mathcal{A}$. □

Following [16], we define real JB*-triples as norm-closed real subtriples of complex JB*-triples. If $\mathcal{E}$ is a complex JB*-triple, then conjugations on $\mathcal{E}$ preserve the triple product of E, and hence the real forms of $\mathcal{E}$ are real JB*-triples. In [16] it is shown that actually every real JB*-triple can be regarded as a real form of a suitable complex JB*-triple.

By a real JBW*-triple we mean a real JB*-triple whose underlying Banach space is a dual Banach space. As in the complex case, the triple product of every real JBW*-triple is separately weak*-continuous [20], and the bidual $\mathcal{E}^{**}$ of a real JB*-triple $\mathcal{E}$ is a real JBW*-triple whose triple product extends the one of $\mathcal{E}$ [16]. Noticing that every real JBW*-triple is a real form of a complex JBW*-triple [16], it follows easily that, if W is a real JBW*-triple and if φ is a norm-one element in W_*, then, for $z \in W$ such that

$\varphi(z) = ||z|| = 1$, the mapping $x \mapsto (\varphi\{x, x, z\})^{\frac{1}{2}}$ is a prehilbert seminorm on W (not depending on z). Such a seminorm will be denoted by $||.||_\varphi$.

The main goal of [21] is to extending Theorem 1.1 to the setting of real JB*-triples. Such an extension is actually obtained in [21, Theorem 4.5] with constant $4\sqrt{2}$ instead of $\sqrt{2}$. However, an easy final touch to the proof of Theorem 4.3 in [21] is letting us to get a better value of the constant.

Proposition 1.4. *Let E be a real JB*-triple, H a real Hilbert space, and $T : E \to H$ a bounded linear operator which attains its norm. Then there exists $\varphi \in S_{E^*}$ satisfying*

$$||T(x)|| \leq (1 + 3\sqrt{2})||T||||x||_\varphi$$

for all $x \in E$.

Proof. Without loss of generality we can suppose $||T|| = 1$. Then, by the proof of [21, Theorem 4.3], there exist $e \in S_{E^{**}}$ and $\psi, \xi \in D(E^{**}, e) \cap E^*$ such that

$$||T(x)|| \leq \sqrt{8}||x||_\psi + (1 + \sqrt{2})\ ||x||_\xi$$

for all $x \in E$. Setting $\rho = \dfrac{2\sqrt{2}}{1+\sqrt{2}}$ and $\varphi := \frac{1}{1+\rho}(\xi + \rho\ \psi)$, φ is a norm-one functional in E^* with $\varphi(e) = 1$, and we have

$$||T(x)|| \leq \sqrt{(1+\sqrt{2})^2 + \frac{8}{\rho}}\sqrt{||x||_\xi^2 + \rho\ ||x||_\psi^2}$$

$$= \left([(1+\sqrt{2})^2 + \frac{8}{\rho}](1+\rho)\right)^{\frac{1}{2}}\ ||x||_\varphi = (1 + 3\sqrt{2})\ ||x||_\varphi$$

for all $x \in E$. □

Keeping in mind Proposition 1.4 above, a new application of [19, Theorem 1] gives us the following improvement of [21, Theorem 4.5].

Theorem 1.5. *Let E be a real JB*-triple and H a real Hilbert space. Then the set of those bounded linear operators T from E to H such that there exists a norm-one functional $\varphi \in E^*$ satisfying*

$$||T(x)|| \leq (1 + 3\sqrt{2})||T||||x||_\varphi$$

for all $x \in E$, is norm dense in $BL(E, H)$.

Let X be a complex Banach space and τ a conjugation on X. We define a conjugation $\tilde{\tau}$ on $BL(X)$ by $\tilde{\tau}(T) := \tau T \tau$. If T is a $\tilde{\tau}$-invariant element of $BL(X)$, then we have $T(X^\tau) \subseteq X^\tau$, and hence we can consider $\Lambda(T) := T|_{X^\tau}$ as a bounded linear operator on the real Banach space X^τ. Since the mapping $\Lambda : BL(X)^{\tilde{\tau}} \to BL(X^\tau)$ is a linear contraction sending I_X to I_{X^τ}, we get

$$V(BL(X^\tau), I_{X^\tau}, \Lambda(T)) \subseteq V(BL(X)^{\tilde{\tau}}, I_X, T)$$

for all $T \in BL(X)^{\tilde{\tau}}$. On the other hand, by the Hahn-Banach Theorem, we have

$$V(BL(X)^{\tilde{\tau}}, I_X, T) = V(BL(X)_{\mathbb{R}}, I_X, T)$$

for every $T \in BL(X)^{\tilde{\tau}}$. It follows

$$V(BL(X^\tau), I_{X^\tau}, \Lambda(T)) \subseteq \Re e\, V(BL(X), I_X, T)$$

for all $T \in BL(X)^{\tilde{\tau}}$.

Let E be a real JB*-triple. Since $E = \mathcal{E}^\tau$ for some complex JB*-triple $\mathcal{E}$ with conjugation τ, it follows from the above paragraph that, for $x \in E$, $V(BL(E), I_E, L(x,x))$ consists only of non-negative real numbers. Therefore, for $\Phi \in D(BL(E), I_E)$, the mapping $(x,y) \to \Phi(L(x,y))$ from $E \times E$ to $\mathbb{R}$ is a positive symmetric bilinear form on E, and hence $|||x|||_\Phi^2 := \Phi(L(x,x))$ defines a prehilbert seminorm on E.

Now, when in the proof of Theorem 1.2 Theorem 1.5 replaces Theorem 1.1, we arrive at a real variant of Theorem 1.2 with constant $(1 + 3\sqrt{2})$ instead of $\sqrt{2}$. However, as we show in Theorem 1.8 below, a better result holds.

Lemma 1.6. *Let X be a complex Banach space with a conjugation τ. Denote by $\mathbf{H}$ the real Banach space of all hermitian operators on X which lie in $BL(X)^{\tilde{\tau}}$. Then, for every $\Phi \in D(BL(X), I_X)$, there exists $\Psi \in D(BL(X^\tau, I_{X^\tau})$ such that $\Phi(T) = \Psi(\Lambda(T))$ for every T in $\mathbf{H}$.*

Proof. It is easy to see that, for T in $BL(X)^{\tilde{\tau}}$, the inequality $||T|| \leq 2||\Lambda(T)||$ holds. Now, let T be in $\mathbf{H}$. Then, for $n \in \mathbb{N}$, T^n lies in $BL(X)^{\tilde{\tau}}$ and, by [5, Theorem 11.17], we have

$$||T||^n = ||T^n|| \leq 2||\Lambda(T^n)|| = 2||\Lambda(T)^n|| \leq 2||\Lambda(T)||^n.$$

By taking n-th roots and letting $n \to +\infty$, we obtain $||T|| \leq ||\Lambda(T)||$. It follows that Λ, regarded as a mapping from $\mathbf{H}$ to $BL(X^\tau)$, is a linear isometry. Therefore, given $\Phi \in D(BL(X), I_X)$, the composition $\Phi|_{\mathbf{H}}\ \Lambda^{-1}$ belongs to $D(\Lambda(\mathbf{H}), I_{X^\tau})$, and it is enough to choose $\Psi \in D(BL(X^\tau), I_{X^\tau})$ extending $\Phi|_{\mathbf{H}}\ \Lambda^{-1}$ to obtain

$$\Phi(T) = \Psi(\Lambda(T))$$

for all $T \in \mathbf{H}$. □

The next corollary follows straightforwardly from Lemma 1.6 above.

Corollary 1.7. *Let $\mathcal{E}$ be a complex JB*-triple with a conjugation τ, and Φ in $D(BL(\mathcal{E}), I_\mathcal{E})$. Then there exists $\Psi \in D(BL(\mathcal{E}^\tau), I_{\mathcal{E}^\tau})$ such that*

$$|||x|||_\Phi = |||x|||_\Psi$$

for all $x \in \mathcal{E}^\tau$.

Theorem 1.8. *Let E be a real JB*-triple, H a real Hilbert space and $T : E \to H$ a bounded linear operator. Then there exists $\Psi \in D(BL(E), I_E)$ such that*

$$||T(x)|| \leq 2\ ||T||\ |||x|||_\Psi$$

for all $x \in E$.

Proof. Let $\mathcal{E}$ be a complex JB*-triple with conjugation τ such that $E = \mathcal{E}^\tau$, let $\mathcal{H}$ be a complex Hilbert space with conjugation ρ such that $\mathcal{H}^\rho = H$, and let $\widehat{T} \in BL(\mathcal{E}, \mathcal{H})$ such that $\widehat{T}|_E = T$. We note that $\|\widehat{T}\| \leq \sqrt{2}\|T\|$. By Theorem 1.2 there exists $\Phi \in D(BL(\mathcal{E}), I_\mathcal{E})$ satisfying

$$\|T(x)\| \leq \sqrt{2}\|\widehat{T}\| |||x|||_\Phi \tag{1}$$

for all $x \in E$. By Corollary 1.7 there exists $\Psi \in D(BL(E), I_E)$ such that

$$|||x|||_\Phi = |||x|||_\Psi \tag{2}$$

for all $x \in E$. Finally combining (1) and (2) we get

$$\|T(x)\| \leq \sqrt{2}\|\widehat{T}\| |||x|||_\Psi \leq 2\|T\| |||x|||_\Psi$$

for all $x \in E$. □

Section 2 of [22] is mainly devoted to obtaining weak*-versions of the "Little Grothendieck's inequality" for real and complex JBW*-triples. In a first approach we prove the following result.

Proposition 1.9. *If $\mathcal{W}$ is a complex (respectively, real) JBW*-triple, if $\mathcal{H}$ a complex (respectively, real) Hilbert space, and if $M = \sqrt{2}$ (respectively, $M > 1 + 3\sqrt{2}$), then the set of weak*-continuous linear operators T from $\mathcal{W}$ to $\mathcal{H}$ such that there exists a norm-one functional $\varphi \in \mathcal{W}_*$ satisfying*

$$\|T(x)\| \leq M \ \|T\| \ \|x\|_\varphi$$

for all $x \in \mathcal{W}$, is norm dense in the space of all weak-continuous linear operators from $\mathcal{W}$ to $\mathcal{H}$.*

Proposition 1.9 above follows from [22, Lemma 3] (respectively, [22, Lemma 4]) and [30]. When the result in [30] is replaced with a finer principle in [25] on approximation of operator by operator attaining their norms, we get the following theorem.

Theorem 1.10. [22, Theorems 3 and 5] *Let $K > \sqrt{2}$ (respectively, $K > 1+3\sqrt{2}$), $\varepsilon > 0$, $\mathcal{W}$ a complex (respectively, real) JBW*-triple, $\mathcal{H}$ a complex (respectively, real) Hilbert space, and $T : \mathcal{W} \to \mathcal{H}$ a weak*-continuous linear operator. Then there exist norm-one functionals $\varphi_1, \varphi_2 \in \mathcal{W}_*$ such that the inequality*

$$\|T(x)\| \leq K \ \|T\| \ \left(\|x\|_{\varphi_2}^2 + \varepsilon^2 \ \|x\|_{\varphi_1}^2\right)^{\frac{1}{2}}$$

holds for all $x \in \mathcal{W}$.

Of course, Theorem 1.10 above has as a corollary the next non-weak* version of "Little Grothendieck's inequality".

Corollary 1.11. *Let $K > \sqrt{2}$ (respectively, $K > 1 + 3\sqrt{2}$) and $\varepsilon > 0$. Then, for every complex (respectively, real) JB*-triple $\mathcal{E}$, every complex (respectively, real) Hilbert space $\mathcal{H}$, and every bounded linear operator $T : \mathcal{E} \to \mathcal{H}$, there exist norm-one functionals $\varphi_1, \varphi_2 \in \mathcal{E}^*$ such that the inequality*

$$\|T(x)\| \leq K \ \|T\| \ \left(\|x\|_{\varphi_2}^2 + \varepsilon^2 \ \|x\|_{\varphi_1}^2\right)^{\frac{1}{2}},$$

holds for all $x \in \mathcal{E}$.

Remark 1.12. Let K, $\mathcal{W}$, $\mathcal{H}$, and T be as in Theorem 1.10. We claim that there exists Φ in $D(BL(\mathcal{W}), I_{\mathcal{W}})$ which lies in the natural non-complete predual $\mathcal{W} \otimes \mathcal{W}_*$ of $BL(\mathcal{W})$ and satisfies

$$\|T(x)\| \leq K \ \|T\| \ |||x|||_\Phi$$

for all $x \in \mathcal{W}$. Indeed, take $\varepsilon > 0$ such that $\frac{K}{\sqrt{1+\varepsilon^2}} > \sqrt{2}$ (respectively $\frac{K}{\sqrt{1+\varepsilon^2}} > 1 + 3\sqrt{2}$) and apply Theorem 1.10 to find $\varphi_1, \varphi_2 \in S_{\mathcal{W}_*}$ satisfying

$$\|T(x)\| \leq \frac{K}{\sqrt{1+\varepsilon^2}} \ \|T\| \ \left(\|x\|_{\varphi_2}^2 + \varepsilon^2 \ \|x\|_{\varphi_1}^2\right)^{\frac{1}{2}}$$

for all $x \in \mathcal{W}$. Then, choosing $e_i \in D(\mathcal{W}_*, \varphi_i)$ $(i = 1, 2)$ and putting

$$\Phi := \frac{1}{1+\varepsilon^2}(\Phi_{\varphi_2, e_2} + \varepsilon^2 \Phi_{\varphi_1, e_1}),$$

Φ lies in $D(BL(\mathcal{W}), I_{\mathcal{W}}) \cap (\mathcal{W} \otimes \mathcal{W}_*)$ and satisfies

$$\|T(x)\| \leq K \ \|T\| \ |||x|||_\Phi$$

for all $x \in \mathcal{W}$. It seems to be plausible that the claim just proved could remain true with $K = \sqrt{2}$ (respectively, $K = 1+3\sqrt{2}$) whenever we allow the element Φ in $D(BL(\mathcal{W}), I_{\mathcal{W}})$ to lie in the natural complete predual $\mathcal{W}\widehat{\otimes}_\pi \mathcal{W}_*$ of $BL(\mathcal{W})$.

The concluding section of the paper [22] deals with some applications of Theorem 1.10, including certain results on the strong*-topology of real and complex JBW*-triples. We recall that, if W is a real or complex JBW*-triple, then the strong*-topology of W, denoted by $S^*(W, W_*)$, is defined as the topology on W generated by the family of seminorms $\{\|.\|_\varphi : \varphi \in W_*, \ \|\varphi\| = 1\}$. It is worth mentioning that, if a JBW*-algebra $\mathcal{A}$ is regarded as a complex JBW*-triple, then $S^*(\mathcal{A}, \mathcal{A}_*)$ coincides with the so-called "algebra-strong* topology" of $\mathcal{A}$, namely the topology on $\mathcal{A}$ generated by the family of seminorms of the form $x \mapsto \sqrt{\xi(x \circ x^*)}$ when ξ is any weak*-continuous positive linear functional on $\mathcal{A}$ [26, Proposition 3]. As a consequence, when a von Neumann algebra $\mathcal{M}$ is regarded as a complex JBW*-triple, $S^*(\mathcal{M}, \mathcal{M}_*)$ coincides with the familiar strong*-topology of $\mathcal{M}$ (compare [28, Definition 1.8.7]).

For every dual Banach space X (with a fixed predual X_*), we denote by $m(X, X_*)$ the Mackey topology on X relative to its duality with X_*.

The following theorem extends to real JBW*-triples some results in [3], [26], and [27] for complex JBW*-triples, and completely solved a gap in the proof of the results of [26].

Theorem 1.13. [22, Corollary 9 and Theorem 9] *Let W be a real or complex JBW*-triple. Then we have:*

1. *The strong*-topology of W is compatible with the duality (W, W_*).*

2. *If V is a weak*-closed subtriple of W, then the inequality $S^*(W, W_*)|_V \leq S^*(V, V_*)$ holds, and in fact $S^*(W, W_*)|_V$ and $S^*(V, V_*)$ coincide on bounded subsets of V.*

3. *The triple product of W is jointly $S^*(W, W_*)$-continuous on bounded subsets of W.*

4. The topologies $m(W, W_*)$ and $S^*(W, W_*)$ coincide on bounded subsets of W.

Moreover, linear mappings between real or complex JBW-triples are strong*-continuous if and only if they are weak*-continuous.*

Remark 1.14. In a recent work L. J. Bunce obtains an improvement of Assertion 2 of Theorem 1.13. Indeed, in [6, Corollary] he proves that, if W is a real or complex JBW*-triple, and if V is a weak*-closed subtriple, then each element of V_* has a norm preserving extension in W_*, and hence $S^*(W, W_*)|_V = S^*(V, V_*)$

From Assertion 4 in Theorem 1.13 we derive in [22, Theorem 10] a Jarchow-type characterization of weakly compact operators from (real or complex) JB*-triples to arbitrary Banach spaces. With Theorems 1.8 and 1.2 instead of [22, Corollaries 5 and 1] in the proof, Theorem 10 of [22] reads as follows.

Theorem 1.15. *Let E be a real (respectively, complex) JB*-triple, X a real (respectively, complex) Banach space, and $T : E \to X$ a bounded linear operator. The following assertions are equivalent:*

1. *T is weakly compact.*
2. *There exist a bounded linear operator G from E to a real (respectively, complex) Hilbert space and a function $N : (0, +\infty) \to (0, +\infty)$ such that*

$$||T(x)|| \leq N(\varepsilon)||G(x)|| + \varepsilon||x||$$

for all $x \in E$ and $\varepsilon > 0$.

3. *There exist $\Phi \in D(BL(E), I_E)$ and a function $N : (0, +\infty) \to (0, +\infty)$ such that*

$$||T(x)|| \leq N(\varepsilon)\ |||x|||_\Phi + \varepsilon||x||$$

for all $x \in E$ and $\varepsilon > 0$.

For a forerunner of the complex case of Theorem 1.15 above the reader is referred to [7] (see also the comment after [22, Theorem 10]).

2. Big Grothendieck's inequality

Big Grothendieck's inequalities for complex JB*-triples appear in the papers [2] and [8]. However, the proofs of such inequalities in both papers contain some gaps, so we are not sure if the statements of those inequalities are true. In any case, putting together facts completely proved in [2] and [8], the complex case of the following theorem follows with minor difficulties (see for instance [22, Section 1]). The real case of the following theorem has no forerunner before [22].

Theorem 2.1. [22, Theorem 1 and Corollary 8] *Let $M > 4(1 + 2\sqrt{3})\ (1 + 3\sqrt{2})^2$ (respectively, $M > 3 + 2\sqrt{3}$) and E, F be real (respectively, complex) JB*-triples. Then the* *set of all bounded bilinear forms* U *on* $E \times F$ *such that there exist norm-one functionals* $\varphi \in E^*$ *and* $\psi \in F^*$ *satisfying*

$$|U(x,y)| \leq M \; \|U\| \; \|x\|_\varphi \; \|y\|_\psi$$

for all $(x,y) \in E \times F$, *is norm dense in the Banach space of all bounded bilinear forms on* $E \times F$.

The complex case of the next theorem follows from Theorem 2.1 above by arguing as in the proof of Theorem 1.2. The real case then follows from the complex one by a suitable application of Corollary 1.7.

Theorem 2.2. *Let* $\mathcal{E}$, $\mathcal{F}$ *be complex (respectively, real) JB*-triples,* $M = 3 + 2\sqrt{3}$ *(respectively,* $M = 2(3 + 2\sqrt{3})$*), and let* U *be a bounded bilinear form on* $\mathcal{E} \times \mathcal{F}$. *Then there are* $\Phi \in D(BL(\mathcal{E}), I_{\mathcal{E}})$ *and* $\Psi \in D(BL(\mathcal{F}), I_{\mathcal{F}})$ *such that*

$$|U(x,y)| \leq M \|U\| |||x|||_\Phi |||y|||_\Psi$$

for all $(x,y) \in \mathcal{E} \times \mathcal{F}$.

The main goal of Section 3 in [22] is to prove weak*-versions of the "Big Grothendieck's inequality" for real and complex JBW*-triples. In this line, the main result is the following.

Theorem 2.3. [22, Theorems 6 and 7] *Let* $M > 4(1 + 2\sqrt{3})\ (1 + 3\sqrt{2})^2$ *(respectively,* $M > 4(1 + 2\sqrt{3})$*) and* $\varepsilon > 0$. *For every pair* (V, W) *of real (respectively, complex) JBW*-triples and every separately weak*-continuous bilinear form* U *on* $V \times W$, *there exist norm-one functionals* $\varphi_1, \varphi_2 \in V_*$, *and* $\psi_1, \psi_2 \in W_*$ *satisfying*

$$|U(x,y)| \leq M \; \|U\| \; \left(\|x\|^2_{\varphi_2} + \varepsilon^2 \; \|x\|^2_{\varphi_1}\right)^{\frac{1}{2}} \; \left(\|y\|^2_{\psi_2} + \varepsilon^2 \; \|y\|^2_{\psi_1}\right)^{\frac{1}{2}}$$

for all $(x,y) \in V \times W$.

Since every bounded bilinear form on the cartesian product of two real or complex JB*-triples has a separately weak*-continuous bilinear extension to the cartesian products of their biduals [22, Lemma 1], Theorem 2.3 above, has a natural non-weak* corollary. However, a better value of the constant M in the complex case of such a corollary can be got by means of an independent argument (see [22, Remark 2]). Precisely, we have the following result.

Corollary 2.4. *Let* $M > 3 + 2\sqrt{3}$ *(respectively,* $M > 4(1 + 2\sqrt{3})\ (1 + 3\sqrt{2})^2$*) and* $\varepsilon > 0$. *Then for every pair* $(\mathcal{E}, \mathcal{F})$ *of complex (respectively, real) JB*-triples and every bounded bilinear form* U *on* $\mathcal{E} \times \mathcal{F}$ *there exist norm-one functionals* $\varphi_1, \varphi_2 \in \mathcal{E}^*$ *and* $\psi_1, \psi_2 \in \mathcal{F}^*$ *satisfying*

$$|U(x,y)| \leq M \; \|U\| \; \left(\|x\|^2_{\varphi_2} + \varepsilon^2 \; \|x\|^2_{\varphi_1}\right)^{\frac{1}{2}} \; \left(\|y\|^2_{\psi_2} + \varepsilon^2 \; \|y\|^2_{\psi_1}\right)^{\frac{1}{2}}$$

for all $(x,y) \in \mathcal{E} \times \mathcal{F}$.

In view of the complex case of Corollary 2.4 above, the question whether the interval $M > 3 + 2\sqrt{3}$ is valid in the complex case of Theorem 2.3 naturally appears. In the rest of this paper we answer affirmatively this question. We recall that if $\mathcal{E}$ and $\mathcal{F}$ are complex JB*-triples, then every bounded bilinear form U on $\mathcal{E} \times \mathcal{F}$ has a (unique) separately weak*-continuous extension, denoted by $\widetilde{U}$, to $\mathcal{E}^{**} \times \mathcal{F}^{**}$.

Lemma 2.5. *Let $M > 3 + 2\sqrt{3}$ and $\varepsilon > 0$. Then for every pair $(\mathcal{E}, \mathcal{F})$ of complex JB*-triples and every bounded bilinear form U on $\mathcal{E} \times \mathcal{F}$ there exist norm-one functionals $\varphi_1, \varphi_2 \in \mathcal{E}^*$ and $\psi_1, \psi_2 \in \mathcal{F}^*$ satisfying*

$$|\widetilde{U}(\alpha, \beta)| \leq M \; \|U\| \; \left(\|\alpha\|_{\varphi_2}^2 + \varepsilon^2 \; \|\alpha\|_{\varphi_1}^2\right)^{\frac{1}{2}} \; \left(\|\beta\|_{\psi_2}^2 + \varepsilon^2 \; \|\beta\|_{\psi_1}^2\right)^{\frac{1}{2}}$$

for all $(\alpha, \beta) \in \mathcal{E}^{**} \times \mathcal{F}^{**}$.

Proof. By Corollary 2.4, there are norm-one functionals $\varphi_1, \varphi_2 \in \mathcal{E}^*$ and $\psi_1, \psi_2 \in \mathcal{F}^*$ satisfying

$$|\widetilde{U}(x, y)| \leq M \; \|U\| \; \left(\|x\|_{\varphi_2}^2 + \varepsilon^2 \; \|x\|_{\varphi_1}^2\right)^{\frac{1}{2}} \; \left(\|y\|_{\psi_2}^2 + \varepsilon^2 \; \|y\|_{\psi_1}^2\right)^{\frac{1}{2}} \tag{3}$$

for all $(x, y) \in \mathcal{E} \times \mathcal{F}$.

Let (α, β) be in $\mathcal{E}^{**} \times \mathcal{F}^{**}$. By Assertion 1 in Theorem 1.13, there are nets $(x_\lambda) \subseteq \mathcal{E}$ and $(y_\mu) \subseteq \mathcal{F}$ converging to α and β in the strong* topology (hence also in the weak* topology) of $\mathcal{E}^{**}$ and $\mathcal{F}^{**}$, respectively. Since, for $i \in \{1, 2\}$, the seminorm $\|.\|_{\psi_i}$ is strong*-continuous on $\mathcal{E}^{**}$, by (3) and the separately weak*-continuity of $\widetilde{U}$ we have

$$|\widetilde{U}(x, \beta)| \leq M \; \|U\| \; \left(\|x\|_{\varphi_2}^2 + \varepsilon^2 \; \|x\|_{\varphi_1}^2\right)^{\frac{1}{2}} \; \left(\|\beta\|_{\psi_2}^2 + \varepsilon^2 \; \|\beta\|_{\psi_1}^2\right)^{\frac{1}{2}}$$

for all $x \in \mathcal{E}$. By taking $x = x_\lambda$ in the last inequality, and arguing similarly, the proof is concluded. □

We can now state the complex case of Theorem 2.3 with constant $M > 3 + 2\sqrt{3}$.

Theorem 2.6. *Let $M > 3 + 2\sqrt{3}$ and $\varepsilon > 0$. For every pair $(\mathcal{V}, \mathcal{W})$ of complex JBW*-triples and every separately weak*-continuous bilinear form U on $\mathcal{V} \times \mathcal{W}$, there exist norm-one functionals $\varphi_1, \varphi_2 \in \mathcal{V}_*$, and $\psi_1, \psi_2 \in \mathcal{W}_*$ satisfying*

$$|U(x, y)| \leq M \; \|U\| \; \left(\|x\|_{\varphi_2}^2 + \varepsilon^2 \; \|x\|_{\varphi_1}^2\right)^{\frac{1}{2}} \; \left(\|y\|_{\psi_2}^2 + \varepsilon^2 \; \|y\|_{\psi_1}^2\right)^{\frac{1}{2}}$$

for all $(x, y) \in \mathcal{V} \times \mathcal{W}$.

Proof. Let $\widetilde{U}$ the unique separately weak*-continuous extension of U to $\mathcal{V}^{**} \times \mathcal{W}^{**}$. By Lemma 2.5 there exist norm-one functionals $\varphi_1, \varphi_2 \in \mathcal{V}^*$ and $\psi_1, \psi_2 \in \mathcal{W}^*$ satisfying

$$|\widetilde{U}(\alpha, \beta)| \leq M \; \|U\| \; \left(\|\alpha\|_{\varphi_2}^2 + \varepsilon^2 \; \|\alpha\|_{\varphi_1}^2\right)^{\frac{1}{2}} \; \left(\|\beta\|_{\psi_2}^2 + \varepsilon^2 \; \|\beta\|_{\psi_1}^2\right)^{\frac{1}{2}} \tag{4}$$

for all $(\alpha, \beta) \in \mathcal{V}^{**} \times \mathcal{W}^{**}$.

Let $\mathcal{U}$ stand for either $\mathcal{V}$ or $\mathcal{W}$. Then $(J_{\mathcal{U}_*})^* : \mathcal{U}^{**} \to \mathcal{U}$ is a weak*-continuous surjective triple homomorphism. Indeed, the map $(J_{\mathcal{U}_*})^* \; J_{\mathcal{U}}$ is the identity on $\mathcal{U}$, $J_{\mathcal{U}}$ is a triple homomorphism, and $J_{\mathcal{U}}(\mathcal{U})$ is weak*-dense in $\mathcal{U}^{**}$. Now $\mathcal{I}(\mathcal{U}) := \ker((J_{\mathcal{U}_*})^*)$ is a weak*-closed ideal of $\mathcal{U}^{**}$, and hence there exists a weak*-closed ideal $\mathcal{J}(\mathcal{U})$ such that $\mathcal{U}^{**} = \mathcal{I}(\mathcal{U}) \oplus^{\ell_\infty} \mathcal{J}(\mathcal{U})$ [15]. Denoting by $\Pi_{\mathcal{U}}$ the linear projection from $\mathcal{U}^{**}$ onto $\mathcal{J}(\mathcal{U})$ corresponding to the decomposition $\mathcal{U}^{**} = \mathcal{I}(\mathcal{U}) \oplus \mathcal{J}(\mathcal{U})$, it follows that the restriction of $(J_{\mathcal{U}_*})^*$ to $\mathcal{J}(\mathcal{U})$ is a weak*-continuous surjective triple isomorphism with inverse mapping $\Psi_{\mathcal{U}} := \Pi_{\mathcal{U}}\ J_{\mathcal{U}} : \mathcal{U} \to \mathcal{J}(\mathcal{U})$.

Now note that, since the bilinear mapping $(\alpha, \beta) \mapsto U((J_{\mathcal{V}_*})^*(\alpha), (J_{\mathcal{W}_*})^*(\beta))$, from $\mathcal{V}^{**} \times \mathcal{W}^{**}$ to $\mathbb{C}$, is separately weak*-continuous and extends U, we have $\widetilde{U}(\alpha, \beta) = U((J_{\mathcal{V}_*})^*(\alpha), (J_{\mathcal{W}_*})^*(\beta))$ for all $(\alpha, \beta) \in \mathcal{V}^{**} \times \mathcal{W}^{**}$. As a consequence, we obtain

$$U(x, y) = \widetilde{U}(\Psi_{\mathcal{V}}(x), \Psi_{\mathcal{W}}(y)) \tag{5}$$

for all $(x, y) \in \mathcal{V} \times \mathcal{W}$.

Since $\mathcal{V}^{**} = \mathcal{I}(\mathcal{V}) \oplus^{\ell_\infty} \mathcal{J}(\mathcal{V})$ and $\mathcal{W}^{**} = \mathcal{I}(\mathcal{W}) \oplus^{\ell_\infty} \mathcal{J}(\mathcal{W})$, we are provided with decompositions $\varphi_i = \varphi_i^1 + \varphi_i^2$ and $\psi_i = \psi_i^1 + \psi_i^2$ $(i \in \{1, 2\})$, where

$$\varphi_i^1 \in (\mathcal{J}(\mathcal{V}))_*,\ \varphi_i^2 \in (\mathcal{I}(\mathcal{V}))_*,\ \|\varphi_i^1\| + \|\varphi_i^2\| = 1,$$

and

$$\psi_i^1 \in (\mathcal{J}(\mathcal{W}))_*,\ \psi_i^2 \in (\mathcal{I}(\mathcal{W}))_*,\ \|\psi_i^1\| + \|\psi_i^2\| = 1.$$

Now, choosing norm-one elements $e_i^1 \in \mathcal{J}(\mathcal{V})$ and $e_i^2 \in \mathcal{I}(\mathcal{V})$ such that $\varphi_i^j(e_i^j) = \|\varphi_i^j\|$ $(i, j \in \{1, 2\})$, taking $\widetilde{\varphi}_i := \dfrac{\varphi_i^1 \Psi_{\mathcal{V}}}{\|\varphi_i^1 \Psi_{\mathcal{V}}\|} \in \mathcal{V}_*$ if $\varphi_i^1 \Psi_{\mathcal{V}} \neq 0$ and $\widetilde{\varphi}_i$ arbitrary in $S_{\mathcal{V}_*}$ otherwise, and keeping in mind that $\mathcal{J}(\mathcal{V})$ and $\mathcal{I}(\mathcal{V})$ are orthogonal, we get

$$\begin{aligned}\|\Psi_{\mathcal{V}}(x)\|_{\varphi_i}^2 &= \varphi_i^1 \left\{\Psi_{\mathcal{V}}(x), \Psi_{\mathcal{V}}(x), e_i^1\right\} + \varphi_i^2 \left\{\Psi_{\mathcal{V}}(x), \Psi_{\mathcal{V}}(x), e_i^2\right\} \\ &= \varphi_i^1 \Psi_{\mathcal{V}} \left\{x, x, \Psi_{\mathcal{V}}^{-1}(e_i^1)\right\} \leq \|x\|_{\widetilde{\varphi}_i}^2\end{aligned}$$

for all $x \in \mathcal{V}$.

Similarly we find norm-one functionals $\widetilde{\psi}_i$ in $\mathcal{W}_*$ such that

$$\|\Psi_{\mathcal{W}}(y)\|_{\psi_i}^2 \leq \|y\|_{\widetilde{\psi}_i}^2$$

for all $y \in \mathcal{W}$, $i \in \{1, 2\}$.

Finally, applying (4) and (5), we get

$$|U(x, y)| \leq M\ \|U\|\ \left(\|x\|_{\widetilde{\varphi}_2}^2 + \varepsilon^2\ \|x\|_{\widetilde{\varphi}_1}^2\right)^{\frac{1}{2}}\ \left(\|y\|_{\widetilde{\psi}_2}^2 + \varepsilon^2\ \|y\|_{\widetilde{\psi}_1}^2\right)^{\frac{1}{2}}$$

for all $(x, y) \in \mathcal{V} \times \mathcal{W}$. □

Remark 2.7. It is shown in [21, Theorem 3.2] that, if A is a JB-algebra, if H is a real Hilbert space, and if $T : A \to H$ is a bounded linear operator, then there is a norm-one positive linear functional φ in A^* such that

$$\|T(x)\| \leq 2\sqrt{2}\ \|T\|\ (\varphi(x^2))^{\frac{1}{2}}$$

for all $x \in A$. Keeping in mind the parallelism between the theories of JB*-triples and JB-algebras (see [14]), the spirit of the arguments in the proof of Theorem 2.6 can be applied to derive from the result in [21] just quoted the following fact which improves [22, Corollary 3].

Fact A: If A is a JBW-algebra, if H is a real Hilbert space, and if $T : A \to H$ is a weak-continuous linear operator, then there is a norm-one positive linear functional φ in A_* such that*

$$\|T(x)\| \leq 2\sqrt{2}\ \|T\|\ (\varphi(x^2))^{\frac{1}{2}}$$

for all $x \in A$.

Now, Lemma 4 in [22] can be improved as follows. Indeed, it is enough to replace in its proof [22, Corollary 3] with Fact A.

Fact B: If W is a real JBW-triple, if H is a real Hilbert space, and if $T : W \to H$ is a weak*-continuous linear operator which attains its norm, then there is a norm-one functional $\varphi \in W_*$ such that*

$$\|T(x)\| \leq (1 + 3\sqrt{2})\ \|T\|\ \|x\|_\varphi$$

for all $x \in W$.

Finally, Fact B above and [30] allow us to take $M = 1 + 3\sqrt{2}$ in the real case of Proposition 1.9.

REFERENCES

1. Alvermann, K.: Real normed Jordan algebras with involution, Arch. Math. **47**, 135-150 (1986).
2. Barton, T. and Friedman Y.: Grothendieck's inequality for JB*-triples and applications, J. London Math. Soc. (2) **36**, 513-523 (1987).
3. Barton, T. and Friedman Y.: Bounded derivations of JB*-triples, Quart. J. Math. Oxford **41**, 255-268 (1990).
4. Barton, T. and Timoney, R. M.: Weak*-continuity of Jordan triple products and its applications, Math. Scand. **59**, 177-191 (1986).
5. Bonsall, F. F. and Duncan, J.: Complete Normed Algebras, Springer-Verlag, New York 1973.
6. Bunce, L. J.: Norm preserving extensions in JBW*-triple preduals, Preprint, 2000.
7. Chu, C-H. and Iochum, B.: Weakly compact operators on Jordan triples, Math. Ann. **281**, 451-458 (1988).
8. Chu, C-H., Iochum, B., and Loupias, G.: Grothendieck's theorem and factorization of operators in Jordan triples, Math. Ann. **284**, 41-53 (1989).
9. Dineen, S.: The second dual of a JB*-triple system, In: Complex analysis, functional analysis and approximation theory (ed. by J. Múgica), 67-69, (North-Holland Math. Stud. 125), North-Holland, Amsterdam-New York, 1986.
10. Goodearl, K. R.: Notes on real and complex C*-algebras, Shiva Publ. 1982.
11. Grothendieck, A.: Résumé de la théorie métrique des produits tensoriels topologiques. Bol. Soc. Mat. Sao Paolo **8**, 1-79 (1956).

12. Haagerup, U.: Solution of the similarity problem for cyclic representations of C*-algebras, Ann. of Math. **118**, 215-240 (1983).
13. Haagerup, U.: The Grothendieck inequality for bilinear forms on C*-algebras, Avd. Math. **56**, 93-116 (1985).
14. Hanche-Olsen, H. and Størmer, E.: Jordan operator algebras, Monographs and Studies in Mathematics 21, Pitman, London-Boston-Melbourne 1984.
15. Horn, G.: Characterization of the predual and ideals structure of a JBW*-triple, Math. Scand. **61** (1987), 117-133.
16. Isidro, J. M., Kaup, W., and Rodríguez, A.: On real forms of JB*-triples, Manuscripta Math. **86**, 311-335 (1995).
17. Kaup, W.: Algebraic characterization of symmetric complex Banach manifolds, Math. Ann. **228**, 39-64 (1977).
18. Kaup, W.: A Riemann mapping theorem for bounded symmetric domains in complex Banach spaces, Math. Z. **183**, 503-529 (1983).
19. Lindenstrauss, J.: On operators which attain their norm, Israel J. Math. **1**, 139-148 (1963).
20. Martínez, J. and Peralta A.M.: Separate weak*-continuity of the triple product in dual real JB*-triples, Math. Z., **234**, 635-646 (2000).
21. Peralta, A. M.: Little Grothendieck's theorem for real JB*-triples, Math. Z., to appear.
22. Peralta, A. M. and Rodríguez, A.: Grothendieck's inequalities for real and complex JBW*-triples, Proc. London Math. Soc., to appear.
23. Pisier, G.: Grothendieck's theorem for non commutative C*-algebras with an appendix on Grothendieck's constant, J. Funct. Anal. **29**, 397-415 (1978).
24. Pisier, G.: Factorization of linear operators and geometry of Banach spaces, Publ. Amer. Math. Soc. CBMS **60**, Am. Math. Soc. 1986.
25. Poliquin, R. A. and Zizler, V. E.: Optimization of convex functions on w*-compact sets, Manuscripta Math. **68**, 249-270 (1990).
26. Rodríguez A.: On the strong* topology of a JBW*-triple, Quart. J. Math. Oxford (2) **42**, 99-103 (1989).
27. Rodríguez A.: Jordan structures in Analysis. In Jordan algebras: Proc. Oberwolfach Conf., August 9-15, 1992 (ed. by W. Kaup, K. McCrimmon and H. Petersson), 97-186. Walter de Gruyter, Berlin, 1994.
28. Sakai, S.: C*-algebras and W*-algebras, Springer-Verlag, Berlin 1971.
29. Upmeier, H.: Symmetric Banach Manifolds and Jordan C*-algebras, Mathematics Studies 104, (Notas de Matemática, ed. by L. Nachbin) North Holland 1985.
30. Zizler, V.: On some extremal problems in Banach spaces, Math. Scand. **32**, 214-224 (1973).

www.ingramcontent.com/pod-product-compliance
Ingram Content Group UK Ltd.
Pitfield, Milton Keynes, MK11 3LW, UK
UKHW020226290926
15470UKWH00001B/408

9 780444 502193